OPTICAL MAGNETOMETRY

Featuring chapters written by leading experts in magnetometry, this book provides comprehensive coverage of the principles, technology, and diverse applications of optical magnetometry, from testing fundamental laws of nature to detecting biomagnetic fields and medical diagnostics.

Readers will find a wealth of technical information, from antirelaxation-coating techniques, microfabrication, and magnetic shielding to geomagnetic-field measurements, space magnetometry, detection of biomagnetic fields, detection of NMR and MRI signals, and rotation sensing. The book includes an original survey of the history of optical magnetometry, and a chapter on the commercial use of these technologies.

The book is supported by extensive online material, containing historical overviews, derivations, side-line discussion, and additional plots and tables, available at www.cambridge.org/9781107010352. As well as introducing graduate students to this field, the book is also a useful reference for researchers in atomic physics.

DMITRY BUDKER is Professor of Physics at the University of California at Berkeley; Faculty Scientist in the Nuclear Science Division, LBNL; and Co-founder and Scientist of Rochester Scientific, LLC. His research interests are related to the study of violation of discrete symmetries and the development and applications of the optical-magnetometry techniques.

DEREK F. JACKSON KIMBALL is Associate Professor and Chair of the Department of Physics at California State University – East Bay. His research focuses on using techniques of experimental atomic physics and nonlinear optics for precision tests of the fundamental laws of physics.

OPTICAL MAGNETOMETRY

Edited by

DMITRY BUDKER

University of California at Berkeley

DEREK F. JACKSON KIMBALL

California State University – East Bay

CAMBRIDGE
UNIVERSITY PRESS

CAMBRIDGE
UNIVERSITY PRESS

University Printing House, Cambridge CB2 8BS, United Kingdom

Published in the United States of America by Cambridge University Press, New York

www.cambridge.org
Information on this title: www.cambridge.org/9781107010352

First published 2013

A catalogue record for this publication is available from the British Library

Library of Congress Cataloguing in Publication data
Optical magnetometry / edited by Dmitry Budker and Derek F. Jackson Kimball.
pages cm
Includes bibliographical references and index.
ISBN 978-1-107-01035-2
1. Magnetic fields–Measurement. 2. Optical measurements. 3. Magnetic instruments.
I. Budker, Dmitry. II. Jackson Kimball, Derek F.
QC754.2.M3O68 2013
538.028′7–dc23
2012038087

ISBN 978-1-107-01035-2 Hardback

Contents

Contributors

V. M. Acosta *Department of Physics, University of California, Berkeley, California 94720-7300, USA.*

E. B. Alexandrov *Ioffe Physical Technical Institute, Russian Academy of Sciences, 26 Polytechnisheskaya, St. Petersburg, 194021, Russia.*

M. V. Balabas *S. I. Vavilov State Optical Institute, St. Petersburg 199034, Russia.*

A. Ben-Amar Baranga *Department of Electrical and Computer Engineering, Ben-Gurion University of the Negev, 84105 Beer-Sheva, Israel.*

D. Bonaccini Calia *Laser Systems Department, European Southern Observatory, D-85748 Garching near Munich, Germany.*

M.-A. Bouchiat *Laboratoire Kastler Brossel, Département de Physique de l'Ecole Normale Supérieure, 24 Rue Lhomond, F-75231 Paris Cedex 05, France.*

A. W. Brown *Polatomic Inc., Richardson, Texas 75081, USA.*

D. Budker *Department of Physics, University of California at Berkeley, Berkeley, California 94720-7300, USA; Nuclear Science Division, Lawrence Berkeley National Laboratory, Berkeley, California 94720, USA.*

T. E. Chupp *Department of Physics, University of Michigan, Ann Arbor, Michigan 48109, USA.*

E. A. Donley *Time and Frequency Division, National Institute of Standards and Technology, Boulder, Colorado 80305-3328, USA.*

W. Gawlik *Institute of Physics, Jagiellonian University, Reymonta 4, 30-059 Kraków, Poland.*

P. R. Hemmer *Department of Electrical and Computer Engineering, Texas A & M University, College Station, Texas 77843, USA.*

J. M. Higbie *Department of Physics and Astronomy, Bucknell University, Lewisburg, Pennsylvania 17837, USA.*

R. Holzlöhner *Laser Systems Department, European Southern Observatory, D-85748 Garching near Munich, Germany.*

D. C. Hovde *Southwest Sciences – Ohio Operations, 6837 Main Street, Cincinnati, Ohio 45244, USA.*

I. Hrvoic *GEM Systems Inc., 135 Spy Court, Markham, Ontario L3R 5H6, Canada.*

D. F. Jackson Kimball *Department of Physics, California State University – East Bay, Hayward, California 94542-3084, USA.*

K. Jensen *Niels Bohr Institute, University of Copenhagen, DK 2100, Denmark; QUANTOP, Danish National Research Foundation Center for Quantum Optics, Denmark.*

R. Johnson *Geometrics Inc., 2190 Fortune Drive, San Jose, California 95131, USA.*

J. Kitching *Time and Frequency Division, National Institute of Standards and Technology, Boulder, Colorado 80305, USA.*

S. Knappe *Time and Frequency Division, National Institute of Standards and Technology, Boulder, Colorado 80305, USA.*

S. K. Lamoreaux *Department of Physics, Yale University, New Haven, Connecticut 06520-8120, USA.*

M. P. Ledbetter *Department of Physics, University of California at Berkeley, Berkeley, California 94720-7300, USA.*

S.-K. Lee *GE Global Research Center, Niskayuna, New York 12309, USA.*

J. R. Maze *Department of Physics, Pontificia Universidad Catolica, Santiago 7820436, Chile.*

D. D. McGregor *Polatomic Inc., Richardson, Texas 75081, USA.*

E. Paperno *Department of Physics, Nuclear Research Center, Negev 84190 Beer-Sheva, Israel*; *Department of Mechanical Engineering, Ben-Gurion University of the Negev, P.O. Box 653, Beer-Sheva 84105, Israel.*

B. Patton *Department of Physics, University of California at Berkeley, Berkeley, California 94720-7300, USA.*

E. S. Polzik *Niels Bohr Institute, University of Copenhagen, DK 2100, Denmark; QUANTOP, Danish National Research Foundation Center for Quantum Optics, Denmark.*

M. D. Prouty *Geometrics Inc., 2190 Fortune Drive, San Jose, California 95131, USA.*

S. Pustelny *Institute of Physics, Jagiellonian University, Reymonta 4, 30-059 Kraków, Poland.*

S. M. Rochester *Rochester Scientific, El Cerrito, California 94530-1757, USA; Department of Physics, University of California at Berkeley, Berkeley, California 94720-7300, USA.*

M. V. Romalis *Department of Physics, Princeton University, Princeton, New Jersey 08544, USA.*

I. Savukov *Los Alamos National Laboratory, Los Alamos, New Mexico 87545, USA.*

S. J. Seltzer *Materials Sciences Division, Lawrence Berkeley National Laboratory, Berkeley, California 94720, USA; Department of Chemistry, University of California, Berkeley, California 94720-7300, USA.*

R. E. Slocum *Polatomic Inc., Richardson, Texas 75081, USA.*

E. J. Smith *Jet Propulsion Laboratory, California Institute of Technology, Pasadena, California 91009, USA.*

A. K. Vershovskiy *Ioffe Physical Technical Institute, Russian Academy of Sciences, 26 Polytechnisheskaya, St. Petersburg, 194021, Russia.*

R. T. Wakai *Department of Medical Physics, University of Wisconsin-Madison, 1111 Highland Avenue, Madison, WI 53705, USA.*

T. G. Walker *Department of Physics, University of Wisconsin, 1150 University Avenue, Madison, Wisconsin 53706, USA.*

R. L. Walsworth *Harvard-Smithsonian Center for Astrophysics, Cambridge, Massachusetts 02138, USA.*

V. V. Yashchuk *Advanced Light Source Division, Lawrence Berkeley National Laboratory, Berkeley, California 94720, USA.*

Preface

Optical magnetometry, in which a magnetic field is measured by observing changes in the properties of light interacting with matter immersed in the field, is not a new field. It has its origins in Michael Faraday's discovery in 1845 of the rotation of the plane of linearly polarized light as it propagated through a dense glass in the presence of a magnetic field. Faraday's historic discovery marked the first experimental evidence relating light and electromagnetism.

A century later, atomic magnetometers based on optical pumping were introduced and gradually perfected by such giants as Alfred Kastler, Hans Dehmelt, Jean Brossel, William Bell, Arnold Bloom, and Claude Cohen-Tannoudji, to name but a few of the pioneers. Recent years have seen a revolution in the field related to the development of tunable diode lasers, efficient antirelaxation wall coatings, techniques for elimination of spin-exchange relaxation, and, most recently, the advent of optical magnetometers based on color centers in diamond. Today, optical magnetometers are pushing the boundaries of sensitivity and spatial resolution, and, in contrast to their able competition from superconducting quantum interference device (SQUID) magnetometers, they do not require cryogenic temperatures. Numerous novel applications of optical magnetometers have flourished, from detecting signals in microfluidic nuclear-magnetic resonance chips to measuring magnetic fields of the human brain to observing single nuclear spins in a solid matrix.

The remarkable progress of optical magnetometry during recent years called for a single-source reference to help those entering the field, including students and practitioners interested in applications, to get a "jump-start" on the principles, techniques, conventions, and the latest achievements. Toward this goal, we have assembled a remarkable group of authors, who have teamed up to prepare twenty pedagogical chapters packed with information. We are excited to offer the result of this effort for the perusal and judgement of the reader. Of course, we welcome and appreciate feedback on the content of the book, which can be sent to us via e-mail (dbudker@gmail.com or derek.jacksonkimball@csueastbay.edu). Please note that the book web site www.cambridge.org/9781107010352 is a repository for additional material related to the subjects discussed in the chapters.

In addition to our deep gratitude to the Cambridge University Press editors for their encouragement and support, and to all the contributors for their hard work writing and rewriting the chapters based on our feedback, we particularly thank Professor Michael V. Romalis who helped us define the scope and direction of this project and who helped edit several chapters.

Part I

Principles and techniques

1

General principles and characteristics of optical magnetometers

D. F. Jackson Kimball, E. B. Alexandrov, and D. Budker

1.1 Introduction

Optical magnetometry encompasses the wide range of experimental techniques in which light is used to measure the response of atomic angular momentum to magnetic fields. Atomic magnetic moments $\boldsymbol{\mu}$ arise due to the magnetic moments associated with the intrinsic spins of constituent electrons and nuclei as well as electronic orbital motion. In the presence of a magnetic field $\mathbf{B}$, a torque

$$\boldsymbol{\tau} = \boldsymbol{\mu} \times \mathbf{B}$$

acts on the atoms. For sufficiently small magnetic fields the atomic magnetic moment $\boldsymbol{\mu}$ can be assumed to be independent of $\mathbf{B}$, in which case $\boldsymbol{\tau}$ causes the component of angular momentum transverse to $\mathbf{B}$ to precess about $\mathbf{B}$ at the Larmor frequency $\Omega_L = \gamma B$, where γ is the gyromagnetic ratio of the atomic species. When light propagates through and interacts with the atomic medium, angular momentum is exchanged between the atoms and the light field. Thus the angular momentum state of the atoms affects the angular momentum state of the light. In this way, precession of atomic angular momentum can be observed via the induced changes in the polarization and intensity of light interacting with the atoms, allowing optical measurement of Larmor precession.

The exchange of angular momentum between the atomic medium and light field that enables optical detection of Larmor precession can similarly be used to polarize the atomic medium through the process of *optical pumping* (see the comprehensive reviews [1]–[3]). Optical pumping refers to the process whereby photons absorbed by atoms transfer their angular momentum to the atoms, thereby creating polarization in both ground and excited states. Ground-state spin polarization is achieved through both depopulation of selected Zeeman sublevels and repopulation of Zeeman sublevels by spontaneous transitions from

Optical Magnetometry, ed. Budker, D. and Jackson Kimball, D. F. Published by Cambridge University Press.

the excited states polarized by the incident light. In most optical magnetometers, the signals used to detect Larmor precession scale with the ground-state spin polarization, and consequently, optical pumping dramatically enhances magnetometer sensitivity.

Optical magnetometers, presently unmatched in both absolute accuracy and magnetometric sensitivity [4, 5], are universally based on these aforementioned principles: (1) optical pumping of atomic spin polarization, (2) time-evolution of atomic spin polarization due to the torque exerted on atomic magnetic moments by the magnetic field, and (3) optical detection of the evolved atomic spin polarization state through the effect of the polarized atoms on light propagating through the atomic medium. Within this basic framework, there is a remarkable diversity of experimental techniques and physical effects, many of which are explored throughout this book, in the cited references, and in online supporting material (available at www.cambridge.org/9781107010352), and undoubtedly still others that remain to be discovered.

1.1.1 Fundamental sensitivity limits

The fundamental quantum-mechanical uncertainty in the measurement of atomic spin projection constrains the potential sensitivity of optical magnetometers. The *spin-projection-noise-limited* (or *atomic shot-noise-limited*) sensitivity $\delta B_{\rm SNL}$ of a polarized atomic sample to magnetic fields[1] is determined by the total number of atoms N and the spin-relaxation rate $\Gamma_{\rm rel}$ for measurement times $\tau \gg \Gamma_{\rm rel}{}^{-1}$ [6]:

$$\delta B_{\rm SNL} \approx \frac{1}{\gamma}\sqrt{\frac{\Gamma_{\rm rel}}{N\tau}}\,. \tag{1.1}$$

Equation (1.1) can be understood by noting that a measurement of a single atomic spin for a time $1/\Gamma_{\rm rel}$ determines the Larmor precession angle with an uncertainty on the order of 1 radian. If the measurement is performed with N atoms, the uncertainty is reduced by $\sqrt{N}$, and if the measurement is repeated multiple times, the uncertainty is reduced by the square root of the number of measurements, which most efficiently is $\sqrt{\Gamma_{\rm rel}\tau}$.

As can be seen from Eq. (1.1), to achieve the highest possible precision in magnetometric measurements, it is advantageous to have the longest possible relaxation time for the atomic polarization, i.e., the smallest possible $\Gamma_{\rm rel}$, as well as the largest possible N. Therefore, optical magnetometers typically are based on measurements of long-lived ground-state spin polarization [an exception is the ^{4}He magnetometer (Chapter 10), in which polarization of the metastable $2\,^3S_1$ state is used]. A variety of techniques can be employed to minimize $\Gamma_{\rm rel}$: antirelaxation coating the walls of the vapor cell containing the atoms to reduce spin-depolarizing wall collisions [7–10] (see Chapter 11), filling the vapor cell with buffer gas to slow diffusion of atoms to the cell walls (see, for example, Refs. [11–15]), and even atom trapping and cooling [16, 17] (Chapter 9). Also of note are optical magnetometers

[1] Here we ignore factors of order unity that depend on the details of the atomic system and optical magnetometer scheme, such as the total atomic angular momentum and relative contributions of different Zeeman sublevels.

using particular condensed matter systems, such as nitrogen-vacancy centers in diamonds (Chapter 8) and alkali atoms trapped in condensed (superfluid or solid) helium [18–20], that have small Γ_{rel}.

There is also a contribution to optical-magnetometer noise from the quantum uncertainty of measurements of light properties (photon shot noise). Optical detection of atomic spin precession is usually performed by measuring either the intensity or polarization of light transmitted through the atomic sample. There are certain intrinsic advantages to measuring the light polarization, in particular, reduced sensitivity to noise due to laser-intensity fluctuations. If, for example, atomic spin precession is detected by measuring optical rotation of the plane of transmitted light polarization [21], the photon-shot-noise-limited sensitivity to the optical rotation angle φ is

$$\delta\varphi \approx \frac{1}{2}\sqrt{\frac{1}{\Phi\tau}}\,, \tag{1.2}$$

where Φ is the probed photon flux (photons/s) detected after the atomic sample and $\delta\varphi$ is measured in $\mathrm{rad}/\sqrt{\mathrm{Hz}}$. It should also be noted that light–atom coupling via AC Stark shifts can generate additional noise (see Ref. [22] and Sec 1.4.3). In optimal operation, the contribution of photon shot noise to overall magnetometric noise does not exceed the contribution from atomic spin-projection noise [6, 23]. Upon optimization of the atomic density n for a given volume V of the sample, where $N = nV$, for an atomic-vapor-based optical magnetometer the dominant spin-relaxation mechanism becomes either spin-exchange or spin-destruction collisions (depending on the details of the magnetometry scheme), in which case $\Gamma_{\mathrm{rel}} = \xi n$, and the optimum magnetometric sensitivity becomes

$$\delta B_{\mathrm{opt}} \approx \frac{1}{\gamma}\sqrt{\frac{\xi}{V\tau}}\,. \tag{1.3}$$

The relaxation constant ξ ranges between $\sim 10^{-9}\ \mathrm{cm}^3/\mathrm{s}$ and $\sim 10^{-13}\ \mathrm{cm}^3/\mathrm{s}$ for alkali atoms, depending on the details of the collisions [1]. Thus for optical magnetometers using alkali vapors, the optimal magnetometric sensitivity for a $V = 1\ \mathrm{cm}^3$ magnetic sensor ranges between 10^{-11} and $10^{-13}\ \mathrm{G}/\sqrt{\mathrm{Hz}}$ (1 to 0.01 $\mathrm{fT}/\sqrt{\mathrm{Hz}}$). Issues related to noise in optical magnetometers, including exploration of squeezed states and quantum nondemolition (QND) measurements, are addressed in detail in Chapters 2 and 3.

1.1.2 Zeeman shifts and atomic spin precession

The language of atomic spectroscopy provides a complementary description of optical magnetometry: the light field propagating through the atomic medium measures the Zeeman shifts of atomic states. The Hamiltonian describing the Zeeman shift is

$$H_Z = -\boldsymbol{\mu}\cdot\mathbf{B}\,. \tag{1.4}$$

For an elementary particle such as the electron, the magnetic moment is directly proportional to the intrinsic spin $\mathbf{S}$, $\boldsymbol{\mu} = g_e \mu_B \mathbf{S}$, where g_e is the electron's Landé g-factor and μ_B is the Bohr magneton,

$$\mu_B = \frac{e\hbar}{2mc} = \frac{\alpha}{2} e a_0 \approx (2\pi\hbar) \times 1.4\ \text{MHz/G}\ , \tag{1.5}$$

where e is the absolute value of the electron charge, $\hbar$ is Planck's constant, m is the electron mass, c is the speed of light, $\alpha = e^2/(\hbar c)$ is the fine structure constant, and $a_0 = \hbar^2/(me^2)$ is the Bohr radius. The Landé g-factor for the electron, $g_e \approx 2.002\,32$, can be determined from extraordinarily precise experimental measurements [24–26] and from the theory of quantum electrodynamics [27], and thus, measurements of the Zeeman shifts of electron spin states constitute absolute measurements of the magnetic field.

For an atom the situation becomes more complicated. For sufficiently small magnitudes of $\mathbf{B}$, the atomic magnetic moment $\boldsymbol{\mu}$ is determined by internal atomic interactions and is approximately independent of $\mathbf{B}$,

$$\boldsymbol{\mu} = g_F \mu_B \mathbf{F}\ , \tag{1.6}$$

where $\mathbf{F} = \mathbf{I} + \mathbf{J}$ is the total atomic angular momentum ($\mathbf{I}$ is the nuclear spin, and $\mathbf{J} = \mathbf{L} + \mathbf{S}$ is the electronic angular momentum, $\mathbf{L}$ is the orbital angular momentum, and $\mathbf{S}$ is the electron spin) and g_F is the Landé factor corresponding to a state with a particular value of F. It is evident from symmetry that $\boldsymbol{\mu}$ is parallel to the total atomic angular momentum $\mathbf{F}$ (this follows, for example, from the Wigner–Eckart theorem – see the discussion in Refs. [28] and [29]). However, for larger magnitudes of $\mathbf{B}$ where the Zeeman shifts become significant compared to energy splittings caused by hyperfine and fine interactions, the magnetic moment $\boldsymbol{\mu}$ becomes dependent on the external magnetic field $\mathbf{B}$ due to mixing of atomic states (this is the nonlinear Zeeman effect, see Fig. 1.1 and the detailed discussion in Ref. [29]). Ultimately, then, the energy splitting ΔE between Zeeman sublevels in an atom, and consequently, the frequency Ω associated with time-evolution of the atomic angular momentum, is described by an expansion in powers of B:

$$\Delta E = \hbar\Omega = \gamma_0 + \gamma_1 B + \gamma_2 B^2 + \gamma_3 B^3 + \cdots = \sum_{n=0}^{\infty} \gamma_n B^n\ , \tag{1.7}$$

where the expansion coefficients γ_n depend on the particular atomic states and Zeeman sublevels involved. As seen in the lower plot of Fig. 1.1, for alkali atoms subjected to magnetic fields in the geophysical range of ~0.5 G, there are nonlinearities of the Zeeman shifts with magnitudes of tens of hertz, often larger than the magnetic resonance linewidths, that must be accounted for in sensitive optical magnetometry schemes.

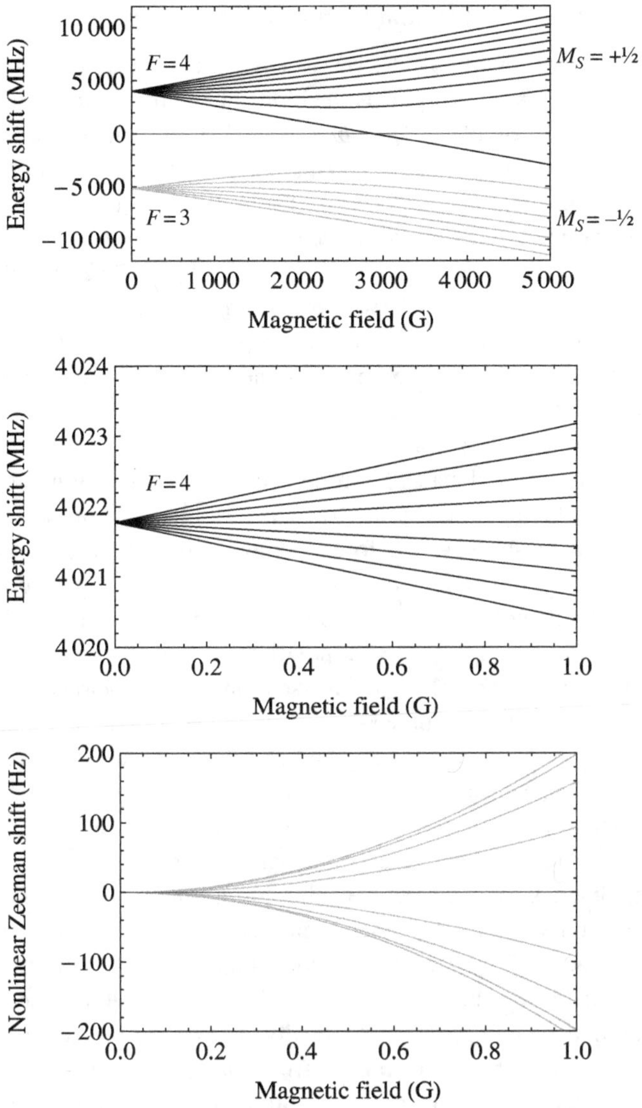

Figure 1.1 Top and middle plots: energies of the ground-state 6s $^2S_{1/2}$ hyperfine manifold of cesium ($I = 7/2$) as a function of applied magnetic field (the Breit–Rabi diagram). At low fields (middle plot), the atomic states are well described by the coupled basis where F and M_F are good quantum numbers and the Hamiltonian (Eq. 1.4) representing the interaction with the applied magnetic field is treated as a perturbation. At high fields, the atomic states are well described by the uncoupled basis where M_I and $M_J = M_S$ are good quantum numbers and the hyperfine interaction is treated as a perturbation. Bottom plot: nonlinear Zeeman shifts for the Cs ground state, computed by subtracting the linear term describing the low-field Zeeman shifts from the exact Breit–Rabi solutions.

1.1.3 Quantum beats and dynamic range

Since under all laboratory conditions Ω is very small compared to optical frequencies, and in many cases Ω is small compared to optical transition linewidths, Zeeman shift measurements typically involve observation of *quantum beats* [30–33]: the time-evolution of a coherent superposition of nondegenerate energy eigenstates at a frequency determined by the energy splittings. If a coherent superposition is created between two quantum states separated in energy by ΔE, according to the time-dependent Schrödinger equation, the complex phase between the states will evolve according to $\varphi(t) = \Delta E t/\hbar$, or in the case of Zeeman shifts, as described by Eq. (1.7), $\varphi(t) = m\Omega t$ where m is an integer. The time-evolution of $\varphi(t)$ causes the complex index of refraction for light propagating though the atomic medium to evolve in time as well, which can be observed in the time-dependence of the intensity and polarization of the light field.

If all fields, magnetic and optical, applied to the atoms are constant or vary in time slowly compared to the ground-state spin-relaxation rate $\Gamma_{\rm rel}$, the atomic spin polarization reaches a steady state. For spin-precession frequencies $\Omega \lesssim \Gamma_{\rm rel}$, atomic spins are unable to undergo an entire quantum-beat cycle before relaxing. In this case, although spin precession angles accrued between pump and probe interactions are random, because the angles are small compared to π the polarization does not average to zero. This produces a residual, magnetic field-dependent steady-state spin polarization that can be optically probed. For spin-precession frequencies $\Omega \gg \Gamma_{\rm rel}$, transverse atomic spin polarization is averaged out because probed atomic spins have precessed by random angles greater that π with respect to one another. In other words, atoms polarized at different times will, in general, beat out of phase with each other, canceling out the overall atomic spin polarization. For this reason optical magnetometers with constant or slowly varying fields have dynamic range limited by $\Gamma_{\rm rel}$ and find application as near-zero-field magnetometers.

Optical magnetometers used to measure fields corresponding to $\Omega \gg \Gamma_{\rm rel}$ therefore generally employ some type of time-dependent field: magnetic field pulses to induce transient responses, radiofrequency (RF) fields to take advantage of magnetic-resonance techniques, or modulated light fields to synchronously optically pump the atomic medium. For example, synchronous optical pumping (or "optically driven spin precession" [34]) creates time-dependent macroscopic atomic spin polarization for high quantum-beat frequencies ($\Omega \gg \Gamma_{\rm rel}$) by modulating the light at the quantum-beat frequency Ω or a subharmonic thereof. Polarization is produced in phase with that of atoms pumped on previous cycles, each optical pumping cycle contributing coherently to the atomic spin polarization, and as a result the ensemble beats in unison.

1.2 Model of an optical magnetometer

The basic principles of optical magnetometry are illustrated by considering an ensemble of two-level atoms with total angular momentum $F = 1/2$ in the ground state and $F' = 1/2$ in the upper state. Suppose that the atoms are immersed in a uniform magnetic field $\mathbf{B} = B\hat{\mathbf{z}}$

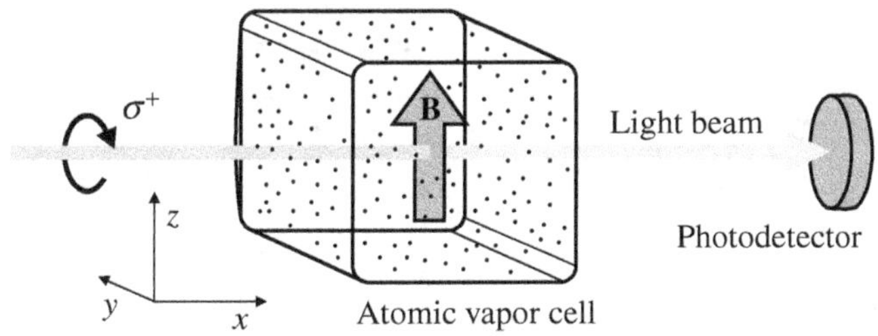

Figure 1.2 Example of a basic experimental setup for optical magnetometry.

and left-circularly polarized (σ^+) light,[2] near resonant with the (allowed electric dipole) $F \to F'$ transition, propagates along $\hat{\mathbf{x}}$ (Fig. 1.2). The strength of the light–atom interaction is parameterized by the *Rabi frequency*

$$\hbar\Omega_R = d\mathcal{E}_0 \,, \tag{1.8}$$

where d is the transition dipole moment [for typical allowed electric-dipole transitions, $d \sim ea_0 \approx 2\pi \times 1.28$ MHz/(V/cm)] and $\mathcal{E}_0$ is the amplitude of the optical electric field. Spontaneous emission from the state F' back to F occurs at a rate Γ_0 (assume a closed transition). The optical pumping rate Γ_{pump} is given by Fermi's golden rule [28]:

$$\Gamma_{\text{pump}} = \frac{\Omega_R^2}{\Gamma_0} \,. \tag{1.9}$$

Let us assume that $\Gamma_{\text{rel}} \ll \Gamma_0$, $\Omega_R \ll \Gamma_0$, and ignore Doppler broadening (assume the atoms are stationary). Furthermore, to simplify our discussion, let us assume that $\Gamma_{\text{rel}} \ll \Gamma_{\text{pump}}$ (but note that for optimized magnetometer performance, typically $\Gamma_{\text{rel}} \gtrsim \Gamma_{\text{pump}}$). Under these conditions, the *saturation parameter* [28, 29]:

$$\mathcal{K} = \frac{\text{excitation rate}}{\text{relaxation rate}} = \frac{\Gamma_{\text{pump}}}{\Gamma_{\text{rel}}} = \frac{\Omega_R^2}{\Gamma_0 \Gamma_{\text{rel}}} \gg 1 \,, \tag{1.10}$$

indicates that the system is in the regime where the populations of ground-state Zeeman sublevels are strongly perturbed by the light field. Since $\Omega_R \ll \Gamma_0$, we can neglect stimulated emission and associated effects.

Suppose that initially $B = 0$ [Fig. 1.3(a)]. Interaction of the atoms with the light field causes ground-state optical pumping: choosing the quantization axis along the light propagation direction $\hat{\mathbf{x}}$, we see that the population of the $M_x = -1/2$ state (denoted $|-\rangle_x$) is depleted relative to the population of the $M_x = +1/2$ state (denoted $|+\rangle_x$), since due to angular-momentum selection rules the light field only causes transitions between the

[2] Here we use the spectroscopists' convention for left and right circular polarization, where a σ^+ photon (one with positive *helicity*, with the photon spin along its direction of propagation) is said to be left-circularly polarized, and a σ^- photon is said to be right-circularly polarized.

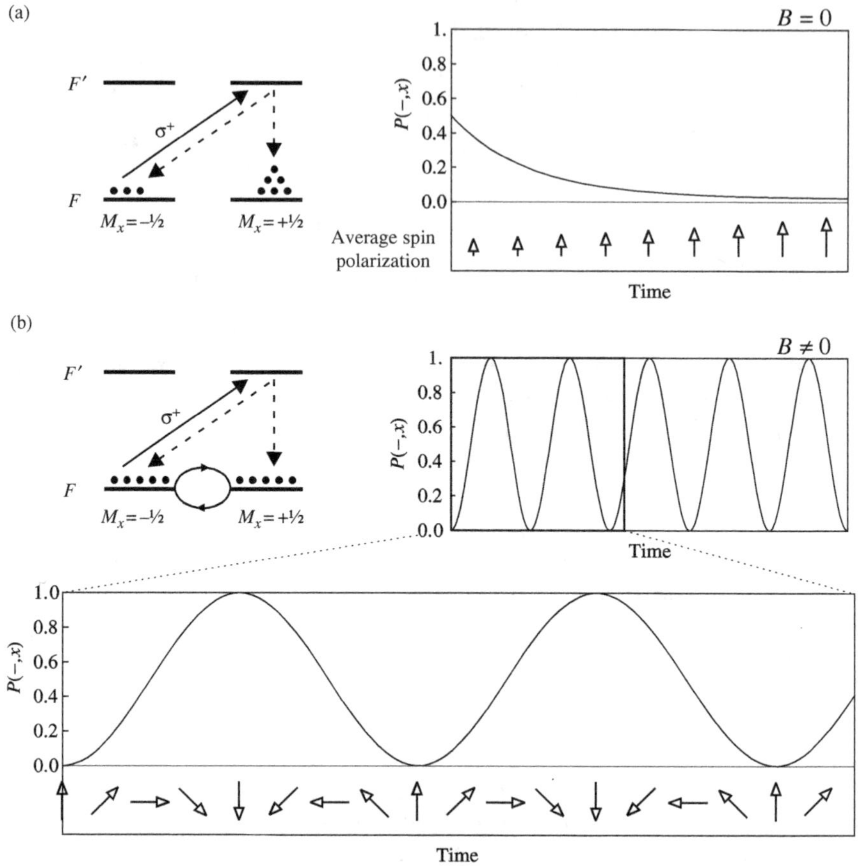

Figure 1.3 (a) Ground-state optical pumping of an ensemble of atoms with $F = 1/2$ in the ground state and $F' = 1/2$ in the upper state for the setup shown in Fig. 1.2 with $\mathbf{B} = 0$. Left-circularly polarized (σ^+) light excites the $M_x = -1/2 \to M'_x = +1/2$ transition, depleting the population of the $M_x = -1/2$ state. This decreases the probability $P(-,x)$ for atoms to be in the $M_x = -1/2$ state and increases the average spin polarization along x, as indicated by the increasing magnitude of the arrows along the bottom of the plot on the right. (b) Optical detection of spin precession with $\mathbf{B} \neq 0$. Larmor precession causes coherent oscillation between the $M_x = -1/2$ and $M_x = +1/2$ states (quantum beats) while the light probes the population of the $M_x = -1/2$ state [proportional to $P(-,x)$]. The coherent oscillation can also be understood as precession of the average spin polarization at the Larmor frequency, as shown by the arrows at the bottom of the lower plot.

$F = 1/2$, $M_x = -1/2$ and $F' = 1/2$, $M'_x = +1/2$ states. The atoms are optically pumped into a pure state $|+\rangle_x$ (since we can neglect ground-state spin relaxation under our assumptions) that does not interact with the light field. States that do not interact with light of particular polarization and/or frequency are commonly known as *dark states*. The atomic medium has acquired net angular momentum along $\hat{\mathbf{x}}$ from the light field.

Suppose that after optically pumping the atomic ensemble into $|+\rangle_x$, the magnetic field is suddenly (nonadiabatically) switched "on" ($B \neq 0$) and the light intensity is simultaneously reduced to a level sufficiently low that the optical pumping rate is slow compared to the measurement time [Fig. 1.3(b)]. Consider the system from the point of view of a quantization axis chosen along $\hat{\mathbf{z}}$, parallel to $\mathbf{B}$: the energy of the $M_z = +1/2$ state (denoted $|+\rangle_z$) is shifted by $+\hbar\Omega_L/2$ and the energy of the $M_z = -1/2$ state (denoted $|-\rangle_z$) is shifted by $-\hbar\Omega_L/2$. Using quantum-mechanical rotation matrices [28,29], it can be shown that

$$|+\rangle_x = \frac{|-\rangle_z + |+\rangle_z}{\sqrt{2}}\,, \tag{1.11}$$

$$|-\rangle_x = \frac{|-\rangle_z - |+\rangle_z}{\sqrt{2}}\,. \tag{1.12}$$

When the magnetic field is switched on at $t = 0$, the time-dependent state of the atomic ensemble (ignoring for now the light–atom interaction) is described by

$$|\psi(t)\rangle = \sqrt{\frac{1}{2}}\left(e^{i\Omega_L t/2}|-\rangle_z + e^{-i\Omega_L t/2}|+\rangle_z\right)\,. \tag{1.13}$$

The time-dependent probability $P(-,x)$ for an atom to be found in $|-\rangle_x$, a *bright state* which interacts strongly with the light field, is given by

$$P(-,x) = |\langle\psi(t)|-\rangle_x|^2 = \frac{1}{2}[1-\cos(\Omega_L t)] = \sin^2(\Omega_L t/2)\,. \tag{1.14}$$

This in turn modulates the light–atom–interaction probability and index of refraction at Ω_L, enabling optical detection of the quantum beats. The observed quantum beats are nothing but Larmor precession of the spin-1/2 atoms in the magnetic field cast in another language. The optical signal contains the information about the magnetic field B.

In the above model of an optical magnetometer, optical pumping occurs in a zero-field environment and optical probing occurs in the presence of the magnetic field to be measured. Practically, it is difficult and inefficient to null the magnetic field during optical pumping, and it is also frequently desirable to operate magnetometers continuously without temporal separation of pumping and probing stages.

An example of a magnetometry scheme enabling continuous optical pumping and probing of the atomic media in the presence of the magnetic field to be measured is the Bell–Bloom method [34]. Consider the same system as in the above model, but in this case $\mathbf{B}$ remains fixed throughout the measurement and the intensity of the light beam is sinusoidally modulated at a frequency Ω_m. Again, to simplify our discussion we assume that Γ_{pump} is fast compared to Ω_m and Γ_{rel}. At times when the light intensity is high, atoms are optically pumped out of $|-\rangle_x$ and into $|+\rangle_x$. The polarized atoms precess about $\hat{\mathbf{z}}$ at Ω_L, which is equivalently understood as coherent oscillation between $|+\rangle_x$ and $|-\rangle_x$ at Ω_L.

If the light intensity is modulated at $\Omega_m \neq \Omega_L$ (with $|\Omega_m - \Omega_L| \gtrsim \Gamma_{\text{rel}}$, which includes the case of unmodulated light where $\Omega_m = 0$), the precessing atomic polarization produced by

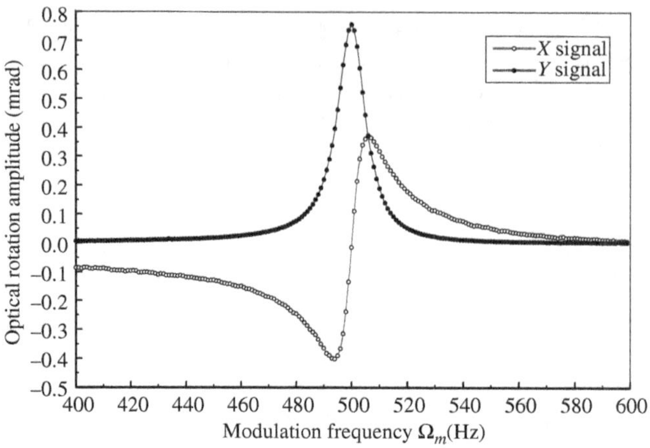

Figure 1.4 Characteristic resonant optical magnetometer signal used to measure the Larmor precession frequency. The X signal represents the signal amplitude in-phase with the frequency modulation and the Y signal represents the signal amplitude out-of-phase with the frequency modulation. Figure reprinted with permission from Ref. [35]. Copyright 2009, American Institute of Physics.

atoms optically pumped from $|-\rangle_x \to |+\rangle_x$ at a particular time is not in-phase with respect to precessing atomic polarization produced at earlier times. However, when $\Omega_m = \Omega_L$, optical pumping from $|-\rangle_x \to |+\rangle_x$ happens synchronously with Larmor precession and precessing atomic polarization produced by atoms optically pumped at time t is reinforced by the atomic polarization optically pumped at $t - 2\pi/\Omega_L$, $t - 4\pi/\Omega_L$, $t - 6\pi/\Omega_L$, etc. Alternatively, one can say that in the "stroboscopic" picture where the atomic polarization is sampled only when the light intensity is at its peak, the atomic spins are always oriented along $\hat{\mathbf{x}}$. Chapter 6 explores such optical-magnetometry techniques that employ modulated light, and similar techniques are also employed in helium magnetometers as discussed in Chapter 10.

Figure 1.4 shows a characteristic lineshape of the amplitude of an optical-magnetometer signal as Ω_m is tuned through resonance. In this particular example, linear polarization instead of circular polarization is used and $\mathbf{B}$ is along instead of perpendicular to the light propagation direction, but the essential physics remains similar (see Refs. [21, 33] and Chapter 6 for more details). For these data (from Ref. [35]), a laser beam tuned to the high-frequency side of the Doppler-broadened $F = 3 \to F'$ component of the D2 transition in atomic ^{85}Rb propagates through an antirelaxation-coated, buffer-gas-free Rb vapor cell. The laser is frequency modulated at Ω_m with a modulation amplitude of 65 MHz, which in turn modulates the light–atom interaction probability at Ω_m. A polarimeter measures the optical rotation of the laser light transmitted through the cell, and the oscillating signal is demodulated at the first harmonic of Ω_m with a lock-in amplifier. The characteristic resonant lineshape can be understood in analogy with a driven, damped harmonic oscillator: the atomic spin polarization produced by optical pumping naturally precesses at Ω_L due to the presence of **B**, and when optical pumping (the driving term for the oscillator) is modulated at $2\Omega_L$ (the factor of 2 arises because of the twofold symmetry of atomic spin polarization generated by linearly polarized light; see Ref. [29] and Section 1.3), there is a resonant enhancement of the atomic polarization leading to an enhancement of the optical-rotation signal. As is the case with driven, damped harmonic oscillators, the phase of the time-dependent optical rotation depends on the detuning of the drive frequency from the natural oscillation frequency ($2\Omega_L - \Omega_m$); thus signals are observed both in-phase (X signal) and out-of-phase (Y signal) with the modulation of the light–atom interaction probability. While the X signal nominally crosses zero at $2\Omega_L = \Omega_m$, the Y signal is maximum when $2\Omega_L = \Omega_m$. This is because when the optical-pumping rate is synchronized with the spin-precession rate, the axis of the atomic spin polarization is parallel with the light polarization at the periodic maxima in the modulated light–atom interaction probability – when the atomic spin polarization axis is parallel with the light polarization, no optical rotation is produced. The maximum optical rotation occurs when the atomic alignment axis is rotated by an angle $\phi = \pi/4$ with respect to the light polarization, which on resonance ($2\Omega_L = \Omega_m$) causes optical rotation out-of-phase with the modulation of the light–atom interaction probability.

For operation in varying ambient magnetic fields, optical magnetometers require a feedback loop to keep Ω_m locked to resonance. One approach is to use a phase-sensitive detector with an external feedback loop and a voltage-controlled oscillator. Another approach, often technically simpler, is to operate the magnetometer in a *self-oscillating* mode where the spin-precession signal is used directly to generate the oscillating RF field (see Ref. [36] and Chapter 4) or modulation of the light (see Refs. [37–39] and Chapter 6).

A brief history of the development of optical magnetometers, ranging from the work of Michael Faraday in the nineteenth century to the recent developments described in this book, is presented in the online supporting material (www.cambridge.org/9781107010352).

1.3 Density matrix and atomic polarization moments

At the heart of all optical magnetometers are polarized atomic samples, whose properties can be modeled and understood using the *density-matrix* formalism. The density matrix makes it possible to describe ensembles[3] of quantum systems that are more general than the ensembles that can be described by single-atom wave functions (referred to as *pure* ensembles). The density matrix is defined in the following way. Consider the ith atom in an ensemble described by the quantum-mechanical state $|\psi_i\rangle$, which can be represented in terms of a set of orthonormal basis states $|m\rangle$:

$$|\psi_i\rangle = \sum_m |m\rangle\langle m|\psi_i\rangle \; . \tag{1.15}$$

[3] An ensemble can either be assembled spatially (e.g., atoms contained in a vapor cell) or consist of sequential measurements separated temporally (under proper circumstances, this could even be a single quantum system).

The elements of the density matrix are given by

$$\rho_{m,n} = \sum_i \langle m|\psi_i\rangle\langle\psi_i|n\rangle \,. \tag{1.16}$$

The density-matrix formalism is discussed in detail in Refs. [29,40].

The density matrix of an ensemble of atoms in a state with angular momentum F has $(2F+1)\times(2F+1)$ components $\rho_{M,M'}$, where M, M' refer to Zeeman sublevels. The density matrix ρ can be thought of as a tensor, and it is often useful to work with the irreducible components of ρ, i.e., the components $\rho_q^{(\kappa)}$ with $\kappa = 0,\ldots,2F$ and $q = -\kappa,\ldots,\kappa$, which transform among themselves under rotations (see, for example, Refs. [41] and [42]). One can represent ρ in the following way:

$$\rho = \sum_{\kappa=0}^{2F}\sum_{q=-\kappa}^{\kappa} \rho_q^{(\kappa)} T_q^{(\kappa)}, \tag{1.17}$$

where $T_q^{(k)}$ are irreducible tensor components represented as $(2F+1)\times(2F+1)$ matrices, and the coefficients $\rho_q^{(\kappa)}$ with $\kappa = 0,\ldots,2F$ and $q = -\kappa,\ldots,\kappa$ are called *state multipoles* or *polarization moments* (PMs). The $\rho_q^{(\kappa)}$ are related to the $\rho_{M,M'}$ by

$$\rho_q^{(\kappa)} = \sum_{M,M'=-F}^{F} (-1)^{F-M'} \langle F,M,F,-M'|\kappa,q\rangle \rho_{M,M'} \,. \tag{1.18}$$

The following terminology is used for the different multipoles: $\rho^{(0)}$, monopole moment (which is equal to the population divided by $\sqrt{2F+1}$); $\rho^{(1)}$, vector moment or *orientation*; $\rho^{(2)}$, quadrupole moment or *alignment*; $\rho^{(3)}$, octupole moment; $\rho^{(4)}$, hexadecapole moment, and so on.[4] Each of the moments $\rho^{(\kappa)}$ has $2\kappa+1$ components. Polarization moments can be visualized by plotting a three-dimensional surface where the distance of the surface from the origin in a given direction is equal to the probability of finding the highest projection of the angular momentum ($M = F$) in that direction [29,44,45].

The term *polarization* is used for the general case of an ensemble that has any moment with $\kappa > 0$. When the Zeeman sublevels are not equally populated, $\rho_0^{(\kappa)} \neq 0$ for some $\kappa > 0$, and the medium is said to have *longitudinal polarization*: polarization along the quantization axis. When there are coherences between the sublevels, $\rho_q^{(\kappa)} \neq 0$ for some $q \neq 0$, and the medium is said to have *transverse polarization*. When the magnetic field is along the quantization axis, transverse polarization precesses about **B**, while longitudinal polarization remains static. For a given quantization axis z, the longitudinal orientation $\mathcal{O}_z$

[4] There are other definitions of the terms "orientation" and "alignment" in the literature. For example, in Ref. [43], alignment designates even moments in atomic polarization (quadrupole, hexadecapole, etc.), while orientation designates the odd moments (dipole, octupole, etc.).

and longitudinal alignment $\mathcal{A}_{zz}$ are given by

$$\mathcal{O}_z \propto \rho_0^{(1)} \propto \langle F_z \rangle,$$
$$\mathcal{A}_{zz} \propto \rho_0^{(2)} \propto \langle 3F_z^2 - F^2 \rangle, \tag{1.19}$$

respectively.

Of particular interest for magnetometry [46–49] are the PMs with the highest possible κ for given electronic angular momenta L and S and nuclear spin I, which exhibit no nonlinear Zeeman effect. This is due to the fact that in the case of transverse polarization, PMs with maximal κ are related to coherences between stretched states, i.e., those states with maximal projections of angular momenta along the quantization axis, which are the same in the low- and high-field regimes (see Fig. 1.1 and surrounding discussion).

Also of importance is the role of spin-exchange collisions in the transfer of PMs between atoms: angular momentum conservation demands that spin-exchange collisions preserve the net orientation in an atomic vapor, whereas higher-order PMs are not conserved in such collisions. Thus spin-exchange collisions inevitably lead to relaxation of PMs with $\kappa \geq 2$, as opposed to oriented atomic vapors where there are techniques to avoid spin-exchange broadening of magnetic resonances [50, 51].

Note that optical pumping with circularly polarized light (in the absence of other external fields) generally creates multipoles of all orders ($\kappa \leq 2F$), while pumping with linearly polarized light creates only even-ordered multipoles. This latter fact is a consequence of the symmetry of linearly polarized light: the optical electric field has no preferred direction in space, only a preferred axis. Also note that electric-dipole transition selection rules based on angular momentum conservation require that a single photon, having spin 1, can only possess PMs with $\kappa = 1$ (orientation) or $\kappa = 2$ (alignment). Thus to either generate or probe PMs with $\kappa > 2$ through light–atom interactions requires multiphoton processes, higher electromagnetic multipole transitions (for example electric quadrupole, etc.), or the presence of some other polarizing effect (such as thermal, spin polarization in strong magnetic fields).

The optical properties of a polarized atomic medium are related to the symmetry of the PMs. A light field with wave vector $\mathbf{k}$ propagating in the $\hat{\mathbf{z}}$ direction can, in general, be written as a superposition of *optical polarization eigenmodes* (i.e., waves that traverse the medium without changing their state of polarization, experiencing only attenuation and phase shifts). The optical eigenmodes are determined by the symmetry of the medium, which in a spin-polarized vapor is given by the symmetry of the PMs. For example, in an oriented atomic vapor with light propagating along the orientation direction, the optical eigenmodes are left- and right-circularly polarized waves, since there is no other preferred axis or direction in the system. In fact, orientation, which is an axial vector,[5] creates a preferred "handedness" in the system, which manifests as a breaking of the symmetry

[5] An *axial vector* is a vector that does not change sign under spatial inversion.

between the two optical eigenmodes. This symmetry breaking creates different complex indices of refraction for the two circular components.

Optical magnetometers often use measurements of optical rotation of linearly polarized light to monitor atomic spin precession. Linearly polarized light can be expressed as a coherent superposition of circular components, for example,

$$\hat{\mathbf{x}} = -\frac{1}{\sqrt{2}}(\hat{\epsilon}_+ - \hat{\epsilon}_-)\,, \tag{1.20}$$

$$\hat{\mathbf{y}} = \frac{i}{\sqrt{2}}(\hat{\epsilon}_+ + \hat{\epsilon}_-)\,, \tag{1.21}$$

where $\hat{\epsilon}_+$ is the unit vector describing left-circular polarization and $\hat{\epsilon}_-$ is the vector describing right-circular polarization (detailed discussion of the parametrization of light polarization is presented, for example, in Refs. [28, 29, 43]). The plane of light polarization is defined by the relative phase between the two circular components. As the light propagates through an oriented atomic vapor, because of the difference in the indices of refraction, the two circular components acquire a relative phase shift and the plane of light polarization is rotated. In this case, Larmor precession of the orientation generates oscillation of the optical rotation at the Larmor frequency. Similar effects occur when polarized light interacts with polarized atoms possessing other higher-rank PMs, such as alignment (see Chapter 6).

1.4 Sensitivity and accuracy

1.4.1 Variational sensitivity (short-term resolution) and long-term stability

The (variational) sensitivity of a magnetometer in a stable field **B** is characterized by the dispersion δB of measurements averaged over a time interval T for a given duration τ of a single measurement (where $T \gg \tau$):

$$\delta B = \sqrt{\langle B(t)^2\rangle - \langle B(t)\rangle^2}\,, \tag{1.22}$$

where $\langle\cdots\rangle$ denotes the average of the measurements over time T.

Figure 1.5 shows an example of the spectrum of magnetic-field noise δB measured by a spin-exchange-relaxation-free (SERF) magnetometer using potassium atoms [51] (see Chapter 5). The magnetic-field-noise spectrum is obtained by recording the magnetometer response for ~100 s, performing a fast Fourier transform (FFT) without windowing, and calculating RMS amplitudes in 1 Hz bins. The SERF magnetometer is a zero- or low-field optical magnetometer, and in this case, it is surrounded by a five-layer mumetal magnetic shield system (see Chapter 12 for a review of magnetic-shielding techniques) and a system of coils null the residual fields and gradients. Residual magnetic field noise at the level of 7 $\mathrm{fT}/\sqrt{\mathrm{Hz}}$ is observed originating from Johnson noise of the innermost magnetic-shield layer. The intrinsic sensitivity of the magnetometer is obtained by canceling the ambient magnetic field noise through operation of the SERF magnetometer in gradient mode, where

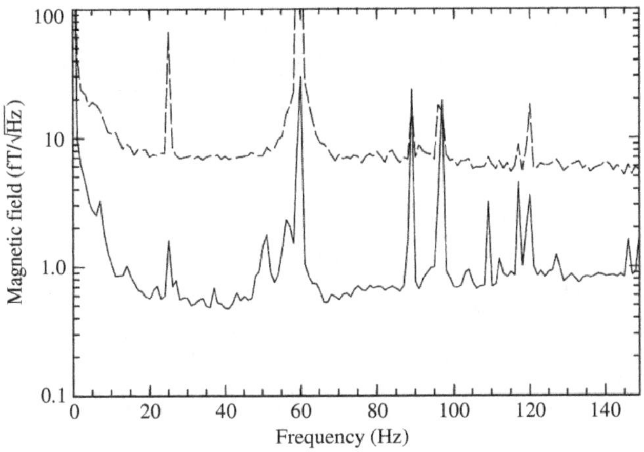

Figure 1.5 Magnetic-field-noise spectrum measured by a single SERF magnetometer channel (dashed line) and intrinsic sensitivity (solid line) obtained by operating the SERF magnetometer in gradiometer mode (taking the difference between adjacent channels). Figure adapted from Ref. [51].

magnetic-field measurements carried out in two adjacent channels are subtracted (diffusion of K atoms is slowed by He buffer gas, enabling localized magnetic-field measurements – for the data in Fig. 1.5 the two channels are spaced 3 mm apart, so magnetic-field fluctuations are common-mode for the two channels).

Fundamentally, δB is limited by quantum fluctuations as noted in Section 1.1. However, in practical applications, technical sources of random noise contribute differently for different measurement durations τ, and thus, as in the study of atomic clocks, it is useful to study δB as a function of the measurement duration τ. The time dependence of sensitivity is usually described in terms of the Allan diagram [52] depicting the dependence of the Allan deviation $\sigma(\tau)$ or the Allan variance $\sigma^2(\tau)$ over the time interval $T = N\tau$:

$$\sigma^2(\tau) = \frac{1}{2(N-1)} \sum_{i=1}^{N} [B_{i+1}(\tau) - B_i(\tau)]^2 , \tag{1.23}$$

where $B_i(\tau)$ represents the results of successive measurements within adjacent time intervals averaged over the duration τ. If the measurement results have a normal distribution, then the Allan deviation $\sigma(\tau)$ coincides with the sensitivity δB as expressed in Eq. (1.22).

Figure 1.6 shows the Allan deviation plot for a self-oscillating Rb magnetometer based on the double radio-optical resonance (DRR) technique [53] (see Chapter 4). The two stable isotopes of rubidium, ^{85}Rb and ^{87}Rb, are contained within a single, antirelaxation-coated cell, and the Larmor precession frequencies f_{85} and f_{87} are simultaneously measured for both isotopes. The self-oscillating DRR Rb optical magnetometer is a high-field magnetometer, in the present example the ambient field is $B = 1170$ nT $= 11.7$ mG, stabilized with a feedback loop using the magnetometric signal from ^{85}Rb. Based on the gyromagnetic ratios for the

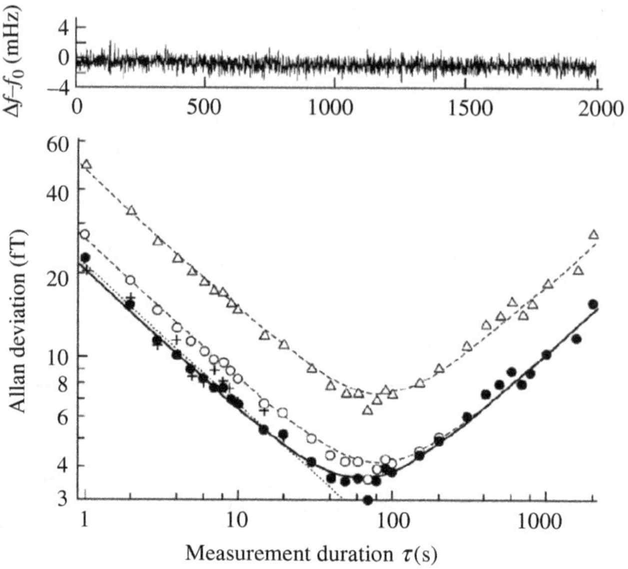

Figure 1.6 Upper plot: record of the difference frequency $\Delta f - f_0$ (see text). Lower plot: Allan deviation of the ^{87}Rb magnetometer signal (open triangles), the magnetic field derived from the difference frequency Δf (open circles), the ^{87}Rb magnetometer signal after subtraction of ambient field noise (filled circles); shot-noise-projected sensitivity (crosses). Figure adapted with kind permission from Springer Science Media from Ref. [53].

Rb isotopes, $3f_{85} \approx 2f_{87}$, and so the difference frequency $\Delta f = 3f_{85} - 2f_{87}$ is near zero with only a small offset $f_0 \approx 12.47$ Hz. This allows straightforward subtraction of the noise due to fluctuations of the ambient magnetic field yielding the intrinsic sensitivity of the optical magnetometer, which agrees well with the projected shot-noise-limited sensitivity for integration times less than 100 s.

A comparison of Allan diagrams of different magnetometric devices indicates that they are characterized (similarly to atomic frequency standards) by the measurement time τ_0 at which the decrease in the Allan deviation as a function of τ is followed by a flat dependence or an increase in $\sigma(\tau)$ as a function of τ (in Fig. 1.6, $\tau_0 \approx 80$ s). Such behavior is due to parametric shifts of the magnetic resonance frequency Ω itself and to errors in the measurement of its value. The variation in $\sigma(\tau)$ over measurement durations $\tau > \tau_0$ determines the long-term stability of the magnetometer.

1.4.2 Parameter optimization

As seen in the preceding examples (Figs. 1.5 and 1.6), the sensitivity of a magnetometer can be determined via direct measurement in a magnetic field environment with magnetic field stability exceeding the magnetometric sensitivity, or via a differential measurement using identical magnetometers measuring the field in approximately the same region of space. For parameter optimization, however, it may be more efficient to determine the response of the optical signal to a known change in the magnetic field and estimate the sensitivity based on the light power reaching the detector. This approach separates the parameter optimization problem from the potentially difficult issues connected with eliminating technical sources of noise limiting the practical magnetometric sensitivity.

In parameter optimization studies, what is typically reported is the *shot-noise-projected* (SNP) sensitivity, given by

$$\delta B_{\rm SNP} = \left(\frac{d\varphi}{dB}\right)^{-1} \delta\varphi \,, \tag{1.24}$$

where $d\varphi/dB$ is the slope of the detected optical signal with respect to the applied magnetic field and $\delta\varphi$ is the photon-shot-noise-projected sensitivity of the optical measurement. It should be noted that in most cases $\delta B_{\rm SNP}$ cannot be reduced below the atomic-shot-noise limit given by Eq. (1.1) (important exceptions are described in Chapters 2 and 3). As seen in Fig. 1.6, $\delta B_{\rm SNP} = \delta B$ when ambient magnetic-field noise can be reduced below the sensitivity of the magnetometer.

1.4.3 Absolute accuracy and systematic errors

One of the key advantages of optical magnetometers as compared to other magnetometry schemes is that, in principle, measurement of the spin-precession frequency Ω yields an intrinsically accurate, calibration-free determination of the magnetic field. Nonetheless, there are several sources of systematic errors that need to be considered in magnetometric measurements (see also Chapter 20 for further discussion of systematic effects).

One important source of systematic error is the effect of the pump and probe light on the precession frequency. The light–atom interaction can shift the energy of Zeeman sublevels through the AC Stark effect, and hence shift Ω. This produces, in effect, a fictitious magnetic field along the axis of light propagation proportional to the degree of circular polarization of the light [54]. There are several techniques that can be used to reduce light shifts: (1) reduce the light intensity, since the AC Stark shifts generally scale proportionally for sufficiently low intensities; (2) employ an experimental geometry where the light propagation direction is orthogonal to the measured magnetic field direction, in which case the fictitious field adds in quadrature with the leading field; (3) employ linearly polarized light, which in principle eliminates the effect, although there can be issues with birefringence of optical elements between the polarizers and the atomic sample, as well as subtle issues related to alignment-to-orientation conversion (AOC) that can produce fictitious fields even from linearly polarized light [35, 55]; and (4) tune the laser light to a frequency where the AC Stark shifts are minimized, taking advantage of zero-crossings in the spectral dependence of the AC Stark effect [56, 57].

While the atomic spins are precessing in the magnetic field they are effectively decoupled from the experimental apparatus, and thus rotations of the apparatus can lead to gyroscopic shifts of the precession frequency (Chapter 19). This is especially important for optical magnetometers on moving platforms. Even gyroscopic effects due to the rotation of the Earth can be significant for stationary optical magnetometers used in high-precision tests of fundamental physics [58, 59] (see Chapter 18).

Because of the vector/tensor nature of the light–atom interaction employed for pumping and probing atomic spins, there are typically certain orientations of the magnetic field relative to the magnetic sensor (in particular relative to the light propagation direction and light polarization axis/direction) for which the magnetometric sensitivity is drastically reduced. Such orientations are known as *dead zones*, and the problem has been addressed by using multiple sensors [60, 61], mechanical rotation of some components [62], use of unpolarized light and spatially varying microwave fields [63], and more recently by simultaneously pumping and probing PMs of different κ [64]. Furthermore, due to shifts, splittings, and lineshape asymmetries of magnetic resonances caused by the nonlinear Zeeman effect, the measured magnetic field magnitude can acquire a false dependence on the relative orientation of the sensor to the field (*heading error*).

Collisions between atoms and with cell walls also can cause systematic shifts of Ω. Collisional frequency shifts are well studied [1–3] and can be accounted for with precise knowledge of the atomic vapor density, while frequency shifts due to wall collisions can be minimized with a spherical cell geometry.

1.5 Vector and scalar magnetometers

Optical magnetometers can be operated in either *scalar mode*, measuring the magnitude of the magnetic field, or *vector mode*, measuring one or more vector components of the field. There are advantages and disadvantages for each mode of operation, and so the choice of configuration largely depends on the application. Scalar magnetometers are insensitive to rotation of the magnetometer in the Earth's magnetic field and rely on frequency measurements which can be made with very high accuracy. For example, a 10 nG change in the Earth's magnetic field corresponds to a fractional frequency shift of 20 ppb, easily within the stability range of even the simplest quartz oscillators. Since the relation between the frequency and the magnetic field [see Eqs. (1.5) and (1.7)] depends only on fundamental constants, scalar atomic magnetometers do not require calibration and are essentially free from aging effects.

Vector magnetometers can measure all three components of the magnetic field and thus can obtain more complete information about the field and provide much better localization information for detection of magnetic-anomaly sources [65]. On the other hand, vector magnetometers are much more sensitive to rotations: in the worst case, a rotation by $\sim$20 nrad will mimic a field change of $\sim$10 nG. Therefore, vector magnetometers have to be mounted on stationary platforms, combined with sensitive gyroscopes, or used in combination to form magnetic gradiometers in order to realize magnetometric sensitivities near the $\mathrm{nG}/\sqrt{\mathrm{Hz}}$ level in practical systems.

Optical magnetometers can naturally operate in either vector or scalar mode using essentially the same setup [66–68]. The technique for conversion between scalar and vector operation relies on the fact that if a small bias field is applied to the sensor in a certain direction in addition to the field to be measured, then the change in the overall field magnitude is linear in the projection of the bias field on the main field, and is only quadratic (and generally negligible) in the projection on the orthogonal plane. Thus, applying three orthogonal bias fields consecutively, and performing three measurements of the overall magnetic field magnitude, one reconstructs the overall field vector.

1.6 Applications

The applications of optical-magnetometry techniques are diverse and expanding. Optical magnetometers are used to carry out a variety of geophysical magnetic field measurements (discussed in Chapter 17) – studying natural and anthropogenic magnetic anomalies, investigating geomagnetic-field dynamics, and measuring the magnetic properties of geological samples. Optical magnetometers are also deployed to measure magnetic fields in space (Chapter 15) and new techniques have recently been proposed to measure geomagnetic fields using the naturally occurring atomic-sodium-rich layer in the mesosphere (Chapter 13). Of considerable interest are direct measurements of magnetic fields of biological origin (Chapter 16), for example those from the heart and brain. A variety of state-of-the-art optical magnetometers with extremely high spatial resolution, in particular those employing nitrogen-vacancy (NV) centers in diamond (Chapter 8) and spinor Bose–Einstein condensates (Chapter 9), are now capable of magnetic microscopy enabling new applications in material science and biology. Optical magnetometers have also been applied to novel nuclear magnetic resonance (NMR) and magnetic resonance imaging (MRI) studies (Chapter 14). Already significant progress has been made in miniaturization of optical magnetometers, to the point where microfabrication techniques allow chip-scale devices (Chapter 7). Related experimental techniques are also used in tests of fundamental physical laws and symmetries of nature (Chapter 18) and for gyroscopes based on atomic spin (Chapter 19). It is exciting to witness and participate in moving optical magnetometers from the laboratory domain to a variety of diverse applications.

Acknowledgments

The authors are grateful to M. V. Romalis for important early contributions to the chapter, to M. Zolotorev for insightful discussions, and to A. K. Vershovskiy for helpful comments.

References

[1] W. Happer, Rev. Mod. Phys. **44**, 169 (1972).
[2] T. G. Walker and W. Happer, Rev. Mod. Phys. **69**, 629 (1997).

[3] W. Happer, Y.-Y. Jau, and T. Walker, *Optically Pumped Atoms* (Wiley-VCH, New York, 2010).
[4] D. Budker and M. V. Romalis, Nature Physics **3**, 227 (2007).
[5] E. B. Aleksandrov and A. K. Vershovskii, Usp. Fiz. Nauk **52**, 573 (2009).
[6] M. Auzinsh, D. Budker, D. F. Kimball, S. M. Rochester, J. E. Stalnaker, A. O. Sushkov, and V. V. Yashchuk, Phys. Rev. Lett. **93**, 173002 (2004).
[7] H. G. Robinson, E. S. Ensberg, and H. G. Dehmelt, Bull. Am. Phys. Soc. **3**, 9 (1958).
[8] M. A. Bouchiat and J. Brossel, Phys. Rev. **147**, 41 (1966); M. A. Bouchiat, Ph. D. Thesis, University of Paris, 1964.
[9] E. B. Aleksandrov, Optiko-Mekh. Promyshl. **55**, 27 (1988) [Sov. J. Opt. Technol. (USA), **55**, 731 (1988)].
[10] M. V. Balabas, T. Karaulanov, M. P. Ledbetter, and D. Budker, Phys. Rev. Lett. **105**, 070801 (2010).
[11] J. Brossel, J. Margerie, and A. Kastler, Compt. Rend. **241**, 865 (1955).
[12] C. Cohen-Tannoudji, J. Brossel, and A. Kastler, Compt. Rend. **244**, 1027 (1957).
[13] H. Dehmelt, Phys. Rev. **105**, 1487 (1957).
[14] T. Skalinsky, Compt. Rend. **245**, 1908 (1957).
[15] F. Hartmann, M. Rambosson, J. Brossel, and A. Kastler, Compt. Rend. **246**, 1522 (1958).
[16] N. Davidson, H. J. Lee, C. S. Adams, M. Kasevich, and S. Chu, Phys. Rev. Lett. **74**, 1311 (1995).
[17] M. Vengalattore, J. M. Higbie, S. R. Leslie, J. Guzman, L. E. Sadler, and D. M. Stamper-Kurn, Phys. Rev. Lett. **98**, 200801 (2007).
[18] M. Arndt, S. I. Kanorsky, A. Weis, and T. W. Hänsch, Phys. Rev. Lett. **74**, 1359 (1995).
[19] A. Weis, S. Kanorsky, M. Arndt, and T. W. Hänsch, Z. Phys. B: Condens. Matter **98**, 359 (1995).
[20] S. I. Kanorsky, S. Lang, S. Lucke, S. B. Ross, T. W. Hänsch, and A. Weis, Phys. Rev. A **54**, R1010 (1996).
[21] D. Budker, W. Gawlik, D. F. Kimball, S. M. Rochester, V. V. Yashchuk, and A. Weiss, Rev. Mod. Phys. **74**, 1153 (2002).
[22] M. Fleischhauer, A. B. Matsko, and M. O. Scully, Phys. Rev. A **62**, 013808 (2000).
[23] M. P. Ledbetter, I. M. Savukov, V. M. Acosta, D. Budker, and M. V. Romalis, Phys. Rev. A **77**, 033408 (2008).
[24] R. S. Van Dyck, Jr., P. B. Schwinberg, and H. G. Dehmelt, Phys. Rev. Lett. **59**, 26 (1987).
[25] B. Odom, D. Hanneke, B. D'Urso, and G. Gabrielse, Phys. Rev. Lett. **97**, 030801 (2006).
[26] D. Hanneke, S. Fogwell, and G. Gabrielse, Phys. Rev. Lett. **100**, 120801 (2008).
[27] T. Aoyama, M. Hayakawa, T. Kinoshita, and M. Nio, Phys. Rev. Lett. **99**, 110406 (2007).
[28] D. Budker, D. F. Kimball, and D. DeMille, *Atomic Physics: An Exploration Through Problems and Solutions*, 2nd edition (Oxford University Press, Oxford, 2008).
[29] M. Auzinsh, D. Budker, and S. Rochester, *Optically Polarized Atoms: Understanding Light–Atom Interactions* (Oxford University Press, Oxford, 2010).
[30] E. Aleksandrov, Opt. Spectros. (USSR) **17**, 522 (1964).
[31] J. N. Dodd, R. Kaul, and D. Warrington, Proc. Phys. Soc. **84**, 176 (1964).
[32] E. B. Alexandrov, M. P. Chaika, and G. I. Khvostenko, *Interference of Atomic States* (Springer, New York, 1993).

[33] E. B. Alexandrov, M. Auzinsh, D. Budker, D. F. Kimball, S. M. Rochester, and V. V. Yashchuk, J. Opt. Soc. Am. B **22**, 7 (2005).
[34] W. E. Bell and H. L. Bloom, Phys. Rev. Lett. **6**, 280 (1961).
[35] D. F. Jackson Kimball, L. R. Jacome, S. Guttikonda, E. J. Bahr, and L. F. Chan, J. Appl. Phys. **106**, 063113 (2009).
[36] A. Bloom, Appl. Opt. **1**, 61 (1962).
[37] A. B. Matsko, D. Strekalov, and L. Maleki, Opt. Commun. **247**, 141 (2005).
[38] P. D. D. Schwindt, L. Hollberg, and J. Kitching, Rev. Sci. Instrum. **76**, 126103 (2005).
[39] J. Higbie, E. Corsini, and D. Budker, Rev. Sci. Instrum. **77**, 113106 (2006).
[40] K. Blum, *Density Matrix Theory and Applications* (Springer-Verlag, New York, 2011).
[41] A. Omont, Prog. Quantum Electron. **5**, 69–138 (1977).
[42] D. A. Varshalovich, A. N. Moskalev, and V. K. Khersonskii, *Quantum Theory of Angular Momentum: Irreducible Tensors, Spherical Harmonics, Vectors Coupling Coefficients, 3nj Symbols* (World Scientific, Singapore, 1988).
[43] R. N. Zare, *Angular Momentum: Understanding Spatial Aspects in Chemistry and Physics* (Wiley, New York, 1988).
[44] M. Auzinsh, Can. J. Phys. **75**, 853 (1997).
[45] S. Rochester and D. Budker, Am. J. Phy. **69**, 450 (2001).
[46] V. V. Yashchuk, D. Budker, W. Gawlik, D. F. Kimball, Yu. P. Malakyan, and S. M. Rochester, Phys. Rev. Lett. **90**, 253001 (2003).
[47] V. M. Acosta, M. Auzinsh, W. Gawlik, P. Grisins, J. M. Higbie, D. F. Jackson Kimball, L. Krzemien, M. P. Ledbetter, S. Pustelny, S. M. Rochester, V. V. Yashchuk, and D. Budker, Optics Express **16**, 11423 (2008).
[48] E. B. Aleksandrov, M. V. Balabas, A. K. Vershovski, and A. S. Pazgalev, J. Tech. Phys. **44**, 1025 (1999).
[49] A. I. Okunevich, Opt. Spectrosc. **91**, 177 (2001).
[50] J. C. Allred, R. N. Lyman, T. W. Kornack, and M. V. Romalis, Phys. Rev. Lett. **89**, 130801, (2002).
[51] I. K. Kominis, T. W. Kornack, J. C. Allred, and M. V. Romalis, Nature **422**, 596 (2003).
[52] D. Allan, Proc. IEEE **54**, 221 (1966).
[53] E. B. Alexandrov, M. V. Balabas, A. K. Vershovski, and A. S. Pazgalev, Tech. Phys. **49**, 779 (2004).
[54] W. Happer and B. Mathur, Phys. Rev. **163**, 12 (1967).
[55] D. Budker, D. F. Kimball, S. M. Rochester, and V. V. Yashchuk, Phys. Rev. Lett. **85**, 2088 (2000).
[56] E. B. Alexandrov, M. V. Balabas, A. S. Pazgalev, A. K. Vershovskii, N. N. Yakobson, Laser Phys. **6**, 244 (1996).
[57] I. Novikova, A. B. Matsko, V. L. Velichansky, M. O. Scully, and G. R. Welch, Phys. Rev. A **63**, 063802 (2001).
[58] J. M. Brown, S. J. Smullin, T. W. Kornack, and M. V. Romalis, Phys. Rev. Lett. **105**, 151604 (2010).
[59] B. J. Venema, P. K. Majumder, S. K. Lamoreaux, B. R. Heckel, and E. N. Fortson, Phys. Rev. Lett. **68**, 135 (1992).
[60] B. Chéron *et al.*, J. Phys. III (France) **7**, 1735 (1997).
[61] B. Chéron, H. Gilles, and J. Hamel, Eur. Phys. J. Appl. Phys. **13**, 143 (2001).
[62] C. Guttin, J. Leger, and F. Stoeckel, J. Phys. IV (France) **4**, C4-655 (1994).
[63] E. B. Aleksandrov, A. K. Vershovskii, and A. S. Pazgalev, Tech. Phys. **51**, 919 (2006).
[64] A. Ben-Kish and M. V. Romalis, Phys. Rev. Lett. **105**, 193601 (2010).
[65] T. R. Clem, Naval Eng. J. **110**, 139 (1998).

[66] S. Seltzer and M. V. Romalis, Appl. Phys. Lett. **85**, 4804 (2004).
[67] E. B. Alexandrov, M. V. Balabas, V. N. Kulyasov, A. E. Ivanov, A. S. Pazgalev, J. L. Rasson, A. K. Vershovskii, and N. N. Yakobson, Meas. Sci. Technol. **15**, 918 (2004).
[68] O. Gravrand, A. Khokhlov, J. L. L. Mouel, and J. M. Leger, Earth Planets Space **53**, 949 (2001).

2

Quantum noise in atomic magnetometers

M. V. Romalis

2.1 Introduction

An atomic magnetometer, as any quantum device, has an intrinsic limit on its sensitivity that is imposed by the Heisenberg uncertainty principle. In this chapter we will discuss such fundamental limits as well as the practical means of approaching and in some cases even overcoming these limits by means of quantum measurements.

As a model system we consider an ensemble of N_a atoms each with spin F and ignore for the moment hyperfine interactions present in alkali-metal atoms. Most atomic magnetometers operate using spin orientation of the atomic ensemble, so we will consider atoms that are initially optically pumped into a fully spin-polarized state, say along the $\hat{\mathbf{x}}$ direction, $\langle F_x \rangle = F$. In a vector spin-exchange-relaxation-free (SERF) magnetometer (Chapter 5) or a radiofrequency magnetometer (Chapter 4) one measures a small transverse component of the spin, $\langle F_z \rangle$, that develops due to the interaction

$$H_B = -\boldsymbol{\mu} \cdot \mathbf{B} = g_F \mu_B \mathbf{B} \cdot \mathbf{F} \tag{2.1}$$

with a weak magnetic field B_y.[1] In a scalar atomic magnetometer the observable is the frequency of spin precession about a large B_y field. This frequency is best measured by finding the time between two zero crossings of an oscillating $\langle F_z \rangle$ component while the spin polarization points in the $\hat{\mathbf{x}}$ direction. Thus, in all cases one measures the component of the spin perpendicular to the direction of the large spin polarization, and the quantum sensitivity limits are nearly the same for vector, RF, and scalar atomic magnetometers.

[1] Note that the definition of μ_B used in this book (Eq. 1.5) uses the absolute value of the electron charge. Thus μ_B is defined to be positive whereas the magnetic moment of the electron is negative, explaining the disappearance of the negative sign on the right-hand side of Eq. (2.1). Also, in our notation the angular momentum F is a number, it does not carry units of $\hbar$.

Optical Magnetometry, ed. Budker, D. and Jackson Kimball, D. F. Published by Cambridge University Press.

2.2 Spin-projection noise

For a fully polarized spin state with $\langle F_x \rangle = F$, the spin wave function $|\psi\rangle$ with quantization along the z axis can be found using Wigner's d matrix to rotate the spin state by $\pi/2$ around the y axis:

$$|\psi\rangle = \sum_{m_F} d^F_{m_F F}\left(\frac{\pi}{2}\right)|F,F\rangle = \sum_{m_F} \sqrt{\frac{(2F)!}{(F+m_F)!(F-m_F)!}} \frac{(-1)^{F-m_F}}{2^F} |F,m_F\rangle\,. \tag{2.2}$$

In a measurement of F_z the probability of obtaining each value of m_F is proportional to $\left[d^F_{m_F F}(\pi/2)\right]^2$. The standard deviation of many measurements on different atoms will then be given by

$$\sigma(F_z) = \sqrt{\sum_{m_F} m_F^2 \left[d^F_{m_F F}\left(\frac{\pi}{2}\right)\right]^2} = \sqrt{\frac{F}{2}}\,. \tag{2.3}$$

The same result can be obtained by using the angular momentum commutation relationship $[F_y, F_z] = iF_x$ to get the minimum quantum-mechanical uncertainty limit $\sigma(F_z)\sigma(F_y) \geq \langle F_x \rangle /2$ and by assuming that the standard deviations in F_z and F_y are the same. If we make a measurement of the mean $\langle F_z \rangle$ for all N_a atoms, each with standard deviation given by Eq. (2.3), the resulting uncertainty will be given by

$$\sigma(\langle F_z \rangle) = \sqrt{\frac{F}{2N_a}}\,. \tag{2.4}$$

This is known as the atomic spin-projection noise or the standard quantum limit (SQL) [1].

The spin-projection noise is approximately the same even if the atoms are not spin polarized. For completely unpolarized atoms $\langle F_x^2 \rangle + \langle F_y^2 \rangle + \langle F_z^2 \rangle = F(F+1)$ and we have

$$\sigma_{\text{unpol}}(\langle F_z \rangle) = \sqrt{\frac{F(F+1)}{3N_a}}\,. \tag{2.5}$$

For spin-1/2 atoms the spin-projection noise is independent of polarization while for larger values of F it changes slowly with polarization.

2.3 Faraday rotation measurements

How can we actually measure the spin state of the atoms to realize this uncertainty? Among different atomic interrogation schemes the one that has been predominantly used for quantum measurements is paramagnetic Faraday rotation of linearly polarized light [2–4]. The rotation is due to the vector atomic polarizability α_V and can be expressed by the following atomic Hamiltonian [5]:

$$H = -\frac{1}{2}\alpha_V \frac{\mathbf{F}}{F} \cdot \frac{\mathbf{E} \times \mathbf{E}^*}{i}, \tag{2.6}$$

where $\mathbf{E}$ is the electric field of the probe light. This interaction causes two closely related effects: rotation of the plane of polarization of linearly polarized probe light and vector AC Stark shifts; they constitute quantum measurement and back-action effects. Note that $(\mathbf{E} \times \mathbf{E}^*)/i = \mathbf{s}E^2 = \mathbf{S}_3$, where $\mathbf{s}$ is a vector parallel to the light propagation direction and proportional to the degree of circular polarization of the probe light; it can also be written in terms of the circular polarization Stokes parameter S_3.

Consider N_a atoms confined to a volume V with a length L along the $\hat{\mathbf{z}}$ direction and cross-sectional area A. The polarization of a linearly polarized probe light propagating along $\hat{\mathbf{z}}$ will be rotated due to the circular birefringence of the atomic vapor with a finite vector polarizability. The rotation angle is equal to

$$\phi = \frac{4\pi^2 L N_a}{\lambda V} \mathrm{Re}[\alpha_V] \frac{\langle F_z \rangle}{F} . \tag{2.7}$$

The uncertainty in the measurement of ϕ is limited by photon shot noise if one uses classical light. If we use a balanced polarimeter to measure the polarization rotation angle ϕ, for small rotations it is given by $\phi = (N_1 - N_2)/2(N_1 + N_2)$, where N_1 and N_2 are the numbers of photons collected in each photodetector. If N_{ph} photons pass through the atomic ensemble with negligible absorption, each photodetector will collect on average $N_{\mathrm{ph}}/2$ photons with shot noise uncertainty given by $\sqrt{N_{\mathrm{ph}}/2}$. Then the uncertainty in the measurement of the angle ϕ is given by

$$\sigma(\phi) = \frac{1}{2\sqrt{N_{\mathrm{ph}}}} . \tag{2.8}$$

From the measurement of the rotation angle, one finds a value of the $\langle F_z \rangle$ spin component with an uncertainty given by

$$\sigma_{\mathrm{ph}}(\langle F_z \rangle) = \frac{\lambda A F}{8\pi^2 N_a \mathrm{Re}[\alpha_V] \sqrt{N_{\mathrm{ph}}}}. \tag{2.9}$$

Although the uncertainty in the measurement of $\langle F_z \rangle$ due to photon shot noise is not a fundamental limit for atomic magnetometers, as we will show below, it is often a practical limitation. On the other hand, it is also possible to reduce this source of noise by using nonclassical probe light with polarization squeezing, as was demonstrated in [6, 7].

2.4 Quantum back-action

The AC Stark shift acts as a quantum back-action mechanism for Faraday rotation spin measurements. Even though the probe light is linearly polarized, it will have quantum fluctuations in the degree of circular polarization, which can limit the ultimate magnetic field sensitivity, as was first considered in Ref. [8]. Due to the interaction (2.6) these fluctuations will have an effect similar to a magnetic field fluctuating along the $\hat{\mathbf{z}}$ direction.

To analyze the effect of AC Stark shift noise it is convenient to use the formalism of stochastic differential equations [9], which we will also use to analyze other types of quantum fluctuations. Consider a short time interval dt during which dN_{ph} probe photons have interacted with the atomic ensemble. The quantum fluctuations of the average circular polarization s of the probe can be calculated by assuming that it consists of $dN_{\text{ph}}/2$ left- and right-circularly polarized photons. The shot-noise fluctuations in the number of these photons will result in fluctuating circular polarization of the probe beam with a zero mean and a standard deviation equal to

$$\sigma(s) = \frac{1}{\sqrt{dN_{\text{ph}}}} . \tag{2.10}$$

Then the evolution of the spin polarization due to the Hamiltonian (2.6) can be calculated using the Bloch equation for spin evolution due to a randomly fluctuating effective B_z magnetic field. We get

$$d\left\langle F_y \right\rangle = \langle F_x \rangle \operatorname{Re}[\alpha_V] \frac{4\pi^2 \nu}{cAF} \frac{dN_{\text{ph}}}{dt} \sigma(s) \xi(t) dt , \tag{2.11}$$

where $\xi(t)$ represents random Gaussian white noise with a root mean square amplitude of unity and a negligibly short correlation time corresponding to a coherent probe laser beam. One can see that $\left\langle F_y \right\rangle$ will execute a random walk and after scattering N_{ph} photons the uncertainty in $\left\langle F_y \right\rangle$ is given by

$$\sigma_{ac}(\left\langle F_y \right\rangle) = \langle F_x \rangle \operatorname{Re}[\alpha_V] \frac{4\pi^2 \nu}{cAF} \sqrt{N_{\text{ph}}} . \tag{2.12}$$

Using Eq. (2.9) we find that $\sigma_{\text{ph}}(\left\langle F_y \right\rangle)\sigma_{ac}(\langle F_z \rangle) = \langle F_x \rangle / 2N_a$, so reduction in the uncertainty of $\langle F_z \rangle$ by a measurement of the probe polarization leads to a corresponding increase in the uncertainty of $\left\langle F_y \right\rangle$ to preserve the fundamental quantum-mechanical uncertainty principle $\sigma(F_z)\sigma(F_y) \geq \langle F_x \rangle / 2$ for N_a atoms. The effects of probe-light quantum noise on spin evolution due to the interaction (2.6) were studied experimentally in [10, 11] and their role in magnetometry was discussed in [12, 13]. Note that the AC Stark shift only introduces extra noise for spin-polarized atoms.

2.5 Time correlation of spin-projection noise

A unique aspect of atomic spin-projection noise is that it has a finite correlation time. Quantum theory tells us that after a measurement the atom will remain in the measured state until some other process, such as spin relaxation, perturbs its state. The first experimental observation of quantum spin noise and its finite correlation time was performed by Aleksandrov

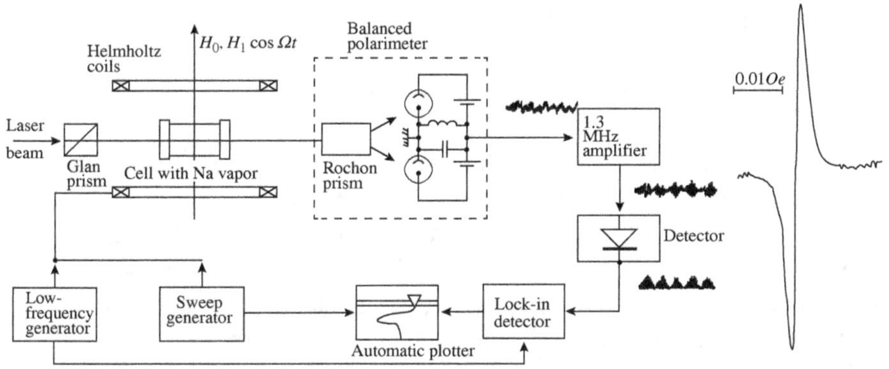

Figure 2.1 The diagram of the experimental setup used by Aleksandrov and Zapasskii for the first observation of atomic spin noise. The bias magnetic field was modulated to measure a change in the polarimeter noise power due to the shift of the Zeeman resonance frequency. The measured derivative of the atom spin noise resonance peak is shown on the right. Adapted from Ref. [14].

and Zapasskii in 1981 [14]. Their experiment is prototypical for much of the more recent work in quantum spin-noise spectroscopy [6, 15, 16] and is illustrated in Fig. 2.1. A linearly polarized laser beam propagates through a vapor cell containing unpolarized sodium atoms and neon buffer gas in a finite magnetic field. The polarization of the laser is analyzed with a balance polarimeter and the output noise power near the Zeeman resonance frequency is measured. The noise power showed a peak at the Zeeman resonance with a width similar to the inverse time of atomic diffusion across the laser beam.

The finite correlation time of atomic spin noise can be modeled quantitatively by a stochastic differential equation. Consider N_a atoms with transverse spin relaxation time T_2. In a time interval dt, $N_a(dt/T_2)$ atoms undergo a spin-relaxation process. We assume that there is no correlation between F_z for each atom before and after such a process, for example, an atomic spin-relaxation collision or a probe-photon absorption. The standard deviation of $\langle F_z \rangle_{\text{col}}$ for the atoms that have collided in time dt is equal to $\sigma(\langle F_z \rangle_{\text{col}}) = \sqrt{FT_2/(2N_a dt)}$, both before and after the collision. Therefore the change in $\langle F_z \rangle$ for the whole ensemble in time dt is a random number with a standard deviation equal to $(dt/T_2)\sqrt{FT_2/(N_a dt)}$. We obtain the following stochastic differential equation:

$$d\langle F_z \rangle = -\frac{\langle F_z \rangle}{T_2}dt + \xi(t)\sqrt{\frac{Fdt}{N_a T_2}}, \tag{2.13}$$

where the first term represents the usual spin relaxation of the average polarization to zero, while the second term represents quantum spin fluctuations. This is an example of a so-called Ornstein–Uhlenbeck process, which describes a random walk with relaxation to the mean [9]. The time correlation of $\langle F_z \rangle$ fluctuations can be expressed by the covariance of two consecutive measurements of $\langle F_z \rangle$ at times t_1 and t_2,

$$\mathrm{cov}[\langle F_z(t_1)\rangle, \langle F_z(t_2)\rangle] = \frac{F}{2N_a} \exp(-|t_1 - t_2|/T_2) , \tag{2.14}$$

and on a time scale much longer than T_2 the standard deviation of $\langle F_z \rangle$ is equal to $\sigma(\langle F_z \rangle) = \sqrt{F/2N_a}$. In the presence of a B_x magnetic field, one can add a precession term to Eq. (2.13), turning it into a system of Bloch equations with noise terms for both F_z and F_y.

The evolution given by the stochastic differential equation (2.13) remains accurate if the measurement error in the Faraday rotation $\sigma_{\mathrm{ph}}(\langle F_z \rangle)$ is smaller than $1/\sqrt{N_a}$ as long as it is much larger than $1/N_a$, which is true for all experiments using large numbers of atoms. If the error $\sigma_{\mathrm{ph}}(\langle F_z \rangle)$ starts to approach $1/N_a$, then higher-order effects of conditional quantum measurement become important and lead to a localization of the wave function to one of the eigenstates of the collective atomic spin $F_z^c = \sum_i^{N_a} F_z^i$ as discussed in more detail in Ref. [17].

It is also convenient to consider the power spectrum of the atomic spin noise. The power spectral density $S(\nu)$ is equal to the Fourier transform of the covariance function (2.14) and is given by a Lorentzian centered at zero frequency with a half-width equal to $\Delta\nu = 1/(2\pi T_2)$,

$$S(\langle F_z \rangle)(\nu) = \frac{F}{N_a \pi \Delta\nu} \frac{1}{1 + (\nu/\Delta\nu)^2} . \tag{2.15}$$

One can check that the integral of the power spectrum $\int_0^\infty S(\langle F_z \rangle) d\nu = \sigma^2(\langle F_z \rangle)$. A more general way to obtain the spectral noise density of atom spin noise is to use the fluctuation-dissipation theorem, which states that the power spectrum of fluctuations, whether they are thermal or quantum in nature, is proportional to the frequency response of the system to a small driving force [18]. Therefore, the power spectrum of quantum spin fluctuations is generally the same as the magnetic resonance spectrum.

2.6 Conditions for spin noise dominance

Under what condition can we observe the finite correlation of atomic spin noise? To be able to resolve the time course of atomic spin evolution we need to have $S(\langle F_z \rangle) > S_{\mathrm{ph}}(\langle F_z \rangle)$, where $S_{\mathrm{ph}}(\langle F_z \rangle)$ is the power spectral density of the photon shot noise given by Eq. (2.9). If the flux of photons through the cell $\Phi = dN_{\mathrm{ph}}/dt$, then the power spectral density due to the photon shot noise is equal to

$$S_{\mathrm{ph}}(\langle F_z \rangle) = \left(\frac{\lambda A F}{8\pi^2 N_a \mathrm{Re}[\alpha_V]} \right)^2 \frac{2}{\eta \Phi} . \tag{2.16}$$

Here we have also introduced the photodetector quantum efficiency η, which gives the fraction of photons actually converted to photoelectrons in the detector. We assume that probe photon absorption in the atomic sample is negligible, which can always be achieved, at least in principle, by detuning the probe laser far from the resonance with a corresponding increase of the photon flux. One can continue to increase Φ until photon absorption begins to dominate the spin resonance linewidth $\Delta\nu$. The photon absorption rate can be calculated in terms of the scalar atomic polarizability α_S. Consider a probe laser tuned in the vicinity of a D1 transition in alkali-metal vapor, which has a dark state with $\langle F\rangle = F$ for circularly polarized probe light. For such a state the total polarizability vanishes, therefore we have

$$\alpha_S(\omega) = \alpha_V(\omega) = \frac{fr_e c^2}{2\omega_0}\frac{(\omega_0-\omega+i\gamma)}{(\omega_0-\omega)^2+\gamma^2}, \tag{2.17}$$

where ω_0 is the optical resonance frequency, γ is the half-width of the optical resonance, f is the oscillator strength, and r_e is the classical electron radius. The photon absorption rate per atom can be calculated from the photon absorption cross-section $\sigma = 4\pi\,\mathrm{Im}(\alpha_S)\omega_0/c$ and the photon flux per unit area. Setting $T_2^{-1} = \sigma\Phi/A$, we get for the ratio of spectral densities

$$\frac{S(\langle F_z\rangle)(0)}{S_{\mathrm{ph}}(\langle F_z\rangle)} = \frac{2\pi fr_e c\eta}{\gamma}\frac{(\omega_0-\omega)^2}{(\omega_0-\omega)^2+\gamma^2}\frac{N_a}{AF}. \tag{2.18}$$

This relationship can be simplified if the detuning $(\omega_0-\omega)$ is much larger than the width of the transition γ. Noting that the photon absorption cross-section on resonance is $\sigma_0 = 2\pi fr_e c/\gamma$ and $N_a = nLA$, we obtain

$$\frac{S(\langle F_z\rangle)(0)}{S_{ph}(\langle F_z\rangle)} = \frac{\sigma_0 nL\eta}{F}. \tag{2.19}$$

Thus, the ratio of atom noise power to the photon noise power is proportional to the optical density $\sigma_0 nL$ of the transition on resonance. Note that the probe laser is actually detuned very far from resonance, so the optical density on resonance can be very large without significant absorption of the probe beam. This can be contrasted with a measurement of spin noise using photon absorption [19], where the ratio of atom noise power to photon noise power is still proportional to the optical density on resonance, but it cannot significantly exceed unity without complete absorption of the probe beam.

For a measurement using a pulse of probe light with N_{ph} photons we can also consider the ratio $\sigma(\langle F_z\rangle)/\sigma_{\mathrm{ph}}(\langle F_z\rangle)$. The number of atoms that absorb a photon during such a pulse is given by $N_{\mathrm{abs}} = \sigma N_a N_{\mathrm{ph}}/A \equiv \varepsilon N_a$, where we define the fraction of atoms ε that absorbed a photon. Combining Eqs. (2.4) and (2.9) with the expression for the polarizability (2.17) we obtain

$$\frac{\sigma(\langle F_z\rangle)}{\sigma_{\mathrm{ph}}(\langle F_z\rangle)} = \sqrt{\frac{\varepsilon\sigma_0 nL\eta}{2F}}. \tag{2.20}$$

If the optical density is sufficiently high, we can have both $\sigma_{\mathrm{ph}}(\langle F_z\rangle)/\sigma(\langle F_z\rangle) \ll 1$ and $\varepsilon \ll 1$. This is an illustration of the quantum-nondemolition (QND) nature of the off-resonant Faraday rotation measurement, which was first pointed out in Refs. [20] and [21];

it allows one to obtain a measure of the spin expectation value with high precision relative to the quantum spin-projection noise $\sigma_{\rm ph}(\langle F_z\rangle) \ll \sigma(\langle F_z\rangle)$, while having only a small fraction of atoms ε in the sample absorb a photon. Even poor photon detection efficiency η can be compensated for by a higher optical depth.

We have assumed a Lorentzian lineshape in Eq. (2.17) for the atomic polarizability. Since the probe beam is usually detuned far from resonance in QND measurements, this is a good approximation even in the presence of Doppler broadening. The optical density $\sigma_0 nL$ corresponds to resonance absorption that would be there in the absence of Doppler broadening. Thus thermal motion of atoms does not detract from QND measurements; in fact, it helps by reducing the real optical density of atoms on resonance and preventing radiation trapping during optical pumping.

Thus we arrive at a set of conditions necessary to realize an atomic magnetometer with sensitivity limited by spin-projection noise. One can use paramagnetic Faraday rotation in an optically thick atomic sample while ensuring that fluctuations of effective magnetic field caused by back-action of the probe laser do not contribute to the measured spin component. If the optical density is sufficiently high, one can reduce the photon flux until the photon scattering no longer contributes to T_2, while still having the spectral density of atom-spin-projection noise exceed the spectral density of photon shot noise.

2.7 Spin projection limits on magnetic field sensitivity

To calculate precisely the best sensitivity for an atomic magnetometer, we first consider a steady-state measurement, as typically used in SERF magnetometers. If the optical density is much greater than unity, the sensitivity will be limited only by the spin-projection noise. Detailed analysis of the sensitivity of the SERF magnetometer is presented in Ref. [22], taking into account the two hyperfine manifolds of alkali-metal atoms in spin-exchange equilibrium. After considering the effects of nuclear spin, it is found that the sensitivity is independent of the value of the nuclear spin and is the same as that of an uncoupled electron with electron spin relaxation time T_{sd}. For continuous operation of the SERF magnetometer the optimal pumping rate is equal to twice the spin relaxation rate. The magnetic field uncertainty after a measurement time $t \gg T_{sd}$ is equal to

$$\sigma_{\rm cont}(B_y) = \frac{\hbar}{g_s \mu_B}\sqrt{\frac{27}{2}\frac{1}{N_a T_{sd} t}}\,. \tag{2.21}$$

The electron-spin relaxation time T_{sd} is usually dominated by atomic collisions. For a sufficiently high alkali-metal density the dominant relaxation mechanism is due to alkali–alkali spin-destruction collisions with a cross-section σ_{sd}, $T_2^{-1} = N_a \bar{v} \sigma_{sd}/V$, where $\bar{v}$ is the average thermal velocity. For these conditions the sensitivity of the magnetometer is determined solely by the measurement volume V and the value of the collisional cross-section:

$$\sigma_{\text{cont}}(B_y) = \frac{\hbar}{g_s \mu_B} \sqrt{\frac{27}{2} \frac{\bar{v}\sigma_{sd}}{Vt}} \,. \tag{2.22}$$

Similar equations for the fundamental sensitivity of a radiofrequency magnetometer operated in a continuous regime are derived in [23] and for a scalar magnetometer in [24]. It is shown that for the radiofrequency magnetometer the sensitivity is determined by the geometric mean of the spin-exchange and spin-destruction cross-sections, while for a scalar magnetometer operated in a continuous regime, it is limited by the spin-exchange cross-section. For magnetometers based on cold atoms (see Chapter 9), the fundamental sensitivity limit per unit volume can be obtained by replacing the collisional spin-relaxation rate constant $\bar{v}\sigma_{sd}$ with the two-body atomic loss rate constant β_2, which is ultimately limited by magnetic dipolar relaxation [25, 26].

The back-action noise due to probe beam AC Stark shifts can also contribute to magnetometer noise, but it does not present a fundamental limit to sensitivity, as it can be avoided by various methods depending on the exact magnetometry scheme. For example, in a magnetometer operated at zero magnetic field the vector AC Stark shift increases the uncertainty in the F_y spin component, which is orthogonal to the measured F_z component. Hence, the magnetometer is naturally back-action evading. For an RF magnetometer operating in a finite B_x magnetic field, the F_y and F_z spin components mix, requiring additional efforts to avoid back-action noise. One possibility is to use two alkali-metal cells with opposite spin polarization to eliminate back-action noise [27], see Chapter 3. Another possibility is to use stroboscopic modulation of the probe beam at twice the Larmor frequency [28]. In a scalar magnetometer using spin orientation, the back-action can be eliminated by operating with both the probe beam and the spin polarization perpendicular to the magnetic field [29]. In alignment-based magnetometers, discussed in more detail in Chapter 6, it should also be possible to realize a back-action-evading measurement by using a double-pass probe beam arrangement proposed in [30].

It is also interesting to consider the sensitivity of the magnetometer as a function of the frequency of the measured magnetic field. For example, in a SERF magnetometer the small-signal frequency response corresponds to that of a first-order low-pass filter with a cut-off frequency equal to $\Delta\nu$ [31]. Therefore, the power spectral density of the signal as a function of the frequency of the magnetic field is given by

$$S(B_y)(\nu) = S(B_y)(0) \frac{1}{1 + (\nu/\Delta\nu)^2} \,, \tag{2.23}$$

which is exactly the same spectrum as that of atomic spin noise (2.15). This is a general consequence of the fluctuation-dissipation theorem and applies to other types of magnetometers as well. Therefore, as long as spin-noise spectral density dominates photon shot-noise spectral density, both the signal and the noise change in the same way as a function of frequency

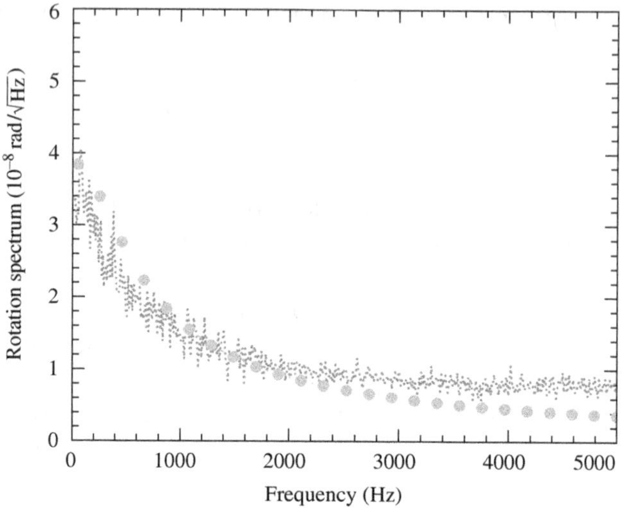

Figure 2.2 Comparison of signal and noise frequency spectrum. Thin dotted line: spectrum of magnetometer rotation noise consisting of a low-frequency peak due to spin-projection noise and broadband photon shot noise. Large dots: magnetometer rotation signal in response to a 22 fT_{RMS} magnetic field as a function of frequency. The signal-to-noise ratio remains approximately the same as long as the spin-projection noise dominates over photon shot noise. Adapted from Ref. [29].

and the sensitivity of the magnetometer remains equal to the sensitivity on resonance even beyond the small-signal bandwidth $\Delta\nu$ [29]. This is illustrated in Fig. 2.2 for a scalar magnetometer, which shows that both optical rotation noise spectral density and the optical rotation signal due to a certain magnetic field decrease at higher frequency in a similar way, maintaining the same signal-to-noise ratio. Hence, QND measurements using Faraday rotation allow one to make high-sensitivity broadband measurements of a magnetic field beyond the natural bandwidth of the magnetometer.

It is also possible to operate atomic magnetometers in a pulsed regime. In this case the atoms are initially prepared in a fully polarized state, the pump beam is turned off, and spins are allowed to evolve in the presence of a magnetic field in the dark. After a certain evolution time a brief high-intensity pulse of probe light measures the spin of the system. The optimal measurement time is set by the spin coherence time. In the case of the SERF magnetometer with equal longitudinal and transverse spin relaxation times, $T_1 = T_2$, the spin evolves as

$$\langle F_z \rangle = \frac{g_s \mu_B B t}{2\hbar} \exp(-t/T_2) \,, \tag{2.24}$$

where $T_2 = q(P)T_{sd}$. Here $q(P)$ is the slowing-down factor that depends on the spin polarization [22, 32]. The maximum response to a small magnetic field is achieved after a measurement time $t_m = T_2$. However, if it is desired to repeat such measurements as fast as possible for a time $t \gg T_2$, then the optimal measurement time is $t_m = T_2/2$. With a high-optical-density vapor it is possible to measure the final spin state with an uncertainty

$\sigma_{ph}(\langle F_z \rangle) \ll \sigma(\langle F_z \rangle)$. However, the initial spin state of the system has an uncertainty $\sigma(\langle F_z \rangle)$ which, for the case of the two alkali-metal hyperfine manifolds participating in the SERF regime, is given by $\sigma(\langle F_z \rangle) = \sqrt{q(P)/4N_a}$ [22]. So the uncertainty in the measurement of B_y after a long time t is given by

$$\sigma_{\text{pulse}}(B_y) = \frac{\hbar}{g_s \mu_B} \sqrt{\frac{2e}{N_a T_{sd} t}}, \tag{2.25}$$

where $e = \exp(1)$. This result for the uncertainty is also obtained for a Ramsey measurement sequence on a two-level system in Ref. [33], and it is quite typical for any frequency measurement in a system with finite relaxation. Compared with a continuous measurement there is an improvement in sensitivity for a pulsed SERF magnetometer by a factor of $\sqrt{27/4e} = 1.57$. Experimentally, however, pulsed measurements are more challenging to implement because they require higher pump-laser power to re-polarize atoms after every measurement in a time much shorter than T_2, an ability to turn the pump laser off while avoiding excitation of transients much larger than quantum spin noise, and high probe laser power with low-noise detection electronics to measure the final spin state in a time much shorter than T_2.

Since for pulsed measurements the magnetic field uncertainty is dominated by the uncertainty in the initial spin state, it is natural to consider what happens if we use QND measurements to determine the initial state of the system. In a sufficiently optically dense sample the initial measurement can be performed without significant perturbation of the atomic state, $\varepsilon \ll 1$, and with an uncertainty $\sigma_{\text{ph}}(\langle F_z \rangle)$ small relative to $\sigma(\langle F_z \rangle)$. To account for an imperfect correlation of the spin state between the initial and final measurement, we consider a combination $C = \langle F_z(t) \rangle - \zeta \langle F_z(0) \rangle$, where we can vary the fraction ζ of the initial state that we subtract to minimize the final uncertainty in C [34] (see Chapter 3). In general one finds that the uncertainty in C is minimized when we choose $\zeta = \text{cov}[\langle F_z(t) \rangle, \langle F_z(0) \rangle]/\sigma^2(\langle F_z \rangle)$. Using the spin covariance given by Eq. (2.14), modified for the spin variance of a real alkali-metal atom in the SERF regime, and the signal given by Eq. (2.24), we find that the uncertainty in B_y based on a measurement of C is given by

$$\sigma_{\text{2-pulse}}(B_y) = \frac{\hbar}{g_s \mu_B} \sqrt{\frac{2}{N_a T_{sd} t}}. \tag{2.26}$$

In this case the optimal measurement time is short compared with T_2. Thus we have further reduced the uncertainty by a factor of $\sqrt{e} = 1.65$. This is the most efficient method of spin-precession measurement for a system with a constant decoherence rate. This result, first obtained in Refs. [33, 35], was recently re-derived in a more general way in Ref. [36]. It was also discussed in relation to magnetometry in Ref. [12], where it was shown that the fundamental sensitivity of any magnetometer scales as $1/\sqrt{N_a T_2 t}$.

2.8 Spin squeezing and atomic magnetometry

This brings us to the question of the role of spin squeezing in atomic magnetometry or more generally in precision spectroscopy. The idea of using multiparticle quantum states with reduced uncertainty was first introduced from quantum optics to spin ensembles by Wineland and co-workers [37]. They introduced a spin-squeezing parameter

$$\xi = \frac{\sigma(\langle F_z \rangle)\sqrt{2N_a F}}{\langle F_x \rangle} \tag{2.27}$$

such that $\xi = 1$ if $\sigma(\langle F_z \rangle)$ is given by the SQL (Eq. 2.4) and the spin state is fully polarized $\langle F_x \rangle = F$. Spin-squeezed states with $\xi < 1$ have unequal spin uncertainties in the two directions orthogonal to their spin polarization, $\sigma(\langle F_z \rangle) < \sigma(\langle F_y \rangle)$, but of course they still obey the Heisenberg uncertainty principle. The easiest way to generate a spin-squeezed state is through QND Faraday rotation measurements [20, 21], as was already discussed above. Spin-squeezed states can also be generated by collective spin self-interactions [38]. A general review of this rich field is outside the scope of this chapter (see, for example, recent reviews [39, 40]); here we will focus only on the potential usefulness of spin squeezing for precision magnetometry.

The squeezing parameter is defined in such a way that it quantifies the improvement in the measurement of the spin precession angle $\phi = \langle F_z \rangle / \langle F_x \rangle$ relative to the SQL, $\sigma(\phi) = \xi\sigma_{\text{SQL}}(\phi)$. If the precession angle is measured with an uncertainty $\sigma(\phi_i)$ and $\sigma(\phi_f)$ at the beginning and at the end, respectively, of a measurement time t_m, the precession frequency is determined with an uncertainty

$$\sigma(\omega) = \frac{\sqrt{\sigma^2(\phi_i) + \sigma^2(\phi_f)}}{t_m} = \sigma_{\text{SQL}}(\phi)\frac{\sqrt{\xi_i^2 + \xi_f^2}}{t_m}. \tag{2.28}$$

Thus reducing the angle measurement error by a factor $\xi < 1$ leads to a corresponding reduction in the frequency error for a given measurement time.

Equation (2.28) is easy to use if the measurement time t_m is fixed by some external requirements and the spin-squeezed state is preserved for the duration of the measurement, with ξ_i, $\xi_f < 1$. However, since the frequency resolution continues to improve without bound with longer measurement time t_m, one has to consider the effects of spin relaxation on ξ_f to determine an intrinsic limit on the frequency resolution of the spin system. The effect of spin relaxation on spin squeezing in spectroscopic measurements was first considered in [33]. Those authors found that in the presence of a constant relaxation rate the improvement in the frequency uncertainty from the use of spin-squeezed states is at most equal to $\sqrt{e}$, as we have reproduced in Eq. (2.26). On the other hand, they found that the optimal measurement time can be shorter than the spin-coherence time, increasing the measurement bandwidth without loss of sensitivity. Thus, spin-squeezing techniques can improve the bandwidth of atomic magnetometers [12], as shown in Fig. 2.2 [29], or equivalently, increase their sensitivity for a measurement time shorter than the atomic spin coherence time [27, 41] (see Chapter 3).

Under what conditions can spin squeezing improve the long-term measurement sensitivity by more than a factor of $\sqrt{e}$? One practical case is if the measurement time is fixed, as is the case, for example, in an atomic fountain, and the density of atoms cannot be increased until collisional effects begin to dominate spin relaxation. Another more fundamental possibility is when the relaxation rate is not a constant but increases in time. A practical example of such a situation is a dense alkali-metal vapor, where the spin-exchange collisions can dominate T_2 relaxation and present a fundamental limit for the magnetic field sensitivity per unit volume of RF and scalar magnetometers, as discussed in relation to Eq. (2.22). Spin exchange is a nonlinear process and the spin-exchange-relaxation rate can be suppressed for a short time by pumping atoms into a high-polarization initial state. Using spin squeezing with a short measurement time one can then take advantage of the smaller initial spin-exchange-relaxation rate [28]. This allows one in principle to achieve magnetic field sensitivity limited by the spin-destruction cross-section even for scalar and RF magnetometers operating in a finite magnetic field. Another possibility, considered recently in [42], is a case of a non-Markovian spin relaxation process with a finite correlation time. It also leads to a suppressed decoherence rate on a time scale shorter than the correlation time and so can also benefit from spin-squeezing measurements.

2.9 Conclusion

We have presented a semi-classical description of quantum noise in atomic magnetometers, deriving relationships necessary for the quantitative analysis of the fundamental noise sources in magnetometry as well as in other spectroscopic measurements. We showed that the spin-projection noise in combination with spin relaxation interactions present a fundamental limit for magnetic field sensitivity per unit volume in a given atomic system. Photon shot noise, while often significant in practical atomic magnetometers, can be eliminated by using optically dense atomic ensembles or polarization-squeezed probe light. The AC Stark shift noise due to probe beam measurement back-action can be eliminated by choosing an appropriate measurement geometry or other back-action-evasion techniques. Spin squeezing can be implemented in atomic magnetometers using quantum-nondemolition measurements based on Faraday rotation. Spin squeezing can improve the bandwidth of magnetic field measurements, extending the frequency range of maximum sensitivity beyond the linewidth of the spin resonance. Under certain conditions spin squeezing can also help to improve the long-term sensitivity of atomic magnetometers, particularly in case when spin relaxation is dominated by a nonlinear process, such as spin-exchange collisions.

References

[1] W. M. Itano, J. C. Bergquist, J. J. Bollinger, J. M. Gilligan, D. J. Heinzen, F. L. Moore, M. G. Raizen, and D. J. Wineland, Phys. Rev. A **47**, 3554 (1993).
[2] A. Kastler, Compt. Rend. **232**, 953 (1951).

[3] J. Manuel and C. Cohen-Tannoudji, Compt. Rend. **257**, 413 (1963).
[4] W. Happer and B. S. Mathur, Phys. Rev. Lett. **18**, 577 (1967).
[5] W. Happer and B. S. Mathur, Phys. Rev. **163**, 12 (1967).
[6] J. L. Sorensen, J. Hald, and E. S. Polzik, Phys. Rev. Lett. **80**, 3487 (1998).
[7] F. Wolfgramm, A. Cerè, F. A. Beduini, A. Predojević, M. Koschorreck, and M. W. Mitchell, Phys. Rev. Lett. **105**, 053601 (2010).
[8] M. Fleischhauer, A. B. Matsko, and M. O. Scully, Phys. Rev. A **62**, 013808 (2000).
[9] C. Gardiner, *Stochastic Methods* (Springer, Berlin, 2009).
[10] C. Schori, B. Julsgaard, J. L. Sorensen, and E. S. Polzik, Phys. Rev. Lett. **89**, 057903 (2002).
[11] J. L. Sorensen, B. Julsgaard, C. Schori, E. S. Polzik, Spectrochim. Acta B **58**, 999 (2003).
[12] M. Auzinsh, D. Budker, D. F. Kimball, S. M. Rochester, J. E. Stalnaker, A. O. Sushkov, and V. V. Yashchuk, Phys. Rev. Lett. **93**, 173002 (2004).
[13] S. M. Rochester, M. P. Ledbetter, T. Zigdon, A. D. Wilson-Gordon, and D. Budker, Phys. Rev. A **85**, 022125 (2012).
[14] E. B. Aleksandrov and V. S. Zapasskii, Sov. Phys. JETP **54**, 64 (1981).
[15] A. Kuzmich, L. Mandel, J. Janis, Y. E. Young, R. Ejnisman, and N. P. Bigelow, Phys. Rev. A **60**, 2346 (1999).
[16] S. A. Crooker, D. G. Rickel, A. V. Balatsky, and D. L. Smith, Nature **431**, 49 (2004).
[17] J. K. Stockton, R. van Handel, and H. Mabuchi, Phys. Rev. A **70**, 022106 (2004).
[18] U. Weiss, *Quantum Dissipative Systems* (World Scientific, Singapore, 2008).
[19] T. Mitsui, Phys. Rev. Lett. **84**, 5292 (2000).
[20] A. Kuzmich, N. P. Bigelow, and L. Mandel, Europhys. Lett. **42**, 481 (1998).
[21] Y. Takahashi, K. Honda, N. Tanaka, K. Toyoda, K. Ishikawa, and T. Yabuzaki, Phys. Rev. A **60**, 4974 (1999).
[22] M. P. Ledbetter, I. M. Savukov, V. M. Acosta, D. Budker, and M. V. Romalis, Phys. Rev. A **77**, 033408 (2008).
[23] I. M. Savukov, S. J. Seltzer, M. V. Romalis, and K. L. Sauer, Phys. Rev. Lett. **95**, 063004 (2005).
[24] S. J. Smullin, I. M. Savukov, G. Vasilakis, R. K. Ghosh, and M. V. Romalis, Phys. Rev. A **80**, 033420 (2009).
[25] J. M. Gerton, C. A. Sackett, B. J. Frew, and R. G. Hulet, Phys. Rev. A **59**, 1514 (1999).
[26] B. Pasquiou, G. Bismut, Q. Beaufils, A. Crubellier, E. Maréchal, P. Pedri, L. Vernac, O. Gorceix, and B. Laburthe-Tolra, Phys. Rev. A **81**, 042716 (2010).
[27] W. Wasilewski, K. Jensen, H. Krauter, J. J. Renema, M. V. Balabas, and E. S. Polzik, Phys. Rev. Lett. **104**, 133601 (2010).
[28] G. Vasilakis, V. Shah, and M. V. Romalis, Phys. Rev. Lett. **106**, 143601 (2011).
[29] V. Shah, G. Vasilakis, and M. V. Romalis, Phys. Rev. Lett. **104**, 013601 (2010).
[30] J. Cviklinski, A. Dantan, J. Ortalo, and M. Pinard, Phys. Rev. A **76**, 033830 (2007).
[31] J. C. Allred, R. N. Lyman, T. W. Kornack, and M. V. Romalis, Phys. Rev. Lett. **89**, 130801 (2002).
[32] I. M. Savukov and M. V. Romalis, Phys. Rev. A **71**, 023405 (2005).
[33] S. F. Huelga, C. Macchiavello, T. Pellizzari, A. K. Ekert, M. B. Plenio, and J. I. Cirac, Phys. Rev. Lett. **79**, 3865 (1997).
[34] J. Appel, P. J. Windpassinger, D. Oblak, U. B. Hoff, N. Kjærgaard, and E. S. Polzik, PNAS **106**, 10960 (2009).
[35] D. Ulam-Orgikh and M. Kitagawa, Phys. Rev. A **64**, 052106 (2001).
[36] B. M. Escher, R. L. de Matos Filho, and L. Davidovich, Nature Physics **7**, 406 (2011).
[37] D. J. Wineland, J. J. Bollinger, W. M. Itano, F. L. Moore, and D. J. Heinzen, Phys. Rev. A **46**, R6797 (1992).

[38] M. Kitagawa and M. Ueda, Phys. Rev. A **47**, 5138 (1993).
[39] J. Maa, X. Wang, C. P. Sun, and F. Nori, Phys. Rep. **509**, 89 (2011).
[40] K. Hammerer, A. S. Sorensen, and E. S. Polzik, Rev. Mod. Phys. **82**, 1041 (2010).
[41] M. Koschorreck, M. Napolitano, B. Dubost, and M. W. Mitchell, Phys. Rev. Lett. **104**, 093602 (2010).
[42] Y. Matsuzaki, S. C. Benjamin, and J. Fitzsimons, Phys. Rev. A **84**, 012103 (2011).

3

Quantum noise, squeezing, and entanglement in radiofrequency optical magnetometers

K. Jensen and E. S. Polzik

3.1 Sources of noise

In optical magnetometry [1], one often uses Faraday rotation,[1] caused by the interaction of light with a spin-polarized atomic ensemble, to measure the strength of a magnetic field (Chapter 1). The precision of the magnetic field measurement is ultimately limited by quantum-mechanical effects. The quantum noise of the measured signal can be due to shot noise (SN) of the probing light, intrinsic atomic-spin-projection noise (PN), and back-action noise arising from the measurement. In this chapter we describe these quantum noise sources and present theoretical and experimental results of radiofrequency (RF) magnetometry where all of them have been significantly suppressed.

3.1.1 Atomic projection noise

The quantum state of an atomic ensemble with N_A atoms can be described using the collective spin $\mathbf{J} = \sum_{k=1}^{N_A} \mathbf{j}_k$, where $\mathbf{j}_k$ is the spin of the kth atom.[2] The spin components are noncommuting $\left[J_y, J_z\right] = iJ_x$. For an atomic ensemble polarized in the x-direction, J_x has a large value and can be considered a classical variable. On the other hand, the transverse spin components J_y and J_z are quantum variables with zero or small mean values. The transverse components are uncertain due to the projection noise arising from the Heisenberg uncertainty relation, which reads $\mathrm{Var}\left(J_y\right)\mathrm{Var}\left(J_z\right) \geq \left|J_x\right|^2/4$, where the variance of an operator equals the square of its standard deviation, i.e., $\mathrm{Var}\left(J_y\right) = \left(\Delta J_y\right)^2$.

For a multilevel atom with a ground-state hyperfine quantum number F and projection along the x-axis m_x, we can consider the collective state consisting of N_A atoms all in the $|F, m_x = F\rangle$ state. Such a "stretched" state is polarized in the x-direction with $J_x = FN_A$.

[1] Historically, the Faraday effect refers to the rotation of the plane of light polarization as it passes through a medium immersed in a longitudinal magnetic field. However, in recent times, this definition has been extended to encompass optical rotation due to the interaction of light with atomic polarization in general.

[2] In this chapter, by "spin" we mean the total angular momentum of an atom. We also adopt a system of units where $\hbar = 1$.

Optical Magnetometry, ed. Budker, D. and Jackson Kimball, D. F. Published by Cambridge University Press.

The transverse spin components J_y and J_z have zero mean values and fluctuations $\Delta J_y = \Delta J_z = \sqrt{|J_x|/2}$. This state and the states which can be obtained from this state by rotations are called coherent spin states (CSS). The fluctuations in the transverse spin components are smaller than the mean spin by a factor $\sqrt{2|J_x|}$, which for centimeter-sized Rb or Cs vapor cells at room temperature is a factor of $\sim 10^6$. The angular uncertainty of the direction of the spin vector is then $\Delta\theta/\theta \approx \Delta J_\perp/J_x = 1/\sqrt{2J_x}$. In a PN-limited magnetometer, the direction of the spin vector can be measured with this high precision.

Coherent states [2] can be thought of as being at the boundary between the classical and quantum worlds. Atomic states prepared in optical magnetometry experiments often approach coherent states (i.e., they are either partially or fully polarized in, for instance, the x-direction), and the noise properties of the prepared states can be compared to those of coherent states.

3.1.2 Photon shot noise

In an optical magnetometer, the components of the atomic spin vector are typically measured by observing the polarization of light transmitted through the atomic sample. A pulse of polarized light propagating in the z-direction can be described in terms of the Stokes vector $\mathbf{S} = (S_1, S_2, S_3)$, where the components are defined by

$$S_1 = \left[N_{\text{ph}}(x) - N_{\text{ph}}(y)\right]/2\,, \tag{3.1}$$

$$S_2 = \left[N_{\text{ph}}(+45°) - N_{\text{ph}}(-45°)\right]/2\,, \tag{3.2}$$

$$S_3 = \left[N_{\text{ph}}(\sigma_+) - N_{\text{ph}}(\sigma_-)\right]/2\,. \tag{3.3}$$

$N_{\text{ph}}(x)$ and $N_{\text{ph}}(y)$ are the number operators for photons polarized in the x- and y-directions, respectively, $N_{\text{ph}}(\pm 45°)$ is the number operator for photons polarized in the directions $\pm 45°$ between the x- and y-directions, and $N_{\text{ph}}(\sigma_\pm)$ is the number operator for $\sigma_\pm$ polarized photons. The Stokes operators satisfy the angular-momentum-like commutation relations, e.g., $[S_2, S_3] = iS_1$ (for a discussion of the quantum Stokes operator see, for example, Appendix A in Ref. [3]). The operator $\mathbf{S}$ counts photons within a certain time duration T and is dimensionless. We can also define a time-dependent Stokes operator $\mathbf{S}(t) = \{S_1(t), S_2(t), S_3(t)\}$ that counts photons per unit time at time t. This operator has the dimension of 1/time. The integrated and the time-dependent operators are related by the formula $\mathbf{S} = \int_0^T \mathbf{S}(t)dt$.

Let us assume that the light is strongly linearly polarized along either the x- or y-direction, such that S_1 has a large amplitude $|S_1| = N_{\text{ph}}/2$ and can be considered a classical variable. We also assume that $S_1(t)$ is time independent such that $S_1 = S_1(t)T$. The Stokes components S_2 and S_3 are small and can be considered quantum variables. The Heisenberg uncertainty relation for the Stokes operators is $\text{Var}(S_2)\,\text{Var}(S_3) \geq |S_1|^2/4$. For coherent light states the uncertainties of S_2 and S_3 are equal and minimal: $\Delta S_2 = \Delta S_3 = \sqrt{|S_1|/2}$. This noise is called the quantum shot noise of light. The angular uncertainty of the direction of the Stokes vector is $\Delta\theta/\theta \approx \Delta S_2/S_1 = 1/\sqrt{2S_1}$, which decreases with the number of photons.

3.1.3 Back-action noise and QND measurements

According to the principles of quantum mechanics, a measurement of a quantum system also disturbs the system. For an atomic ensemble probed by light, back-action noise can come, for example, from the Stark shift imposed by the quantum fluctuations of the circular light polarization on J_y when J_z is measured. When the atoms are located in a magnetic field, both J_y and J_z experience Larmor precession and hence both of them acquire back-action noise. We note that while the atomic projection noise of a spin-polarized ensemble scales as $\mathrm{Var}(J_z) \propto J_x$, the contribution to the atomic noise from the back-action of light onto atoms scales as $\mathrm{Var}(J_z) \propto J_x^2 N_{\mathrm{ph}}$ [see Section 3.3 and the online supplementary material (www.cambridge.org/9781107010352) for details of the light–atom interaction]. Quantum nondemolition (QND) techniques can be utilized to achieve back-action-evading measurements. In this chapter we focus on a QND technique which utilizes two oppositely polarized atomic ensembles [4, 5]. Other techniques are based on stroboscopic probing (at twice the Larmor frequency) of the atoms [6] and probe-pulse trains with alternating linear polarization [7].

3.1.4 Technical (classical) noise

Quantum noise such as PN or SN is fundamental in the sense that it is given by the Heisenberg uncertainty relation. On the other hand, "classical noise" is of technical character and should be reduced if possible. Classical noise could be introduced to the spins by, for instance, lasers with excess intensity noise, imperfectly polarized light, and fluctuations in the bias magnetic field. Technical noise might have a specific frequency dependence, such as $1/f$ noise and noise peaked around a specific frequency (such as 50 or 60 Hz). Shot noise, on the other hand is "white" in the sense that it is independent of frequency. The frequency spectrum of PN is centered around the Larmor frequency and has a width comparable to $1/T_2$, where T_2 is the transverse spin-relaxation time.

3.1.5 Entanglement and spin squeezing

Spin squeezing

Spin-squeezed states (SSSs) [8, 9] are atomic states with reduced uncertainty in one of the transverse spin components. An example of a SSS is a state with $\langle \mathbf{J} \rangle = J_x \hat{\mathbf{e}}_x$ ($\hat{\mathbf{e}}_x$ is a unit vector in the x-direction) and $\mathrm{Var}(J_z) < |J_x|/2$. The squeezed spin component J_z has reduced uncertainty compared to a CSS while the anti-squeezed component J_y has increased uncertainty $\mathrm{Var}\left(J_y\right) > |J_x|/2$ such that the Heisenberg uncertainty relation is still fulfilled.

Spin-squeezed states are useful in magnetometry and quantum sensing because the signal-to-noise ratio of the measurement can be improved in certain cases. However, since squeezed (or entangled) states are more fragile than coherent states, they decay faster. This has the consequence that squeezed states only improve the sensitivity for "short times" as discussed in detail in [10, 11] and in Section 3.2.2. There have been experimental demonstrations of improvement in magnetic field sensing using both squeezed atomic states [12] and squeezed light states [13].

Entanglement between atomic ensembles

Entangled states can be useful for optical magnetometry, since, like squeezed states, entangled states of (light) atoms can have noise below the (SN) PN associated with coherent states. Entanglement can be described as nonlocal and nonclassical correlations between two subsystems. A particular kind of entanglement useful for atomic magnetometry, the EPR entanglement, was first introduced by Einstein, Podolsky, and Rosen [14] for the position and momentum operators of two particles. The quantitative criterion for such entanglement was introduced in Ref. [15].

In this chapter, we focus on entanglement between two atomic ensembles constituting an atomic magnetometer. It turns out that for two oppositely polarized atomic ensembles described by spin vectors $\mathbf{J}_1$ and $\mathbf{J}_2$ with $J_{x1} = -J_{x2} = |J_x|$ the Einstein–Podolsky–Rosen (EPR) entanglement criterion can be presented as[3]

$$\Sigma_{\mathrm{EPR}} \equiv \mathrm{Var}\left(\frac{J'_{z1}+J'_{z2}}{\sqrt{2|J_x|}}\right) + \mathrm{Var}\left(\frac{J'_{y1}+J'_{y2}}{\sqrt{2|J_x|}}\right) < 1. \tag{3.4}$$

Σ_{EPR} is the atomic noise in PN units. The entanglement criterion is easy to remember since for two uncorrelated atomic ensembles, both in a CSS, where the variances of the spin operators J'_{y1}, J'_{z1} and J'_{y2}, J'_{z2} all equal $|J_x|/2$, we have $\Sigma_{\mathrm{EPR}} = 1$. Two uncorrelated CSS are therefore at the boundary for entanglement.

Assuming equal noise in the y- and z-directions, we see that the condition (3.4) implies that $\mathrm{Var}\left(J'_{z1}+J'_{z2}\right) < |J_x|$ and $\mathrm{Var}\left(J'_{y1}+J'_{y2}\right) < |J_x|$. The entanglement criterion $\Sigma_{\mathrm{EPR}} < 1$ is therefore equivalent to having (two-mode) squeezing of the two nonlocal operators $J'_{z1}+J'_{z2}$ and $J'_{y1}+J'_{y2}$. Entanglement between two atomic ensembles obeying the condition (3.4) was first demonstrated using QND measurements [4].

3.2 A pulsed radiofrequency magnetometer and the projection noise limit

We now describe how atomic ensembles can be used as a sensor for radiofrequency magnetic fields [12,16,17]. We analyze both single-cell and two-cell schemes. Two atomic ensembles allow us to suppress the quantum-back-action noise and to generate an entangled atomic state, thus reducing PN of the magnetometer. From a practical perspective the two-ensemble setup is also well suited for differential magnetometry.

3 In the described experiment, the atomic ensembles are located in a bias magnetic field. This leads to Larmor precession of the spins. The notation is here that the primed spin operators correspond to the spin components in a frame rotating at the Larmor frequency.

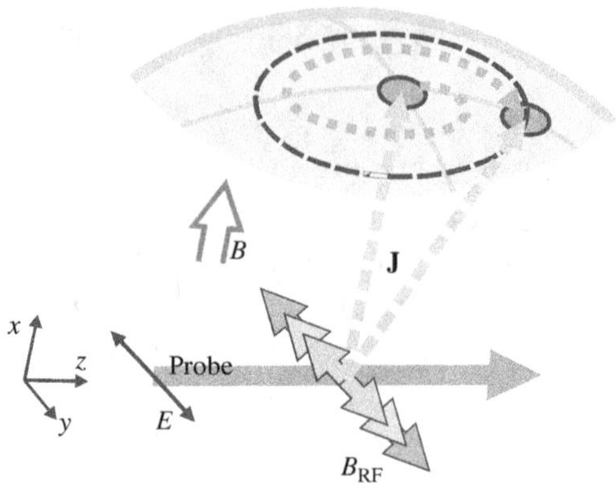

Figure 3.1 Radiofrequency magnetometer. The atomic spin **J** precesses (the dashed spiral) in crossed DC (B) and RF (B_{RF}) magnetic fields. The precessing **J** imposes an oscillating polarization rotation on the probe light.

3.2.1 Pulsed RF magnetometry

Consider first a special case of measuring oscillating (RF) magnetic fields with a pulsed scheme using a single ensemble. The atoms are initially prepared in a CSS polarized in the x-direction by optical pumping with a pulse of resonant light. In this case the atomic spins can be represented as a long spin vector $\mathbf{J}$ with an uncertainty as depicted in Fig. 3.1. The length of the vector is J_x, which is a macroscopic classical value. On the other hand, the transverse spin components J_y and J_z are small quantum variables with equal uncertainties $\Delta J_y = \Delta J_z = \sqrt{|J_x|/2}$. The atoms are located in a bias magnetic field B_x pointing in the x-direction such that the transverse spin components J_y and J_z precess at the Larmor frequency Ω_L.

At time $t = 0$, we apply an RF magnetic field with constant amplitude B_{RF} along the y-direction for a certain duration τ. This creates a transverse spin component

$$\langle J_\perp \rangle = \gamma B_{RF} J_x T_2 [1 - \exp(-\tau/T_2)]/2 , \tag{3.5}$$

where T_2 is the transverse spin coherence time and $\gamma = g_F \mu_B/\hbar = \Omega_L/B_x$ is the gyromagnetic ratio (g_F is the Landé factor and μ_B is the Bohr magneton). For cesium atoms in the $F = 4$ ground state, for example, $\gamma = 2.2 \times 10^{10}$ rad/(T s).

The transverse spin component can be measured with a subsequent pulse of light.[4] Assuming that the dominant noise source is the atomic PN, we find the minimal detectable field by equating Eq. (3.5) to the PN: $\Delta J_\perp = \sqrt{|J_x|/2}$. The result is

$$B_{\min} = \left[\gamma \sqrt{|J_x|/2}\, T_2 \{1 - \exp(-\tau/T_2)\}\right]^{-1} . \tag{3.6}$$

[4] The light–atom interaction is described in Section 3.3 and in the online supplementary material (www.cambridge.org/9781107010352).

We call this the "minimal detectable field," but it should really be understood as the standard deviation ΔB_{RF} of the measurement of the RF magnetic field using the measurement time $\approx \tau$. In a pulsed scheme, where the measurement takes place after the RF magnetic field is applied, one has the advantage of a long coherence time T_2. The atoms decay more slowly during the RF pulse than during the probing pulse because of the absence of probe-light-induced relaxation during the RF pulse. In Eq. (3.5), T_2 should be understood as the coherence time during the RF pulse. For a pulsed scheme, this coherence time equals the decay time "in the dark," T_2^{dark}.

3.2.2 Sensitivity and bandwidth

The PN-limited sensitivity is here[5] defined as $\mathcal{S}_{PN} \equiv B_{\mathrm{min}}\sqrt{\tau}$. It has units of $\mathrm{T}/\sqrt{\mathrm{Hz}}$ (or $\mathrm{G}/\sqrt{\mathrm{Hz}}$) and equals the standard deviation of the measured field if one uses repeated measurements for a total duration of 1 second.[6] Inserting the expression for B_{min} given by Eq. (3.6), we find the full expression for the sensitivity:

$$\mathcal{S}_{PN} = \frac{1}{\gamma}\sqrt{\frac{2}{|J_x|}}\frac{\sqrt{\tau}}{T_2[1-\exp(-\tau/T_2)]}. \tag{3.7}$$

$\mathcal{S}_{PN}$ depends on T_2 and τ, or equivalently, the magnetometer bandwidth $\delta_A \equiv 1/T_2$ and the bandwidth of the RF pulse $\delta_{\mathrm{RF}} \equiv 1/\tau$. Below we examine how the sensitivity scales with T_2 and τ.

We can consider the two limits of short/long RF pulses compared to T_2. We find

$$\begin{aligned} \mathcal{S}_{PN} &= \frac{1}{\gamma}\sqrt{\frac{2}{|J_x|}}\frac{1}{\sqrt{\tau}} \quad \text{for} \quad \tau \ll T_2 \quad \text{or} \quad \delta_{\mathrm{RF}} \gg 1/T_2 \\ &= \frac{1}{\gamma}\sqrt{\frac{2}{|J_x|}}\frac{\sqrt{\tau}}{T_2} \quad \text{for} \quad \tau \gg T_2 \quad \text{or} \quad \delta_{\mathrm{RF}} \ll 1/T_2. \end{aligned} \tag{3.8}$$

The sensitivity given by Eq. (3.7) together with the asymptotic expressions in Eq. (3.8) (all divided by $\frac{1}{\gamma}\sqrt{2/|J_x|}$) are shown in Fig. 3.2. The left panel shows the sensitivity as a function of T_2 for a fixed value of $\tau = 1$ s. The right panel shows the sensitivity as a function τ with a fixed value for $T_2 = 1$ s.

The right panel shows that for an atomic magnetometer with a fixed T_2, there is an optimal RF-pulse duration $\tau \approx T_2$ at which the magnetometer has the best sensitivity. In this sense, the RF bandwidth and the atomic bandwidth should be matched. On the other hand, for a

5 We here relate the pulse duration τ to the bandwidth δ_{RF} by the formula $\delta_{\mathrm{RF}} \approx 1/\tau$. This formula is only true up to a numerical factor, and the definition for sensitivity used in this chapter could therefore differ by a numerical factor (around a factor of $\sqrt{2}$) compared to the definition used in other chapters. See also Chapter 2 for further discussion of sensitivity and bandwidth.

6 Repeating the same measurement M times for a total duration $M\tau$ improves the uncertainty ΔB on the magnetic field by the factor $\sqrt{M}$.

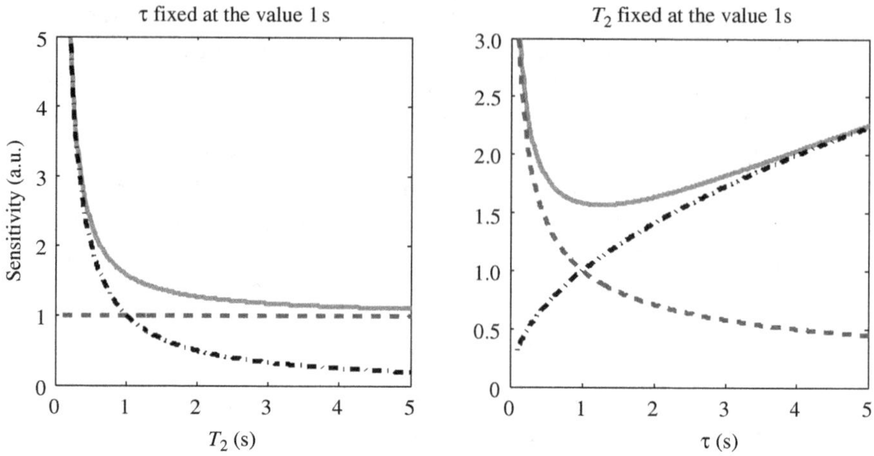

Figure 3.2 Solid lines: projection-noise-limited sensitivity given by Eq. (3.7) as a function of atomic T_2 time and RF duration τ, respectively. Dashed and dash-dotted lines are asymptotic expressions given by Eq. (3.8).

fixed RF-pulse duration τ (left panel), the sensitivity improves with larger T_2. For $T_2 \gg \tau$ the sensitivity becomes independent of T_2, as can also be seen from Eq. (3.8). This is the setting where the magnetometer has the *best sensitivity*. If one would like to demonstrate a magnetometer with the best sensitivity, one would need a magnetometer with as large T_2 as possible, and then to measure the RF pulses with duration $\tau \approx T_2$.

If one uses two atomic ensembles to sense the RF magnetic field, one can suppress the projection noise of the magnetometer by entangling the two ensembles. The entanglement has a certain lifetime T_{2E}. For RF-pulse durations $\tau \geq T_{2E}$, the entanglement decays and thus will not improve the performance of the magnetometer. Usually entangled states decay faster than coherent states such that $T_{2E} < T_2$. Entanglement will therefore only improve the magnetometer in the situation described by the inequalities

$$\tau \leq T_{2E} < T_2 \quad \text{or} \quad \delta_{\text{RF}} \geq 1/T_{2E} > 1/T_2. \tag{3.9}$$

As argued above, this is *not* the limit where the magnetometer has best sensitivity. The best sensitivity (when $\tau \approx T_2$) will not be improved by entanglement. Instead, entanglement improves the magnetometer for short RF pulses. This will also be apparent in the experimental results presented in Section 3.5.

3.3 Light–atom interaction

The interaction between a spin-polarized atomic ensemble and linearly polarized light will now be briefly discussed. For more details, the reader is referred to the online supplementary material (www.cambridge.org/9781107010352) or the review [18].

3.3.1 A spin-polarized atomic ensemble interacting with polarized light

The atomic ensemble is described using the total spin vector $\mathbf{J}$, and it is assumed that the atoms are oriented in the x-direction such that J_x is large compared to the transverse spin components J_y and J_z. Essentially, the atomic ensemble is in an eigenstate of J_x and thus J_x can be regarded as a classical variable, while the transverse spin components J_y and J_z are so small that they are dominated by quantum fluctuations. The polarization of the light is described with the Stokes vector $\mathbf{S}(t)$. It is here assumed that the light propagates in the z-direction and that it is polarized in either the x- or y-direction such that S_1 is large compared to S_2 and S_3, i.e., S_1 is a classical variable and S_2 and S_3 are dominated by quantum fluctuations. It is also assumed that both J_x and S_1 are constant during the interaction. The above assumptions will be valid for the experiment presented later in this chapter.

The interaction will be described in terms of input–output equations for $\mathbf{J}$ and $\mathbf{S}$. Such equations can be derived from a Hamiltonian (see the online supplementary material), which represents the dipole interaction $-\mathbf{d}\cdot\mathbf{E}$, where $\mathbf{d}$ is the induced dipole moment of the atoms and $\mathbf{E}$ is the electric field of the light. It is assumed that the light is detuned from the atomic transition, such that an effective Hamiltonian for the ground states can be derived. This Hamiltonian consists of three terms: scalar, vector, and tensor terms proportional to the scalar (a_0), vector (a_1), and tensor (a_2) polarizabilities of the atoms, which in turn depend on the detuning of laser light from atomic resonance.

The vector polarizability leads to Faraday rotation, where the polarization of the light is rotated by an angle proportional to the spin component along the propagation direction. For small rotation angles (such that S_1 and J_x can be considered constant during the interaction), we have [3]

$$S_2^{\text{out}}(t) = S_2^{\text{in}}(t) + aJ_z(t)S_1(t) \qquad \text{and} \qquad S_3^{\text{out}}(t) = S_3^{\text{in}}(t), \tag{3.10}$$

where a is a coupling constant[7] proportional to a_1. The light entering/exiting the ensemble is described with the Stokes vector $\mathbf{S}^{\text{in/out}}$. We see that the atomic spin component J_z is imprinted on the Stokes component S_2 of the output light. In an atomic magnetometer based on nonlinear Faraday rotation, the magnetic field value is typically extracted from the magnetometer signal $S_2^{\text{out}}(t)$. We emphasize that the equation for $S_3^{\text{in/out}}$ is only true far from resonance where one can neglect circular dichroism.

In zero magnetic field or for interaction times T much shorter than the Larmor precession time $1/\Omega_L$, the z-component of the spin J_z is conserved during the interaction while the y-component is changed due to the AC Stark shift introduced by the light:

$$J_y^{\text{out}} = J_y^{\text{in}} + aS_3^{\text{in}}J_x \qquad \text{and} \qquad J_z^{\text{out}} = J_z^{\text{in}}. \tag{3.11}$$

We use the notation $\mathbf{J}^{\text{in}} = \mathbf{J}(t=0)$ and $\mathbf{J}^{\text{out}} = \mathbf{J}(t=T)$. The equations (3.10) and (3.11) describe a QND measurement since J_z can be measured without being disturbed (it is

[7] The coupling constant a appearing in Eqs. (3.10) and (3.11) is the same. The derivation of the input–output equations and the full expression for a can be found in Ref. [3].

conserved). For linearly polarized light, the mean value of S_3 is zero. However, quantum fluctuations of the light $\mathrm{Var}(S_3) = S_1/2$ will be transferred to J_y. This is the back-action of light on atoms.

3.3.2 Conditional spin squeezing

It is possible to generate spin squeezing using QND measurements. A measurement of S_2^{out} provides information about the atomic operator J_z. This information can be used to reduce the uncertainty on J_z. Suppose we perform two measurements using two separate light pulses[8] and compare the results. Consider the conditional Stokes operator $S_2^{\mathrm{cond}} = S_2^{\mathrm{2nd}} - \alpha S_2^{\mathrm{1st}}$, where the parameter α is chosen to minimize the conditional variance

$$\mathrm{Var}\left(S_2^{\mathrm{cond}}\right) = \mathrm{Var}\left(S_2^{\mathrm{2nd}} - \alpha S_2^{\mathrm{1st}}\right)|_{\alpha=\alpha_{\mathrm{opt}}}. \tag{3.12}$$

The parameter α_{opt} describes how well one can predict the outcome of the second measurement based on the first measurement result. $\alpha_{\mathrm{opt}} = 1$ corresponds to perfect correlation between the first and second measurement results, while $\alpha_{\mathrm{opt}} = 0$ corresponds to no correlation. How close α_{opt} is to 0 or 1 depends on, for instance, the interaction strength and decoherence rates.

If the QND measurement is good, the first and the second measurement results should be equal, leading to a low conditional variance. This is called conditional spin squeezing since the atomic part of the noise (on the output light) is reduced conditioned on the first measurement result. Squeezing can also be generated unconditionally if the atomic state is changed using a feedback proportional to the first measurement result.

3.3.3 Larmor precession, back-action noise, and two atomic ensembles

Located in a magnetic field, the atomic spins will precess at the Larmor frequency. The magnetometer signal is therefore encoded in the cosine and sine components S_{2c}^{out} and S_{2s}^{out} of the Stokes vector defined by $S_{2c}^{\mathrm{out}} \propto \int_0^T S_2^{\mathrm{out}}(t)\cos(\Omega_L t)\,dt$ (S_{2s}^{out} is similarly defined). These operators are here normalized such that $\mathrm{Var}\left(S_{2c}^{\mathrm{in}}\right) = \mathrm{Var}\left(S_{2s}^{\mathrm{in}}\right) = N_{\mathrm{ph}}/2$, where N_{ph} is the number of photons. For a single ensemble, the back-action noise will pile up in both J_y and J_z due to the Larmor precession. In this case, the measurement of the atomic spins is not of the QND type. However, with two oppositely oriented ensembles ($J_{x1} = -J_{x2} = |J_x|$), see Fig. 3.3, one can achieve a QND interaction since the back-action noise contributions on the two ensembles cancel each other.

[8] For a single ensemble, J_z can be measured twice using two light pulses. For two ensembles (Fig. 3.3), where the light interacts with one after the other, $J'_{y1} + J'_{y2}$ and $J'_{z1} + J'_{z2}$ [see Eq. (3.13)] can be measured twice using two light pulses.

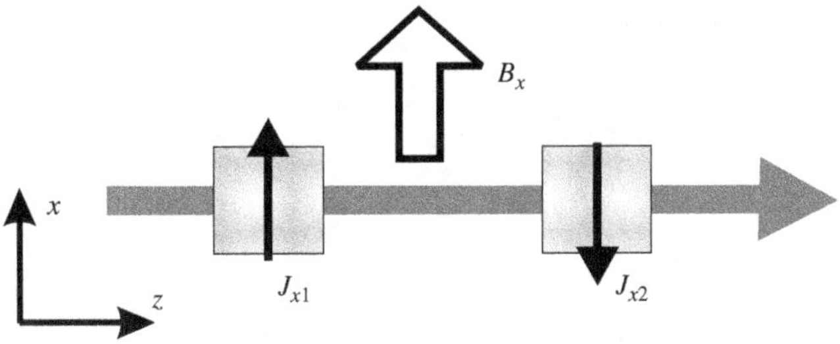

Figure 3.3 Illustration of light interacting with two oppositely polarized atomic ensembles one after the other.

For the case with two oppositely oriented atomic ensembles, we have the input–output equations

$$S_{2c}^{\text{out}} = S_{2c}^{\text{in}} + \kappa\sqrt{\frac{N_{\text{ph}}}{2\left|J_x\right|}}\left[J_{z1}^{'\text{in}} + J_{z2}^{'\text{in}}\right] \quad \text{and} \quad S_{2s}^{\text{out}} = S_{2s}^{\text{in}} + \kappa\sqrt{\frac{N_{\text{ph}}}{2\left|J_x\right|}}\left[J_{y1}^{'\text{in}} + J_{y2}^{'\text{in}}\right]. \tag{3.13}$$

The dimensionless coupling constant κ^2 is defined as $\kappa^2 = a^2\left|J_x\right|\left|S_1\right|$. The value of $\kappa^2 = 1$ corresponds to the situation where the fluctuations on the magnetometer signal $\text{Var}\left(S_{2c}^{\text{out}}\right)$ are equally due to the shot noise of the input light $\text{Var}\left(S_{2c}^{\text{in}}\right) = N_{\text{ph}}/2$ and the projection noise of the atoms $\text{Var}\left(J_{z1}^{'\text{in}} + J_{z2}^{'\text{in}}\right) = \left|J_x\right|$. The two nonlocal combinations $J_{y1}'(t) + J_{y2}'(t)$ and $J_{z1}'(t) + J_{z2}'(t)$ are conserved during the interaction. It is therefore possible to realize QND measurements of $J_{y1}'(t) + J_{y2}'(t)$ and $J_{z1}'(t) + J_{z2}'(t)$.

3.3.4 Swap and squeezing interaction

The tensor polarizability of the atoms (corresponding to the term in the Hamiltonian proportional to a_2) can also be included in the description of the light–atom interaction. For long interaction times, we obtain the following input–output equations for the Stokes operators [19]:

$$S_{2c}^{\text{out}} = \zeta\sqrt{\frac{N_{\text{ph}}}{2\left|J_x\right|}}\left(J_{z1}^{'\text{in}} + J_{z2}^{'\text{in}}\right) \tag{3.14}$$

and

$$S_{3c}^{\text{out}} = -\frac{1}{\zeta}\sqrt{\frac{N_{\text{ph}}}{2\left|J_x\right|}}\left(J_{y1}^{'\text{in}} - J_{y2}^{'\text{in}}\right), \tag{3.15}$$

and the nonlocal spin operators

$$J_{y1}^{'\text{out}} - J_{y2}^{'\text{out}} = +\zeta\sqrt{\frac{2|J_x|}{N_{\text{ph}}}}S_{3c}^{\text{in}} \tag{3.16}$$

and

$$J_{z1}^{'\text{out}} + J_{z2}^{'\text{out}} = -\frac{1}{\zeta}\sqrt{\frac{2|J_x|}{N_{\text{ph}}}}S_{2c}^{\text{in}}. \tag{3.17}$$

For atomic Cs, the parameter $\zeta^2 = -14a_2(\Delta)/a_1(\Delta)$, where Δ is the light detuning. For the experiments presented below, $\zeta^2 \approx 6.4$.

Equations (3.14)–(3.17) describe a swap between the Stokes and spin operators. The term "swap" refers to the fact that the output Stokes operator is proportional to the input spin operator and vice versa. For instance,

$$S_{2c}^{\text{out}} \propto \left(J_{z1}^{'\text{in}} + J_{z2}^{'\text{in}}\right)$$

and

$$\left(J_{z1}^{'\text{out}} + J_{z2}^{'\text{out}}\right) \propto S_{2c}^{\text{in}},$$

which implies that *one can perform a back-action-noise-free, shot-noise-free, projection-noise-limited measurement of the magnetic field.*

We also see from Eqs. (3.14)–(3.17) that the output operators are squeezed by the amount ζ^2 in the variance. For instance,

$$\text{Var}\left(J_{z1}^{'\text{out}} + J_{z2}^{'\text{out}}\right) = \frac{1}{\zeta^2}\frac{2|J_x|}{N_{\text{ph}}}\text{Var}\left(S_{2c}^{\text{in}}\right) = \frac{1}{\zeta^2}|J_x|, \tag{3.18}$$

where we assumed the input light noise equals the shot noise [i.e., $\text{Var}\left(S_{2c}^{\text{in}}\right) = N_{\text{ph}}/2$]. In other words, the variance of the output spin operator $\left(J_{z1}^{'\text{out}} + J_{z2}^{'\text{out}}\right)$ is a factor ζ^2 below the projection noise $|J_x|$. The "swap and squeezing" interaction can generate two-mode squeezing and entangled light as demonstrated in Ref. [19] and steady-state entanglement between the two atomic ensembles as demonstrated in Refs. [20–22].

3.4 Demonstration of high-sensitivity, projection-noise-limited magnetometry

3.4.1 Setup, pulse sequence, and procedure

A sketch of a two-cell magnetometer is shown in Fig. 3.4(a). The RF magnetic field B_{RF} is generated by sending a current through a coil. In the sketch, the coil is arranged such that both atomic ensembles are located in the same RF magnetic field. The ensembles, which are located in a bias magnetic field, are oppositely oriented using optical pumping methods, and probed by linearly polarized light.

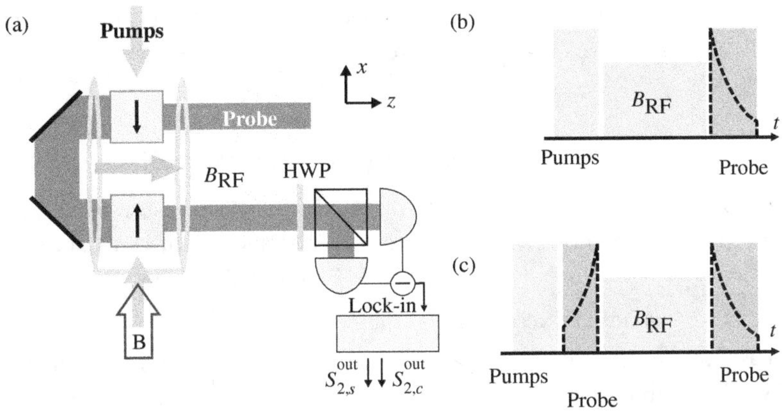

Figure 3.4 (a) Sketch of the experimental setup. Cubic 22 mm paraffin-coated cells filled with cesium are placed in the DC bias field B in two separate magnetic shields. The atoms are optically pumped so that the directions of the collective spins in the two cells are opposite. A pulse of B_{RF} at the frequency Ω_L is applied orthogonally to the B field. The same RF source (not shown) is used for the B_{RF} and as a reference to the lock-in amplifier. The polarization rotation of the top-hat-shaped probe beam (diameter 21 mm) is detected by two detectors (HWP, half-wave plate). The lock-in amplifier measures the $\cos(\Omega_L t)$ and $\sin(\Omega_L t)$ components of the photocurrent. (b) Pulse sequence for projection-noise-limited magnetometry. The temporal mode function for the probe is shown with a dashed curve. (c) Pulse sequence and temporal modes for entanglement-assisted magnetometry.

The Stokes operator $S_2^{out}(t)$ is measured by polarization homodyning. The outputs of the detectors are sent to a lock-in amplifier where the $\cos(\Omega_L t)$ and $\sin(\Omega_L t)$ components of the photocurrent are extracted. The outputs of the lock-in amplifier are measured using a field programmable gate array (FPGA) data acquisition card. On the computer, the cosine and sine components are integrated over the pulse duration to obtain the signals S_{2c}^{out} and S_{2s}^{out}.

Figure 3.4(b) shows the pulse sequence for the experiment. First, optical pumping light polarizes the two atomic ensembles. Then, the RF magnetic field is applied for a certain duration τ. During the RF magnetic pulse, no light is on the atoms. The decay time in the dark (when no light is on the atoms) is $T_2^{dark} > 30$ ms. This long coherence time is needed for high-sensitivity RF magnetic-field measurements as discussed in Section 3.2.2. Then a probe pulse is used to measure the atomic spins and thereby the RF-field amplitude B_{RF}. Due to decay of the spin state during the probe pulse, the atomic signal is encoded in the output-light mode with an exponentially falling temporal profile (see the online supplementary material), as illustrated in Fig. 3.4(b). Experimentally, this temporal light mode can be measured by multiplying the output of the lock-in amplifier with a temporal mode function $\propto e^{-\gamma_{opt} t}$, where γ_{opt} is some suitable decay constant of size $\approx 1/T_2$.

The pulse sequence for the entanglement-assisted measurements is shown in Fig. 3.4(c). A first probe pulse prior to the RF pulse is used to entangle the two ensembles. Afterwards the RF pulse is applied, and finally a second probe pulse is used for measuring the magnetic field amplitude B_{RF}. As seen in Fig. 3.4(c), we choose the temporal mode function of the first probe to be exponentially rising and the mode function of the second probe pulse to be exponentially falling. These choices optimize the conditional variance.

3.4.2 The projection-noise-limited magnetometer

We now present the results of high-sensitivity magnetic field measurements very close to the PN limit. Figure 3.5 shows two series (data sets A and B) of raw data points of the integrated signals S_{2c}^{out} and S_{2s}^{out} normalized to PN units. This means that we have plotted $S_{2c}^{\text{out}}/\sqrt{(N_{\text{ph}}/2)\kappa^2}$, such that the standard variation $\Delta\left[S_{2c}^{\text{out}}/\sqrt{(N_{\text{ph}}/2)\kappa^2}\right]=1$ for the PN-limited measurement. In order to normalize the raw data to PN units, one needs to know $N_{\text{ph}}/2$ and the coupling constant κ^2. N_{ph} is proportional to the light power P and κ^2 is a function of the light power and the Faraday angle θ_F (the polarization rotation angle of a weak, linearly polarized probe propagating in the x-direction, not shown in Fig. 3.4). Both P and θ_F are continuously monitored.

Prior to the magnetic field measurements it was checked that the homodyne detection was shot-noise limited by verifying that the input-light noise $\text{Var}\left(S_{2c}^{\text{in}}\right)$ was proportional

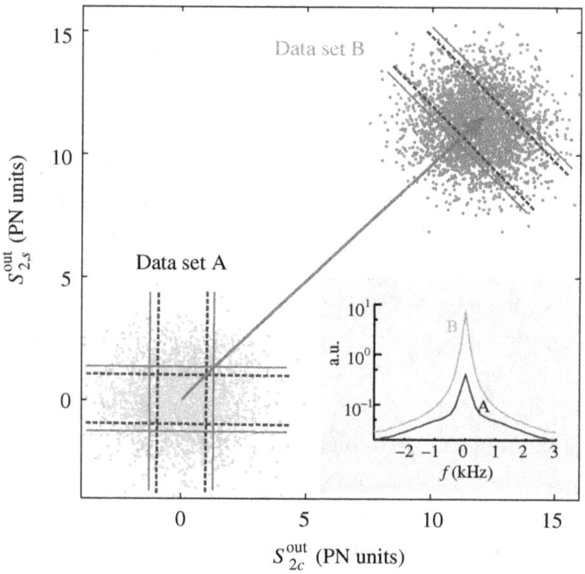

Figure 3.5 Results of a series of measurements of $\hat{S}_{2s,2c}$ without B_{RF} (data set A) and after a $\tau=15$ ms RF pulse with frequency 322 kHz and amplitude $B_{\text{RF}}=30(7)$ fT (data set B). The solid and the dashed lines show the standard deviation of the experimental points and of the PN-limited measurement, respectively. The spin displacement in the rotating frame is indicated by the arrow. (Inset) Power spectrum of the photocurrent (arbitrary units, centered at Ω_L). The narrow peak of the spectrum is the atomic-spin signal (projection noise). The broad part of the spectrum is the shot noise of light with the width set by the detection bandwidth.

to the light power. The coupling constant κ^2 was calibrated using the method described in [12, 19, 23]. Data set A in Fig. 3.5 corresponds to measurements of S_{2c}^{out} and S_{2s}^{out} where no B_{RF} is applied to the atoms. The mean values $\langle S_{2c}^{\text{out}}\rangle$ and $\langle S_{2s}^{\text{out}}\rangle$ are therefore zero. The distance between the horizontal (vertical) solid lines equals twice the measured standard deviation of S_{2c}^{out} (S_{2s}^{out}). The dashed lines represent the PN-limited measurement. The horizontal (or vertical) dashed lines are separated by two PN units. The measured standard deviations represented by the solid lines are a bit larger than the dashed lines, which correspond to the PN of atoms. This is expected because of the contribution from the shot noise of light to the total magnetometer noise. By comparing the solid lines with the dashed lines, one can directly see that the dominant source of noise is the PN.

For the measurements presented in Fig. 3.5, we measured the Faraday angle $\theta_F = 17.1°$ at the beginning of the probe pulse. From the measured Faraday angle, the numbers of atoms in the cell[9] can be estimated to be $N_A = 2 \times 7.8(7) \times 10^{11}$. The atomic decoherence time in the dark was measured to be $T_2^{\text{dark}} = 32$ ms.

Figure 3.5 also shows results (data set B) of measurements of a RF magnetic field (here $B_{\text{RF}} = 30(7)$ fT at the frequency of 322 kHz was applied for $\tau = 15$ ms prior to the measurement). The RF magnetic field tilts the spin vector and results in a nonzero mean value in the integrated signal $S_{2c}^{\text{out}} + iS_{2s}^{\text{out}}$.

An important figure of merit is the signal-to-noise ratio defined by

$$\text{SNR} = \frac{\left|\left\langle S_{2c}^{\text{out}} + iS_{2s}^{\text{out}}\right\rangle\right|}{\sqrt{\frac{1}{2}\left[\text{Var}\left(S_{2c}^{\text{out}}\right) + \text{Var}\left(S_{2s}^{\text{out}}\right)\right]}}. \tag{3.19}$$

The SNR of the data presented in Fig. 3.5 is given by the length of the arrow divided by the standard deviation (1/2 of the distance between the solid lines), yielding SNR = 12.3.

The minimal detectable field can be calculated from the applied RF magnetic field and the signal-to-noise ratio: $B_{\min} = B_{\text{RF}}/\text{SNR} = 30\ \text{fT}/12.3 = 2.4(5)$ fT. The sensitivity is then calculated as $\mathcal{S}_\tau = B_{\min}\sqrt{\tau} = 2.9(6) \times 10^{-16}\ \text{T}/\sqrt{\text{Hz}}$ where we neglected the time used for optical pumping and probing.

The measured sensitivity can be compared with the PN-limited sensitivity of $\mathcal{S}_{PN} = 2.6(1) \times 10^{-16}\ \text{T}/\sqrt{\text{Hz}}$ which can be calculated using Eq. (3.7) and the experimental values $N_A = 2 \times 7.8(7) \times 10^{11}$, $T_2^{\text{dark}} = 32$ ms, and $\tau = 15$ ms. Since the measured sensitivity is close to the PN-limited sensitivity, the magnetometer is mainly limited by the projection noise of atoms. The small difference between the two sensitivities is due to the residual shot noise, atomic decay, and classical noise of the atomic spins. We emphasize that the measurements have been done utilizing the back-action-evading swap and squeezing interaction and that the contribution from the shot noise to the total magnetometer noise is reduced.

The best experimental sensitivity $\mathcal{S}_{\text{tot}} = B_{\text{RF}}\sqrt{T_{\text{tot}}}/\text{SNR}$ reached was $3.5(8) \times 10^{-16}\ \text{T}/\sqrt{\text{Hz}}$, which was calculated using the total cycle time T_{tot} (RF duration $\tau =$

[9] In the above experiment, the probe has a top-hat-shaped beam profile with 21 mm diameter close to the cubic cell length of 22 mm. The probe light therefore fills up almost the whole cell.

22.5 ms, probe duration $T = 1.5$ ms and optical pumping duration $T_{\text{pump}} = 6$ ms). This best experimental sensitivity is close to the best-to-date sensitivity obtained by an atomic magnetometer [24]. The magnetometer [24] is a so-called spin-exchange-relaxation-free (SERF) atomic magnetometer (Chapter 5) and operates with orders of magnitudes more atoms. The magnetometer described in this chapter achieves a similar sensitivity with fewer atoms because the measurement of the magnetic field is limited only by the quantum projection noise.

3.5 Demonstration of entanglement-assisted magnetometry

One can reduce atomic noise below the PN limit by the use of entanglement between two atomic ensembles. Entanglement is quantified in terms of the EPR variance Σ_{EPR} defined in Eq. (3.4), and the criterion for entanglement between the ensembles is that $\Sigma_{\text{EPR}} < 1$.

The two atomic ensembles can be entangled prior to the RF pulse [Fig. 3.4(c)] using the first probe pulse. The magnetic field B_{RF} is subsequently measured with a second probe pulse. Using the measurement result obtained with the first probe pulse, we can reduce the uncertainty on the second probe measurement. This leads to a better SNR for the measurement of B_{RF}.

When utilizing the first pulse measurement to reduce the noise on the second pulse measurement, we use the notion of conditional variance. The conditional cosine Stokes operator is defined as

$$S_{2c}^{\text{cond}} = S_{2c}^{\text{out,2nd}} - \alpha S_{2c}^{\text{out,1st}}, \tag{3.20}$$

and the conditional sine operator S_{2s}^{cond} is defined similarly. The parameter α is optimized in the experiment in order to minimize the conditional Stokes variance defined by $\left[\text{Var}\left(S_{2c}^{\text{cond}}\right) + \text{Var}\left(S_{2c}^{\text{cond}}\right)\right]/2$. From the conditional Stokes variance, we can calculate the conditional atomic variance $\Sigma_{\text{EPR}}^{\text{cond}}$. The criterion for conditional entanglement is $\Sigma_{\text{EPR}}^{\text{cond}} < 1$. For entanglement-assisted magnetometry, we define the signal-to-noise ratio as

$$\text{SNR} = \frac{\left|S_{2c}^{\text{out,2nd}} + iS_{2s}^{\text{out,2nd}}\right|}{\sqrt{\frac{1}{2}\left[\text{Var}\left(S_{2c}^{\text{cond}}\right) + \text{Var}\left(S_{2s}^{\text{cond}}\right)\right]}}. \tag{3.21}$$

Entanglement improves the signal-to-noise ratio since $\text{Var}\left(S_{2c}^{\text{cond}}\right) + \text{Var}\left(S_{2c}^{\text{cond}}\right) < \text{Var}\left(S_{2c}^{\text{out}}\right) + \text{Var}\left(S_{2c}^{\text{out}}\right)$.

We now present experiments demonstrating that entanglement can improve magnetometric measurements. The results presented here are obtained at room temperature, corresponding to the Faraday angle of $\theta_F \approx 8°$. The field B_{RF}, applied to one of the ensembles for a variable duration τ, is measured using two approaches. In the first approach [Fig. 3.4(b)], the RF magnetic field is applied directly after the optical-pumping stage and measured with a probe pulse. In the second approach [Fig. 3.4(c)], the two atomic ensembles

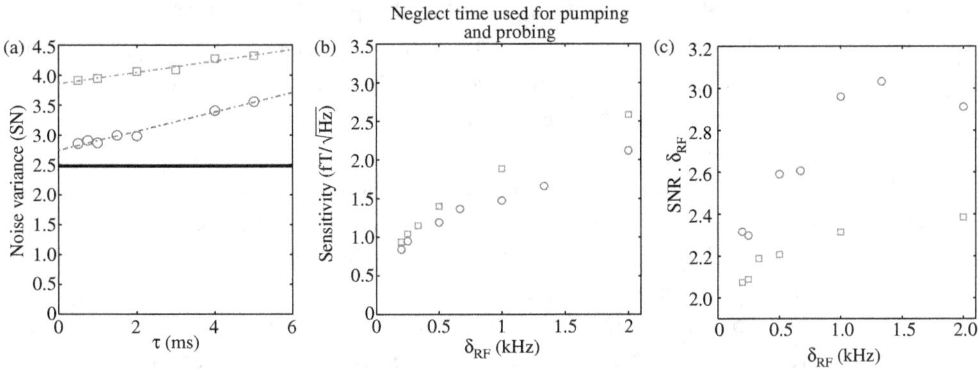

Figure 3.6 Entanglement-assisted magnetometry results. Circles and squares represent data with and without entanglement, respectively. (a) Magnetometer read-out noise in units of shot noise of light. Dashed lines are a linear fit. The solid line at 2.5 is the limit for the PN-limited measurement of the magnetic field given by Eq. (3.22). (b) Sensitivity $\mathcal{S}_\tau = B_{\mathrm{RF}}\sqrt{\tau}/\mathrm{SNR}$ of the magnetic-field measurement. (c) Signal-to-noise ratio SNR times the RF bandwidth δ_{RF} for the magnetometer. Since SNR depends linearly on the applied magnetic-field amplitude B_{RF}, SNR has been normalized to the value $B_{\mathrm{RF}} = 138$ fT.

are entangled using a first probe pulse that occurs after optical pumping. Following this, the RF magnetic field is applied and measured with a second probe pulse. The data sets for the first and second approaches are denoted "without" and "with" entanglement, respectively. The data without entanglement serve as reference to which the data with entanglement can be compared.

The results of the measurements are shown in Fig. 3.6. In the three plots, circles represent data with entanglement and squares represent data without entanglement. We start by considering Fig. 3.6(a). The square data points without entanglement show the variance of the signal $\left[\mathrm{Var}\left(S_{2c}^{\mathrm{out}}\right) + \mathrm{Var}\left(S_{2s}^{\mathrm{out}}\right)\right]/2$ normalized to $N_{\mathrm{ph}}/2$ as a function of RF pulse duration τ. The solid line represents the results one would obtain with a PN-limited measurement of the spin. If we include a detection efficiency η_{det} in the analysis, we have the following criterion for the PN-limited measurement:

$$\frac{1}{2}\left[\mathrm{Var}\left(S_{2c}^{\mathrm{out}}\right) + \mathrm{Var}\left(S_{2s}^{\mathrm{out}}\right)\right] \leq \eta_{\mathrm{det}}\kappa^2\frac{N_{\mathrm{ph}}}{2}. \tag{3.22}$$

Using the measured value of $\kappa^2 = 3.1$ (valid for the specific experimental settings used for obtaining the results plotted in Fig. 3.6) and the estimated detection efficiency $\eta_{\mathrm{det}} \approx 0.8$, we find that the PN-limited measurement has

$$\left[\mathrm{Var}\left(S_{2c}^{\mathrm{out}}\right) + \mathrm{Var}\left(S_{2s}^{\mathrm{out}}\right)\right]/2 \approx 0.8 \times 3.1 \times \left(N_{\mathrm{ph}}/2\right) \approx 2.5 \times \left(N_{\mathrm{ph}}/2\right) .$$

The measured data without entanglement has $\left[\mathrm{Var}\left(S_{2c}^{\mathrm{out}}\right)+\mathrm{Var}\left(S_{2s}^{\mathrm{out}}\right)\right]/2 \approx 3.9 \times \left(N_{\mathrm{ph}}/2\right)$ (for short RF durations). In terms of the standard deviation, the measurement has only $\sqrt{3.9/2.5} \approx 1.25$ times higher standard deviation than the PN-limited measurement.

We now turn to the data with entanglement plotted as circles in Fig. 3.6(a). The entanglement between the ensembles is visible in the conditional variances $\left[\mathrm{Var}\left(S_{2c}^{\mathrm{cond}}\right)+\mathrm{Var}\left(S_{2s}^{\mathrm{cond}}\right)\right]/2$ which are plotted with circles. We see that the conditional variances are significantly lower than the variances obtained without entanglement. The difference is largest for short RF durations where the conditional variance is 25–30% lower than the variance obtained without entanglement.

For RF magnetic pulse durations $\tau \leq 2$ ms, the measured conditional variances are $\left[\mathrm{Var}\left(S_{2c}^{\mathrm{cond}}\right)+\mathrm{Var}\left(S_{2s}^{\mathrm{cond}}\right)\right]/2 \approx 2.9 \cdot \left(N_{\mathrm{ph}}/2\right)$. In terms of the standard deviation, the entanglement-assisted measurement has $\sqrt{2.9/2.5} \approx 1.08$ higher standard deviation than the PN-limited measurement. This is even closer to the PN-limited measurement than when no entanglement was utilized (which had the standard deviation 25% higher than the PN-limited measurement). We conclude that the entanglement-assisted measurement of B_{RF} is close to being PN limited. It should be possible to achieve a better than PN-limited measurement of B_{RF} by improving the detection efficiency and increasing the optical depth. Since the data were taken at room temperature, the optical depth can easily be increased by a factor 2 or 3 by increasing the temperature of the cells.

In order to determine whether the two ensembles are entangled or not, we should calculate the atomic noise in PN units. Using a full swap and squeezing interaction theory which includes decoherence [25], the atomic noise is calculated from the measured output variances and the measured κ^2. For short RF durations, we calculate the atomic noise $\Sigma_{\mathrm{EPR}} = 1.10(8)$ for the data without entanglement and the conditional atomic noise $\Sigma_{\mathrm{EPR}}^{\mathrm{cond}} = 0.70(5)$ for the data with entanglement. Since $\Sigma_{\mathrm{EPR}}^{\mathrm{cond}} < 1$, we conclude that the two atomic ensembles were entangled by the first probe pulse. For longer RF pulses the entanglement degrades due to atomic decoherence.

With increasing RF pulse duration, the measured output-light noise increases as seen in Fig. 3.6(a). This is true both for the case with and for that without entanglement. To illustrate this, we make linear fits to the data (squares and circles). The fits are shown in Fig. 3.6(a) as dash-dotted lines. From the slopes of the linear fits, it is seen that the conditional variance decays faster than the variance obtained without entanglement, as discussed in Section 3.2.2.

Since the signal is the same with or without entanglement (if one neglects the insignificant decay of the mean spin during the first probe pulse), the reduction of the noise using entanglement is therefore directly transferred into an improved SNR. Figure 3.6(c) shows the signal-to-noise ratio times the bandwidth, SNR $\times\, \delta_{\mathrm{RF}}$, as a function of the RF magnetic field bandwidth here defined as $\delta_{\mathrm{RF}} = 1/\tau$. For large bandwidths the product SNR $\times\, \delta_{\mathrm{RF}}$ reaches an approximately constant value. Using entanglement, this product is improved due to the lower noise of the entangled atomic state.

From the measured SNR, the applied B_{RF} and the duration of the RF field τ, we calculate the sensitivity $\mathcal{S}_\tau = B_{\mathrm{RF}}\sqrt{\tau}/\mathrm{SNR}$ for the case with and without entanglement. Note that here we neglect the time it takes for the pumping T_{pump} and probing T_{probe}. When neglecting T_{pump} and T_{probe}, the improvement in *SNR* due to entanglement is directly translated into an improvement of the sensitivity. This sensitivity as a function of RF bandwidth is plotted in Fig. 3.6(b) for the cases with and without entanglement. The best sensitivity is reached for long RF pulses or equivalently small bandwidths. We also see that entanglement improves the magnetometer sensitivity in particular for large bandwidths. This is expected since the conditional variance decays faster than the variance obtained without entanglement, as discussed in Section 3.2.2.

3.6 Conclusions

In conclusion, we have discussed an atomic RF magnetometer in which all three main sources of quantum noise – the light shot noise, the projection noise of atoms, and the back-action noise of the measurement – are significantly suppressed. This is achieved with a back-action-evading measurement, the reduction of the shot noise from the probing light by the swap interaction and by entanglement of atoms. Experimentally, the magnetometer demonstrated a sensitivity $\mathcal{S}_{\text{tot}} = 3.5(8) \times 10^{-16}\ \text{T}/\sqrt{\text{Hz}}$, which in absolute units is comparable with the sensitivity obtained with state-of-the-art (SERF) magnetometers (see Ref. [24] and Chapter 5), and exceeds the best magnetometers to date in terms of sensitivity per atom. Since the PN-limited sensitivity increases with the number of atoms, it is possible to further increase the sensitivity by using larger cells or heating the cells further. It is also possible to obtain higher sensitivity to long RF pulses by improving the atomic coherence time (in the dark). This could be done by improving the bias magnetic-field homogeneity and by improving the antirelaxation coating of the cells (see Ref. [26] and Chapter 11).

We also presented results of entanglement-assisted metrology with the highest-to-date numbers of atoms. The degree of entanglement can in principle reach $\Sigma_{\text{EPR}} = \zeta^{-2} \approx 1/6.4$ for the specific experimental probe detuning. This corresponds to 8 dB of spin squeezing. This expected degree of entanglement can be even larger for a further detuned probe. Due to various decoherence effects including spontaneous emission, the entanglement generated in our experiment is lower than this value $1/6.4$. With a higher optical depth, we expect that the degree of entanglement can be improved.

Using the entanglement-assisted magnetometer presented here, it was shown that it is possible to get close to the PN-limited measurement of magnetic fields, and it should be possible to surpass the limit with only slightly higher optical depth. It was also demonstrated that using entanglement, the signal-to-noise ratio of the magnetometer could be improved. However, for a short RF-pulse duration, finite pumping and probing times limit the magnetometer sensitivity. This is true both for the measurement with and that without entanglement. The magnetometer can therefore be improved by shortening these times.

There has been some discussion of whether entanglement can improve an atomic magnetometer [10, 11]. Here we showed results demonstrating that the magnetic-field sensitivity is indeed improved by using entanglement when one is detecting RF magnetic-field pulses which are shorter than the entanglement lifetime.

Entanglement generated by QND measurements becomes better as the number of atoms grows. Therefore, scaling of the precision of metrology and magnetometry can exceed the projection-noise-limited $1/\sqrt{N}$ scaling and approach the $1/N$ Heisenberg scaling. In the experiment presented in this chapter, the $1/\sqrt{N}$ scaling was not surpassed since an increase in N was accompanied by growing decoherence that limited the degree of entanglement. However, in a cold atom system [27], the $1/\sqrt{N}$ scaling was surpassed and scaling approaching the $1/N$ Heisenberg scaling was demonstrated.

The atomic magnetometer described here utilizes two ensembles, one of which is optically pumped into an energetically inverted state of the magnetic-sublevel manifold. Such a polarized ensemble is similar to an inverted harmonic oscillator, or an oscillator with a "negative mass." As demonstrated in [28], an inverted atomic ensemble can be entangled with a real mechanical oscillator and this system can be used to measure a force applied to the oscillator with a sensitivity beyond the standard quantum limit, much in the same way that two atomic ensembles are used to measure the magnetic field beyond the projection-noise level. The two-cell magnetometer is thus the first example of a more general approach where the concept of a negative mass is used for quantum measurements [29].

Acknowledgments

The authors would like to thank Wojciech Wasilewski and Hanna Krauter. This research was supported by EU grants Q-ESSENCE and MALICIA and the QUASAR program of DARPA. K. J. acknowledges support from the Danish Research Council.

References

[1] D. Budker and M. Romalis, Nature Physics **3**, 227 (2007).
[2] C. C. Gerry and P. L. Knight, *Introductory Quantum Optics* (Cambridge University Press, Cambridge, 2005).
[3] B. Julsgaard, Ph.D. thesis, University of Aarhus (2003) [www.quantop.nbi.dk/pdfs/theses/brianthesis.pdf].
[4] B. Julsgaard, A. Kozhekin, and E. S. Polzik, Nature **413**, 400 (2001).
[5] B. Julsgaard, J. Sherson, J. I. Cirac, J. Fiurášek, and E. S. Polzik, Nature **432**, 482 (2004).
[6] G. Vasilakis, V. Shah, and M. V. Romalis, Phys. Rev. Lett. **106**, 143601 (2011).
[7] M. Koschorreck, M. Napolitano, B. Dubost, and M. W. Mitchell, Phys. Rev. Lett. **105**, 093602 (2010).
[8] M. Kitagawa and M. Ueda, Phys. Rev. A **47**, 5138 (1993).
[9] D. J. Wineland, J. J. Bollinger, W. M. Itano, F. L. Moore, and D. J. Heinzen, Phys. Rev. A **46**, R6797 (1992).
[10] S. F. Huelga, C. Macchiavello, T. Pellizzari, A. K. Ekert, M. B. Plenio, and J. I. Cirac, Phys. Rev. Lett. **79**, 3865 (1997).
[11] M. Auzinsh, D. Budker, D. F. Kimball, S. M. Rochester, J. E. Stalnaker, A. O. Sushkov, and V. V. Yashchuk, Phys. Rev. Lett. **93**, 173002 (2004).

[12] W. Wasilewski, K. Jensen, H. Krauter, J. J. Renema, M. V. Balabas, and E. S. Polzik, Phys. Rev. Lett. **104**, 133601 (2010).
[13] F. Wolfgramm, A. Cerè, F. A. Beduini, A. Predojević, M. Koschorreck, and M. W. Mitchell, Phys. Rev. Lett. **105**, 053601 (2010).
[14] A. Einstein, B. Podolsky, and N. Rosen, Phys. Rev. **47**, 47 (1935).
[15] L. M. Duan, G. Giedke, J. I. Cirac, and P. Zoller, Phys. Rev. Lett. **84**, 2722 (2000).
[16] I. M. Savukov, S. J. Seltzer, M. V. Romalis, and K. L. Sauer, Phys. Rev. Lett. **95**, 063004 (2005).
[17] M. P. Ledbetter, V. M. Acosta, S. M. Rochester, D. Budker, S. Pustelny, and V. V. Yashchuk, Phys. Rev. A **75**, 023405 (2007).
[18] K. Hammerer, A. S. Sorensen, and E. S. Polzik, Rev. Mod. Phys. **82**, 1041 (2010).
[19] W. Wasilewski, T. Fernholz, K. Jensen, L. S. Madsen, H. Krauter, C. Muschik, and E. S. Polzik, Opt. Express **17**, 14444 (2009).
[20] H. Krauter, C. A. Muschik, K. Jensen, W. Wasilewski, J. M. Petersen, J. I. Cirac, and E. S. Polzik, arXiv:1006.4344 (2010).
[21] H. Krauter, C. A. Muschik, K. Jensen, W. Wasilewski, J. M. Petersen, J. I. Cirac, and E. S. Polzik, Phys. Rev. Lett. **107**, 080503 (2011).
[22] C. A. Muschik, H. Krauter, K. Jensen, J. M. Petersen, J. I. Cirac, and E. S. Polzik, J. Phys. B: At. Mol. Opt. Phys. **45**, 124021 (2012).
[23] K. Jensen, W. Wasilewski, H. Krauter, T. Fernholz, B. M. Nielsen, M. Owari, M. B. Plenio, A. Serafini, M. M. Wolf, and E. S. Polzik, Nature Physics **7**, 13 (2011).
[24] S.-K. Lee, K. L. Sauer, S. J. Seltzer, O. Alem, and M. V. Romalis, Appl. Phys. Lett. **89**, 214106 (2006).
[25] K. Jensen, Ph.D. thesis, University of Copenhagen (2011) [http://www.nbi.ku.dk/english/research/phd theses/phd theses 2011/kasper jensen].
[26] M. V. Balabas, T. Karaulanov, M. P. Ledbetter, and D. Budker, Phys. Rev. Lett. **105**, 070801 (2010).
[27] J. Appel, P. J. Windpassinger, D. Oblak, U. B. Hoff, N. Kjærgaard, and E. S. Polzik, Proc. Nat. Acad. Sci. **106**, 10960 (2009).
[28] K. Hammerer, M. Aspelmeyer, E. S. Polzik, and P. Zoller, Phys. Rev. Lett. **102**, 020501 (2009).
[29] M. Tsang, H. M. Wiseman, and C. M. Caves, Phys. Rev. Lett. **106**, 090401 (2011).

4

M_x and M_z magnetometers

E. B. Alexandrov and A. K. Vershovskiy

4.1 Dynamics of magnetic resonance in an alternating field

4.1.1 Bloch equations and Bloch sphere

In a typical setup of an optically pumped quantum magnetometer (OPQM), the working medium is placed within a superposition of two magnetic fields: a slowly changing field $\mathbf{B}$ being measured, and a field $\mathbf{B}_1$ oscillating (or rotating) at frequency ω that excites the atomic transition. Equations describing the evolution of the magnetic moment $\mathbf{M}$ subject to such fields were first formulated by F. Bloch [1]:

$$\frac{d\mathbf{M}}{dt} = \gamma \mathbf{M} \times \mathbf{B}\,, \tag{4.1}$$

where γ is the gyromagnetic ratio of a system, i.e. the ratio of its magnetic dipole moment to its angular momentum (see Chapter 1). Equation (4.1) describes the evolution of a magnetic moment vector $\mathbf{M}$ in an arbitrary magnetic field $\mathbf{B}$, without taking into consideration the relaxation processes. In the case of a constant field $\mathbf{B}_0$ directed along the Z axis, the moment component along the Z axis stays constant, and the transverse component rotates around Z with frequency $\omega_0 = \gamma B_0$.

Let us introduce a coordinate system $X'Y'Z$, rotating around the Z axis with angular velocity $\boldsymbol{\omega}$, so $\mathbf{B}_1 \parallel \mathbf{Y}'$ (Fig. 4.1). Transverse components of $\mathbf{M}$ in this system are traditionally denoted by $u \equiv M_y'$ and $v \equiv M_x'$. Applying vector transformation rules to (4.1) in the rotating frame yields:

$$\frac{d\mathbf{M}}{dt} = \gamma \left[\mathbf{M} \times \left(\mathbf{B} - \frac{\boldsymbol{\omega}}{\gamma}\right)\right] = \gamma \mathbf{M} \times \mathbf{B}'\,. \tag{4.2}$$

Thus the magnetic moment vector in a rotating coordinate system experiences the influence of an effective magnetic field $\mathbf{B}' = \mathbf{B} - \boldsymbol{\omega}/\gamma$. This result is known as Larmor's theorem.

Suppose that at time $t = 0$ the magnetic moment $\mathbf{M}_0$ is directed along the Z axis. The magnetic field $\mathbf{B}'$, acting on $\mathbf{M}_0$ in the rotating frame, is constant and equal to the vector

Optical Magnetometry, ed. Budker, D. and Jackson Kimball, D. F. Published by Cambridge University Press.

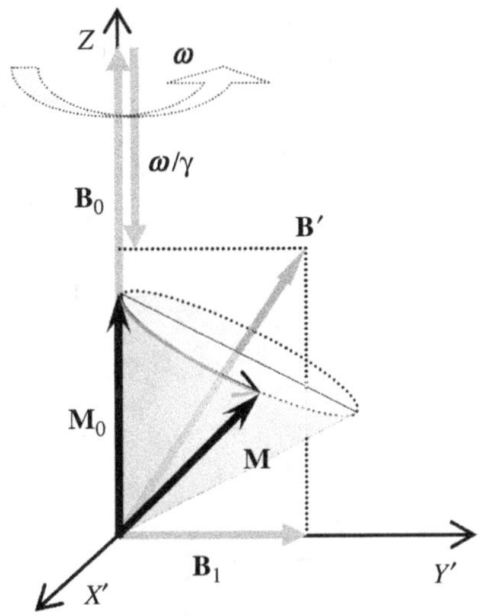

Figure 4.1 Evolution of a magnetic moment **M** in the presence of constant ($\mathbf{B}_0$) and rotating ($\mathbf{B}_1$) magnetic fields in the rotating coordinate system.

sum of the field $\mathbf{B}_0 - \omega/\gamma$ parallel to the Z axis and the field $\mathbf{B}_1$ parallel to $\mathbf{Y}'$. Under these conditions the moment **M** precesses without changing its length around $\mathbf{B}'$ with frequency

$$\omega' = \sqrt{(\Delta\omega)^2 + \Omega^2}\,, \tag{4.3}$$

where $\Delta\omega = \omega_0 - \omega$ is the detuning of the rotating field frequency, $\omega_0 = \gamma B_0$ is the magnetic resonance frequency, and $\Omega = \gamma B_1$ is the Rabi frequency, determined by the amplitude of the rotating field.

It turns out that in the laboratory coordinates near resonance, the magnetic moment vector moves along a spiroid curve laying on a spheric surface. The area comprising the totality of the solutions of the Bloch equations (4.1) is called a Bloch sphere (Fig. 4.2). In quantum mechanics it is often used to describe the state of a two-level system; it is customary to call the upper pole of the sphere "north," the bottom pole "south," and the $Z = 0$ plane the "equator."

Thus, in the laboratory frame the projection of the moment **M** on the equatorial plane rotates at the frequency ω of the rotating field $\mathbf{B}_1$. Its movement back and forth between the north and south poles is also harmonic with the nutation frequency ω' [Eq. (4.3)]. In exact resonance the vector hodograph of the magnetic moment encloses the entire sphere from the north pole to the south, and the nutation frequency equals the Rabi frequency $\Omega = \gamma B_1$.

If relaxation processes are taken into consideration, the system settles to a new equilibrium state with reduced longitudinal magnetization. In the general case, longitudinal and transverse components of the magnetic moment vector may relax in different ways: relaxation processes tend to bring the longitudinal component to the stationary value $\mathbf{M}_0$

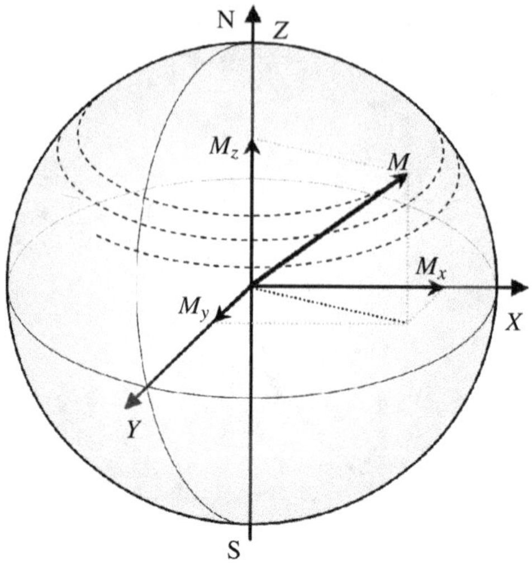

Figure 4.2 The Bloch sphere – the area comprising the totality of solutions of the Bloch equations; the dashed line represents the typical evolution of the magnetic moment on resonance.

with time constant T_1, and transverse components to zero with time constant T_2 (generally $T_1 \geq 2T_2$).

The stationary solution of the Bloch equation modified under these assumptions in the rotating coordinate system is described by the following expressions:

$$\begin{aligned} v &= -M_0 \frac{\Omega T_2}{1+(\Delta\omega T_2)^2+\Omega^2 T_1 T_2}, \\ u &= M_0 \frac{\Delta\omega\Omega T_2^2}{1+(\Delta\omega T_2)^2+\Omega^2 T_1 T_2}, \\ M_z &= M_0 \frac{1+(\Delta\omega T_2)^2}{1+(\Delta\omega T_2)^2+\Omega^2 T_1 T_2}. \end{aligned} \tag{4.4}$$

Absorption of the alternating field energy by the system per unit volume is proportional to $\langle \mathbf{B} \cdot d\mathbf{M}/dt \rangle$ (brackets here denote time averaging), and it can be shown that it is determined by the v component, while the u component determines dispersion. We note that there is no component u on exact resonance, meaning that the transverse component of the angular momentum precesses with a 90° phase shift relative to the $\mathbf{B}_1$ field vector.

4.1.2 Types of magnetic resonance signals: M_z and M_x signals

Two types of magnetic-resonance signals are distinguished according to the projections of the magnetic moment:

1. M_z signals, associated with the longitudinal component of the magnetic moment, i.e., constant magnetization M_z;

2. M_x signals, associated with any transverse component of the magnetic moment, described in the rotating coordinate system by variables u and v.

In the vicinity of the resonance these signals behave quite differently; this applies to the dependence of these values on the amplitude and frequency of the resonant field, as well as to the dynamics of these values.

4.2 M_z and M_x magnetometers: general principles

All magnetometers based on magnetic resonance phenomena are divided into two types, classified according to the method used for the resonance detection:

1. M_z magnetometers: relatively slow, since their time response is principally limited by T_1, but highly accurate instruments in which a quantity proportional to the longitudinal component of the magnetic moment serves as a signal;
2. M_x magnetometers: fast (their time response is not limited by T_1 or T_2) instruments that detect the phase of the oscillating transverse component of the magnetic moment.

All commercially produced OPQMs use electronic paramagnets (alkali metals in the gas state or metastable helium) as working substances. A typical level structure of an alkali atom is given in Fig. 4.3. The transitions $\Delta F = 0$, $\Delta m_F = \pm 1$ between magnetic sublevels belonging to the same hyperfine level F are called *Zeeman transitions*; their frequencies in the Earth's magnetic field (EMF) are in the radiofrequency range. The transitions $\Delta F = \pm 1$, $\Delta m_F = 0, \ \pm 1$ are called *hyperfine transitions*, and their frequencies are in the microwave range.

The D1 and D2 transitions connecting the atomic ground state S with the fine-structure sublevels ($P_{1/2}$, $P_{3/2}$) of the first excited level P are optical; they are generally used for pumping the reference transition.

The idea behind an M_z magnetometer is as follows. A resonant RF field, applied to an atomic transition, reduces the population difference created by optical pumping; one may modulate the RF field frequency and observe changes in the absorption of pumping light. The feedback system tunes the mean frequency of the RF field to the frequency of the resonance line center.

The history of the optical M_x magnetometer dates back to the work of H. Dehmelt [2,3], who was the first to use a transverse light beam to detect magnetic resonance and to show that light absorption can be used to create self-generating systems. Dehmelt's idea was experimentally verified in the same year, 1957, by W. Bell and A. Bloom [4].

Let us consider an alkali atom with spin 1/2, being excited by left-circularly (σ^+) polarized light propagating along the Z axis (supposing that $\mathbf{B}$ is parallel to Z). This light induces σ^+ transitions only, resulting in depletion of the $m = -1/2$ sublevel and reduction of pump light absorption. Let us apply to the oriented atoms a transverse magnetic field $\mathbf{B}_1$ oscillating with frequency ω, which is close to the resonance frequency $\omega_0 = \gamma B_0$. As follows from Eq. (4.4), this results in a reduction of longitudinal magnetization, as

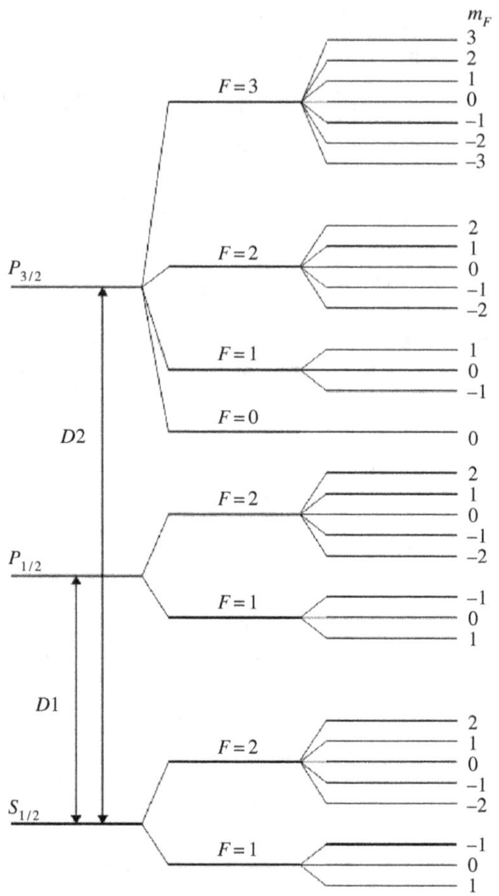

Figure 4.3 Level diagram (ground and first excited states) of an alkali atom with $I = 3/2$ (isotopes ^{7}Li, ^{23}Na, ^{39}K, ^{41}K, ^{87}Rb). The splittings are not to scale.

well as in the appearance of a transverse component of magnetization precessing with frequency ω.

In order to detect the transverse magnetization, one can use an auxiliary circularly polarized light beam directed, for example, along the X axis (perpendicular to $\mathbf{B}$). Then, in a coordinate system with the quantization axis parallel to X, there is a periodic (with frequency ω) change of population of the $m = -1/2$ sublevel, and therefore, a periodic change of the absorption coefficient for the transverse light beam. This AC signal can be used for phase locking to the resonance.

For real alkali atoms, the two-level model of the ground state has to be replaced with the multilevel one. The question of the Bloch model's applicability to alkali atoms was discussed by M.-A. Bouchiat in Ref. [5], and it was shown that this model stays true in the case of a detecting beam having a wide spectrum over the range of the hyperfine structure of one line of the fine-structure doublet (D1 or D2). The frequencies of the different radiofrequency transitions between adjacent magnetic sublevels are supposed to be equal, so the signals of these transitions could be summed together. This model fits very well the case of cesium in EMF, where the frequency difference of the adjacent resonances inside one hyperfine level is smaller than the resonance width.

Another extreme case corresponds to the potassium isotopes ^{39}K and ^{41}K, because their radiofrequency spectrum is almost completely resolved in the whole EMF range. Therefore, each transition can be considered separately, and the multilevel system can be replaced with a set of independent two-level systems described by Bloch equations.

There are several types of "classical" alkali OPQMs [i.e., using single Zeeman transitions, Fig. 4.4(a)], such as the rubidium magnetometer, the first one historically, but rarely used now, the much more widespread cesium magnetometer, and the relatively new potassium magnetometer. There are also devices that use hyperfine resonances [so-called HFS magnetometers, Fig. 4.4(b)], or multiphoton transitions in the radiofrequency range [Fig. 4.4(c)], or multiphoton transitions in the microwave and optical range [Fig. 4.4(d)].

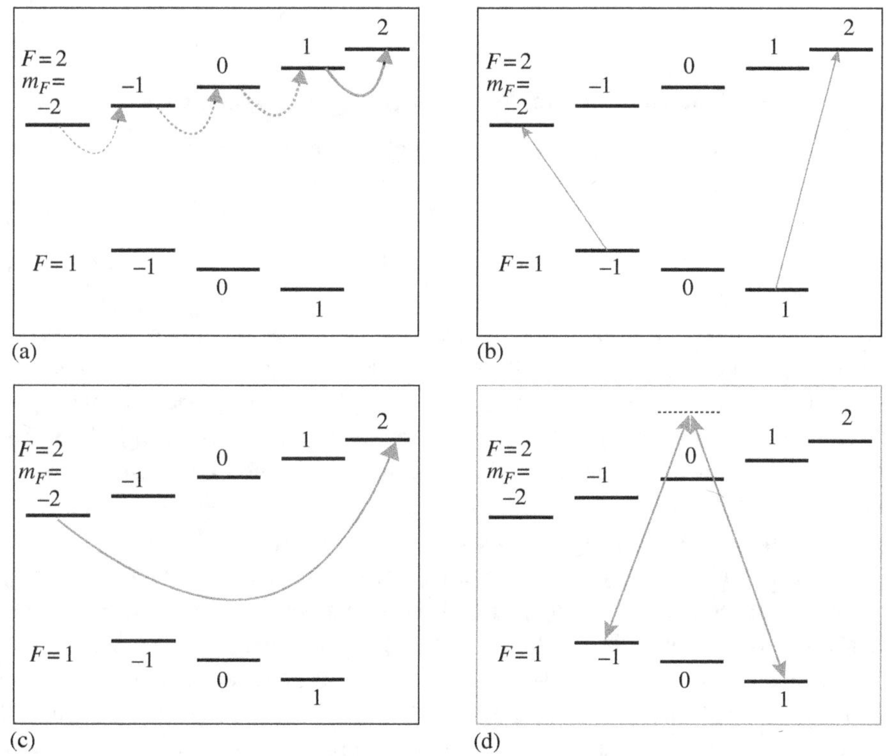

Figure 4.4 Detection schemes of the resonance in the alkali atom with $I = 3/2$ ground state $nS_{1/2}$: (a) Zeeman transition; (b) hyperfine transition; (c) four-photon transition; (d) coherent Λ-transition (microwave, optical).

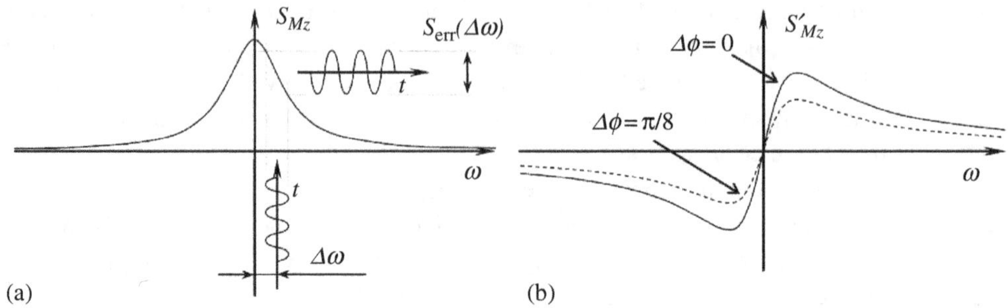

Figure 4.5 (a) M_z signal as recorded with a photodetector, and its response to slow frequency modulation; (b) M_z signal after the synchronous detection. The dashed line corresponds to the phase error $\Delta\phi = \pi/8$.

The "classical" scheme can be realized as an M_x magnetometer with fast response, whereas other schemes such as HFS, Zeeman multiphoton, or pure optical OPQMs are implemented in magnetometers of the M_z type.

4.2.1 Advantages and disadvantages of M_z magnetometers

At low intensity of the field $\mathbf{B}_1$ (the criterion being given by the condition $\Omega^2 T_1 T_2 \ll 1$) the dependence $\Delta M_z(\Delta\omega)$ has a characteristic Lorentzian form

$$\Delta M_z(\Delta\omega) = M_0 \Omega^2 T_1 T_2 \mathfrak{L}(\Delta\omega T_2) \tag{4.5}$$

where $\mathfrak{L}(x) = 1/(1+x^2)$ is the Lorentzian contour with the extremum at the resonance point (Fig. 4.5). In order to determine the sign and value of the mean detuning $\Delta\omega$ one has to use modulation of ω at low frequency Ω_m ($\Omega_m < 1/T_1,\ 1/T_2$), accompanied with the synchronous detection of the signal at the same frequency Ω_m [Fig. 4.5(a)]. This procedure limits the OPQM time response, which is especially noticeable in the case of the narrow resonances.

These drawbacks are compensated by the main advantage of an M_z magnetometer – its high accuracy, conditioned by two factors: (1) the center of the resonance line position [Fig. 4.5(b)] does not depend on the phase of the signal observation; (2) if more than one resonance line is present in the working substance radiofrequency spectrum [Fig. 4.4(a)], the reference line center position is shifted much less in the case of a Lorentzian line shape than in the case of a dispersive line shape which is observed in M_x schemes. The shift of the frequency of the selected resonance by the neighboring resonance wing can be estimated as $\Delta\omega \approx \alpha\Gamma^2/\Delta$ for M_x signals, and $\Delta\omega \approx \alpha\Gamma^4/\Delta^3$ for M_z signals – here $\alpha < 1$ is the relative magnitude of the adjacent resonance, Δ is the detuning from it, and Γ is its width ($\Gamma/\Delta \ll 1$).

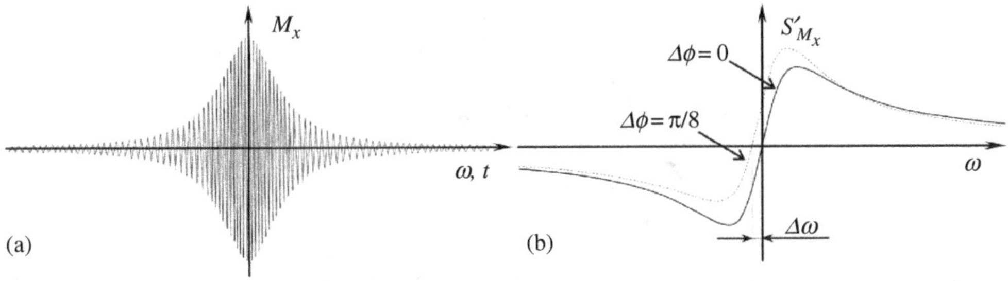

Figure 4.6 (a) M_x signal as recorded with a photodetector in the case of slow linear scan of the RF frequency – the oscillations are at the Larmor frequency with a Lorentz-shaped envelope; (b) M_x signal after phase detection. The dashed line denotes a signal with a $\Delta\phi = \pi/8$ phase error; $\Delta\omega$ is the corresponding error in the frequency domain.

Another M_z magnetometer advantage is the pumping scheme simplicity and its axial symmetry: only one optical beam, directed along the magnetic field vector **B**, is needed to pump the atoms in a "classical" Zeeman scheme and to detect the M_z signal.

It follows from the above considerations that the orientation of this device relative to **B** cannot be chosen arbitrarily: no resonance pumping or detection is possible with the optical beam perpendicular to **B**. This means that "classical" M_z magnetometers have what are usually called *angular dead zones*. There are more complicated M_z schemes, however, able to function at any angle to the magnetic field vector – i.e., hyperfine magnetometers that use isotropic spectral optical pumping together with a divergent resonant field distribution. The frequency of the modulation used in M_z magnetometers is usually limited by the resonance width, and therefore M_z schemes are relatively slow.

4.2.2 Advantages and disadvantages of M_x magnetometers

Registration of M_x signals is usually more convenient because they oscillate even under stationary conditions (Fig. 4.6). It is very important that whereas the populations' rate of change is restricted by the relaxation constants, the oscillation frequency of the coherence reacts to the field module changes immediately. As a result, M_x magnetometers can be very fast.

The principal factor placing an upper limit on the bandwidth of M_x magnetometers is the Larmor frequency $\omega = \gamma B$; furthermore, the operational speed of any magnetometer is limited by the fundamental uncertainty of the measurement of a noise-containing harmonic signal frequency within a given time interval [6, 7].

The synchronous phase detection method allows selective extraction of the components u and v. A signal proportional to u, with a dispersion-shaped function of the detuning [Fig. 4.6(b)], is most convenient as a frequency discriminator. However, it is phase dependent, so the point where it crosses zero, and where the feedback system locks to the

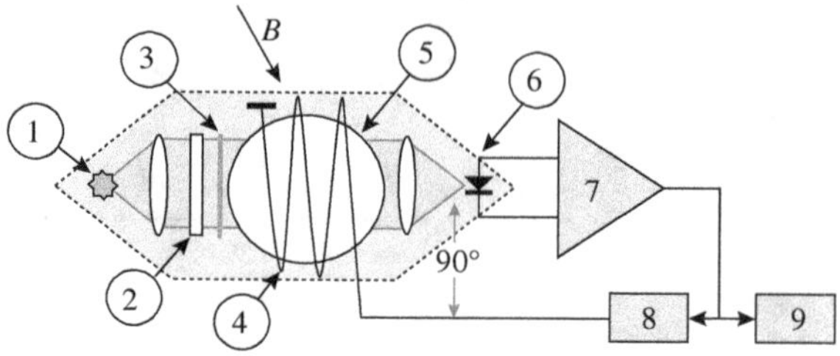

Figure 4.7 The block scheme of a self-oscillating M_x OPQM: 1 spectral lamp, 2 D1 filter, 3 circular polarizer, 4 RF coil, 5 vapor cell, 6 photo detector, 7 amplifier, 8 phase shifter, 9 frequency meter.

resonance, changes with phase [Fig. 4.6(b)]. Phase error $\Delta\phi$ leads to the frequency error $\delta\omega = \Gamma \tan(\Delta\phi)$.

Two optical beams are needed for an M_x scheme: one pumping beam directed along the **B** vector, and one detecting beam perpendicular to it, so the M_x scheme has no axial symmetry. In practice, these two beams are often combined into one beam tilted by $\theta = \pi/4$ to the magnetic field vector, and the magnetometer has dead zones corresponding to the orientations when (a) the beam is parallel to **B**, and therefore does not contain a perpendicular component needed for the signal detection, and (b) the beam is perpendicular to **B**, and therefore does not contain a pumping component. In the simplest case, the M_x signal is proportional to $\sin\theta \sin(\pi/2 - \theta)$, and dead zones include angles where it is not big enough to provide normal feedback operation. As mentioned above, if more than one magnetic resonance line is present in the spectrum, the shift of the chosen line center is larger in M_x schemes.

M_x-resonance registration techniques: self-oscillating and non-self-oscillating schemes

The simplest M_x magnetometer works as an atomic frequency generator (also called a *spin generator*) with an amplifier in a feedback loop. Its main elements are a resonant *Bell–Bloom* lamp, a cell, a photodetector, and electronic circuits (see Fig. 4.7). The lamp contains atoms of the same type as in the magnetometer cell; it is placed in microwave field, causing an electronic discharge accompanied by emission of resonant light. A number of attempts to replace the lamp with a frequency-stabilized semiconductor laser have been made [7–11], but none have led to the production of a commercial magnetometer yet.

The maximal sensitivity of the scheme is achieved if the light is directed at an angle of $\pi/4$ with respect to **B**. After passing through a D1 interference filter (in the simplest schemes it may be omitted) and a circular polarizer, this light is used both for optical pumping of the alkali vapor in a cell [filled with buffer gas or an antirelaxation (paraffin) coating to prevent spin relaxation] and for the resonance detection. The amplified photodetector signal is transmitted to the RF inductor (the coil). At system start-up the noise components of the photo-signal trigger a priming signal at the Larmor precession frequency. This signal is then amplified, phase-shifted, and again transferred to the cell through a feedback loop.

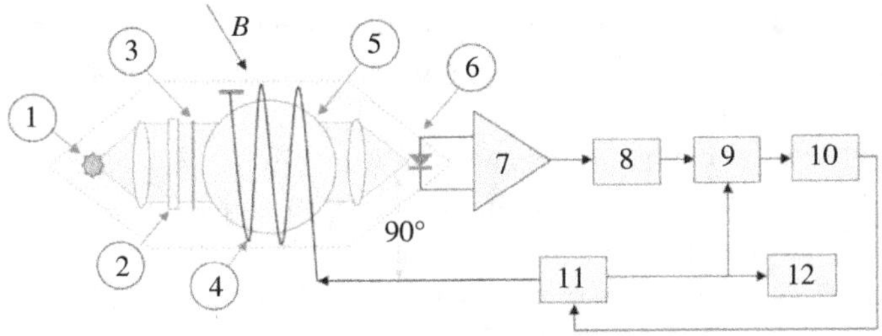

Figure 4.8 PLL M_x magnetometer block-scheme: 1–8 the same as in Fig. 4.7, 9 phase detector, 10 integrating amplifier, 11 voltage-controlled oscillator, 12 frequency meter.

The advantage of this scheme is it simplicity. Its drawbacks include the following.

1. It requires a very fast and precise frequency meter able to measure frequency with 10^{-8} resolution for 10^{-2} to 10^{-1} s measuring time.
2. Apart from shifting phase by $\pi/2$, the phase shifter must also compensate for frequency-dependent phase shifts arising in electronic circuits. Phase compensation should be provided in the full range of working frequencies which can be spread over an octave or more.
3. If the atomic spectrum contains more than one line in the radiofrequency spectrum, it is very difficult to achieve single-frequency generation on the selected transition.

Accordingly, nowadays the use of non-self-oscillating schemes (Fig. 4.8) has become more frequent. Their optical part is the same, and their electronic part (based on a phase-locked loop, PLL) contains a phase detector (PD) and tunable voltage-controlled oscillator (VCO). The VCO frequency is controlled by the PD, which compares the M_x signal phase to the VCO output phase; in this way the frequency of the reference generator can be locked to any resonance into the vicinity of which it is forcibly brought, which solves the problem of resonance line selection. This scheme also makes it potentially possible to control the phase shifter with the VCO controlling voltage. Still, the scheme requires a precise frequency meter.

The advent of high-frequency digital frequency synthesizers (DFS) provided a basis for the creation of digital PLL circuits, making it possible to exclude the frequency-measuring procedure altogether. In the scheme presented in [12], the VCO is replaced with a quartz-stabilized DFS; its frequency is locked to the selected resonance by a digital control loop. Digital codes which control the DFS carry information about the resonance frequency. The M_x signal is detected with two quadrature phase detectors. Their digitized outputs S_1, S_2 are mixed in a digital phase shifter with weights depending on variable angle ϕ:

$$S(\phi) = S_1 \cos(\phi) - S_2 \sin(\phi)\ ; \tag{4.6}$$

this way the signal phase can be rotated with very high precision. The dependence of ϕ on the Larmor frequency ω, providing exact compensation of the phase shifts, is pre-built into the device for the whole range of ω.

Still, the problem of temperature-dependent phase shifts as well as all other phase drifts persists. Periodic phase setup can be done by recording and symmetrizing the resonance shape with the PLL loop open; for this procedure the magnetometer must be placed in a stabilized magnetic field, which is rarely possible in normal field conditions.

Several solutions have been proposed for this problem, for example, modulation of the resonance linewidth Γ. Then (according to $\Delta\omega = \Gamma\Delta\phi$) in the case of non-zero phase error, the signal frequency modulation would appear; it can be measured and used for phase error minimization. In Refs. [13] and [14] it was proposed to use a two-dimensional M-signal representation (so-called Nyquist plots) for different PLL parameters; but all these methods require interruption of the measurement procedure. Schemes allowing dynamical phase correction during the measurement are described in Section 4.2.3 on M_x–M_z tandems.

M_x-*magnetometer sensor optimization*

The main parameters of quantum magnetometers are accuracy, variational sensitivity (or short-time resolution), and time response. As discussed in Chapter 1, they are basically limited by the parameters of the magnetic resonance, such as the resonance linewidth and shift, symmetry, and the signal/noise ratio.

Of all possible types of OPQM noise, only two of a quantum nature are unavoidable in principle: the shot noise of the light and the quantum noise of an atomic ensemble (see Chapter 2). In "classical" pumping schemes, the light noise dominates over the atomic quantum noise. If all technical noises are suppressed below the light shot-noise level, the minimally measured field variation in a frequency band Δf corresponding to the measurement time τ is written (see, for example, [15]) as

$$\delta B_{\min} = \frac{k_F}{\gamma}\frac{\rho_N \Gamma_{\text{full}}}{S}\sqrt{\Delta f} = \frac{1}{Q}\sqrt{\Delta f}\,, \tag{4.7}$$

where Γ_{full} is the total resonance linewidth (taking in account all broadening factors), ρ_N is the shot-noise spectral density, S is the signal amplitude (for an M_x-signal it can be defined as $S = \max\left[S'(\omega)\right]$, where $S'(\omega)$ is the signal detected at the Larmor frequency ω, see Fig. 4.6), and $k_F \approx 1$ is the resonance form factor:

$$k_F = \frac{S}{\Gamma_{\text{full}}}\left(\left.\frac{dS'}{d\omega}\right|_{\Delta\omega=0}\right)^{-1}. \tag{4.8}$$

Introducing the magnetic resonance quality factor, or Q factor (as was done, for example, in Ref. [16]):

$$Q = \frac{\gamma}{k_F}\frac{S}{\rho_N \Gamma_{\text{full}}} \tag{4.9}$$

one can rewrite Eq. (4.7) as $\delta B_{\min} = (1/Q)\sqrt{\Delta f}$.

Other important metrological characteristics of magnetometers include the absolute accuracy, long-term stability, reproducibility of readings, and so on. Most of these parameters are in turn related to the stability of the pumping light spectral profile, the cell coating aging, and so on. These errors are not easy to formalize, but most of them are proportional to the resonance width Γ.

In a paraffin-coated cell with a side-arm (stem) containing an alkali metal sample, Γ can be represented (disregarding broadening by the RF field) as a sum of several terms:

$$\Gamma = \underbrace{\left[\underbrace{(\Gamma_{\text{wall}} + \Gamma_{\text{stem}})}_{\Gamma_0} + \Gamma_{\text{coll}} \right]}_{\Gamma_d} + \Gamma_{\text{light}} , \qquad (4.10)$$

where Γ_{wall} is the contribution of relaxation due to collisions with the walls; Γ_{stem} is the absorption rate of polarized atoms in the stem; Γ_{coll} is the line broadening due to atom–atom collisions (spin-exchange broadening) proportional to the atomic vapor concentration n:

$$\Gamma_{\text{coll}} = \sigma_{\text{se}} n \langle v_{\text{rel}} \rangle , \qquad (4.11)$$

where σ_{se} is the cross-section of the spin-exchange process; $\langle v_{\text{rel}} \rangle = \sqrt{2} \langle v \rangle$ is the relative velocity of colliding particles; Γ_{light} is light-induced broadening proportional to the pumping light intensity I_{ph}. The square brackets in Eq. (4.10) are introduced to group terms of a similar nature. That is, the first two terms denote the intrinsic resonance width Γ_0, the first three terms the "dark" width Γ_d.

Expressions for the dependence of the signal amplitude on the pumping light intensity were first proposed by H. G. Dehmelt [2, 3]; A. L. Bloom [15] was the first to point out the necessity of Q-factor optimization based on light intensity measurements. Typical dependences of the signal, slope, and quality factor on the pumping intensity $I = \Gamma_{\text{light}}/\Gamma_d$ (expressed in terms of the relative line broadening by the pumping light) are presented in Fig. 4.9.

One can see that the quality factor reaches a maximum at $I_{\text{opt}} = 3$, which corresponds to a fourfold light broadening of the resonance line. However, all light and related parametric shifts of the resonance line (hence, their variations) are proportional to the pumping intensity, and the quality factor dependence on I is weak in the vicinity of the optimum; therefore, it makes sense to choose the working intensity to be somewhat below I_{opt}.

Relaxation of optically oriented rubidium atoms on coated walls was investigated by M. A. Bouchiat and others in Refs. [17–19]. Investigations of the relaxation mechanisms in alkali metals were continued in Refs. [20–22] by M. V. Balabas and colleagues in the group headed by E. B. Alexandrov. As a result, very narrow (sub-Hz) atomic lines were obtained. These and other investigations of wall coatings are discussed in Chapter 11.

According to a relatively simple model of the M_x resonance in a coated cell proposed in Ref. [23] by A. K. Vershovskiy and A. S. Pazgalev, the dependence of the quality factor

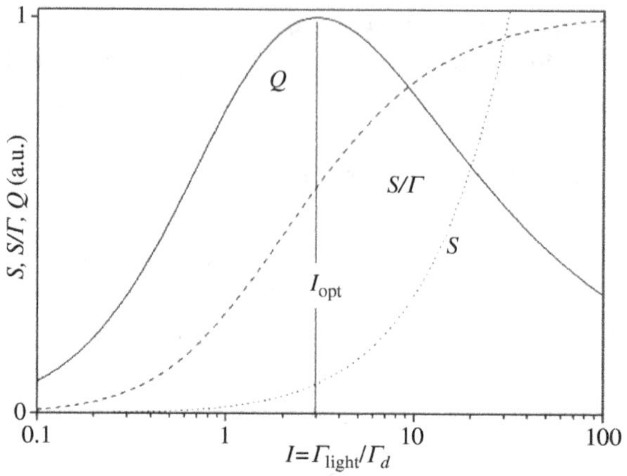

Figure 4.9 Plots of signal S, slope (S/Γ), and Q-factor of magnetic resonance vs. light broadening $I = \Gamma_{\text{light}}/\Gamma_d$.

on the absorption within the cell imposes a stringent limitation on its optical thickness ($x_{\text{opt}} \geq 1.6$) and thereby on the spin-exchange broadening. For example, optimal collisional broadening for the cell with the effective length ℓ filled with ^{39}K and pumped on the D1 line is determined by the expression

$$\Gamma_{\text{coll}}\ell \geq 2\pi \times 16 \text{ Hz} \cdot \text{cm} , \tag{4.12}$$

and for $\ell = 5$ cm the optimal collisional width $\Gamma_{\text{coll}} = 3.2$ Hz. Therefore, the advantages of using the ultra-narrow ($\ll 1$ Hz) lines that can be obtained in coated cells are to a considerable degree neutralized by spin-exchange broadening required to obtain optimal optical thickness of the cell. Still, the parameters of an OPQM must be optimized for a concrete purpose: if accuracy and long-term stability have priority over sensitivity, it can be reasonable to reduce the linewidth by simultaneous reduction of the optical pumping intensity and the vapor density.

4.2.3 Attempts to combine advantages of M_x and M_z magnetometers: M_x–M_z tandems

To meet the contradictory requirements that a metrological instrument should possess – both high accuracy and high operational speed – two different devices can be combined into a single system in which the measurements of the faster device are corrected by the slower but more accurate device. This approach was proposed by A. H. Allen and P. L. Bender in 1972 [24]; a magnetometer of this type consists of M_x and M_z OPQM modules. The M_z OPQM uses a magnetic resonance spectrum with the resolved line structure, thereby ensuring high accuracy; in contrast, the M_x OPQM operates on a wide unresolved line.

Certainly, this approach has weak points: first, the M_x and M_z OPQMs measure the field in different spatial locations; second, tandems are complicated, bulky, and expensive, therefore only suitable for observatory applications. Later in this chapter, schemes intended to overcome these drawbacks are presented.

4.3 Applications: radio-optical M_x and M_z magnetometers

4.3.1 Alkali M_z magnetometers

Alkali–helium magnetometer

An alkali–helium magnetometer can only be implemented in an M_z scheme. It uses the effect of spin-polarization transfer during atomic collisions in an alkali–helium gas discharge plasma. The working substance of this magnetometer is helium in the metastable 2^3S_1 state – i.e., orthohelium – polarized by collisions with optically oriented alkali atoms; it is exposed to an RF field, causing resonant transitions between magnetic sublevels of the 2^3S_1 state. Reduction in orthohelium polarization affects the equilibrium polarization of alkali atoms, and it can be detected due to the increase in absorption of alkali pumping light. This indirect registration of magnetic resonance in helium entails a noticeable loss of polarization signal; on the other hand, it helps avoid the light shift of helium sublevels. The process of orientation transfer from optically oriented atoms of alkali metals to orthohelium was first discovered by G. M. Keiser and co-workers [25]. The most important contribution to the development of the alkali–helium OPQM was made by R. A. Zhitnikov and co-workers [26,27].

Typical systematic errors of the device do not exceed 0.15 nT across the entire EMF range, and random error is about 0.02 nT with an integration time of 1 s. The alkali–helium OPQM was quite popular in the 1980s as a reference instrument, but today it is inferior to the potassium narrow-line magnetometer with respect to sensitivity, accuracy, and time response. Apart from the alkali–helium magnetometer, there are many types of pure helium optical magnetometers, which are discussed in Chapter 10.

Balanced K and Rb HFS magnetometers

An HFS magnetometer is an OPQM based on the resonance between magnetic sublevels of different hyperfine levels of the atomic ground state. The first HFS magnetometer was proposed in the 1970s by E. B. Alexandrov *et al.* [28–30]. At first sight, the idea of using the field dependence of the HF transition frequency seems strange, because it contains a very large field-independent term equal to the HF splitting. However, this inconvenience is outweighed by many advantages, both technical and fundamental, over the traditional OPQM design.

First, the structure of most HF transitions is resolved even in ultra-weak fields of the order of 100 nT, because the g-factors of the lower and upper hyperfine levels of the ground state of alkali metals have opposite signs (see Figs. 4.3 and 4.10). This accounts for the HFS magnetometer's ability to operate in the magnetic field range 10^{-7}–10^{-3} T.

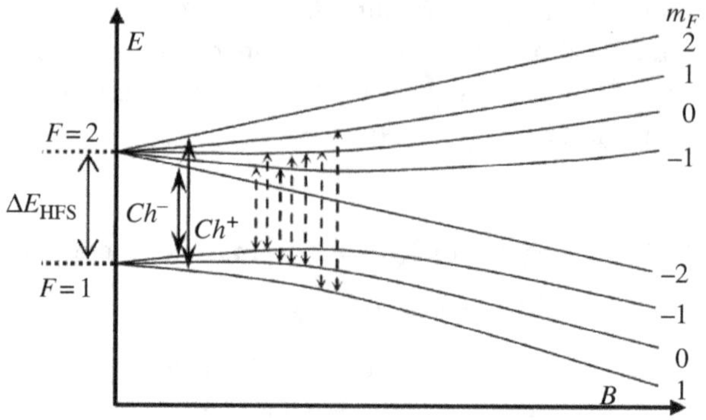

Figure 4.10 Hyperfine transitions in the ground state of an alkali atom with $I = 3/2$ (isotopes ^{7}Li, ^{23}Na, ^{39}K, ^{41}K, ^{87}Rb). Transitions not used in the balanced HFS scheme are denoted by dashed lines.

Second, a spectrum-selective pumping of HF transitions is possible instead of polarization pumping. This permits the use of unpolarized D1 and D2 lines, directed arbitrarily with respect to **B**. Hence, "dead zones" can only appear due to the vector character of the microwave field, and it is possible to form an appropriate microwave field configuration in the cell to prevent these dead zones.

Third, the systematic errors of such a magnetometer can be reduced to the pT level by using a so-called balanced modification, which implies measuring the difference between frequencies of two symmetric transitions. The maximal sensitivity can be reached by use of the two outermost transitions Ch^- and Ch^+ (e.g., $|F = 1, m_F = -1\rangle \leftrightarrow |F = 2, m_F = -2\rangle$ and $|F = 1, m_F = 1\rangle \leftrightarrow |F = 2, m_F = 2\rangle$, see Fig. 4.10) with corresponding frequencies $\omega^-(B)$ and $\omega^+(B)$. The useful information is carried by the frequency difference $\omega(B) = \omega^+(B) - \omega^-(B)$. In the case of pumping with unpolarized light, this difference is virtually free of light and collisional shifts; it is given by

$$\frac{\omega(B)}{2\pi} = f(B) = \frac{6B\mu_B(g_J - g_I)}{4h} + \frac{2Bg_I\mu_B}{h} \approx B \times (42\ \text{Hz/nT}). \tag{4.13}$$

This means that the sensitivity of a balanced magnetometer is 6 times higher than that of a "classical" magnetometer. The field dependence of the output frequency contains no quadratic term, i.e., it is practically linear because all higher terms are negligibly small.

The HFS magnetometer was first investigated experimentally by E. B. Alexandrov *et al.*, in the 1970s with potassium [28, 31]. The structure of the optical HF spectrum of the D1 lines of two stable isotopes of K is shown in Fig. 4.11. Both the isotopic shift (i.e., the difference between centers of mass of the contours belonging to different isotopes) and HF splitting in K are smaller than the Doppler half-width ($\Delta\nu_D \approx 800$ MHz at 300 K), and therefore high efficiency of hyperfine spectral pumping in K cannot be reached easily.

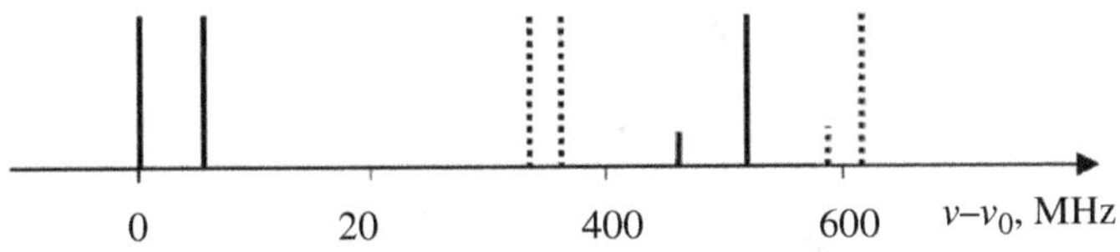

Figure 4.11 Structure of the HF spectrum of the D1 lines of ^{39}K (solid lines) and ^{41}K (dashed lines).

In the experiment discussed in Refs. [28, 31] a ^{39}K cell filled with argon as a buffer gas was optically pumped by a potassium lamp light filtered by ^{41}K. With proper choice of the ^{41}K filter density, the contours of the D1 and D2 lines (769.9 nm and 766.5 nm) in the pumping-light spectrum were distorted in such a way that ^{39}K atoms became excited mostly from the $4^2S_{1/2}$, $F = 2$ state. A sensitivity of 1 pT/$\sqrt{\text{Hz}}$ has been reached. A potassium HFS magnetometer has now been produced in a one-channel version (see Refs. [32, 33]).

Much higher sensitivity was expected from an HFS magnetometer based on ^{87}Rb, since (1) its HF splitting is approximately 15 times higher, and (2) an efficient method of isotopic filtration providing selective HF pumping exists for ^{87}Rb and ^{85}Rb. Moving from K to Rb was anticipated to result in significant growth of the signal amplitude; but experiments showed that in this case a new noise source, related to the poor short-time stability of existing microwave sources, arose [30]. In case of K this noise did not exceed photon shot noise, but the frequency increase from 462 MHz (K) to 8634 MHz (^{87}Rb) resulted in microwave noise dominating. A similar problem exists in atomic frequency standards; it must be solved by using high-Q microwave sources. Taking into account the microwave noise-limiting sensitivity of Rb-HFS schemes, one can state that the K-HFS scheme is still attractive, especially in combination with efficient laser pumping.

In order to excite and separate signals in the channels Ch^+ and Ch^-, frequencies $\omega^+(B)$ and $\omega^-(B)$ are usually modulated [28, 30] at different low frequencies Ω^+ and Ω^-; as a result, the two channels acquire different time responses, which deteriorates their balance. In Ref. [34] a perfectly symmetric balanced ^{87}Rb HFS magnetometer is described, using modulation at a single frequency Ω with a $\pi/2$ phase difference between channels Ch^+ and Ch^-.

4.3.2 M_x magnetometers

Self-oscillating Cs magnetometer

The self-oscillating Cs magnetometer appears to be the simplest and most widely used quantum magnetometer, although among the alkalis Cs has the most complicated structure of ground-state magnetic sublevels. The ground state of ^{133}Cs comprises two hyperfine levels, $F = 3$ and $F = 4$, subject to respective splitting to 7 and 9 Zeeman sublevels, respectively, in a magnetic field; the difference between the frequencies of adjacent Zeeman sublevels in the average EMF does not exceed 7 Hz. Therefore, many resonances actually blend into a single asymmetric resonance hybrid line about 50 Hz in width, roughly corresponding to 15 nT on the magnetic field scale ($\gamma/(2\pi) \approx 3.5$ Hz/nT). This line is schematically shown

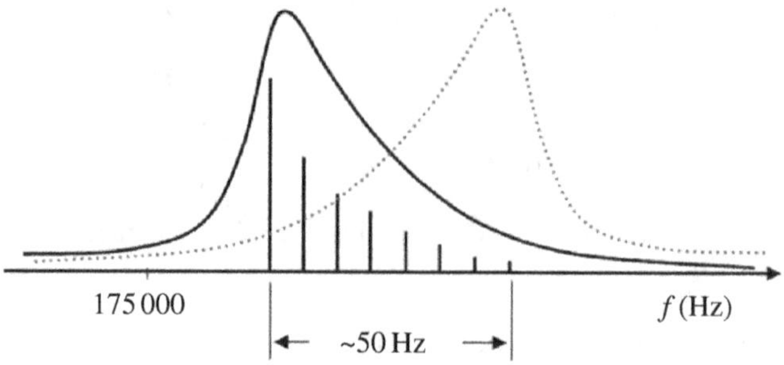

Figure 4.12 Cs magnetic resonance line in Earth's magnetic field (EMF). The dotted line represents the line contour when the direction of **B** is reversed.

in Fig. 4.12 for $F = 4$ and $B = 50$ μT. The line corresponding to $F = 3$ is much weaker and shifted by ~280 Hz to the high frequencies.

A typical Cs magnetometer is built according to the self-oscillating scheme described in Section 4.2.2. Under typical laboratory conditions the magnetometer may be operated without a thermal stabilization system, Cs vapor pressure at room temperature being close to optimum. Pumping is feasible without light filters (i.e., with a mixture of two resonance lines) in order to simplify the optical scheme of the instrument.

The contribution of various partial resonances strongly depends on many factors, such as the light and radio field intensity, the direction of the pumping light relative to the direction of the constant field, and the density of Cs vapor; all of these factors contribute to uncertainty in the position of the resonance maximum. Consequently, the accuracy of a Cs magnetometer does not typically exceed several nT, which manifests itself in both the irreproducibility of the readings in different runs and in slow drifts of the readings even in an ideally stabilized field.

The so-called orientational shifts of the magnetometer frequency upon a change of the angle between the optical axis and the magnetic field vector are especially undesirable for magnetic prospecting. It can be concluded from Fig. 4.12 that maximal orientation error is close to the distance between the outermost spectral lines in the $F = 4$ state, i.e., ~15 nT at 50 μT.

On the other hand, a resolution of 1 pT/$\sqrt{\text{Hz}}$ (or even better) can be reached within a short observation time, $\tau = 10$–100 s, if the magnetometer is fixed in space. The slow drifts may be diminished by thorough stabilization of the pumping light intensity, the working volume temperature, and the RF field strength.

Orientational shifts and drifts can be controlled by special resonance symmetrization techniques. A configuration for this purpose was proposed by A. L. Bloom [15] in 1961 in application to a Rb-OPQM; the idea was to pump two cells in opposite directions and to sum up the two output signals. The result was the superposition of the two asymmetric contours, giving rise to a single wide quasi-symmetric contour. This made it possible to reduce measurement errors by a factor of 10; however, it was still insufficient for many applications.

Subsequent decades witnessed considerable efforts to reduce systematic errors of Cs magnetometers by the groups headed by M. Arditi, F. A. Franz, A. N. Kozlov, and many others [35–40]. Magnetometers with six and more cells were constructed, but, as was shown in Ref. [41], even these devices failed to resolve the problem because complete compensation of the shifts is precluded by variations of partial pressures in individual cells, pumping conditions, and so on. Moreover, even small field gradients around a multicell magnetometer serve as additional sources of errors. Attempts to use airborne magnetometers with an active spatial orientation system were undertaken in the USA in the 1960s. These instruments had (at the expense of being much more costly and complicated) better characteristics than multicell magnetometers, but even they were not totally free of errors related to parametric shifts when operated with a broad asymmetric resonance line.

The next step toward improving the absolute accuracy of Cs magnetometers was the development of a sensor with a two-compartment cell pumped by left- and right-circularly polarized beams by T. Yabuzaki and T. J. Ogawa [42]. Such a configuration makes it possible to reduce orientational shifts to just several tenths of a nT – to the extent of the two signal channels balancing in accuracy.

For all that, Cs magnetometers are still the simplest, most reliable, and widespread OPQMs. They are also widely used in gradientometric systems [41]; Cs-OPQMs are finding increasingly extensive applications due to the advent of readily available sources of laser pumping. An example is the cardiogradiometer with Cs-M_x sensors and laser pumping, developed by the group of A. Weis. This instrument with a 20 mm cell has a sensitivity of $< 100\ \mathrm{fT}/\sqrt{\mathrm{Hz}}$ [10, 13].

The disadvantages of Cs-OPQMs related to the resonance line asymmetry are practically negligible in magnetic fields not exceeding 10% of EMF (i.e., 5 mT), in which Cs line broadening associated with the quadratic Zeeman splitting becomes negligibly small. Precision metrology of sub-Earth magnetic fields is relevant, in particular, for the fundamental problem of the search for a neutron electric dipole moment – see, for example, Chapter 18 and also Ref. [43], where multisensor cesium systems for magnetic field control in a multilayer shield are described.

For the same purpose a cesium magnetometer with laser pumping with sensitivity of $15\ \mathrm{fT}/\sqrt{\mathrm{Hz}}$ in a 2 mT field was also developed by the group of A. Weis [11]. The light driving the magnetometer was produced by a tunable extended-cavity diode laser, actively locked to the 4-3 hyperfine component of the Cs D1 transition ($\lambda = 894$ nm) in an auxiliary cesium vapor cell by means of the dichroic atomic vapor laser lock (DAVLL) technique [44]. The stabilization to a Doppler-broadened resonance provided continuous stable operation over weeks.

Non-self-oscillating K magnetometer

A natural mixture of potassium consists of ^{39}K ($\sim 93\%$) and ^{41}K ($\sim 7\%$). The RF spectrum of these potassium isotopes in EMF ($\sim$50 nT) is presented in Fig. 4.13. The strongest resonance

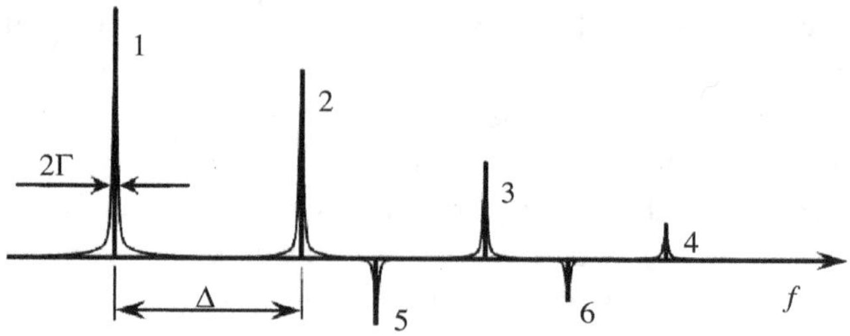

Figure 4.13 Magnetic resonance spectrum of potassium in the Earth's magnetic field (EMF). $\Delta \approx$ 0.5 kHz for ^{39}K and $\Delta \approx 1$ kHz for ^{41}K.

line 1 corresponds to the $|F=2, m_F=2\rangle \leftrightarrow |F=2, m_F=1\rangle$ transition. Transitions between the sublevels of the $F=2$ HF level correspond to the peaks 1–4, transitions within the $F=1$ HF level correspond to the peaks 5–6. The peaks 5–6 are inverted because of the negative sign of the g-factor for the $F=1$ level (see Fig. 4.3). The spectrum is completely resolved; it is therefore possible to isolate a single line and largely disregard others.

The strongest resonance line position shift as a function of pumping conditions is measured in pT rather than nT, in contrast to the compound cesium line (see Fig. 4.12). This motivates the use of potassium for precision measurement of magnetic fields. The first attempts to use this effect for OPQMs dating to the 1960s [45] were not quite successful due to the difficulty of separating a single line from the complex structure given on Fig. 4.13 using a self-oscillating scheme. The first potassium narrow-line M_z magnetometer was described by G. S. Vasyutochkin in Ref. [46], and in its most promising M_x modification by E. B. Alexandrov *et al.* in Refs. [20, 47]. Both self-oscillating and non-self-oscillating variants of a potassium narrow-line magnetometer were realized by the E. B. Alexandrov group [8, 9, 47, 48]. Self-oscillation on a single line was achieved by introducing automatic gain control in the feedback loop.

Each individual resonance position in the potassium Zeeman spectrum is to a greater or lesser extent subject to the effect of adjacent resonances. The distance for ^{39}K and ^{41}K in an average EMF is on the order of 0.5 and 1 kHz, respectively. At $\Gamma \approx 1$ Hz, the frequency shift of the main resonance (see Section 4.2.1) does not exceed 2 mHz ($\sim$0.3 pT) even for ^{39}K, which is four orders of magnitude smaller than the systematic error of Cs magnetometers.

Since the errors due to the overlap of resonance lines are highly suppressed in a K-OPQM, errors due to light shift and phase errors become most significant. As mentioned before, these errors do not exceed a small part of the resonance width, and therefore, it makes sense to work with the narrowest linewidth possible. It was shown in Ref. [23] that the optimum quality factor with respect to the collision-broadened resonance linewidth in a paraffin-coated cell is inversely proportional to the cell dimension; this means that one has to increase the cell diameter in order to achieve both accuracy and sensitivity. The biggest cells used in potassium magnetometers were spherical with a diameter of 15 cm [8, 9];

according to Eq. (4.12), the optimal collision-broadened resonance linewidth in such a cell is expected to be about 1.1 Hz.

The ultimate signal and noise parameters of the magnetic resonance in potassium in a 15-cm spherical paraffin-coated cell using laser pumping were demonstrated by E. B. Alexandrov and his group together with V. L. Velichanskiy [9]. An external cavity diode laser was stabilized using the saturation-absorption technique to a narrow (~10 MHz) "cross-over" resonance situated near the D1 line center; special measures were undertaken to reduce laser intensity noise down to the shot-noise level. The ultimate sensitivity of this resonance was estimated as 1.8 $\mathrm{fT}/\sqrt{\mathrm{Hz}}$.

It was mentioned in Section 4.2.2 that the principal factor placing a limit on the bandwidth of M_x schemes is the Larmor frequency. In the case of potassium, the phase-locked loop bandwidth should not exceed the resonance spacing Δ, otherwise the sidebands of the varying RF field will excite neighboring transitions. Still, the ^{41}K line spacing allows a 1 kHz bandwidth in EMF.

The metrological potential of K-OPQM covers the full range of EMF with a record-breaking resolving power of the order of 100 $\mathrm{fT}/\sqrt{\mathrm{Hz}}$ for commercial devices; use of laser pumping and big cells allows increase of the resolving power up to ~1 $\mathrm{fT}/\sqrt{\mathrm{Hz}}$. Systematic errors of the K-OPQM are small; today, the absolute accuracy of K-OPQMs in EMF is estimated to be 10–20 pT. This roughly corresponds to the accuracy with which fundamental constants that are necessary to convert the resonance frequency to the EMF induction have been determined.

4.3.3 M_x–M_z tandems

Rb M_x–M_z tandem

The first tandem magnetometer, built by A. H. Allen and P. L. Bender [24], used a ^{87}Rb vapor in both sensors. The Zeeman spectrum of ^{87}Rb is similar to that of K, but the spacing between adjacent lines in each group in EMF is less than 50 Hz. Therefore, special precautions were taken by the authors to keep the resonance width below this splitting: the light and RF intensities were kept low in the M_z part of the device, which led to significant signal-to-noise reduction.

Since the same working substance was used in both parts of the tandem, these parts had to be placed at a distance >0.5 m from each other to avoid mutual influence through bias DC and AC fields. Both magnetometers used the same RF field, and the resonant RF frequency of the M_x scheme was locked to that of the M_z scheme by applying a weak bias magnetic field to the M_x cell. In order to obtain the M_z signal without RF-field modulation, the magnetic field inside the M_z cell was modulated at a low frequency (5 Hz). Stability tests of the device indicated reproducibility of ± 0.02 nT, and a standard deviation of individual 10 s measurements of about 0.03 nT.

Cs-K M_x–M_z tandem using a 4-quantum resonance

The first M_x–M_z tandem had several drawbacks, and the worst were determined by the choice of the working substance: even though the Rb spectrum can be resolved in EMF, there is still significant overlap between the resonant lines, and each resonance position depends on the parameters of its neighbors. This disadvantage was overcome after many years by a group of researchers headed by E. Pulz [49], who proposed a tandem in which Cs was used in the M_x scheme and a 1-photon (and latterly 4-photon) resonance in K was used in the M_z scheme.

The idea of using a 4-photon resonance in an M_z scheme was presented by E. B. Alexandrov *et al.* in Refs. [50,51]. The unique properties of a 4-quantum resonance were studied, and it was shown that among n-quantum transitions belonging to the $\Delta F = 0$, $m_F = |n|$ type, the highest ($n = 2F$) transition frequency is practically independent of the amplitude of the alternating field $\mathbf{B}_1$. Besides, at optimal $B = B_1(4-\text{opt})$ the linewidth of the 4-quantum resonance is four times smaller than that of the 1-quantum resonance, whereas at the same $B_1(4-\text{opt})$ all low-order resonances (1-, 2-, and 3-quantum) are over-broadened; this avoids the problem of seeking and maintaining the desired resonance.

The narrowness of the 4-quantum resonance necessitates very slow modulation of the RF frequency. Therefore, its advantages can only be used in a tandem scheme. The tandem described by E. Pulz *et al.* in Ref. [49] used the signal from the 4-photon resonance in such a way that it gradually corrected the feedback loop parameters of the faster Cs M_x OPQM. The system had a response time of 1 ms, a resolving power 10 pT, and an absolute accuracy 0.1 nT.

The main distinction of the Cs-K tandem presented by E. B. Alexandrov and coauthors [51] from its precursor is that it contains a single sensor with the Cs-K cell instead of two: this solution allows one to overcome another drawback of the original tandem scheme – the uncertainty of the spatial point where the field is measured. Experiments have shown that while short-term sensitivity of the tandem corresponds to that of a standard Cs M_x magnetometer, its readings remained stable within 10 pT under twofold variation of the optical pumping intensity.

M_x–M_R tandem

A further development of the idea of the M_x–M_z tandem using only one sensor and one magnetic resonance in both M_x and M_z parts was presented by A. K. Vershovskiy and A. S. Pazgalev in Ref. [52]. As a reference signal for the slow and precise part of the scheme, the authors suggest using the signal produced due to the radial component of the rotating magnetic moment, which occurs in the equatorial (XY) plane and is called the M_R signal.

The squared M_R signal can be measured as a sum of squared outputs of two quadrature detectors [12]. Its shape is symmetric with respect to the frequency detuning δ:

$$M_R^2 = \left(M_z^{(0)}\right)^2 V^2 \frac{\Gamma_2^2+\delta^2}{\left(\Gamma_2^2+\delta^2+4V^2\frac{\Gamma_2}{\Gamma_1}\right)^2}, \tag{4.14}$$

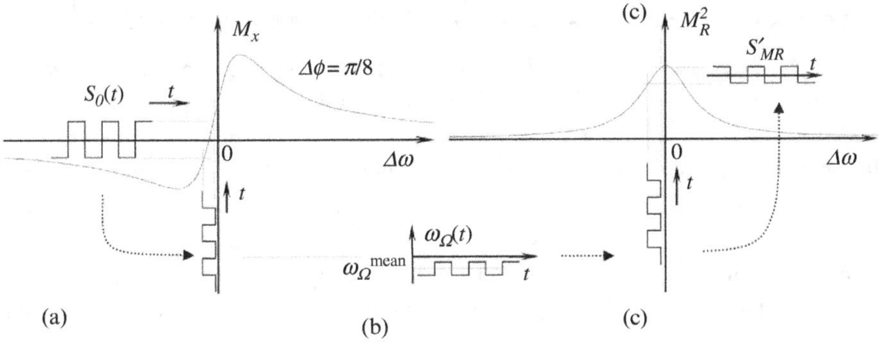

Figure 4.14 M_R signal generation scheme: M_x and M_R signals' dependence on the RF frequency detuning $\Delta\omega$.

where $4V^2\Gamma_2/\Gamma_1$ is the RF-induced broadening. Basically, the M_R signal looks and behaves very similarly to the M_z signal – with the exception that it can be detected at the Larmor frequency. Particularly, the M_R signal shape does not depend on the detection phase. Moreover, its dependence on the neighboring resonance parameters is as weak as that of the M_z signal.

Unfortunately, the drawbacks of the M_z and the M_R signals are also similar. First, one needs low-frequency modulation in order to lock to the M_R signal. Second, the M_R signal does not react to the field change immediately – its reaction is determined by relaxation. Both of these factors limit the time response of the M_R scheme, making it suitable mostly for tandem applications.

An M_x–M_R tandem can be realized using the same magnetic resonance in both parts. In a standard M_x scheme, the feedback loop tends to bring the M_x signal value $S(\Delta\omega)$ to the value $S_0 = 0$ by tuning the RF field frequency ω (Fig. 4.6). If we slowly modulate S_0, ω will also be modulated, as shown in Fig. 4.14(a,b): $\omega(t) = \omega[B(t)] + \omega_\Omega(t)$. Apart from the field-dependent component $\omega[B(t)]$ it will contain a periodically changing summand ω_Ω. Being applied to the M_R channel, the modulated RF field produces an AC signal S'_{MR} [Fig. 4.14(c)], which can be detected at frequency Ω and used for continuous phase adjustment in the M_x scheme.

The term $\omega[B(t)]$ can be separated by measuring $\omega(t)$ for time τ_i, integrating its components repeating with period $1/\Omega$, and subtracting them. As a result, an M_x–M_R OPQM loses sensitivity to the field components in the range $\Omega \pm 1/\tau_i$ and to their high harmonics; but by choosing a sufficiently large τ_i, these "blind zones" can be made negligibly small.

The proposed concept of the M_x–M_R magnetometer was experimentally verified using a standard potassium M_x sensor [52]. At the modulation frequency $\Omega = 1$ Hz, and $\tau_i \approx$ 100 s, the prototype device showed a response speed of 10 readings per second and a reproducibility of measurements on a level of 2–3 pT RMS.

4.4 Summary: M_x and M_z scheme limitations, prospects, and application areas

Both M_x and M_z types of OPQMs have numerous and diverse applications. Devices of the M_x type are more suitable for geodesic, archeological, and geological tasks such as magnetic mapping, search, and exploration, because their fast response provides high spatial resolution when these devices are installed on a moving carrier. For tasks where a particularly simple, robust, small, and sensitive device is needed, and drifts are not critical, a Cs-OPQM is the best choice.

On the other hand, magnetic-mapping tasks require high long-term stability: slow drifts of the magnetometer during the mapping time should not exceed measurement resolution, otherwise it will be difficult to stitch the flight data into one map. Therefore, magnetometers using isolated narrow lines (such as the K-OPQM) are preferable for such tasks.

For some mapping and exploration applications, the HFS version of the M_z magnetometer can be more attractive than the M_x scheme, since it has no dead zones. This is especially important in the equatorial zone where EMF has a very small vertical component.

Because of their fast response, M_x magnetometers are best suited for use in the schemes of measuring magnetic field vector components [53–55], though the first optically pumped component devices of this kind, designed by J. L. Rasson [56] and O. Gravrand *et al.* [57] were based on M_z sensors. Most of these devices use a scalar sensor integrated into a system of magnetic coils which create a succession of fields perpendicular to the field being measured, so this field can be calculated after a sequence of scalar measurements. But, as was shown back in the 1970s by A. J. Fairweather and M. J. Usher [58], the M_x signal angular dependences can also be directly used for vector measurements; this idea was further developed in Ref. [59].

M_z OPQMs are preferable as reference devices in magnetic observatories because of their high accuracy. M_z devices using isolated narrow lines [such as the M_z K-OPQM, or the K (Rb) HFS OPQMs] have a considerable advantage in accuracy over devices using combined wide lines. Thus, for all the variety of methods available to excite and detect magnetic resonances, the basic trend toward the improvement of field measuring systems is the narrowing of the magnetic-resonance line.

Problems requiring both high accuracy and high speed can be successfully solved by combining different types of quantum measuring devices. Some of these combined instruments, like the M_x–M_z tandem with a single cell, or (especially) the M_x–M_R magnetometer, are not much larger or more difficult to use than a "single" device.

Further progress can be expected by using lasers in M_x and M_z schemes, especially in super-compact schemes [60], schemes with suppressed spin-exchange broadening [61–63], and principally new pumping schemes which will be described in the following chapters of this book.

References

[1] F. Bloch, Phys. Rev. **70**, 460 (1946).
[2] H. G. Dehmelt, Phys. Rev. **105**, 1487 (1957).

[3] H. G. Dehmelt, Phys. Rev. **105**, 1924 (1957).
[4] W. Bell and A. Bloom, Phys. Rev. **107**, 1559 (1957).
[5] M. A. Bouchiat, J. Phys. **26**, 415 (1966).
[6] H. Cramér, *Mathematical Methods of Statistics* (Princeton University Press, Princeton, NJ, 1946).
[7] S. Groeger, G. Bison, P. E. Knowles, and A. Weis, Eur. Phys. J. Appl. Phys. **33**, 221 (2006).
[8] E. B. Aleksandrov, M. V. Balabas, A. K. Vershovskii, A. E. Ivanov, N. N. Yakobson, V. L. Velichanskii, and N. V. Senkov, Opt. Spektrosk. **78**, 325 (1995) [Opt. Spectrosc. **78**, 292 (1995)].
[9] E. B. Alexandrov, M. V. Balabas, A. S. Pazgalev, A. K. Vershovskii, and N. N. Yakobson, Laser Phys. **6**, 244 (1996).
[10] G. Bison, R. Wynands, and A. Weis, Appl. Phys. B **76**, 325 (2003).
[11] S. Groeger, G. Bison, J.-L. Schenker, R. Wynands, and A. Weis, Eur. Phys. J. D **38**, 239 (2006).
[12] A. K. Vershovskii and A. S. Pazgalev, Zh. Tekh. Fiz. **76**, 108 (2006) [Tech. Phys. **51**, 924 (2006)].
[13] G. Bison, R. Wynands, and A. Weis, J. Opt. Soc. Am. B **22**, 77 (2005).
[14] A. K. Vershovskii and E. B. Aleksandrov, Opt. Spektrosk. **100**, 17 (2006) [Opt. Spectrosc. **100**, 12 (2006)].
[15] A. L. Bloom, Appl. Optics **1**, 61 (1962).
[16] E. B. Aleksandrov, A. K. Vershovskii, and N. N. Yakobson, Zh. Tekh. Fiz. **56**, 970 (1986) [Tech. Phys. **31**, 592 (1986)].
[17] M. A. Bouchiat and J. Brossel, Compt. Rend. Acad. Sci. **254**, 3828 (1962).
[18] M. A. Bouchiat and J. Brossel, Phys. Rev. **147**, 41 (1966).
[19] V. Liberman and R. J. Knize, Phys. Rev. A **34**, 5115 (1986).
[20] E. B. Aleksandrov, V. A. Bonch-Bruevich, S. V. Provotorov, and N. N. Yakobson, Opt. Spektrosk. **58**, 953 (1985) [Opt. Spectrosc. **58**, 439 (1985)].
[21] M. V. Balabas and V. A. Bonch-Bruevich, Pis'ma Zh. Tekh. Fiz. **19**, 6 (1993) [Tech. Phys. Lett. **19**, 191 (1993)].
[22] M. V. Balabas and S. G. Przhibel'sky, Chem. Phys. Rep. **14**, 882 (1995).
[23] A. K. Vershovskii and A. S. Pazgalev, Zh. Tekh. Fiz. **78**, 116 (2008) [Tech. Phys. **53**, 646 (2008)].
[24] A. H. Allen and P. L. Bender, J. Geomagn. Geoelectr. **24**, 105 (1972).
[25] G. M. Keiser, H. G. Robinson, and C. E. Johnson, Phys. Lett. A **51**, 5 (1975).
[26] S. P. Dmitriev, R. A. Zhitnikov, and A. I. Okunevich, Zh. Eksp. Teor. Fiz. **70**, 69 (1976) [Sov. Phys. JETP **43**, 35 (1976)].
[27] E. V. Blinov, R. A. Zhitnikov, and P. P. Kuleshov, Pis'ma Zh. Tekh. Fiz. **2**, 305 (1976) [Tech. Phys. Lett. **2**, 117 (1976)].
[28] E. B. Aleksandrov, A. B. Mamyrin, and A. P. Sokolov, Opt. Spektrosk. **34**, 1216 (1973).
[29] E. B. Aleksandrov, A. B. Mamyrin, and N. N. Yakobson, Inventor's certificate No. 438345; Bull. Izobret. **45**, 230 (1976).
[30] E. B. Aleksandrov, A. B. Mamyrin, and N. N. Yakobson, Zh. Tekh. Fiz. **51**, 607 (1981) [Sov. Phys. Tech. Phys. **26**, 363 (1981)].
[31] E. B. Aleksandrov, S. L. Izotova, A. B. Mamyrin, M. S. Frish, Opt. Spektrosk. **38**, 818 (1975) [Opt. Spectrosc. **38**, 466 (1975)].
[32] V. A. Mogilevkin and V. I. Shepshelevich, Razvedka i ohrana nedr (Exploration and preservation of natural resources) **9**, 74 (2001).

[33] V. I. Shepshelevich and V. Yu. Yackaer, Razvedka i ohrana nedr (Exploration and preservation of natural resources) **12**, 36 (2002).

[34] E. B. Aleksandrov, A. K. Vershovskii, and A. S. Pazgalev, Zh. Tekh. Fiz. **76**, 103 (2006) [Tech. Phys. **51**, 919 (2006)].

[35] S. Ando, Japan Appl. Phys. **4**, 793 (1965).

[36] K. Ruddock, W. Bell, A. Bloom, J. Arnold, and R. Sarles, Patent USA, N3252081 (1966).

[37] M. Arditi, US Patent, N3281663 (1966).

[38] F. A. Franz, Rev. Sci. Instrum. **34**, 589 (1963).

[39] B. A. Andrianov, I. E. Grin'ko, A. F. Lukoshin, and P. S. Ovcharenko, Izmer. Tekh. **10**, 85 (1976).

[40] A. N. Kozlov, Geofizicheskaya Apparatura **24**, 86 (1965).

[41] C. D. Hardwick, Geophysics **49**, 2024 (1984).

[42] T. Yabuzaki and T. J. Ogawa, Appl. Phys. **45**, 1342 (1974).

[43] E. B. Aleksandrov, M. V. Balabas, S. P. Dmitriev, N. A. Dovator, A. I. Ivanov, M. I. Karuzin, V. N. Kulyasov, A. S. Pazgalev, and A. P. Serebrov, Pis'ma Zh. Tekh. Fiz **32**, 58 (2006) [Tech. Phys. Lett. **32**, 627 (2006)].

[44] V. V. Yashchuk, D. Budker, and J. R. Davis, Rev. Sci. Instrum. **71**, 341 (2000).

[45] J. Mosnier, Ann. Geophys. **113**, 22 (1966).

[46] G. S. Vasyutochkin, in *Melody Razvedochnoi Geofiziki. Para-shchelochnye Kvantovye Magnitometry i Ikh Primenenie* (*Methods of Prospecting Geophysics. Para-Alkaline Quantum Magnetometers and Their Applications*) **19** (NPO *Geofizika*, Leningrad, 1976).

[47] E. B. Aleksandrov, M. V. Balabas, and V. A. Bonch-Bruevich, Pis'ma Zh. Tekh. Fiz. **13**, 749 (1987) [Sov. Tech. Phys. Lett. **13**, 312 (1987)].

[48] E. B. Aleksandrov and V. A. Bonch-Bruevich, Opt. Eng. **31**, 711 (1992).

[49] E. Pulz, K.-H. Jäckel, and H.-J. Linthe, Meas. Sci. Tech. **10**, 1025 (1999).

[50] E. B. Aleksandrov and A. S. Pazgalev, Opt. Spektrosk. **80**, 534 (1996) [Opt. Spectrosc. **80**, 473 (1996)].

[51] E. B. Aleksandrov, A. S. Pazgalev, and J. L. Rasson, Opt. Spektrosk. **82**, 14 (1997) [Opt. Spectrosc. **82**, 10 (1997)].

[52] A. K. Vershovskii and A. S. Pazgalev, Pis'ma Zh. Tekh. Fiz. **37**, 48 (2011) [Tech. Phys. Lett. **37**, 23 (2011)].

[53] E. B. Alexandrov, M. V. Balabas, V. N. Kulyasov, A. E. Ivanov, A. S. Pazgalev, J. L. Rasson, A. K. Vershovski, and N. N. Yakobson, Meas. Sci. Technol. **15**, 918 (2004).

[54] A. K. Vershovskii, M. V. Balabas, A. E. Ivanov, V. N. Kulyasov, A. S. Pazgalev, and E. B. Aleksandrov, Zh. Tekh. Fiz. **76**, 115 (2006) [Tech. Phys. **51**, 112 (2006)].

[55] A. K. Vershovskii, Opt. Spektrosk. **101**, 2 (2006) [Opt. Spectrosc. **101**, 309 (2006)].

[56] J. L. Rasson, Geophys. Trans. **36**, 187 (1991).

[57] O. Gravrand, A. Khokhlov, J. L. Le Mouël, and J. M. Léger, Earth Planets Space **53**, 949 (2001).

[58] A. J. Fairweather and M. J. Usher, J. Phys. E: Scient. Instrum. **5**, 986 (1972).

[59] A. K. Vershovskii, Zh. Tekh Fiz. **37**, 93 (2011) [Tech. Phys. **37**, 140 (2011)].

[60] P. D. D. Schwindt, B. Lindsen, S. Knappe, V. Shah, and J. Kitching, Appl. Phys. Lett. **90**, 81102 (2007).

[61] J. Allred, R. Lyman, T. Kornack, and M. Romalis, Phys. Rev. Lett. **89**, 130801 (2002).

[62] I. K. Kominis, T. W. Kornack, J. C. Allred, and M. V. Romalis, Nature **422**, 596 (2003).

[63] W. C. Griffith, R. Jimenez, V. Shah, S. Knappe, and J. Kitching, Appl. Phys. Lett. **94**, 023502 (2009).

5

Spin-exchange-relaxation-free (SERF) magnetometers

I. Savukov and S. J. Seltzer

5.1 Introduction

In this chapter, we describe the spin-exchange-relaxation-free (SERF) atomic magnetometer. As the name suggests, this is an atomic magnetometer (AM) in which the effects of spin-exchange (SE) collisions on spin relaxation are effectively turned off, even at a very high rate of such collisions. The relaxation rate is an important parameter affecting the operation of an atomic magnetometer, often limiting its sensitivity. In particular, the transverse relaxation time, T_2, is related to the AM sensitivity. Because of the dramatic reduction in bandwidth when the effect of SE collisions on T_2 relaxation is "turned off," the SERF magnetometer has superior sensitivity, well below the level of 1 $\mathrm{fT}/\sqrt{\mathrm{Hz}}$. To better explain the SERF magnetometer, especially why it has superior sensitivity to other detectors, in this chapter we first discuss SE collisions and their effect on atomic spins. We will focus on the effects of SE collisions in the SERF and near-SERF regimes in atomic cells with high density of atoms, typically on the order of 10^{14} per cm^3.

For ideal operation in the SERF regime, the magnetic field must be exactly zero. In practice, the field can only be zeroed with some level of accuracy, and sometimes a finite bias field is applied to tune the magnetometer to operate at a nonzero frequency of interest. Tuning the field in the SERF regime can be necessary because the bandwidth is often as small as 10–100 Hz, while some applications require a substantial spectral range. For example, if a SERF magnetometer is used to detect nuclear magnetic resonance (NMR) at ultra-low frequency (for example 2 kHz) it will need to be tuned far beyond several bandwidths and the broadening due to SE will be significant [1]. If a SERF magnetometer is applied to magnetoencephalography (MEG), it might need to cover a 100 Hz frequency range of the brain neural activity spectrum, and bias-field tuning can help to extend sensitive operation over this range as an alternative to broadening the AM resonance. Thus it is interesting for practical applications to cover a range of AM parameters where the SE contribution to relaxation is not negligible. Frequency dependence of the AM response, tuning, and zeroing will become easy to understand with the aid of the Bloch equations discussed below. After

Optical Magnetometry, ed. Budker, D. and Jackson Kimball, D. F. Published by Cambridge University Press.

discussion of theoretical aspects of SERF magnetometers, we will turn to practical questions such as a typical design of SERF magnetometers and basic experimental principles of operation.

5.2 Spin-exchange collisions

Among the various collisions in magnetometer vapor cells, SE collisions deserve special attention because their cross-section is very large. For example, the SE cross-section exceeds that of potassium–potassium spin-destruction (SD) collisions by four orders of magnitude [2]. SE collisions can be the dominant contribution to the T_2 relaxation of the atomic spins and limit the sensitivity of AMs. The effects of SE collisions can be analyzed with the density-matrix-equation (DME) formalism.

Abundant literature exists on the calculations, measurements, and applications of spin-exchange collisions. Of particular interest to SERF development is the discovery by Happer and Tang [3] of the narrowing of magnetic-resonance lines at high densities of alkali-metal vapors when placed in very small magnetic fields. This unusual behavior of magnetic resonances was later analyzed by Happer and Tam [4], using the equation for the rate of change of the atomic density matrix due to SE collisions given by Grossetête [5] and Anderson *et al.* [6, 7]. Happer and Tam [4] also derived an analytical expression for the frequency shift and width of magnetic resonances for an arbitrary SE rate in the limiting case of low polarization. This equation predicts zero broadening at large SE rates at zero field, essentially the SERF regime, although low polarization is not optimal for SERF operation. Another interesting effect – light narrowing of magnetic resonances, more precisely the reduction of the SE contribution to transverse relaxation rate at high polarization levels – was discovered much later in 1998 by Appelt *et al.* [8]. The analysis of SE effects at low magnetic field for an arbitrary spin polarization was first performed in Ref. [2], where it was shown that SE relaxation is completely eliminated at zero field for arbitrary spin polarization. The density-matrix equations that contain SE collisions, optical pumping, and other terms for complete description of the SERF magnetometer have been formulated [2, 9]. The numerical solution of this density-matrix equation for an extensive range of AM parameters (such as SE rate, pumping rate, and magnetic field) has been obtained and compared with experimental measurements to establish a firm basis for the analysis of SERF and other high-density AMs [10].

5.2.1 The density-matrix equation

The density-matrix equation (DME) is the basis for quantitative analysis of atomic spin behavior in SERF and other magnetometers. In the DME formalism, the expectation value of an observable A (for example, the projection of the electron spin) is given by $\langle A\rangle = \mathrm{Tr}(\rho A)$, where the density matrix ρ satisfies the conditions $\mathrm{Tr}(\rho) = 1$ and $\rho = \rho^\dagger$. The DME contains various terms responsible for interactions in the atomic spin system. Terms that do not arise from collisions can be obtained from the Hamiltonian H of the corresponding interaction using Liouville's theorem:

$$\frac{d\rho}{dt} = -\frac{i}{\hbar}[H,\rho]. \tag{5.1}$$

Relaxation terms due to collisions come from additional considerations, and are added to the right-hand side of this equation, which becomes [2,9]:

$$\frac{d\rho}{dt} = A_{hf}\frac{[\mathbf{I}\cdot\mathbf{S},\rho]}{i\hbar} + \mu_B g_S\frac{[\mathbf{B}\cdot\mathbf{S},\rho]}{i\hbar} + \frac{\varphi(1+4\langle\mathbf{S}\rangle\cdot\mathbf{S})-\rho}{T_{SE}} + \frac{\varphi-\rho}{T_{SD}} + R[\varphi(1+2\mathbf{s}\cdot\mathbf{S})-\rho] + D\nabla^2\rho. \tag{5.2}$$

Here $\varphi = \rho/4 + \mathbf{S}\cdot\rho\mathbf{S}$ is the purely nuclear part of the density matrix ρ (meaning that $\langle\varphi\mathbf{S}\rangle = 0$), A_{hf} is the hyperfine constant of the magnetic-dipole coupling, $g_S \approx 2.002$ is the electron g-factor, $\mu_B \approx 9.274\times10^{-24}$ J/T is the Bohr magneton, T_{SD} is the spin-destruction relaxation time, T_{SE} is the spin-exchange collision time, and $\mathbf{s}$ is the optical-pumping vector defining the direction and the degree of circular polarization of the laser beam $\mathbf{s} = \boldsymbol{\epsilon}\times\boldsymbol{\epsilon}^*$. The first two terms on the right-hand side of Eq. (5.2) are due to the hyperfine- and magnetic-interaction Hamiltonians in the Schrödinger equation. The remaining terms are due to SE, SD, optical pumping (OP), and diffusion and will be explained below.

Spin-exchange term: The effects of SE collisions are included with the third term of the DME given in Eq. (5.2). The magnitude of this term is proportional to the SE rate given by

$$R_{SE} = 1/T_{SE} = n\bar{v}\sigma_{SE}, \tag{5.3}$$

where σ_{SE} is the SE cross-section, $\bar{v}$ is the average interatomic velocity, and n is the density of colliding atoms. This term is nonlinear in the spin density ρ, making an analytical solution of the DME impossible in general and a numerical solution time-consuming. Several cases can be considered separately. When the SE term is much smaller than the hyperfine term but substantially exceeds all other terms, the SE contribution to spin relaxation is greatly reduced; this is another way to define the SERF regime. Under the effect of fast SE collisions, the hyperfine sublevels are populated according to the spin-temperature distribution $\sim\exp(\beta m_F)$, where the spin-temperature parameter $\beta = \ln[(1+P)/(1-P)]$ is a function of the polarization P [6,11,12], as illustrated in Fig. 5.1. The multilevel system is characterized by a single polarization vector. When a bias magnetic field is applied, the SE rate can become comparable to the spin-precession frequency, and then SE collisions will contribute to relaxation. For the case of small polarization it is possible to use perturbation theory to derive analytical expressions for the relaxation rate and precession frequency [4].

Relaxation and optical pumping terms: The third, fourth, and fifth terms in the DME are bulk SD, OP, and diffusion terms, respectively. These terms affect both longitudinal and transverse spin relaxation. The SD term causes evolution of the spin density toward the

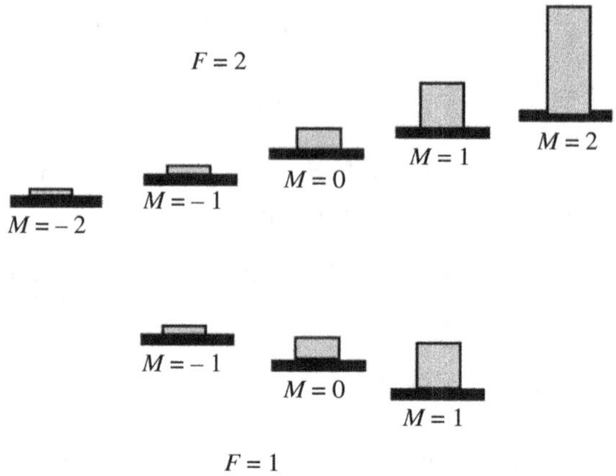

Figure 5.1 Spin-temperature distribution for an alkali atom with nuclear spin $I = 3/2$.

zero-polarization state φ with the rate $R_{SD} = 1/T_{SD}$. There are multiple atomic species in a SERF cell, and they all contribute to the SD rate:

$$R_{SD} = \sum_i n_i \bar{v}_{ji} \sigma_{ji} \,, \tag{5.4}$$

where n_i is the density of atoms of the species i, and

$$\bar{v}_{ji} = \sqrt{8k_B T(1/m_j + 1/m_i)/\pi} \tag{5.5}$$

is the relative thermal velocity between an alkali-metal atom j of mass m_j and an atom i of mass m_i. In particular, a typical SERF potassium magnetometer uses a cell containing potassium vapor, helium, and nitrogen, so

$$R_{SD} = n_K \bar{v}_{KK} \sigma_{KK} + n_{He} \bar{v}_{KHe} \sigma_{KHe} + n_{N_2} \bar{v}_{KN_2} \sigma_{KN_2}. \tag{5.6}$$

The duration of collisions between atoms is much shorter than the time needed for nuclear spins to evolve, so during a collision the nuclear spin orientation is assumed to remain preserved. The form of the SD operator reveals this property explicitly by containing the pure nuclear part of the density matrix φ. SE collisions have a similar property that the nuclear part is conserved during the collision, and the SE term likewise contains the projection operator. However, SE collisions are different in that the "pumping" action is directed toward the average spin, while SD collisions "pump" to the zero spin state. Similarly, optical pumping evolves electron spins toward the state where they are oriented in the same direction as the photon spin, which for circularly polarized light is parallel to the pump-beam direction.

If the cell does not have an antirelaxation coating, as is typical for SERF magnetometers operating at high temperature, then collisions with the walls also contribute to the SD rate via the diffusion term, which in general depends on the spatial distribution of polarization. Note that the diffusion carries the total angular momentum of the atoms, while bulk collisions act on the electron spin part only, because it is assumed that the total angular momentum is lost during collisions with the walls, not just electron-spin polarization. This assumption is validated by surface scattering experiments [13]. For simple geometries, the diffusion-relaxation rate can be estimated from analytical expressions. For example, in the approximation of the fundamental diffusion mode for a spherical cell and relaxation of the total atomic angular momentum to zero at the walls, the diffusion relaxation rate will be $R_{SD,D} = qD(\pi/a)^2$, where D is the diffusion coefficient and a is the radius of the cell. The slowing-down factor q, which defines how much slower the atomic spin precesses in a magnetic field compared to an electron spin, depends on polarization and accounts for the fact that collisions with the walls destroy not only the electron spin but also the nuclear spin component.

The spin-destruction collisions broaden magnetic resonances and determine the equilibrium polarization for a given pumping rate. In addition, the diffusion term leads to spatial redistribution of polarization and some other interesting effects such as motional narrowing of the magnetic gradient broadening.

Exclusion of the hyperfine term in the density-matrix equation: Some simplification can be achieved by exclusion of the large hyperfine term which leads to fast oscillations of matrix elements nondiagonal in F, the total electron and nuclear spin angular momentum. Such oscillations considerably slow down numerical computation. The oscillations are large when hyperfine transitions are excited, for example in atomic clocks, but in SERF atomic magnetometers the oscillations are negligible. In high field – such that the magnetometer would not be in the SERF regime – multiple resonances due to the nonlinear Zeeman splitting can be observed, and it is necessary to include the hyperfine term to explain this effect.

Spin-exchange collisions redistribute populations between the two sets of ground-state hyperfine sublevels $F = I \pm 1/2$, so these sublevels need to be included in the analysis. The density matrix can be expanded in a complete basis of coupled hyperfine states $|i\rangle = |Fm_F\rangle$,

$$\rho = \sum_{ij} \rho_{ij} |i\rangle\langle j|. \tag{5.7}$$

In this basis the hyperfine term becomes very simple:

$$\left(A_{hf}/i\hbar\right)[\mathbf{I}\cdot\mathbf{S},\rho] = \left(\Delta E_{hf}/i\hbar\right)(F - F')\rho_{ij} , \tag{5.8}$$

where ΔE_{hf} is the splitting between hyperfine levels; for example, in the case of $I = 3/2$ this splitting is equal to $2A_{hf}$. If the density matrix elements ρ_{ij} are not diagonal in F, then large factors ΔE_{hf} appear, leading to fast oscillations at the hyperfine frequency. It is convenient to use a projection operator

$$\Pi\left\{\begin{bmatrix} \rho_a & * \\ * & \rho_b \end{bmatrix}\right\} = \begin{bmatrix} \rho_a & 0 \\ 0 & \rho_b \end{bmatrix} \tag{5.9}$$

to exclude these density matrix elements. Here $a \equiv (F = I + 1/2)$ and $b \equiv (F = I - 1/2)$, and applying the projection operator we can redefine $\rho \equiv \Pi\{\rho\}$ and write $\rho = \rho_a + \rho_b$, where

$$\rho_a \equiv \begin{bmatrix} \rho_a & 0 \\ 0 & 0 \end{bmatrix} \tag{5.10}$$

and

$$\rho_b \equiv \begin{bmatrix} 0 & 0 \\ 0 & \rho_b \end{bmatrix}. \tag{5.11}$$

The DME after the projection will have the hyperfine term identically equal to zero. It has been shown numerically that the complete solution with the hyperfine term included agrees closely with the projected solution in the case where the hyperfine term greatly exceeds the other terms [10]. In SERF magnetometers, this is usually the case; however, a magnetometer with a bias field on the order of 0.1 mT will start to exhibit nonlinear Zeeman splitting, and to account for this the hyperfine term should be retained and the projection operator removed.

The numerical solution of the density-matrix equation: Numerical simulation is the only approach to quantitatively investigating the spin behavior for an arbitrary range of parameters. Examples of numerical simulations are given in the supplementary online material (www.cambridge.org/9781107010352).

5.2.2 Simple model of spin exchange

Once we have partitioned ρ into a and b components, we can define the upper and lower hyperfine components of the total angular momentum and the electron spin:

$$F_{a,b} = \mathrm{Tr}(\rho_{a,b} F) \tag{5.12}$$

and

$$S_{a,b} = \mathrm{Tr}(\rho_{a,b} S). \tag{5.13}$$

Taking the trace of the magnetic field term and the time derivative term in the DME [Eq. (5.2)], we obtain

$$\frac{d}{dt}\mathbf{F}_{a,b} = (\mu_B g_S/\hbar)\,\mathbf{B} \times \mathbf{S}_{a,b}. \tag{5.14}$$

Using the proportionality between the matrix elements of the total angular momentum and spin, $\mathbf{S}_{a,b} = \pm\mathbf{F}_{a,b}/(2I+1)$, when these matrix elements are diagonal in F we obtain the time evolution of the $F_{a,b}$ components of the total angular momentum:

$$\frac{d\mathbf{F}_a}{dt} = \gamma \mathbf{B} \times \mathbf{F}_a\,, \tag{5.15}$$

$$\frac{d\mathbf{F}_b}{dt} = -\gamma \mathbf{B} \times \mathbf{F}_b\,, \tag{5.16}$$

where $\gamma = \mu_B g_S/(2I+1)\hbar$ is the gyromagnetic ratio of the atom.

We can also combine the two equations:

$$\frac{d}{dt}(\mathbf{F}_a + \mathbf{F}_b) = \gamma \mathbf{B} \times (\mathbf{F}_a - \mathbf{F}_b). \tag{5.17}$$

In the SERF regime, the a and b components are "forced" to precess together by rapid SE, so

$$\mathbf{F}_a = k_a \mathbf{F} \tag{5.18}$$

and

$$\mathbf{F}_b = k_b \mathbf{F}\,, \tag{5.19}$$

where $k_{a,b}$ are scalar factors determined by the spin-temperature parameter β, which is a function of polarization (see Section 5.2.1). Thus,

$$\frac{d\mathbf{F}}{dt} = \left(\frac{k_a - k_b}{k_a + k_b}\right) \gamma \mathbf{B} \times \mathbf{F} = g\gamma \mathbf{B} \times \mathbf{F}. \tag{5.20}$$

This equation is simply the standard Bloch equation, which is the basis for the analysis of NMR dynamics, although unlike nuclear spins in NMR, atomic spins have a variable gyromagnetic ratio $g\gamma$. Actually, g is a function of spin temperature or polarization:

$$g = \frac{\sum_{m=-a}^{a} m \exp[m\beta] - \sum_{m=-b}^{b} m \exp[m\beta]}{\sum_{m=-a}^{a} m \exp[m\beta] + \sum_{m=-b}^{b} m \exp[m\beta]}\,, \tag{5.21}$$

where m is the projection of F.

Here we have separately summed contributions for the a and b components, since they enter expression (5.20) with opposite signs. This equation can be simplified for specific values of the nuclear spin. In the case of $I = 3/2$, of direct interest for a magnetometer using K or ^{87}Rb atoms, $g = (2\cosh\beta)/(1+2\cosh\beta)$, or in terms of polarization $P = 2S$, $g = 2 - 4/\left(3 + P^2\right)$. For other nuclear spins, similar expressions can be obtained that give the gyromagnetic ratio as a function of polarization [10]. Note that due to the polarization dependence of the precession rate, the Bloch equation is nonlinear in polarization or total spin.

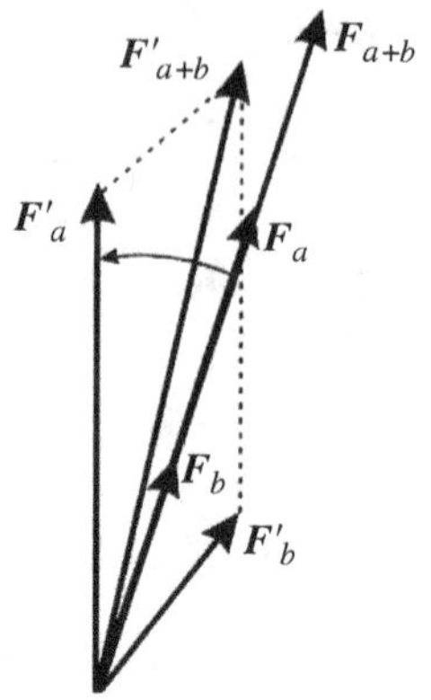

Figure 5.2 Vector diagram illustrating mechanisms of broadening and variation of the gyromagnetic factor due to the actions of SE and precession in a magnetic field. Initially $\mathbf{F}_a$ and $\mathbf{F}_b$ are aligned, and their vector sum is $\mathbf{F}_{a+b}$. The magnetic field, applied perpendicular to the plane of the picture, rotates the two vectors during the time between SE collisions in opposite directions; $\mathbf{F}_a$ becomes $\mathbf{F}'_a$, and $\mathbf{F}_b$ becomes $\mathbf{F}'_b$. The SE collisions bring $\mathbf{F}'_a$ and $\mathbf{F}'_b$ together along a new direction $\mathbf{F}'_{a+b}$ with the conservation of the total angular momentum $\mathbf{F}'_{a+b} = \mathbf{F}'_a + \mathbf{F}'_b$. This cycle leads to the reduction of the total angular momentum by $\mathbf{F}_{a+b} - \mathbf{F}'_{a+b}$ and hence of polarization.

When the precession frequency approaches the spin-exchange rate, the angle between $\mathbf{F}_a$ and $\mathbf{F}_b$ becomes significant, as shown in Fig. 5.2. Although SE collisions conserve the total angular momentum F, which can be shown analytically, the combined action of the magnetic field and of SE collisions does not. In Fig. 5.2, the length of the initial vector before rotation by the magnetic field is greater than the length of the vector after rotation followed by SE realignment. The magnetic field rotates only the transverse component of the spin and hence only transverse polarization is changed. In other words, SE collisions contribute to T_2 but not T_1. Using this geometrical interpretation, it can be shown that the SE T_2 relaxation rate depends quadratically on the magnetic field, as demonstrated in Fig. 5.3. This was observed experimentally [2, 3, 10] and was analytically shown for the case of small polarization [2, 4]. The diagrammatic model works for arbitrary polarization level. In particular, this diagram helps to demonstrate why at high polarization, when the a component of the angular momentum is much larger than the b component, the SE broadening is reduced, an effect termed light narrowing. This two-vector diagram provides a simple visualization of the SE effects.

5.3 Bloch equation description

In the SERF regime, it is convenient to describe the evolution of the alkali spin $\mathbf{S}$ in a magnetic field $\mathbf{B}$ using the phenomenological Bloch equation, which is consistent with numerical simulations [10] and experimental observations of spin precession. One nontrivial feature that is necessary for the Bloch equation to be an adequate basis for theory is the

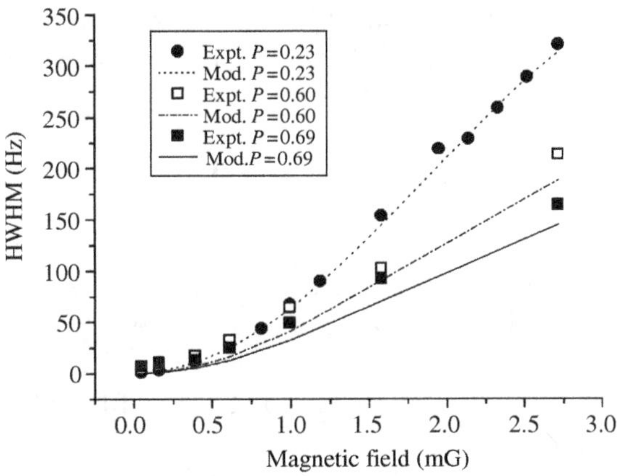

Figure 5.3 Dependence of magnetometer linewidth (HWHM) on the amplitude of the magnetic field in the SERF regime, showing a quadratic dependence near zero field. Experimental data are compared to the results of numerical simulation as a function of polarization P, as described in the supplemental material. From Ref. [10].

presence of a single resonance in a rotating field, which has been verified by experiments. Another feature is radiofrequency (RF) broadening, which also was observed in the SERF regime [10].

In Section 5.2.2, we showed that the essential term of the Bloch equation,

$$\frac{d\mathbf{S}}{dt} = (\gamma^e/q)\mathbf{B} \times \mathbf{S}\,, \tag{5.22}$$

can be derived from the DME in the spin-temperature distribution regime. The optical pumping and spin-destruction terms can be similarly derived by taking the trace of corresponding terms in the DME, and they account for equilibrium spin orientation and spin relaxation:

$$\frac{d\mathbf{S}}{dt} = \frac{1}{q}\left[\gamma^e\mathbf{B} \times \mathbf{S} + R_{\mathrm{OP}}\left(\frac{1}{2}s\hat{\mathbf{z}} - \mathbf{S}\right) - R_{\mathrm{rel}}\mathbf{S}\right]. \tag{5.23}$$

Here q is the nuclear slowing-down factor, $\gamma^e = 2\pi \times 2.8$ MHz/G is the gyromagnetic ratio of the electron, R_{OP} is the optical pumping rate, and R_{rel} is the spin relaxation rate in the absence of optical pumping. The polarization of the pump beam, s, is defined to be along the z direction. In the absence of any magnetic field, the equilibrium spin polarization $\mathbf{S}_0 = S_0\hat{\mathbf{z}}$ is given by setting $d\mathbf{S}/dt = 0$ and finding the steady-state solution,

$$S_0 = \frac{sR_{\mathrm{OP}}}{2(R_{\mathrm{OP}} + R_{\mathrm{rel}})}\,. \tag{5.24}$$

From here on, it will be assumed that $s = 1$, corresponding to a pump beam with circular polarization.

Following Ref. [14], for notational simplicity we introduce the dimensionless magnetic field parameter $\boldsymbol{\beta}$ (not to be confused with the spin-temperature parameter),

$$\beta_i = \left(\frac{\gamma^e}{R_{\mathrm{OP}} + R_{\mathrm{rel}}} \right) B_i. \tag{5.25}$$

In the presence of a slowly changing magnetic field, the quasi-steady-state solution to Eq. (5.23) can be found. Most SERF magnetometers use a probe laser beam orthogonal to the pump beam that is tuned off the absorption line, with the probing axis defined here as $\hat{\boldsymbol{x}}$. Thus the signal of the SERF magnetometer will be proportional to

$$S_x = S_0 \frac{\beta_y + \beta_x \beta_z}{1 + \left(\beta_x^2 + \beta_y^2 + \beta_z^2 \right)}. \tag{5.26}$$

If the magnetic field is also small ($|\boldsymbol{\beta}| \ll 1$), then the magnetometer will be most sensitive to β_y (the component perpendicular to both the probe and pump directions), $S_x \approx S_0 \beta_y$. From Eq. (5.26) it follows that the sensitivity is optimal when $R_{\mathrm{OP}} = R_{\mathrm{rel}}$ giving S_0=1/4, or 50% polarization. At higher pumping rates the polarization saturates while the linewidth continues to broaden, reducing the magnetometer signal, as shown in Fig. 5.4.

Application of a small bias field β_z permits detection of a slowly oscillating field (the frequency is ultimately limited by the temperature of the vapor) $\boldsymbol{\beta}_1(t) = \beta_1 \cos(\omega t) \hat{\mathbf{y}}$.

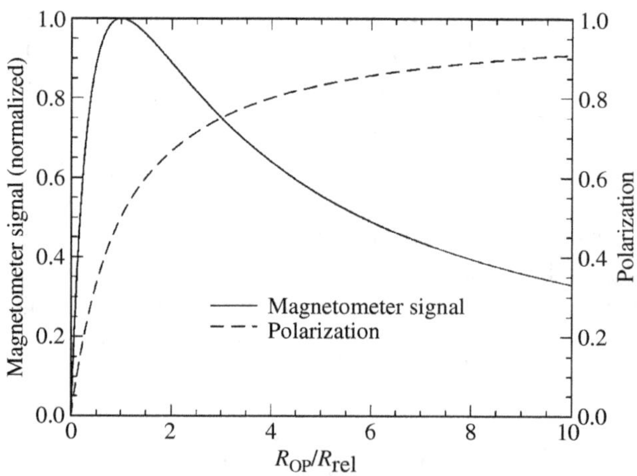

Figure 5.4 SERF magnetometer signal, proportional to S_x as given by Eq. (5.26), as a function of alkali spin polarization. Optimum signal occurs at 50% polarization, corresponding to $R_{\mathrm{OP}} = R_{\mathrm{rel}}$. Adapted from Ref. [15].

Solving Eq. (5.23) for the time-dependent magnetometer signal created by β_z and β_1 gives

$$S_x = \frac{1}{2} S_0 \beta_1 \Delta\omega \left[\frac{\Delta\omega \cos(\omega t) + (\omega - \omega_z)\sin(\omega t)}{(\Delta\omega)^2 + (\omega - \omega_z)^2} + \frac{\Delta\omega \cos(-\omega t) + (\omega + \omega_z)\sin(-\omega t)}{(\Delta\omega)^2 + (\omega + \omega_z)^2} \right], \tag{5.27}$$

where the resonance frequency is written as $\omega_z = \gamma^e B_z/q$, and the magnetic linewidth $\Delta\omega = R_{\rm OP} + R_{\rm rel}/q = 1/T_2$. The response takes the form of two Lorentzians centered at $\pm\omega_z$, with both in-phase (adsorptive) and out-of-phase (dispersive) components. Under conditions such that $|\omega_z| \lesssim \Delta\omega$, these Lorentzians overlap at least partially, and it is necessary to include the effects of the resonance at $-\omega_z$ when considering the magnetometer signal. In the supplementary material we discuss the application of this analysis to measurement of all three vector components of the magnetic field.

5.4 Experimental realization

5.4.1 Classic SERF atomic magnetometer arrangement

While it is possible to design a large number of schemes for atomic magnetometry, the scheme shown in Fig. 5.5 can be considered as the "classic" SERF design and is particularly suitable for illustrative purposes due to its simplicity. Various modifications will be discussed

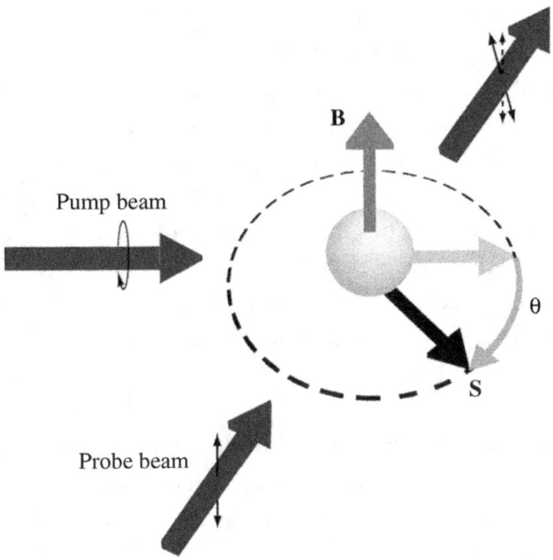

Figure 5.5 Typical arrangement of a SERF magnetometer, with a circularly polarized pump beam tuned to the D1 line and a linearly polarized probe beam tuned slightly off the D1 line. Adapted from Ref. [15].

briefly with their advantages and disadvantages. This scheme has one particular advantage in that the probe and pump beams are perpendicular to each other, minimizing the noise due to fluctuations in the intensity of the pump beam. The magnetometer operates as follows: The circularly polarized pump beam orients spins along its propagation direction. The magnetic field rotates the spins from the pump direction into the probe direction, and their acquired projection along the probe-beam direction is detected as a rotation of the polarization of a linearly polarized probe beam. This detection is based on the Faraday effect, and the spin component along the probe beam is essentially the magnetometer signal. Equation (5.26) gives the dependence of the magnetometer signal on the applied magnetic field. The most efficient pumping is achieved at the center of the D1 line (for instance, 770 nm for K), while the most sensitive detection is achieved when the probe beam is detuned a few linewidths from the center of this line, the latter for two reasons: first, the optical rotation induced by the vapor has a dispersion-like Lorentzian profile, with zero value in the center and extrema at one linewidth away; second, absorption by the vapor decreases the probe beam intensity near resonance and further increases the detuning required for the maximum magnetometer signal. Broadening of magnetic resonances by partial absorption of the probe beam, so-called probe spin destruction, can also increase the optimal probe detuning. Thus it might be necessary to empirically adjust the wavelength by observing simultaneously signal and noise to achieve the best sensitivity.

Owing to conflicting requirements on the wavelength for the pump and probe beams, the best optimization of the sensitivity is achieved with two separate lasers, but it is still possible to achieve high sensitivity using a single laser, as demonstrated in [16]. In order to use a single laser, the laser beam polarization was made elliptical, as a compromise between circular polarization for the pump and linear polarization for the probe. Furthermore, because the single laser beam effectively functioned in a parallel pump-and-probe beam configuration, in order to obtain signal from the magnetometer the atomic spins were tilted by applying a relatively strong modulated magnetic field. The modulation also allowed reduction of $1/f$ noise. The modulated field broadened the magnetic resonance, but because optical noise was substantially reduced by modulation, high sensitivity of 7 $\mathrm{fT}/\sqrt{\mathrm{Hz}}$ was demonstrated. A one-laser magnetometer design has the advantages of price reduction and convenience. Another modification on the optical detection scheme is the use of absorption instead of Faraday rotation [17]. If the cell is small or its optical density is low, then the probe beam can be tuned closer to the center of the D1 line, and this can be convenient for constructing a single-laser magnetometer [18].

Polarization rotation measurement schemes: Optical detection of the atomic spin state is normally based on the measurement of polarization rotation of the probe beam using laser polarimetry. A conventional polarimeter (see for example Ref. [12]) consists of a polarizer before the cell and a polarizing beam splitter (PBS) after the cell rotated at an angle $\sim 45°$ with respect to the first polarizer. The intensities of the split beams are $I_0 \cos^2\theta$ and $I_0 \sin^2\theta$, where θ is the angle between light polarization and the axis of the beam-splitter cube. By measuring the difference with the angle set to 45°, it is possible to subtract noise arising from laser intensity fluctuations. Polarization modulators can be also employed to reduce noise arising from intensity fluctuations in the probe beam and other factors, such as $1/f$ noise (see for example Ref. [2]). Polarization modulation techniques usually employ only one photodiode, facilitating multichannel magnetic field measurements.

Atomic cell: The atomic vapor cell is the central element of an alkali-metal atomic magnetometer. Here the measurement of the magnetic field is realized via the interaction of the atomic spins with the magnetic field and readout of spin states via laser light. In a SERF magnetometer, spin-destruction collisions due to the walls must be suppressed in order to maintain high sensitivity. For this purpose a buffer gas is usually added to slow the diffusion. The best choice of buffer gas is helium, which causes minimal spin destruction and thus provides highest sensitivity (see, for example, Table I in Ref. [2]). Other noble gases and nitrogen can also be used to slow atomic diffusion. Another function of the nitrogen gas, which is generally added in about 30 mtorr quantities, is to quench excited states, so radiation trapping does not occur and spontaneous re-emission of light does not destroy spin polarization. A typical SERF cell contains a small droplet of K, 3 atm of He as a buffer, and 30 mtorr of N_2 for quenching, although some SERF cells have been constructed differently in order to realize particular applications. For example, for purposes of safety and ease of construction, it is sometimes desirable to use less than 1 atm of helium. On the other hand, cells with pressure as high as 12 atm have been used in experiments where it was necessary to achieve uniform polarization; dealing with such high-pressure cells requires caution since they can explode. As discussed in Section 5.4.3, research has also recently focused on realizing a SERF magnetometer using antirelaxation coatings rather than buffer gas.

Initially, SERF cells were made of special aluminosilicate glass which minimized helium diffusion and interaction of alkali-metal with the walls. This type of glass is expensive, especially for large cells, and difficult to shape. Pyrex (borosilicate glass) has also been successfully used for SERF magnetometer cells, although at high temperature, diffusion of helium through the glass is significant, so the atomic cell may change its properties over time. In this case neon can be used as a buffer gas to slow diffusion through the glass.

Oven construction: A SERF magnetometer typically operates at atomic densities on the order of 10^{14} cm^{-3}, which require heating to relatively high temperatures (for example, 160–180°C for K). To avoid electromagnetic noise, ovens should be constructed out of nonmagnetic and nonmetallic components. An oven can be heated with hot air, in which case it is connected to a supply of compressed air that flows through a heater. The heater power is fairly high for typical SERF ovens ($\sim$1 kW for cm-sized cells), which is inconvenient in portable applications. It is much more efficient to heat electrically, which can be arranged easily in the oven. Unfortunately, the electric heating elements produce magnetic noise due to the heater current, coupling to external circuits, and Johnson noise. To reduce current noise, the heater is disconnected during the measurements with an electronic switch. Alternatively, high-frequency AC current can be used that has minimal effect on the atomic spins and continuous operation can be implemented. The noise can also be reduced by twisting the heater wires or by moving the electric heating elements away from the cell and transferring heat with heat-conducting, nonmagnetic materials. The Johnson noise is reduced by using high-resistivity heating elements. In biomagnetic applications, it is also necessary to cool efficiently the outer walls of the oven to allow contact with the body of the subject, which can be done with running-water pads.

5.4.2 Zeroing the magnetic field

For sensitive operation of a SERF magnetometer, SE relaxation must be suppressed by zeroing the ambient magnetic field. In a SERF magnetometer, the field is usually minimized by enclosing the magnetometer in a multiple-layer mumetal shield and by degaussing the shield (see Chapter 12). Small residual fields are further compensated using field coils. In an unshielded environment, compensation is more difficult [14]. In either case, Eq. (5.26) suggests a straightforward method of zeroing the field with the use of compensation coils: First, B_y is set so that the DC component of the magnetometer signal is zero (it is possible to have an offset in the magnetometer signal due to misbalance of the polarimeter, but we assume that this offset is zeroed in advance). Next, we apply a small modulation to B_x and adjust B_z so that the oscillation of the magnetometer signal disappears. Then, B_z is modulated and B_x is adjusted to zero the signal. Repeating this process iteratively will result in making all three components of the magnetic field as small as necessary for optimal SERF operation. This procedure should be used whenever a SERF magnetometer is first turned on, and it should be repeated as necessary to correct for drifts in the ambient magnetic field.

5.4.3 Use of antirelaxation coatings

Most implementations of the SERF magnetometer thus far have used buffer gas to prevent depolarization due to collisions of the alkali atoms with the walls of the glass cell, providing the side-benefit of allowing easy multiple-channel measurements [19, 20]. However, there are a number of advantages to using antirelaxation surface coatings instead of buffer gas for this purpose, such as suppression of gradient broadening, larger optical rotation signals, and lower laser-power requirements (see Chapter 11). Previously, magnetometry with surface coatings was limited to relatively low temperatures due to lack of suitable coating materials, necessitating the use of buffer gas for SERF magnetometry. However, recent experiments have demonstrated SERF magnetometry with antirelaxation coatings in two different operating regimes, not just at high temperature but also near room temperature.

A SERF magnetometer was demonstrated at temperatures up to about 170°C using a multilayer coating of octadecyltrichlorosilane (OTS) with both potassium and rubidium [21]. Observed optical rotation signals were much larger than for buffer-gas cells under the same operating conditions, although the measured alkali density in all cells was smaller than the saturated vapor density. Certain paraffin materials have also been demonstrated to maintain spin polarization above 100°C with observed lifetimes comparable to those given by OTS (M. V. Balabas *et al.*, personal communication, 2012). One concern that was discovered is that the need for sufficient quenching gas pressure at conditions of high optical density is exacerbated in coated cells. Buffer-gas cells may contain an excess of quenching gas without significantly affecting the spin-polarization lifetime because the buffer gas pressure is still typically much greater. However, it is ideal in coated cells to include just enough quenching gas to permit high alkali polarization [22] without needlessly broadening the optical resonance, reducing the polarization lifetime, or restricting motion of the atoms throughout the cell. The determination of optimal quenching gas pressure for a given cell geometry and alkali species remains to be studied. It is possible to increase alkali polarization without the inclusion of quenching gas by employing a hybrid optical pumping scheme: by pumping on the optically thin vapor of a secondary alkali species present in trace amounts in the cell as an impurity, and allowing the secondary species to transfer polarization to the primary species through spin-exchange collisions, the primary species can attain high spin polarization without direct optical excitation and the associated emission of resonant fluorescence. As an example, it was demonstrated that polarization in a particular OTS-coated potassium cell without quenching gas could be increased from 3% using standard optical pumping to 14% using hybrid pumping [23].

Coated cells also open up the possibility of operating in the SERF regime at low alkali density near room temperature, although the requirement for operating at zero field in the SERF regime will be much more stringent at such temperatures. At low alkali density in a typical cell (with characteristic size on the order of 0.1–10 cm), the rate of spin-destruction collisions with the buffer gas is much greater than the spin-exchange rate. In the low-density regime, the narrowing of the magnetic linewidth that occurs at zero field can thus only be taken advantage of in a cell without buffer gas and with a coating of sufficient quality that relaxation at the wall is much slower than relaxation due to spin-exchange collisions. This was demonstrated in a cell coated with an alkene material that allowed about 10^6 collisions with the surface without depolarization, with observed T_2 lifetimes as long as 77 s for rubidium at natural abundance [24]. This SERF magnetometer maintained its ultra-narrow resonance linewidth, dominated by spin-exchange relaxation near zero field, from room temperature up to the melting point of the coating material at 33°C, significantly lower than for any other implementation of SERF magnetometry.

5.4.4 Comparison with SQUIDs

The most sensitive magnetometers at low frequencies are the SERF atomic magnetometer and the superconducting quantum interference device, or SQUID. The technology of SQUID detectors has had decades to mature [25], with careful engineering having produced commercially available devices for a wide range of applications, while SERF magnetometers are only at the beginning stages of their development. Judging by sensitivity only, SERF magnetometers should in principle be capable of replacing SQUIDs in most applications. However, a comparison of SERF magnetometers and SQUIDs shows that they each have advantages and disadvantages for specific applications. Noncryogenic operation is the primary advantage of SERF magnetometers over SQUIDs. Liquid helium is expensive, its supply is often very limited, and the required availability of helium infrastructure is restrictive for many applications. SQUID systems also require constant maintenance, such as liquid-helium-level checks and refills. Frequent interruptions in maintenance are not recommended, since warming of a SQUID system can lead to component failure. High-T_c SQUIDs that only require liquid nitrogen for cooling are less sensitive than their low-T_c counterparts. Finally, conductive thermal shields – needed for improving the efficiency of cryogenic cooling – can produce substantial noise.

However, by avoiding cryogens, the SERF magnetometer is deprived of the convenience of superconducting flux transformers (SFTs) and superconducting shields. The ability to use such transformers is an important advantage of SQUIDs that facilitates many applications. For example, an SFT can be configured as a gradiometer of arbitrary order (1–3 orders are frequently implemented), and with precise winding it is highly efficient in removing uniform fields, noise, and low-order gradients. Typical uniform-field suppression factors in SFTs can be as large as 1000. SERF magnetometers can be also configured as gradiometers either by using multichannel detection or by using several magnetometers, but the uniform field suppression in this case is much lower and is limited by the stability of the SERF channels. The bias field and gradients can negatively affect the SERF sensitivity and stability of operation, while SQUID systems can operate in a relatively large DC field. Note that other types of atomic magnetometers can operate in a large ambient field, such as the related high-density scalar magnetometer demonstrated in Ref. [26].

Another key difference is that SERF magnetometers are absolute field sensors, while SQUIDs only measure relative field changes. Thus, SERF magnetometers are particularly useful for absolute monitoring of all three field components, while SQUIDs operate better in the presence of a large constant magnetic field. The SERF regime itself requires careful zeroing of all three components of the magnetic field to a high precision. For example, the Earth's field must be compensated for in order to achieve the SERF regime. This problem can be partially solved as was shown in Ref. [14], but the performance of a SERF magnetometer in an unshielded environment is not necessarily stable. With SFTs, it is also much more convenient to use SQUIDs for MRI detection, where a large bias field and gradients need to be applied. In principle SFTs can be connected to a SERF magnetometer, but the advantage of noncryogenic operation will be lost and it is less convenient to combine cryogenic and high-temperature systems. Often, magnetometers have to operate reliably in an unshielded environment in applications such as submarine detection, mine detection, geophysical surveys, etc. With an SFT, SQUIDs can operate in an open environment with high sensitivity and gradiometric noise reduction.

Another substantial difference between SERF atomic magnetometers and SQUIDs is their bandwidths. A SERF magnetometer has a relatively narrow bandwidth of operation, from a few Hz to hundreds of Hz, but a SQUID can operate in a much larger frequency range. Bandwidth and sensitivity are related for a SERF magnetometer, and so while it is possible to increase the bandwidth of a magnetometer, this leads to reduction of overall sensitivity. On the other hand, the narrow frequency response of a SERF device gives it immunity to high-frequency noise, which is a major problem for SQUIDs: it disturbs their operation and leads to the loss of feedback lock. To reduce the effect of this noise, SQUIDs require careful grounding, wrapping in Mylar foil, and optical decoupling of the signal and control circuits. SERF magnetometers, on the other hand, do not need special protection against this noise.

SERF magnetometers generally have a heated cell, and it is necessary to insulate and potentially cool the external surface of its oven, which increases the separation between the magnetometer and a sample. SQUIDs have a similar problem because they operate at cryogenic temperatures. Both detectors therefore require special design considerations to achieve the highest sensitivity to small magnetic objects, which is needed in such applications as nanoparticle detection, single-nerve magnetic-field measurements, and nondestructive evaluation. Although SERF magnetometers have many problems, the absence of cryogens is such a compelling advantage that there is no doubt that atomic magnetometers will be replacing SQUIDs in the future for many applications.

5.5 Fundamental sensitivity

The sensitivity of a SERF magnetometer is usually limited by technical sources of noise, such as thermal noise from magnetic shielding and fluctuations in pump/probe laser power or wavelength. In the absence of such experimental imperfections, the sensitivity of the magnetometer is ultimately limited by quantum noise associated with the finite number of alkali atoms and probe beam photons used for measurement. A full derivation of the fundamental sensitivity of a SERF magnetometer is presented in Ref. [27], based on a similar derivation of the fundamental sensitivity of a high-density radiofrequency magnetometer [28]; here we only present the results, assuming a cell filled with buffer gas and the use of a separate probe beam orthogonal to the pumping axis. The total relaxation rate in the absence of pumping can be written as $R_{\mathrm{rel}} = R_{\mathrm{pr}} + R_{\mathrm{SD}}$, where R_{pr} is the rate due to absorption of probe beam photons, and R_{SD} is the rate due to spin-destruction collisions with the wall, the buffer and quenching gases, and other alkali atoms. The spin-projection noise δB_{spn} of the alkali vapor is

$$\delta B_{\mathrm{spn}} = \frac{1}{R_{\mathrm{OP}}\gamma^e}\sqrt{\frac{2\left(R_{\mathrm{OP}} + R_{\mathrm{pr}} + R_{\mathrm{SD}}\right)^3}{nVt}}, \tag{5.28}$$

where t is the measurement time, n is the density of the alkali vapor, and V is the active measurement volume defined by the intersection of the pump and probe beams. The photon shot noise δB_{psn} of the probe beam is given by

$$\delta B_{\mathrm{psn}} = \frac{\left(R_{\mathrm{OP}} + R_{\mathrm{pr}} + R_{\mathrm{SD}}\right)^2}{R_{\mathrm{OP}}\gamma^e}\sqrt{\frac{1}{nVtR_{\mathrm{pr}}OD\eta}}, \tag{5.29}$$

where η is the quantum efficiency of the photodiodes used for detection of the probe beam, and OD is the optical depth of the probe beam if it were tuned on resonance. The probe beam is here assumed to be tuned near the D1 resonance; δB_{psn} is larger by a factor of $\sqrt{2}$ for probing on the D2 resonance.

Note that the spin-projection noise is minimized for $R_{\text{OP}} = 2(R_{\text{pr}} + R_{\text{SD}})$, corresponding to a polarization of 67%, while the photon shot noise is minimized for $R_{\text{OP}} = R_{\text{pr}} + R_{\text{SD}}$, corresponding to 50% polarization. The total quantum noise δB of the magnetometer is

$$\begin{aligned}\delta B &= \sqrt{\delta B_{\text{spn}}^2 + \delta B_{\text{psn}}^2} \\ &= \frac{1}{R_{\text{OP}}\gamma^e\sqrt{nVt}}\sqrt{2\left(R_{\text{OP}} + R_{\text{pr}} + R_{\text{SD}}\right)^3 + \frac{\left(R_{\text{OP}} + R_{\text{pr}} + R_{\text{SD}}\right)^4}{R_{\text{pr}}OD\eta}}.\end{aligned} \tag{5.30}$$

As an example, the fundamental sensitivity limit is 12 aT/$\sqrt{\text{Hz}}$ in volume $V = 1$ cm^3 for a SERF magnetometer operating with $n = 1\times10^{14}$ cm^{-3}, $\frac{1}{2}R_{\text{OP}} = R_{\text{pr}} = R_{\text{SD}} = 2\pi \times$ 1 Hz, $OD = 20$, and $\eta = 0.8$. Currently, the best sensitivity demonstrated with a SERF magnetometer is 160 aT/$\sqrt{\text{Hz}}$ using a measurement volume of 0.45 cm^3 [29], so there still remains significant potential for further improvement.

Acknowledgments

The authors sincerely thank M. V. Romalis for his important early contributions to the chapter. The work of Igor Savukov was sponsored by NIH Grant 5 R01 EB009355. Scott Seltzer thanks Alexander Pines for his generous support.

References

[1] I. M. Savukov, V. S. Zotev, P. L. Volegov, M. A. Espy, A. N. Matlashov, J. J. Gomez, and R. H. Kraus, Jr., J. Magn. Reson. **199**, 188 (2009).
[2] J. C. Allred, R. N. Lyman, T. W. Kornack, and M. V. Romalis, Phys. Rev. Lett. **89**, 130801 (2002).
[3] W. Happer and H. Tang, Phys. Rev. Lett. **31**, 273 (1973).
[4] W. Happer and A. C. Tam, Phys. Rev. A **16**, 1877 (1977).
[5] F. Grossetête, J. Phys. (Paris) **25**, 383 (1965); **29**, 456 (1968).
[6] L. W. Anderson, F. M. Pipkin, and J. C. Baird, Phys. Rev. **120**, 1279 (1960).
[7] L. W. Anderson and A. T. Ramsey, Phys. Rev. **132**, 712 (1963).
[8] S. Appelt, A. B.-A. Baranga, A. R. Young, and W. Happer, Phys. Rev. A **59**, 2078 (1999).
[9] S. Appelt *et al.*, Phys. Rev. A **58**, 1412 (1998).
[10] I. M. Savukov and M. V. Romalis, Phys. Rev. A **71**, 023405 (2005).
[11] A. Abragam, *Principles of Nuclear Magnetism* (Oxford University Press, Oxford, 1961).
[12] D. Budker, D. F. Kimball, and D. P. DeMille, *Atomic Physics: An Exploration through Problems and Solutions, 2nd Edition* (Oxford University Press, Oxford, 2008).
[13] H. N. de Freitas, M. Oria, and M. Chevrollier, Appl. Phys. B, **75**, 703 (2002).
[14] S. J. Seltzer and M. V. Romalis, Appl. Phys. Lett. **85**, 4804 (2004).
[15] S. J. Seltzer, Ph.D. thesis, Princeton University (2008).

[16] V. Shah and M. V. Romalis, Phys. Rev. A **80**, 013416 (2009).
[17] J. Dupont-Roc, S. Haroche, and C. Cohen-Tannoudji, Phys. Lett. A **28**, 638 (1969).
[18] V. Shah, S. Knappe, P. D. D. Schwindt, and J. Kitching, Nature Photonics **1**, 649 (2007).
[19] I. K. Kominis, T. W. Kornack, J. C. Allred, and M. V. Romalis, Nature **422**, 596 (2003).
[20] H. Xia, A. B.-A. Baranga, D. Hoffman, and M. V. Romalis, Appl. Phys. Lett. **89**, 211104 (2006).
[21] S. J. Seltzer and M. V. Romalis, J. Appl. Phys. **106**, 114905 (2009).
[22] M. A. Rosenberry, J. P. Reyes, D. Tupa, and T. J. Gay, Phys. Rev. A **75**, 023401 (2007).
[23] M. V. Romalis, Phys. Rev. Lett. **105**, 243001 (2010).
[24] M. V. Balabas, T. Karaulanov, M. P. Ledbetter, and D. Budker, Phys. Rev. Lett. **105**, 070801 (2010).
[25] J. Clarke and A. I. Braginski, *The SQUID Handbook* (Wiley-VCH, New York, 2004).
[26] S. J. Smullin, I. M. Savukov, G. Vasilakis, R. K. Ghosh, and M. V. Romalis, Phys. Rev. A **80**, 033420 (2009).
[27] M. P. Ledbetter, I. M. Savukov, V. M. Acosta, D. Budker, and M. V. Romalis, Phys. Rev. A **77**, 033408 (2008).
[28] I. M. Savukov, S. J. Seltzer, M. V. Romalis, and K. L. Sauer, Phys. Rev. Lett. **95**, 063004 (2005).
[29] H. B. Dang, A. C. Maloof, and M. V. Romalis, Appl. Phys. Lett. **97**, 151110 (2010).

6

Optical magnetometry with modulated light

D. F. Jackson Kimball, S. Pustelny, V. V. Yashchuk, and D. Budker

6.1 Introduction

Soon after the development of optical magnetometers based on the radio-optical double resonance method (see Chapter 4), it was realized by Bell and Bloom [1] that an alternative method for optical magnetometry was to modulate the light used for optical pumping at a frequency resonant with the Larmor precession of atomic spins. In a Bell–Bloom optical magnetometer, circularly polarized light resonant with an atomic transition propagates through an atomic vapor along a direction transverse to a magnetic field **B**. Atomic spins immersed in **B** precess at the Larmor frequency Ω_L, and when the light intensity is modulated at $\Omega_m = \Omega_L$, a resonance in the transmitted light intensity is observed.[1] The essential ideas of the Bell–Bloom optical magnetometer are reviewed in Chapter 1 (Section 1.2), and can be summarized in terms of what Bell and Bloom termed *optically driven spin precession*: in analogy with a driven harmonic oscillator, in a magnetic field **B** atomic spins precess at a natural frequency equal to Ω_L and the light acts as a driving force oscillating at the modulation frequency Ω_m. From another point of view, the Bell–Bloom optical magnetometer can be described in terms of *synchronous optical pumping*: when $\Omega_m = \Omega_L$, there is a "stroboscopic" resonance in which atoms are optically pumped into a spin state stationary in the frame rotating with Ω_L. Depending on the details of the atomic structure, the spin state stationary in the rotating frame can be either a *dark state* that does not interact with the modulated light or a *bright state* for which the strength of the light–atom interaction is increased. A dark state resonance is signified by a peak in the transmitted light intensity, while a bright state resonance is signified by a maximum in light absorption. Measurement of the resonant modulation frequency gives an accurate determination of Ω_L and therefore directly determines the magnitude of **B**.

[1] At high light powers, there can appear resonances at higher harmonics of Ω_L associated with high-order polarization moments (Section 1.3). For nonsinusoidal light modulation, resonances appear at subharmonics of Ω_L.

Optical Magnetometry, ed. Budker, D. and Jackson Kimball, D. F. Published by Cambridge University Press.

In parallel with the extensive research on optical pumping that gave rise to the Bell–Bloom optical magnetometer, numerous experimental and theoretical studies of *nonlinear magneto-optical rotation* (NMOR, also known as nonlinear Faraday rotation, discovered in 1974 [2,3] and reviewed in Ref. [4]) were being carried out. The effect occurs when linearly polarized light propagates through an atomic medium along the direction of an applied magnetic field. When the light is near-resonant with an atomic transition, and of sufficient power to perturb the equilibrium population of atomic states, light-power-dependent rotation of the plane of light polarization is observed. Research on optical pumping and studies of NMOR often overlapped, but for optical magnetometry perhaps the most important intersection of these two lines of inquiry was the discovery of narrow ($\sim$1 Hz) NMOR resonances in antirelaxation-coated cells [5,6]. These narrow NMOR resonances are related to optical pumping of long-lived ground-state atomic polarization, and are analogous to the narrow lines observed in M_x and M_z magnetometers which also employ cells with antirelaxation coating (see Chapter 4 and Chapter 11). A detailed study [7] of the shot-noise-projected magnetometric sensitivity of NMOR in a paraffin-coated rubidium (Rb) cell demonstrated that, with proper choice of laser light power and detuning, it is possible, in principle, to achieve sensitivities close to the fundamental limit described by Eq. (1.1), establishing NMOR as a highly efficient method of probing precession of atomic polarization.

Shortly after the discovery of narrow NMOR resonances in antirelaxation-coated cells, it was realized that there were considerable practical advantages for magnetometry if modulated light was used in the experimental scheme. The first implementation of this idea [8] employed a single, frequency-modulated light beam for optical pumping and detection of NMOR resonances (FM NMOR). This initial work was, in fact, inspired by frequency-modulation techniques employed in measurements of parity-violating optical rotation and linear Faraday rotation [9, 10], yet it is clear that FM NMOR magnetometry bears a close resemblance to the work of Bell and Bloom [1].

Another approach to optical magnetometry is the measurement of NMOR resonances with amplitude-modulated light (AM NMOR or AMOR), which combines the high efficiency of the synchronous optical pumping technique of Bell and Bloom with sensitive probing of spin precession dynamics using optical rotation [11]. Several experiments (see, for example, Ref. [12] and references therein) utilize a two-beam pump-probe arrangement in which the pump beam is modulated synchronously with the Larmor precession and an unmodulated probe beam monitors the spin-precession frequency via optical rotation. There are notable advantages to this scheme in terms of sensitivity optimization, since experimental parameters governing optical pumping and probing can be independently adjusted.

Anything that changes the light–atom interaction probability can be used as a source of modulation to generate a magnetic resonance signal, and so one can construct optical magnetometers based on amplitude, frequency, or polarization modulation of the light, modulation of an applied magnetic field [13, 14], or even modulation of the atomic spin relaxation rate [15–17]. In this chapter we focus on optical magnetometry based on light modulation, and explore the properties of the magnetic-field resonances, effects at high magnetic fields and high light powers, spectral dependence of signals, and practical magnetometric measurement techniques employing modulated light.

6.2 Typical experimental arrangements

Examples of one- and two-beam optical magnetometer setups based on NMOR with modulated light are shown in Fig. 6.1. The single-beam arrangement depicted in Fig. 6.1(a) uses the Faraday geometry [8, 18–22][2] in which a linearly polarized laser beam propagates along the direction ($\hat{\mathbf{z}}$) of a magnetic field $\mathbf{B} = B_z\hat{\mathbf{z}}$. Figure 6.1(b) shows a two-beam arrangement where the unmodulated probe is in the Faraday geometry and a modulated pump beam propagates along $\hat{\mathbf{x}}$, with initial linear polarization of both beams along y [26]. The resonant (or near-resonant) interaction of the pump beam with a sample of atoms generates ground-state atomic polarization via optical pumping, producing an optically anisotropic medium. The optically pumped ground-state atomic polarization evolves in the presence of the magnetic field **B**, changing the optical anisotropy. Measurement of optical rotation, which is sensitive to optical anisotropy in the form of *linear dichroism* (corresponding to a difference between the imaginary parts of the refractive index for different linear polarizations) and *circular birefringence* (corresponding to a difference between the real parts of the refractive index for different circular polarizations), is performed by placing the atomic sample between a polarizer and an analyzer. In the setup shown, a balanced polarimeter is used where the polarizer and analyzer are oriented at $\sim$45° with respect to each other. In the experiments considered in this chapter, the atomic sample is a gas of alkali atoms contained in an antirelaxation-coated cell (Chapter 11) operating at or near room temperature, typically between $\sim$20°C and 60°C; some of the most common antirelaxation coatings melt at higher temperatures, and alkali vapor pressures are below optimum at lower temperatures (although multipass optical cells could permit use of lower temperatures and vapor pressures while retaining high magnetometric sensitivity [27]). The difference signal between the current outputs of the two photodiodes after the analyzer is measured at a given harmonic of Ω_m with a lock-in amplifier. The difference signal, upon normalization by twice the time-averaged sum of the photodiode outputs, represents a measurement of the amplitude of the optical rotation angle (for small rotation angles).

A tunable diode laser is typically used as the light source. For FM NMOR, the laser frequency can be modulated via the laser diode current or through grating feedback. Optimum sensitivity is generally achieved when the modulation amplitude $\Delta\omega$ is on the order of the transition linewidth. For room-temperature alkali vapor cells, this is the Doppler width of a few hundred MHz. This choice of $\Delta\omega$ maximizes the modulation amplitude of the light–atom interaction probability which yields the best efficiency for synchronous optical pumping. For AM NMOR, pure amplitude modulation (without accompanying frequency modulation) can be achieved by employing an external opto-electronic device such as an

[2] This experimental geometry is the same as that employed over 160 years ago by Michael Faraday in the original discovery of magneto-optical rotation [23–25].

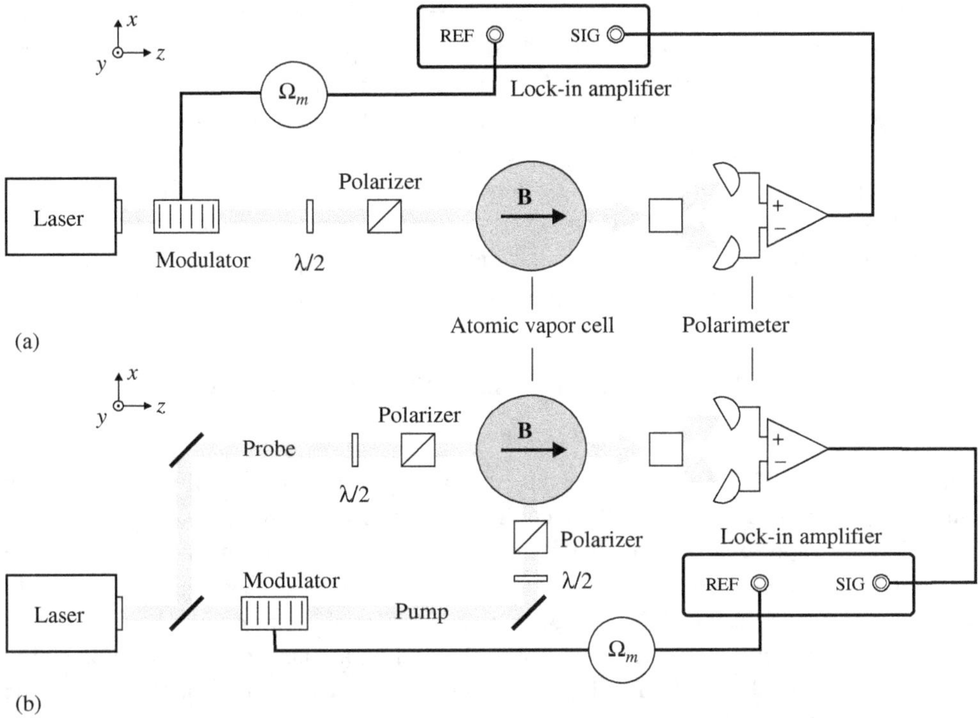

Figure 6.1 Schematic diagrams of (a) one-beam and (b) two-beam experimental arrangements for measurement of nonlinear magneto-optical rotation (NMOR) with modulated light. The half-wave plates ($\lambda/2$) are for control of overall light intensity. Many variations on these basic setups can be employed: for example, in (b) two separate laser systems can be employed for the pump and probe beams.

acousto-optic modulator (AOM). The range of possible Ω_m values achievable with these methods extends beyond several MHz, sufficient for geophysical applications as well as low-field laboratory investigations. Arbitrary waveform modulation can be easily achieved with AOMs, offering another parameter (beyond laser power, detuning, and $\Delta\omega$) for optimization of magnetometric sensitivity [11, 28–30]. For an isolated, Doppler-broadened transition, the highest efficiency of synchronous optical pumping is achieved for relatively short light pulses (duty cycle on the order of 20%) and light tuned to the center of the transition [29,31].[3] On the other hand, the highest efficiency for optical probing is achieved

[3] Note that when multiple optical transitions overlap within their Doppler widths (e.g., the $F = 3 \rightarrow F'$ transitions of the ^{85}Rb D2 line), the efficiency of synchronous optical pumping for generating a given polarization moment may not be the highest for light tuned to the center of the Doppler-broadened resonance. In such cases, excitations of different transitions pump the same ground-state polarization moments but, potentially, with opposite signs. This can cause the contributions to ground-state polarization from optical pumping along different transitions to cancel for certain laser detunings. Magneto-optical effects with partially resolved hyperfine structure are discussed in great detail in Refs. [32, 33].

for CW light (100% duty cycle). As noted in Section 6.1, a significant advantage of the two-beam arrangement shown in Fig. 6.1(b) is that the parameters controlling the pump and probe light can be separately optimized, and the pumping and probing stages can even be temporally separated by employing a shutter system. This allows flexibility in both optimization of magnetometric sensitivity and suppression of systematic effects in precision measurements. For example, pumping can be performed near-resonance and at high light power (where there can be significant light shifts and nonlinear evolution of atomic polarization) to maximize atomic polarization, while probing can be performed far-detuned from the resonance and/or at low light power to minimize light shifts and broadening.

6.3 Resonances in the magnetic field dependence

6.3.1 Frequency modulation

Figure 6.2 shows FM NMOR data obtained in a one-beam geometry [Fig. 6.1(a)] using an antirelaxation-coated vapor cell containing a natural isotopic mixture of rubidium (Rb) [34]. The Rb vapor cell is contained within a five-layer magnetic shield and surrounded by a system of coils used for precise control of the magnetic field applied to the cell. For the data shown in Fig. 6.2, the light power $P \approx 20\ \mu\text{W}$ and the light beam diameter ≈ 2 mm. Frequency modulation is produced by adding a sinusoidal current modulation at $\Omega_m = 2\pi \times 500$ Hz to the DC laser diode current, which produced an FM amplitude $\Delta\omega = 2\pi \times 65$ MHz. The cell temperature was $T \approx 20°\text{C}$, for which the Rb vapor density was measured to be $\sim 4 \times 10^9$ atoms/cm^3 by fitting a low light power ($\sim 1\ \mu\text{W}$) absorption spectrum to a Voigt profile.

The dependence of the optical-rotation amplitude demodulated at the first harmonic of Ω_m as a function of the longitudinal magnetic field B_z (along the direction of light propagation $\mathbf{k}$) is shown for two different laser light detunings: for the upper plot the light is tuned to an ^{85}Rb transition and for the lower plot the light is tuned to an ^{87}Rb transition. In general, for NMOR with modulated light, resonances in the magnetic field dependence of the optical rotation amplitude measured at the first harmonic of Ω_m are observed when

$$n\Omega_m = K\Omega_L\ , \tag{6.1}$$

where K and n are integers and the most prominent signals at low light power occur for $n = 0, 1$ and $K = 2$. In Fig. 6.2 the high-field resonances corresponding to $\pm B_z$ are labeled the $n = 1\pm$ resonances.

In many cases of interest, the parameter K is equal to the rank κ of the atomic polarization moment (PM) detected (see Section 1.3). The spatial distribution of angular momentum for the q component of a PM has $|q|$-fold symmetry about the quantization axis (see Fig. 6.3), and the maximum possible q is equal to the rank κ. In the case where $\mathbf{B}$ and $\mathbf{k}$ are parallel and define the quantization axis, a resonance related to a PM with rank κ and component q occurs when the optical pumping interaction is modulated at the frequency $|q|\Omega_L$. Typically the strongest contribution to the signal comes from the $|q| = \kappa$ components (which corresponds

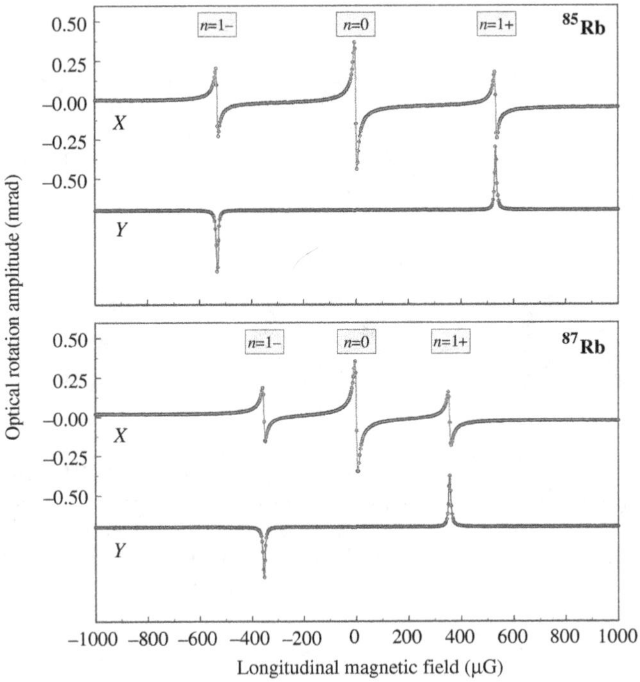

Figure 6.2 (Reprinted with permission from Ref. [34]. Copyright 2009, American Institute of Physics.) Optical rotation amplitude for a one-beam FM-NMOR magnetometer as a function of B_z, demodulated at the first harmonic of Ω_m. Upper plot: in-phase (X, data offset above) and out-of-phase (Y, or quadrature, data offset below) FM NMOR signal for laser tuned near the ^{85}Rb D2 $F = 3 \to F'$ transition. Lower plot: FM NMOR signal for laser tuned near the ^{87}Rb D2 $F = 2 \to F'$ transition.

to pure transverse atomic polarization). Atomic polarization precession at Ω_L causes the atomic medium to acquire a time-dependent optical anisotropy that is periodic at a frequency equal to $|q|\Omega_L$ [18]. Linearly polarized light pumps PMs with even κ, and the lowest-order PM has $\kappa = 2$ (alignment), which, as Fig. 6.3 shows, possesses the axial symmetry of the light polarization. The aligned atomic vapor has a preferred axis but no preferred direction. The medium thus possesses different indices of refraction for light polarized parallel and orthogonal to the alignment axis (linear dichroism/birefringence). When light interacts with the polarized atoms, it is preferentially transmitted either along or perpendicular to the alignment axis depending on whether for the resonant optical transition the ground-state polarization corresponds to a dark or bright state, respectively – see Ref. [4]. Because the atomic alignment precesses due to **B**, the alignment axis is generally not along the initial light polarization and the light–atom interaction can generate optical rotation.

Smaller-amplitude resonances in the FM NMOR signal demodulated at the first harmonic of Ω_m with $n > 1$ can be observed due to nonsinusoidal modulation of the light–atom interaction probability [8, 19] and misalignment between **B** and the wave vector **k** of the light beam [22]. Larger amplitude resonances with $n > 1$ can be observed when the FM

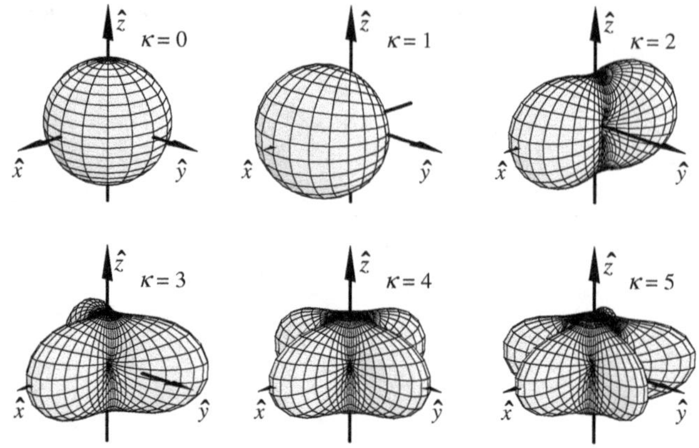

Figure 6.3 (From Ref. [35]. Copyright 2005, The Optical Society.) Angular momentum spatial distribution for states composed only of population (ρ_0^0) and the maximum possible values of the components $\rho_{\pm\kappa}^{\kappa}$ for a particular κ. $\kappa = 0$: monopole moment (isotropic state with population only); $\kappa = 1$: dipole moment (oriented state); $\kappa = 2$: quadrupole moment (aligned state); $\kappa = 3$: octupole moment; $\kappa = 4$: hexadecapole moment; $\kappa = 5$: triakontadipole moment. See Chapter 1, Section 1.3 for more information on PMs.

NMOR signal is demodulated at higher harmonics of Ω_m [8, 19]. (Because of the different Landé g-factors for ^{87}Rb and ^{85}Rb, the FM NMOR resonances are observed at different magnetic fields for the two isotopes).

The $n = 0$ resonance occurs under the conditions where $\Omega_L \ll \Omega_m$ and optical rotation achieves a maximum amplitude when $\Omega_L \approx \Gamma_{\rm rel}$. At $B_z = 0$, there is no Larmor precession and therefore no optical rotation. As B_z departs from zero the atomic alignment axis rotates, but if $\Omega_L \lesssim \Gamma_{\rm rel}$, optically pumped alignment relaxes before a full period of rotation can be completed. Therefore in this regime, the average atomic alignment in the cell has its axis tilted away from the axis of the incident light polarization by some angle $\phi < \pi/4$. The tilted alignment causes optical rotation, and the optical rotation is modulated at Ω_m because of the frequency modulation of the light. For the $n = 0$ resonance, it should be noted that frequency modulation for the pump interaction, which creates the initial atomic alignment, is not essential for the effect, whereas frequency modulation for the probe interaction, which causes optical rotation, *is* essential to generate the time-dependent signal at the first harmonic of Ω_m. As a consequence, the $n = 0$ resonance is absent in the two-beam arrangement with unmodulated probe [Fig. 6.1(b)]. As B_z is increased so that $\Omega_L \gtrsim \Gamma_{\rm rel}$, the optical rotation amplitude decreases since the alignment for atoms optically pumped at different times dephases, and eventually the magneto-optical rotation averages to zero. The shape of the resonance is well described by a dispersive Lorentzian profile, as discussed in detail in Ref. [19]. The time-dependent optical rotation for the $n = 0$ resonance is in-phase with the modulation of the probe interaction, and defines the phase of what is denoted the X signal in Fig. 6.2. The $n = 0$ resonance can be used as the basis for a zero-field optical magnetometer.

It is of interest to note that PMs corresponding to any value of κ can contribute to the $n=0$ signal (however, the lineshapes of the $n=0$ signals from different k can be different).

The $n=1$ resonance occurs under the conditions where $\Omega_L \gg \Gamma_{\text{rel}}$. In this case, in contrast to the $n=0$ case, the FM NMOR resonance can also be observed in the Ω_m-dependence of the demodulated optical rotation signal (see Fig. 1.4). The physical mechanism giving rise to the $n=1$ resonance can be generally understood in the same way as the Bell–Bloom resonance: atomic alignment precesses at Ω_L and both the optical properties and quantum state of the atomic vapor are modulated at twice that frequency on account of the twofold symmetry of the polarization. Therefore, when the optical-pumping rate is modulated at $2\Omega_L$ there is resonant enhancement of the precessing atomic alignment and the optical rotation signal. As is the case with driven, damped harmonic oscillators, the phase of the time-dependent optical rotation acquires a dependence on the detuning of the drive frequency from the natural oscillation frequency ($\Omega_m - 2\Omega_L$). Thus signals are observed both in-phase (X signal) and out-of-phase (Y signal, quadrature) with the modulation of the light–atom interaction probability. While the X signal nominally [36] crosses zero at $2\Omega_L = \Omega_m$, the Y signal is maximum when $2\Omega_L = \Omega_m$. This is because when the optical pumping rate is synchronized with the atomic alignment precession rate, the axis of the atomic alignment is parallel with the light polarization at the periodic maxima in the modulated light–atom interaction probability – when the atomic alignment axis is parallel with the light polarization, no optical rotation is produced. The maximum optical rotation occurs when the atomic alignment axis is rotated by an angle $\phi = \pi/4$ with respect to the light polarization. On resonance ($2\Omega_L = \Omega_m$) this causes optical rotation out of phase with the modulation of the light–atom interaction probability.

6.3.2 Amplitude modulation

Figure 6.4 presents AM NMOR (AMOR) signals demodulated at Ω_m and raw polarimeter signals (insets) measured in the (a) one-beam and (b) two-beam arrangements. As noted in Section 6.2, optical pumping is optimized for a 20% duty cycle and optical probing is optimized for CW (100%) duty cycle. For a one-beam arrangement, a reasonable compromise between these competing requirements is sinusoidal modulation of the light intensity. In this case, however, the polarimeter signal is a convolution of the light intensity modulation and alignment rotation,

$$\varphi(t) = A\left(\frac{I_0}{2}[1+\cos(\Omega_m t)]\right)\sin(2\Omega_L t)\,, \tag{6.2}$$

where $A(I)$ is the light-intensity-dependent amplitude of the AM NMOR signal and I_0 is the maximum instantaneous value of the intensity (see insets to Fig. 6.4).

The most pronounced difference between the data acquired for the two arrangements is the appearance of the $n=0$ resonance for the one-beam case and the absence of the $n=0$ resonance for the two-beam case. As noted earlier in the discussion of FM NMOR

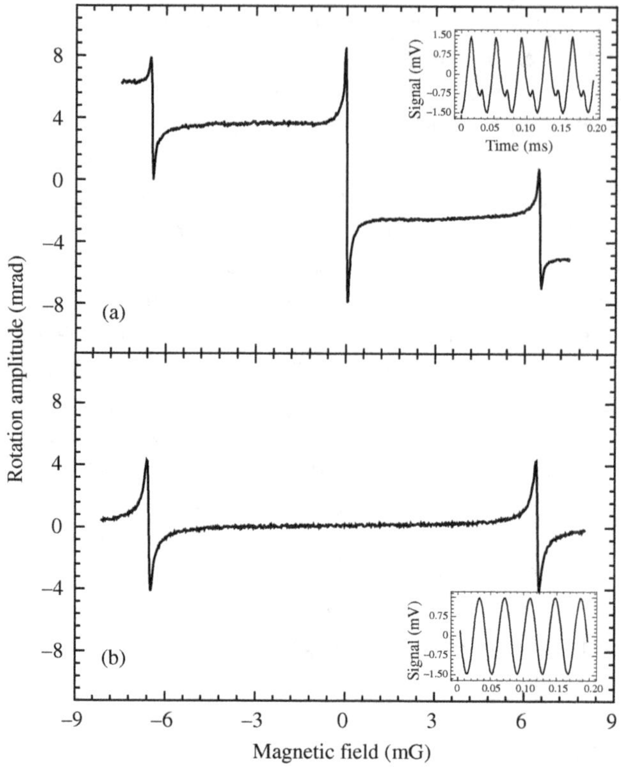

Figure 6.4 Typical AM NMOR (AMOR) signals measured at the first harmonic of the modulation frequency in (a) one-beam and (b) two-beam arrangements. The insets show the raw polarimeter signals (without demodulation) for the two cases. The signals were recorded for sinusoidally modulated pump light with 100% modulation depth, $\Omega_m \approx 2\pi \times 8.5$ kHz, and net intensity of 5 μW/mm^2. The probe-light intensity in the two-beam arrangement (b) was equal to 2 μW/mm^2.

resonances, observation of the $n = 0$ resonance requires modulation of the probe light (since the ensemble-averaged PMs are static for the $\Omega_L \ll \Omega_m$ case).

Another notable difference is that for the two-beam case the optical rotation far from the resonances, where $|K\Omega_L - n\Omega_m| \gg \Gamma_{\rm rel}$, is near zero, whereas for the one-beam case there is a significant offset from zero optical rotation on either side of the resonances. This nonzero optical rotation far from the $n = 0, 1$ resonances in the one-beam case [29] arises from the transit effect: atoms are pumped, precess in the field, and are probed during a single transit through the laser beam. The background rotation seen in Fig. 6.4(a) is from the $n = 0$ transit effect resonance, and the width of this feature, $\sim$10 mG, is determined by the transit rate $\Gamma_t \sim 2\pi \times 10$ kHz. The transit effect is more prominent in the data plotted in Fig. 6.4(a) than for the data plotted in Figs. 6.2 and 6.6 because of the larger range of B_z scanned. Due to nonuniform light-intensity distribution within the light beam, the shape of the transit effect resonance deviates significantly from a Lorentzian shape [29, 37, 38]. In particular, the wings of the transit effect resonance extend to larger magnetic fields, causing offsets of the $n = 1$ resonances. In a single-beam AM NMOR magnetometer, the background rotation signals from the transit effect wings need to be accounted for and compensated to avoid systematic errors. The absence of the $n = 0$ resonances for the two-beam arrangement, along with the capacity for separate optimization of pump and probe parameters, offer advantages for optical magnetometry.

6.3.3 Polarization modulation

The polarization of the light can also be modulated to generate a parametric magnetic resonance [35, 39]. Figure 6.5 shows data for the case of light-polarization rotation generated by mechanical rotation of a $\lambda/2$ plate. Polarization rotation is equivalent to out-of-phase amplitude modulation of the light polarization components, providing a conceptual link between this effect and AM NMOR. For rotating linear polarization at Ω_L, the light polarization is always parallel to the alignment of the atomic ensemble. In this case, optical pumping contributes continuously and coherently to all of the precessing, even-rank PMs (in essence, the $n = 0$ resonance has been shifted to higher frequencies). A bright or dark resonance (a decrease or increase in transmitted light intensity), normally centered at zero field when light polarization is fixed [Fig. 6.5(b)], is shifted by the polarization-rotation frequency in Fig. 6.5(c).

Of particular interest is recent work [40] demonstrating how optical magnetometry using another type of polarization modulation can eliminate the problem of *dead zones* – certain orientations of the magnetic field where the magnetometric sensitivity is lost. Dead zones are an inherent feature of the vector/tensor interactions involved in optical pumping and probing of atomic polarization precession: for certain magnetic field orientations the interaction terms drop to zero. If the light polarization is modulated at Ω_L with a linear polarizer followed by an electro-optic modulator (EOM) with its principal axes at 45° relative to the polarizer, the light polarization oscillates between linear and left/right circular polarization states. A single-beam arrangement can be used where the transmitted light intensity is detected. In such a scheme, both $\kappa = 1$ and $\kappa = 2$ PMs (orientation and alignment) can be simultaneously pumped and probed. Because of the different vector nature of the light interaction with the $\kappa = 1$ and $\kappa = 2$ PMs, the transmission of the laser beam turns out to be sensitive to the magnetic field in any orientation.

6.4 Effects at high light powers

At higher light powers, three notable effects become important: (1) AC Stark shifts of the ground-state Zeeman sublevels due to the optical electric field $\mathcal{E}$ in combination with the Zeeman shifts due to $\mathbf{B}$ lead to more complicated time evolution of PMs than simple precession [33, 34, 41], (2) the AC Stark effect caused by $\mathcal{E}$ can lead to broadening, shifting, and splitting of FM NMOR resonances [20, 42]; and (3) higher-order ($\kappa > 2$) PMs can be

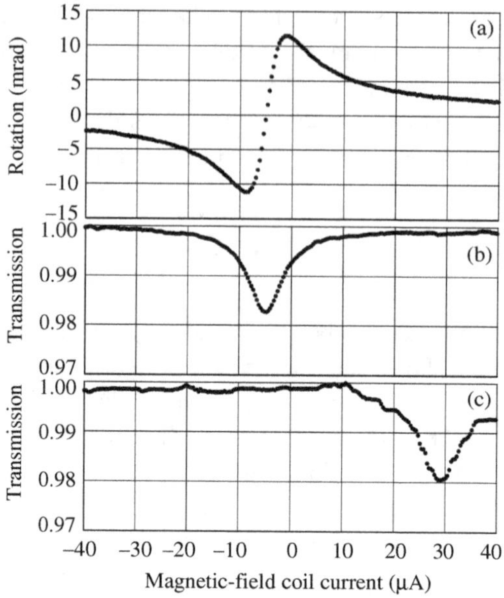

Figure 6.5 (From Ref. [35]. Copyright 2005, The Optical Society.) Magnetic-field dependence of the Faraday rotation angle (a) and transmission (b) recorded with stationary linear polarization of light in the one-beam geometry. Trace (c) shows light transmission as a function of magnetic field when the linear polarization is rotated at $\Omega_m = 2\pi \times 14$ Hz. Transmission was observed with lock-in detection at the rotation frequency while the longitudinal magnetic field strength was swept. Coil current of 1 μA corresponds to a magnetic field of approximately 1 μG.

generated through multiple light–atom interactions within the ground-state coherence time, leading to new FM/AM NMOR resonances [18,26].

The most prominent example of the first effect is alignment-to-orientation conversion (AOC, see Refs. [34,41]), where the combined action of **B** and $\mathcal{E}$ act to convert the optically pumped $\kappa = 2$ PM (alignment) into a $\kappa = 1$ PM (orientation). This effect is of special importance for optical magnetometry since NMOR-based magnetometers typically achieve optimum sensitivity at light powers where AOC is a dominant mechanism causing optical rotation [7,34]. For both $n = 0$ and $n = 1$ resonances, AOC creates atomic orientation parallel to **B**, and thus the atomic orientation does not precess in the magnetic field. In a one-beam arrangement, time-dependent optical rotation can be detected at the first harmonic of Ω_m as a result of modulation of the probe interaction. Two-beam arrangements where the probe beam is unmodulated are thus generally insensitive to NMOR due to AOC and as a result can have lower overall magnetometric sensitivities.

The third high-light-power effect is illustrated in the data plotted in Fig. 6.6, which shows the appearance of FM NMOR resonances corresponding to the $\kappa = 4$ multipole moment (hexadecapole moment) at higher light powers. High multipole moments

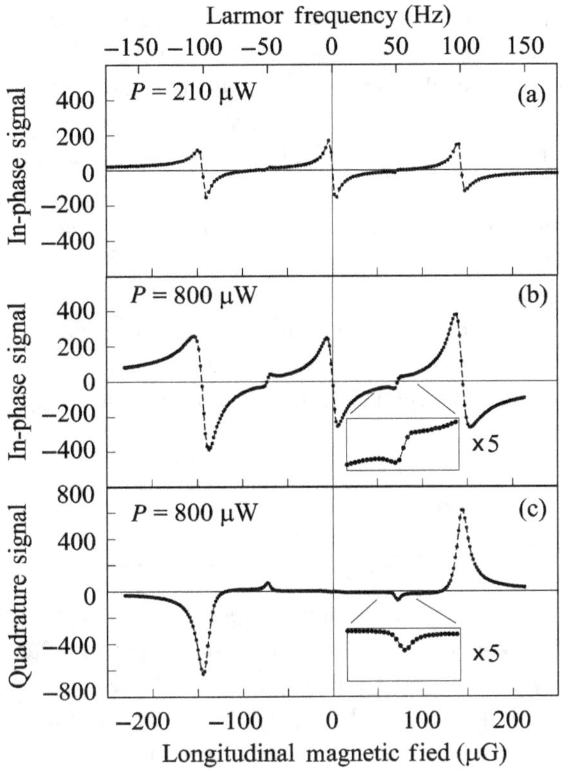

Figure 6.6 (From Ref. [18].) Optical rotation amplitude for a one-beam FM NMOR magnetometer as a function of B_z, demodulated at the first harmonic of $\Omega_m = 2\pi \times 200$ Hz. The laser is tuned near the ^{87}Rb $F = 2 \rightarrow F' = 1$ transition. Modulation amplitude $\Delta\omega = 2\pi \times 40$ MHz; light-powers P are shown in upper left corner of plots; and other experimental parameters are similar to those for Fig. 6.2. $K = 2$ resonances appear at $B = \pm 143.0$ μG, and $K = 4$ resonances appear at ± 71.5 μG. The insets show zooms on the $K = 4$ hexadecapole resonances.

are of interest for optical magnetometry in the geophysical field range where there can be broadening and splitting of FM NMOR resonances due to the nonlinear Zeeman effect (see Chapter 1 and Section 6.5 below), since the highest multipole moment for maximum total atomic angular momentum $F = I + J$, $\kappa = 2F = 2(I + J)$, is unaffected by nonlinear Zeeman splitting [43].[4] Note that due to angular-momentum conservation, creation of a $\kappa = 4$ $q = \pm 4$ PM requires two pump photons and detection requires two probe photons, whereas for the $\kappa = 2$ $q = \pm 2$ PM a single pump interaction and single probe interaction are sufficient. Thus the light-power dependence, of FM/AM NMOR signal amplitudes are different for PMs with different κ. Furthermore,

[4] The highest multipole moment $\kappa = 2F$ is related to coherence between the $M = \pm F$ Zeeman sublevels, for which the energy separation is linear with the magnetic field magnitude in the case of the $F = I + J$ level (see Fig. 1.1).

when pumping at $\Omega_m = 4\Omega_L$, both photons must interact with the atoms within the same "pulse" in order to create the $\kappa = 4$ PM, since separate single-photon processes occurring during successive pumping cycles create $\kappa = 2$ PMs that are orthogonal to each other. As B increases, and, consequently, Ω_L and the Ω_m necessary to satisfy the resonance condition (6.1) increase, the duration of a single optical pumping pulse reduces as $1/\Omega_m$. For a fixed light power, this means that when Ω_m exceeds the transit rate Γ_t of atoms through the light beam, Ω_m becomes the limiting factor in determining the optical pumping saturation parameter. Past a certain point as Ω_L increases, the optical pumping efficiency for a single cycle decreases. This effect can drastically reduce the optical pumping efficiency for high-κ PMs due to the requirement of multiple pump interactions during a single cycle [43]. This effect can be ameliorated by working, for example, on the $n = 2$, $K = 4$ resonance where pump interactions in different cycles can contribute additively to produce the $\kappa = 4$ PM [43].

6.5 Nonlinear Zeeman effect

At magnetic fields in the practically important geophysical range, corresponding to the strength of the Earth's magnetic field of ~500 mG, splitting of the FM/AM/PM NMOR resonances due to the nonlinear Zeeman effect (see Section 1.1.2) is observed [20]. Figure 6.7 shows the $n = 1$ FM NMOR in-phase (X) signal in a geophysical-range field, demonstrating that the $n = 1$ resonance takes on a more complicated shape, the result of three separate dispersive curves centered at slightly different modulation frequencies, plotted separately in the lower plot of Fig. 6.7. The three resonances correspond to the three different $\Delta M = 2$ Zeeman coherences created in the ^{87}Rb $F = 2$ ground state: coherences between $M = 2$ and $M = 0$, between $M = 1$ and $M = -1$, and between $M = 0$ and $M = -2$. Under the conditions for the data shown in Fig. 6.7, the quantum-beat frequency associated with each coherence is shifted due to the nonlinear Zeeman effect, causing the FM NMOR resonance to split into three resolved components – a central resonance centered at $\approx 2\Omega_L$ and two satellites shifted by $\pm\delta$. Note that the central frequencies for the two different alkali ground-state hyperfine levels, in contrast to the low-field case, are shifted from one another due to the nuclear magnetic moment. In the geophysical field range this effect is nonnegligible, resulting in a modification of the Landé factors according to [44]:

$$
\begin{aligned}
g_{F=I+\frac{1}{2}} &= \frac{2}{2I+1} - g_I \frac{m_e}{M_p} \frac{2I}{2I+1} \\
g_{F=I-\frac{1}{2}} &= -\frac{2}{2I+1} - g_I \frac{m_e}{M_p} \frac{2(I+1)}{2I+1} ,
\end{aligned}
\tag{6.3}
$$

where g_I is the nuclear Landé factor, m_e is the electron mass, and M_p is the proton mass ($m_e/M_p \approx 5 \times 10^{-4}$, setting the scale of this effect).

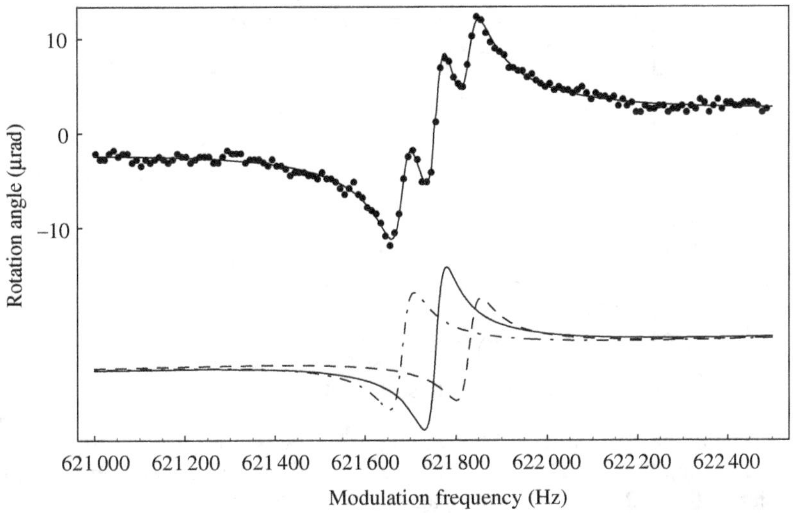

Figure 6.7 Upper curve shows data (dots) and fit (solid line) demonstrating the dependence of the in-phase (X) component of the FM NMOR signal on modulation frequency in the vicinity of the $n = 1$, $K = 2$ resonance for a longitudinal magnetic field of ~444 mG. Lower curves (dashed, solid, and dot-dashed lines) show the individual contributions from the three separate $\Delta M = 2$ resonances whose frequencies are shifted relative to one another by the nonlinear Zeeman effect (and, to a lesser extent, the AC Stark effect caused by the optical field). The laser is tuned to the low-frequency side of the $F = 2 \rightarrow F' = 1$ hyperfine component of the D1 line of ^{87}Rb.

The nonlinear Zeeman effect arises because a sufficiently strong magnetic field mixes Zeeman sublevels in different ground state hyperfine levels. Neglecting the nuclear magnetic moment, in moderate magnetic fields where $\Omega_L \ll \Delta_{\rm HF}$ (with $\Delta_{\rm HF}$ being the energy separation between the hyperfine levels), the energy $E(F,M)$ of a particular ground state Zeeman sublevel of an alkali atom with nuclear spin $I = 3/2$ is approximately given by a perturbative expansion of the Breit–Rabi formula (see, for example, Ref. [45]):

$$E(F,M) \approx E_F + (-1)^F \left[M\,\Omega_L + \left(4 - M^2\right) \frac{\Omega_L^2}{\Delta_{\rm HF}} \right], \tag{6.4}$$

where E_F is the energy of the hyperfine level. Equation (6.4) shows that the resonance frequencies $\Omega_0(M,M')$ for the FM NMOR features corresponding to different $\Delta M = 2$ Zeeman coherences are given by

$$\begin{aligned} \Omega_0(2,0) &= 2\Omega_L - 4\frac{\Omega_L^2}{\Delta_{\rm HF}}, \\ \Omega_0(-1,1) &= 2\Omega_L, \\ \Omega_0(0,-2) &= 2\Omega_L + 4\frac{\Omega_L^2}{\Delta_{\rm HF}}. \end{aligned} \tag{6.5}$$

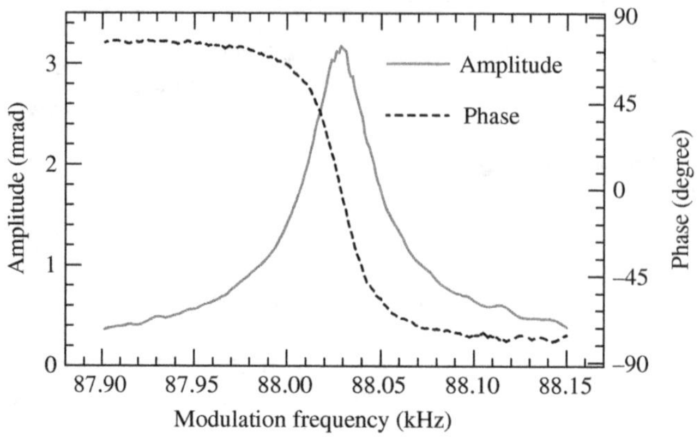

Figure 6.8 Amplitude and phase of the AM NMOR signal measured as a function of Ω_m (demodulated at the first harmonic). The signals were measured in the two-beam arrangement with pump- and probe-light intensities of 2 μW/mm^2 and a magnetic field of 63 mG.

The nonlinear Zeeman effect causes systematic effects known as *heading errors* in practical magnetometric measurements: because of the splittings and shifts of the different components of the magnetic resonances, as well as lineshape asymmetries, the magnetic field acquired from electronic or algorithmic determination of Ω_L tends to depend on the orientation of the sensor with respect to the field. Light shifts due to the AC Stark effect caused by the optical electric field produce a similar pattern of magnetic resonance shifts, and it has been demonstrated that light shifts can be used to directly compensate for the splitting of magnetic resonances due to the nonlinear Zeeman effect [42]. Also, as noted in Section 6.4, PMs of the highest possible κ with $q = \pm\kappa$ are immune to the nonlinear Zeeman effect. Furthermore, the light polarization modulation scheme described in Ref. [40] largely eliminates heading errors.

6.6 Magnetometric measurements with modulated light

Figure 6.8 shows the amplitude and phase of a $K = 2$, $n = 1$ AM NMOR signal measured as a function of Ω_m. The AM NMOR signal amplitude follows a Lorentzian curve reaching a maximum at $\Omega_m = 2\Omega_L$ where the phase, which follows arctangent dependence, reaches zero.

For practical operation at finite fields, optical magnetometers require a feedback loop to keep Ω_m locked to resonance as the magnetic field is changing. One method of magnetometer operation, known as the passive mode, is to track the point of maximum amplitude and zero phase shift using lock-in detection in combination with an external feedback loop implemented through electronics or a computer algorithm, and a voltage-controlled oscillator (VCO) [29]. Passive-mode operation using a fully digital feedback loop was demonstrated

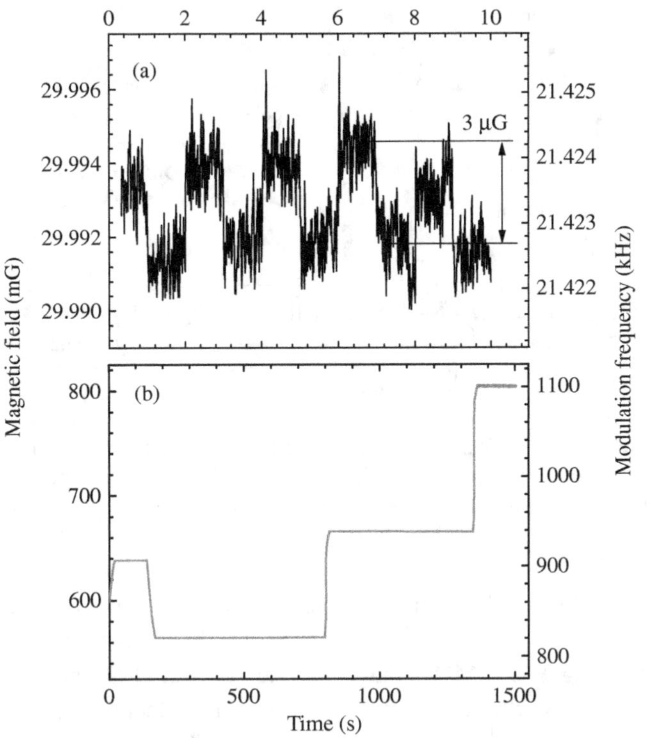

Figure 6.9 A passive-mode AM NMOR magnetic-field tracking signal for relatively small (a) and large (b) magnetic-field changes. The lower plot (b) shows the tracking signal for significant changes of the magnetic field (on the order of 100 mG). In this case the magnetometer finds new resonance conditions ($\Omega_m = 2\Omega_L$) within a few seconds (time limited by the iterative algorithm used for magnetic-field tracking). Both measurements were performed in the one-beam arrangement with light intensity of 5 μW/mm^2 and 100% sinusoidal modulation.

in Ref. [46]. Although lock-in detection limits the magnetometer response time, it also enables magnetometric measurements with relatively high signal-to-noise ratios.

Magnetic-field-tracking signals in the geophysical field range using AM NMOR are shown in Fig. 6.9. In the data shown in Fig. 6.9(a) the magnetic field was changed by 3 μG every second, and the tracking signal responded within a time of $\sim$10 ms, equal to the lock-in integration time. For more significant changes in the field value, on the order of 100 mG [Fig. 6.9(b)], a time of a few seconds, limited by an iterative computer algorithm, was needed to find the new resonance condition ($\Omega_m = 2\Omega_L$).

Another approach to magnetometry, which is often simpler, is a *self-oscillating magnetometer* that uses the measured spin-precession signal to directly generate the modulation in a positive feedback loop [47]. Several optical self-oscillating atomic magnetometers have been demonstrated, utilizing transitions between hyperfine [48] and Zeeman sublevels [12, 49]. A key advantage of a self-oscillating magnetometer is that the

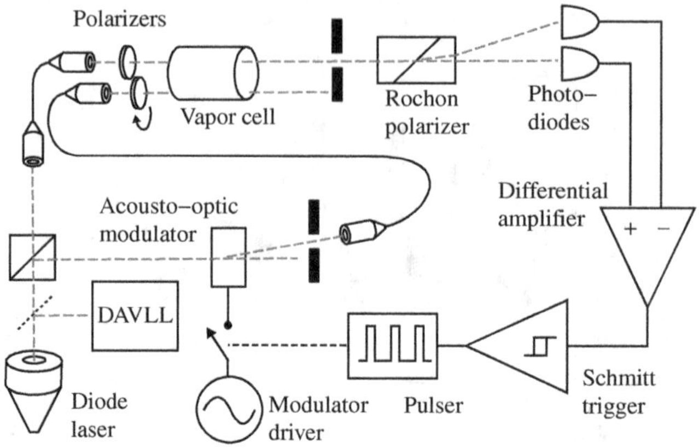

Figure 6.10 (From Ref. [12].) Diagram of a self-oscillating Rb magnetometer based on a two-beam AM NMOR arrangement. A zero-crossing detector (Schmitt trigger) and pulser control the pump power via the AOM, closing the loop and sustaining self-oscillation.

response time can be nearly instantaneous, limited only by the electronic feedback loop.

The self-oscillating FM NMOR magnetometer of Ref. [49] utilized a single-beam arrangement and, as a result, the detected signal was a product both of the rotating atomic alignment and of the modulated detuning. This resulted in a complicated waveform that required significant electronic processing before being suitable for feeding back to the laser modulation, as required in the self-oscillating scheme.

In the work of Ref. [12], a self-oscillating AM NMOR magnetometer based on a two-beam arrangement was used that avoided many of the difficulties encountered in the single-beam experiment, since the optical-rotation signal could be made accurately sinusoidal, avoiding the complexity of digital or other variable-frequency filters in the feedback loop. The use of two beams also permits optical adjustment of the relative phase of the detected signal and the driving modulation by changing the angle between their respective linear polarizations. For magnetometry at large bias field and requiring a wide range of fields, this optical tuning of the feedback-loop phase promises both good long-term stability and much greater uniformity with respect to frequency than can readily be obtained with an electronic phase shift.

The experimental apparatus of Ref. [12] is shown schematically in Fig. 6.10. The linearly polarized pump and probe beams were supplied by a single extended-cavity diode laser operating on the D2 line of rubidium, frequency stabilized ~300 MHz below the center of the $F = 2 \longrightarrow F'$ Doppler-broadened ^{87}Rb line by means of a dichroic atomic vapor laser lock (DAVLL) [50,51]. The probe beam was left unmodulated, while the pump was amplitude modulated with an AOM. Pump and probe were delivered to the cell by separate polarization-maintaining fibers. After exiting the cell, the pump beam was blocked and the probe was analyzed by a balanced polarimeter consisting of a Rochon polarizing beam-splitter and a pair of photodiodes. The difference photocurrent was amplified and passed through a resonant LC filter centered at 20 kHz with a bandwidth of 11 kHz, much wider than either the NMOR resonance (~80 Hz) or the desired magnetometer bandwidth (~1 kHz). This filter reduced jitter in the frequency-counter readings, but is not necessary in principle. The pump modulation was derived from this amplified signal, closing the feedback loop, by triggering a pulse generator on the negative-going zero-crossings of the signal, and allowing these pulses to switch on and off the radiofrequency power delivered to the AOM. The pulse duty cycle was approximately 15%. For characterization of the magnetometer in the laboratory, the vapor cell was placed in a three-layer cylindrical magnetic shield, provided with internal coils for the generation of a stable, well-defined magnetic bias field and gradients. The Rb density in the cell was maintained at an elevated value ($\sim 5 \times 10^{10}$ cm^{-3}) by heating the interior of the magnetic shields to around 40°C with a forced-air heat exchanger. The photodiode signal was fed into a frequency counter.

Provided the trigger threshold of the pulse generator was close enough to zero (i.e., within a few times the noise level of the signal), oscillation would occur spontaneously when the loop was closed at a frequency set by the magnetic field. Optimum settings for the magnetometer sensitivity were found to be approximately 7 μW mean incident pump power, 7 μW continuous incident probe power, and optical absorption of around 60% at the lock point. A sensitivity of 3 nG was achieved for a measurement time of 1 s at these settings (Fig. 6.11), somewhat lower than anticipated based on general considerations [see Eq. (1.1)] – possibly due to alignment-to-orientation conversion [34]. Field operation of

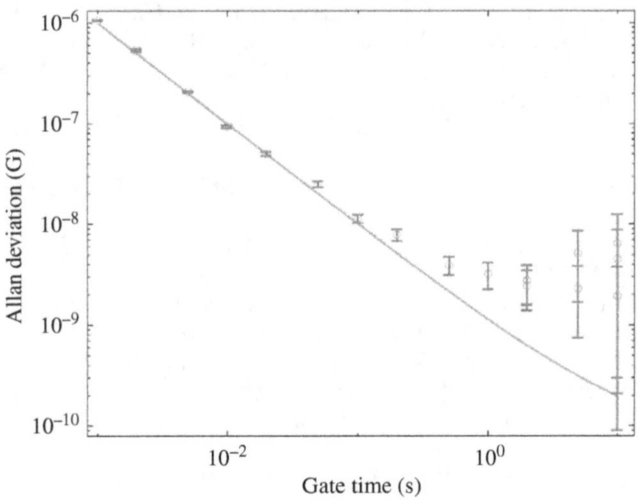

Figure 6.11 (From Ref. [12].) Allan deviation of self-oscillating magnetometer depicted in Fig. 6.10 as obtained from the frequency counter. The solid line indicates the projected sensitivity based on measured experimental parameters, showing good agreement for short measurement times.

a portable self-oscillating magnetometer based on the design of Ref. [12] is reported in Ref. [52].

6.7 Conclusion

Optical magnetometers based on modulated light have already been applied to measurements of nuclear magnetism [53–55], magnetic resonance imaging (MRI) [46, 56, 57], geophysical field measurements [20,52], magnetic particle detection [58], and biomagnetic fields [59]. Practical schemes for chip-scale optical magnetometers based on modulated light have been developed [30, 49]. These techniques have also enabled selective creation and detection of high-order atomic polarization moments [18,26,43] that are of interest not only for magnetometry but also for quantum computation [60, 61]. Modulated-light techniques also find application in remote sensing, both with remotely located vapor cells and using naturally occurring sodium atoms in the atmosphere (see Chapter 13). After evolution for half a century, optical magnetometers based on modulated light continue to improve and become more useful.

References

[1] W. E. Bell and H. L. Bloom, Phys. Rev. Lett. **6**, 280 (1961).
[2] W. Gawlik, J. Kowalski, R. Neumann, and F. Träger, Opt. Commun. **12**, 400 (1974).
[3] W. Gawlik, J. Kowalski, R. Neumann, and F. Träger, Phys. Lett. **48A**, 283 (1974).
[4] D. Budker, W. Gawlik, D. F. Kimball, S. M. Rochester, V. V. Yashchuk, and A. Weiss, Rev. Mod. Phys. **74**, 1153 (2002).
[5] S. I. Kanorsky, A. Weis, and J. Skalla, Appl. Phys. B, **60**, S165 (1995).
[6] D. Budker, V. Yashchuk, and M. Zolotorev, Phys. Rev. Lett. **81**, 5788, (1998).
[7] D. Budker, D. F. Kimball, S. M. Rochester, V. V. Yashchuk, and M. Zolotorev, Phys. Rev. A **62**, 043403 (2000).
[8] D. Budker, D. F. Kimball, V. V. Yashchuk, and M. Zolotorev, Phys. Rev. A **65**, 055403 (2002).
[9] L. M. Barkov and M. S. Zolotorev, Pis'ma v ZhETF **28**, 544 (1978) [JETP Lett. **28**, 503 (1978)].
[10] L. M. Barkov, M. Zolotorev, and D. Melik-Pashaev, Pis'ma v ZhETF **48**, 134 (1988) [JETP Lett. **48**, 144 (1988)].
[11] W. Gawlik, L. Krzemien, S. Pustelny, D. Sangla, J. Zachorowski, M. Graf, A. O. Sushkov, and D. Budker, Appl. Phys. Lett. **88**, 131108 (2006).
[12] J. Higbie, E. Corsini, and D. Budker, Rev. Sci. Instrum. **77**, 113106 (2006).
[13] J. Dupont-Roc, Rev. Phys. Appl. **5**, 853 (1970).
[14] H. Failache, P. Valente, G. Ban, V. Lorent, and A. Lezama, Phys. Rev. A **67**, 043810 (2003).
[15] A. I. Okunevich, Zh. Eksp. Teor. Fiz. **66**, 1578 (1974).
[16] L. Novikov, Comptes Rendus B **278**, 1063 (1974).
[17] A. I. Okunevich, Zh. Eksp. Teor. Fiz. **67**, 881 (1974).
[18] V. V. Yashchuk, D. Budker, W. Gawlik, D. F. Kimball, Yu. P. Malakyan, and S. M. Rochester, Phys. Rev. Lett. **90**, 253001 (2003).

[19] Yu. P. Malakyan, S. M. Rochester, D. Budker, D. F. Kimball, and V. V. Yashchuk, Phys. Rev. A. **69**, 013817 (2004).
[20] V. Acosta, M. P. Ledbetter, S. M. Rochester, D. Budker, D. F. Jackson Kimball, D. C. Hovde, W. Gawlik, S. Pustelny, and J. Zachorowski, Phys. Rev. A **73**, 053404 (2006).
[21] S. Pustelny, D. F. Jackson Kimball, S. M. Rochester, V. V. Yashchuk, D. Budker, Phys. Rev. A **74**, 063406 (2006).
[22] S. Pustelny, S. M. Rochester, D. F. Jackson Kimball, V. V. Yashchuk, D. Budker, and W. Gawlik, Phys. Rev. A **74**, 063420 (2006).
[23] M. Faraday, Philos. Trans. R. Soc. London **136**, 1-20 (1846).
[24] M. Faraday, Philos. Magn. **28**, 294 (1846).
[25] M. Faraday, *Experimental researches in Electricity* **III** (Taylor, London, 1855).
[26] S. Pustelny, D. F. Jackson Kimball, S. M. Rochester, V. V. Yashchuk, W. Gawlik, and D. Budker, Phys. Rev. A **73**, 023817 (2006).
[27] S. Li, P. Vachaspati, D. Sheng, N. Dural, and M. V. Romalis, Phys. Rev. A **84**, 061403 (2011).
[28] S. Pustelny, A. Wojciechowski, M. Kotyrba, K. Sycz, J. Zachorowski, W. Gawlik, A. Cingoz, N. Leefer, J. M. Higbie, E. Corsini, M. P. Ledbetter, S. M. Rochester, A. O. Sushkov, and D. Budker, Proc. SPIE **6604**, 660404 (2007).
[29] S. Pustelny, A. Wojciechowski, M. Gring, M. Kotyrba, J. Zachorowski, and W. Gawlik, J. Appl. Phys. **103**, 063108 (2008).
[30] M. V. Balabas, D. Budker, J. Kitching, P. D. D. Schwindt, and J. E. Stalnaker, J. Opt. Soc. Am. B, **23** 1001 (2006).
[31] S. Pustelny, Ph.D. dissertation, Jagiellonian University (2007).
[32] M. Auzinsh, D. Budker, and S. M. Rochester, Phys. Rev. A **80**, 053406 (2009).
[33] M. Auzinsh, D. Budker, and S. Rochester, *Optically Polarized Atoms: Understanding Light–Atom Interactions* (Oxford University Press, Oxford, 2010).
[34] D. F. Jackson Kimball, L. R. Jacome, S. Guttikonda, E. J. Bahr, and L. F. Chan, J. Appl. Phys. **106**, 063113 (2009).
[35] E. B. Alexandrov, M. Auzinsh, D. Budker, D. F. Kimball, S. M. Rochester, and V. V. Yashchuk, J. Opt. Soc. Am. B **22**, 7 (2005).
[36] D. Budker, D. F. Kimball, S. M. Rochester, and V. V. Yashchuk, Phys. Rev. A **65**, 033401 (2002).
[37] Y. Xiao, I. Novikova, D. F. Phillips, and R. L. Walsworth, Phys. Rev. Lett. **96**, 043601 (2006).
[38] A. V. Taichenachev, A. M. Tumaikin, V. I. Yudin, M. Sthler, R. Wynands, J. Kitching, and L. Hollberg, Phys. Rev. A **69**, 024501 (2004).
[39] E. B. Aleksandrov, Sov. Phys. – Usp. **15**, 436 (1972).
[40] A. Ben-Kish and M.V. Romalis, Phys. Rev. Lett. **105**, 193601 (2010).
[41] D. Budker, D. F. Kimball, S. M. Rochester, and V. V. Yashchuk, Phys. Rev. Lett. **85**, 2088 (2000).
[42] K. Jensen, V. M. Acosta, J. M. Higbie, M. P. Ledbetter, S. M. Rochester, and D. Budker, Phys. Rev. A **79**, 023406 (2009).
[43] V. M. Acosta, M. Auzinsh, W. Gawlik, P. Grisins, J. M. Higbie, D. F. Jackson Kimball, L. Krzemien, M. P. Ledbetter, S. Pustelny, S. M. Rochester, V. V. Yashchuk, and D. Budker, Optics Express **16**, 11423 (2008).
[44] E. B. Alexandrov, M. P. Chaika, and G. I. Khvostenko, *Interference of Atomic States* (Springer, Berlin, 1993).

[45] I. I. Sobelman, *Atomic spectra and Radiative Transitions* (Springer-Verlag, Berlin, 1992).
[46] S. Xu, S. M. Rochester, V. V. Yashchuk, M. H. Donaldson, and D. Budker, Rev. Sci. Instrum. **77**, 083106 (2006).
[47] A. L. Bloom, Appl. Opt. **1**, 61 (1962).
[48] A. B. Matsko, D. Strekalov, and L. Maleki, Opt. Comm. **247**, 141 (2005).
[49] P. D. D. Schwindt, L. Hollberg, and J. Kitching, Rev. Sci. Inst. **76**, 126103 (2005).
[50] K. L. Corwin, Z.-T. Lu, C. F. Hand, R. J. Epstain, and C. E. Wieman, Appl. Opt. **37**, 3295 (1998).
[51] V. V. Yashchuk, D. Budker, and J. Davis, Rev. Sci. Instrum. **71**, 341 (2000).
[52] C. Hovde, B. Patton, E. Corsini, J. Higbie, and D. Budker, Proc. SPIE **7693**, 769313 (2010).
[53] V. V. Yashchuk, J. Granwehr, D. F. Kimball, S. M. Rochester, A. H. Trabesinger, J. T. Urban, D. Budker, and A. Pines, Phys. Rev. Lett. **93**, 160801 (2004).
[54] C. W. Crawford, S. Xu, E. J. Siegel, D. Budker, and A. Pines, Appl. Phys. Lett. **93**, 092507 (2008).
[55] M. P. Ledbetter, C. W. Crawford, A. Pines, D. E. Wemmer, S. Knappe, J. Kitching, and D. Budker, J. Magn. Reson. **199**, 25 (2009).
[56] S. Xu, V. V. Yashchuk, M. H. Donaldson, S. M. Rochester, D. Budker, and A. Pines, Proc. Nat. Acad. Sci. **103**, 12668 (2006).
[57] S. Xu, C. W. Crawford, S. Rochester, V. Yashchuk, D. Budker, and A. Pines, Phys. Rev. A **78**, 013404 (2008).
[58] S. Xu, M. H. Donaldson, A. Pines, S. M. Rochester, D. Budker, and V. V. Yashchuk, Appl. Phys. Lett. **89**, 224105 (2006).
[59] E. Corsini, V. Acosta, N. Baddour, J. Higbie, B. Lester, P. Licht, B. Patton, M. Prouty, and D. Budker, J. Appl. Phys. **109**, 074701 (2011).
[60] T. Opatrny and J. Fiurasek, Phys. Rev. Lett. **95**, 053602 (2005).
[61] S. Pustelny, M. Koczwara, L. Cincio, and W. Gawlik, Phys. Rev. A **83**, 043832 (2011).

7

Microfabricated atomic magnetometers

S. Knappe and J. Kitching

7.1 Introduction

In this chapter, we discuss miniaturized atomic magnetometers, and the technology and applications relevant to this somewhat unusual direction in magnetometer research and development [1]. By "miniaturized," we mean, in addition to their small size, magnetometers that have associated desirable qualities such as low power consumption, low cost, high reliability, and the potential for mass fabrication. Together with the high sensitivity usually obtained from the use of atoms, these properties result in magnetic sensors that fill a unique application space and may in fact enable new applications for which atomic magnetometers have not before been used.

It is perhaps surprising that atomic magnetometers in general are not more widely used in the world today. The main application areas at present are geophysical surveying and magnetic anomaly detection. Geophysical surveying is important in oil and mineral exploration, archeology, and unexploded ordnance detection and is typically carried out by moving one or more atomic magnetometers over the area to be surveyed. The magnetic "map" generated from this data can show the locations and in some cases the size and shape of magnetic objects or structures buried beneath the surface of the earth. Magnetic anomalies include vehicles, ships, and submarines and are typically detected via magnetic gradiometry. There are, however, only three major companies in North America, employing perhaps a few hundred people, that manufacture and sell atomic magnetometers. This effort represents a rather small fraction of the worldwide yearly market for magnetic sensors, which was estimated in 2005 to be about $1 billion [2]. Commercial atomic magnetometers are described in Chapter 20.

One major impediment to more widespread use of atomic magnetometers is their high cost and instrumental complexity relative to other types of magnetic sensor technology. Hall

Optical Magnetometry, ed. Budker, D. and Jackson Kimball, D. F. Published by Cambridge University Press.

probes and magnetoresistive sensors, for example, have captured the bulk of the magnetometer market share mainly because they are manufactured as low-cost integrated circuits that can be integrated as electronic components into automobiles and other consumer or industrial products. Thus miniaturization and associated low-power operation and low-cost production are critical factors in determining whether a technology achieves commercial success and widespread use in real-world applications.

This sentiment is a key driving force behind our work, and the work of others, to develop highly miniaturized atomic magnetometers. Research in the field of atomic magnetometers has traditionally focused largely on sensitivity and the underlying physics that determines this aspect of the performance of magnetic sensors. However, the development of novel fabrication techniques that enable a high degree of miniaturization may be as important in determining how atomic magnetometers are used.

We discuss here work that was initiated at NIST in 2004 as an offshoot of a program to develop highly miniaturized "chip-scale" atomic clocks [3]. We begin with an analysis of how the sensitivity of a vapor-cell magnetic sensor scales with the size of the cell under certain reasonable assumptions about how the sensor is operated. We then discuss aspects of the design and fabrication of chip-scale atomic magnetometers with emphasis on the use of micromachining processes. We conclude with a discussion of applications of chip-scale atomic magnetometers, and in particular focus on biomagnetic imaging and the detection of nuclear magnetic resonance.

7.2 Sensitivity scaling with size

The sensitivity of an atomic magnetometer based on N uncorrelated atoms is limited fundamentally by quantum projection noise in the measurement of individual atomic spins. Under these conditions, the sensitivity can be written as (see Ref. [4] as well as Chapter 1, Section 1.1.1, and Chapter 2):

$$\delta B = \frac{1}{\gamma}\frac{1}{\sqrt{N\tau T}}, \tag{7.1}$$

where γ is the atom's gyromagnetic ratio, τ is the atom relaxation time, and T is the measurement period. We consider here how the sensitivity scales with the size of the vapor cell. The cell size influences the sensitivity in two main ways: through the relaxation time τ and through the atom number N. The atom number is clearly determined by the alkali vapor density n_{Al} and the cell volume V as

$$N = n_{\mathrm{Al}} V . \tag{7.2}$$

In the limit of high alkali atom density, where the relaxation time is dominated by collisions of alkali atoms with other alkali atoms, the relaxation time can be written

$$\tau_{\mathrm{Al\text{-}Al}} \approx \frac{1}{n_{\mathrm{Al}}\sigma_{\mathrm{col}} v}, \tag{7.3}$$

where σ_{col} is the collision cross-section and v is the mean relative velocity of the atoms. Under these conditions, the field sensitivity is given by [5]

$$\delta B \approx \frac{1}{\gamma}\sqrt{\frac{\sigma_{\text{col}} v}{VT}}, \tag{7.4}$$

and scales as the inverse one-half power of the cell volume. When spin-exchange collisions dominate the relaxation rate, $\sigma_{\text{col}} \approx \sigma_{\text{se}}$, and the sensitivity scales with cell dimension approximately as $1\ \text{fT}/\sqrt{\text{Hz}\cdot\text{cm}^3}$ for alkali atoms, where the spin-exchange collision cross-section is about $2\times10^{-14}\ \text{cm}^2$. Under some circumstances the effects of spin-exchange collisions can be suppressed [6, 7], in which case weaker spin-destruction processes dominate. In this case the collision cross-section can be as low as $2\times10^{-18}\ \text{cm}^2$ for ^{39}K, and a sensitivity near $10\ \text{aT}/\sqrt{\text{Hz}}$ may be possible in a cell 1 cm in size.

The above analysis assumes relaxation dominated by alkali–alkali collisions. However, additional relaxation processes are also usually present. These processes include collisions with the cell walls, collisions with buffer-gas atoms if a buffer gas is present, and relaxation due to magnetic field gradients. These other processes do not depend on the alkali atom density but do, in general, depend directly or indirectly on the cell volume. We may therefore write

$$\frac{1}{\tau} = \frac{1}{\tau_{\text{bg}}} + \frac{1}{\tau_w} + \frac{1}{\tau_{\text{oth}}} + \frac{1}{\tau_{\text{Al-Al}}}, \tag{7.5}$$

where the wall-induced relaxation rate, $1/\tau_w$, depends on the cell volume, the buffer-gas-induced relaxation rate is $1/\tau_{\text{bg}}$, and the relaxation due to other processes, $1/\tau_{\text{oth}}$, can be made small with appropriate environmental controls. As discussed below, the buffer-gas pressure is usually adjusted to minimize the relaxation rate for a given cell volume, and is hence indirectly dependent on this parameter.

In an otherwise evacuated cell containing alkali atoms, the atom density is typically small enough that the atoms fly ballistically between wall collisions. Since a single collision with an uncoated glass wall depolarizes an atom completely, the relaxation time is roughly equal to the transit time across the cell and is on the order of 30 ms for a cell with a characteristic size of 1 cm. Wall coatings such as paraffin are known to lengthen considerably the wall-collisional relaxation time (see Chapter 11); typically a thousand wall collisions or more are obtained before the atoms relax [8], although recent results have indicated the possibility that some unique coatings may allow considerably more bounces [9]. In either case the wall-collision-induced relaxation rate can be written as $1/\tau_w = \alpha v/L$, where α is the depolarization probability for one collision and $L \sim V^{1/3}$ is the characteristic size of the cell. Highly miniaturized glass-blown vapor cells with high-quality wall coatings have been recently demonstrated [10].

Buffer gases can also be used to reduce wall-induced relaxation by forcing the alkali atoms to diffuse through the cell and therefore collide with the walls less often. In this case, the relaxation rate is a balance between the diffusion-mediated wall collisions and the relaxation due to collisions with the buffer-gas atoms themselves [11]. Since the wall-related component is inversely proportional to the buffer-gas pressure, while the buffer-gas relaxation is proportional to this quantity, an optimal buffer-gas pressure exists for a given cell size that minimizes the relaxation. At the optimum buffer-gas pressure, the combined relaxation rate due to buffer gas and walls is equal to

$$\frac{1}{\tau_{\mathrm{bg}}} + \frac{1}{\tau_w} = 2\frac{\sqrt{D_0\gamma_0}}{\beta L}\,, \tag{7.6}$$

where D_0 is the diffusion constant and γ_0 is the buffer-gas-induced relaxation rate, both at a buffer-gas density of 1 amagat, and β is a constant of order unity that depends on the cell geometry. We do not consider the effect of nuclear slow-down factors on relaxation here.

We therefore find that both a wall-coated cell and a cell containing a buffer gas whose pressure is optimized for the size of the cell generate relaxation rates that are inversely proportional to the linear dimension of the cell. In fact, for most of the commonly used wall coatings and buffer gases, the coefficients relating the cell size to the relaxation time are quite similar and approximately equal to $2\pi \times 100$ Hz · mm.

At low cell temperatures, where the alkali–alkali collisions do not contribute to the linewidth, the magnetometer sensitivity improves as the alkali density (cell temperature) is increased because of the improved signal-to-noise ratio. When alkali–alkali collisions dominate the linewidth at high cell temperatures, any gains in sensitivity due to improved signal-to-noise are offset by a broader resonance linewidth. The best operating point therefore occurs when the relaxation due to alkali–alkali collisions is approximately equal to the relaxation due to other processes. These processes are collisions with the cell walls and buffer-gas atoms, if a buffer gas is present, and the relaxation time scales linearly with the cell size under optimal conditions [12]. Figure 7.1(a), adapted from Shah *et al.* [1], shows the magnetometer sensitivity as a function of cell size assuming the linewidth is optimized at each cell size as above and assuming spin projection as the dominant noise source. We also assume two different relaxation regimes: one in which spin-exchange collisions dominate (lines labeled A) and another in which spin-exchange relaxation is suppressed and spin-destruction relaxation dominates (lines labeled B). Spin-exchange-relaxation-free (SERF) magnetometry, corresponding to the latter case (B lines), is discussed in Chapter 5. Clearly better sensitivity is obtained at a given cell size when spin-exchange relaxation is suppressed. This is because higher cell temperatures can be used in this case without broadening the natural cell linewidth, and hence more atoms can be interrogated, resulting in a higher signal-to-noise ratio. The difference in sensitivity between atoms of different species at a given cell size is determined almost entirely by the difference in collision cross-sections.

The analysis in Fig. 7.1(a) assumes spin projection as the only noise source. A number of noise contributions other than spin projection noise are also present in atomic magnetometers. These include photon shot noise, light-shift noise, laser intensity noise, and frequency-to-amplitude conversion noise. The effects of these noise terms on the sensitivity of atomic magnetometers can be evaluated somewhat heuristically by describing them

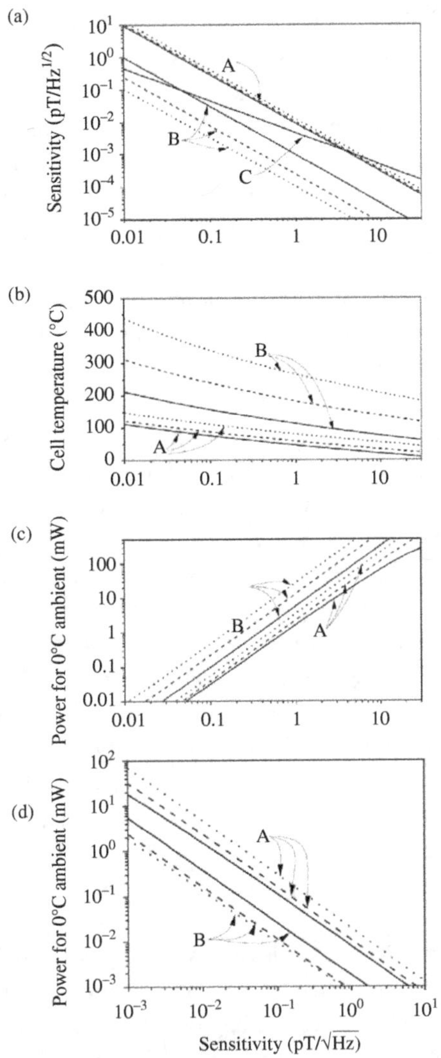

Figure 7.1 (Adapted from Ref. [1].) Scaling of magnetometer parameters: (a) sensitivity, (b) cell temperature, and (c) power as a function of size. (d) Electrical power required to reach a given sensitivity, under assumptions outlined in the text. Traces A and B show the scaling for linewidths limited by spin-exchange collisions and spin-destruction collisions, respectively. Solid lines refer to Cs, dashed lines to ^{87}Rb, and dotted lines to ^{39}K. Solid line C in plot (a) refers to a signal-to-noise ratio of $5 \times 10^6\ \sqrt{\text{Hz}}$.

in terms of a detection signal-to-noise ratio (S/N), which limits how well the center of the magnetic resonance can be determined. Photon shot noise on 1 mW of light results in a S/N of $5 \times 10^7\ \sqrt{\text{Hz}}$. In practice, a S/N between 10^4 and 10^5 at one second can be achieved without too much difficulty, although a S/N as high as $\sim 2 \times 10^7\ \sqrt{\text{Hz}}$ has been achieved in some experiments [13]. The solid line C in Fig. 7.1(a) shows the expected sensitivity as a function of cell size for a fixed S/N of $5 \times 10^6 \sqrt{\mathrm{Hz}}$ for typical buffer-gas vapor cells.

In practice, it seems reasonable that alkali vapor cells could be produced ranging in size from a few centimeters down to about 10 μm. Over this size range, the fundamental magnetometer sensitivity may vary from a few $\mathrm{aT}/\sqrt{\mathrm{Hz}}$ for the largest cells to a few $\mathrm{pT}/\sqrt{\mathrm{Hz}}$ for the smallest. This scaling implies an interesting and important technical trade-off between sensitivity and size, particularly when considering instrumentation for use in real-world applications. By making their vapor cells larger, atomic magnetometers can certainly be made more sensitive, but with loss of overall utility, and ultimately value, due to the bulkiness of the apparatus. The value of high sensitivity is compromised even further when considering magnetic noise that is present in all shielded and unshielded environments. Field fluctuations of geomagnetic origin, for example, have a magnitude of about $1\ \mathrm{pT}/\sqrt{\mathrm{Hz}}$ at 1 Hz, while thermal noise from magnetic shields is approximately a few $\mathrm{fT}/\sqrt{\mathrm{Hz}}$ for shield diameters of about 1 m. For biomagnetic imaging, the finite conductance of the human body generates magnetic noise at a level of a fraction of a $\mathrm{fT}/\sqrt{\mathrm{Hz}}$. Even though gradiometry can be used in some cases, the benefits of improving the magnetometer sensitivity are therefore substantially reduced once the femtotesla regime is reached. From a purely instrument engineering perspective, this viewpoint motivates the development of very small magnetometers, which trade off high sensitivity for low power, ease of fabrication and operation, and low cost.

The optimal sensitivity in Fig. 7.1(a) is reached only when the relaxation rate is dominated by alkali–alkali collisions. Since the alkali density, and hence the relaxation rate, is determined by the cell temperature, the curves in Fig. 7.1(a) define a minimum temperature at which the cell must operate to achieve the desired sensitivity. The optimal alkali density is

$$n_{\mathrm{Al}} = \frac{2\sqrt{D_0 \gamma_0}}{\beta L \sigma_{\mathrm{col}} v}, \tag{7.7}$$

and the cell temperature required to achieve this density is plotted in Fig. 7.1(b). Even for rather small cells, approaching 10 μm in size, the required temperatures are practical to achieve, at least for ^{87}Rb and Cs. Cesium begins diffusing rapidly into Pyrex at temperatures near 350°C, and operation of a Pyrex cell containing Cs is not practical above that temperature.

Finally, it is possible to estimate the power required to heat the cell to its operating temperature. We assume here that the cell can be thermally isolated from the environment to a point where radiation is the dominant source of heat loss. Using the Stefan–Boltzmann law, and assuming a surface emissivity of unity, the power required to maintain the cell temperature in a 0°C ambient is shown in Fig. 7.1(c) as a function of cell size. The results of Fig. 7.1(a) and Fig. 7.1(c) can be combined to determine how much power is required to reach a given sensitivity level. This is plotted in Fig. 7.1(d). Clearly, higher sensitivity requires more power. However, sensitivities near $1\ \mathrm{pT}/\sqrt{\mathrm{Hz}}$ can in principle be achieved with far less than 1 mW of heating power, suggesting that moderately sensitive but highly autonomous battery-operated sensors are possible. While the electronics required to control such a sensor must be made to operate on comparably little power, it is likely that this could be achieved with moderate investment in an appropriate low-power application-specific integrated circuit.

7.3 Sensor fabrication

The design and fabrication of atomic magnetometers with millimeter-scale dimensions has presented a number of interesting challenges over the last decade. Research on this topic grew out of previous work to develop chip-scale atomic clocks [3, 12] and many of the elements of the two types of instruments are similar. In fact, the first chip-scale atomic magnetometer to be operated was the same device as an earlier chip-scale atomic clock.

The main components of an optical magnetometer physics package are the following: a light source, a vapor cell, optics to direct and polarize the light, heaters to heat the vapor cell, and photodetectors. In conventional commercial alkali magnetometers an alkali lamp excited with an RF discharge is used as the light source. The vapor cell is fabricated by use of glass-blowing techniques and contains the alkali-metal atoms and a buffer gas to reduce the effects of wall collisions as described above. Commercial atomic magnetometers have a sensor volume of approximately 1 L and require about 10 W of electrical power to operate.

Highly miniaturized versions of these instruments have incorporated two main improvements: the first is to use a low-power semiconductor laser as the light source. The RF discharges needed to activate the lamp require 1 W of electrical power or more to operate. Lasers, on the other hand, are considerably more efficient. Vertical-cavity surface emitting lasers (VCSELs), for example, can have a wall-plug efficiency above 10% and can therefore run on a few milliwatts of electrical power. The second improvement is the use of microfabricated alkali vapor cells, which allow the alkali atoms to be contained in a much smaller volume than is achieved with glass-blown cells. The smaller cell size in turn allows the power needed to heat the cell to its operating temperature to be much lower. These two improvements, along with a number of more minor changes, have allowed the demonstration of atomic magnetometer sensor heads with volumes below 10 mm^3, as well as some novel designs that are outlined below.

To address applications with different requirements, two generally different approaches to miniaturization have been tried: a fully integrated chip-scale magnetic sensor and microfabricated remote sensor heads fiber-optically coupled to a central control unit.

The first approach to miniaturization of optical magnetometers closely followed that developed for chip-scale atomic clocks. The goal of this design, in addition to small size and low power, was to allow for wafer-level fabrication and assembly in order to make parallel fabrication of large numbers possible at reasonable cost. Figure 7.2 shows a photograph of one of the first chip-scale atomic magnetometer physics packages, as well as a sketch of the separate components. The physics package, based on the standard M_x configuration (see Chapter 4) and pumped on the D1 line of ^{87}Rb at 795 nm, had a total volume of 25 mm^3 and required about 200 mW at an ambient temperature near room temperature. The device housed a VCSEL bonded to the bottom baseplate that created the light tuned to the D1

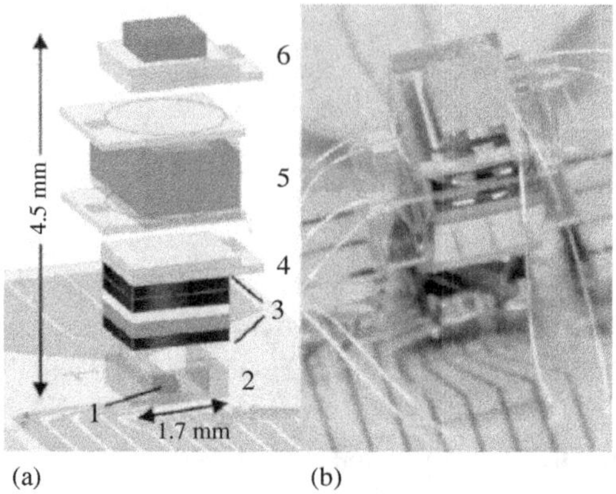

Figure 7.2 (Adapted from Ref. [14].) (a) Design and (b) photograph of a chip-scale atomic magnetometer. A laser (1), surrounded by a spacer (2), emits light through some optics (3) to illuminate a vapor cell (5); the light is detected with a photodiode (6). Ohmic heaters (4) are used to heat the cell.

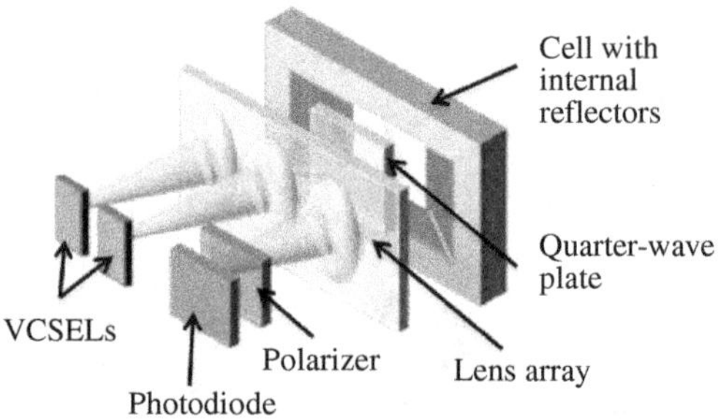

Figure 7.3 Reflectors internal to the cell can be used to redirect laser beams and create perpendicular pump/probe geometries.

transition of ^{87}Rb to optically pump the atoms and probe the precessing atomic spins. The light emitted from the laser was detected by the photodiode on top of the stack, after it was attenuated, collimated, circularly polarized and transmitted through an alkali vapor cell. The cell was heated to 90°C by two heaters on either side of the cell, to achieve an optical depth of 0.7 on resonance. Two H-field coils were added to create an oscillating field parallel to the laser beam. Electrical connections were made through wire bonds to the baseplate. More complicated MEMS designs for multibeam geometries have been proposed. One of them, as shown in Fig. 7.3, uses vapor cells with internal reflectors, and dielectric coatings applied on angled surfaces inside the cell [15]. Another design simply tilted the cell at an angle of 45° [1,16]. A third design used a diverging laser beam to optically pump and probe atoms within the same cell in two different directions [17].

While this type of design enables inexpensive fabrication of many individual physics packages, the proximity of the laser and many electrical connections to the detection volume

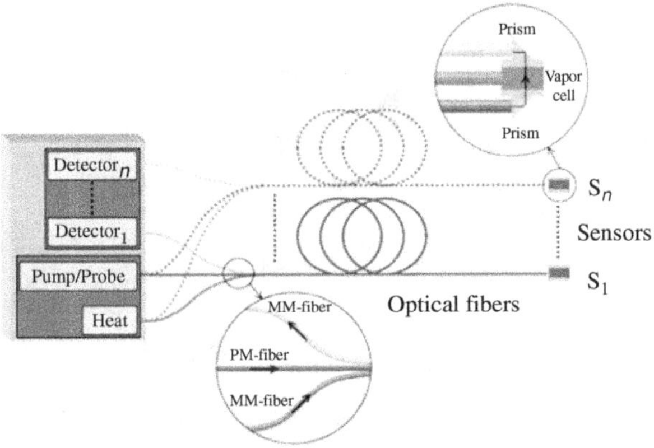

Figure 7.4 Concept of fiber-optically coupled magnetometer array. A series of sensor heads ($S_1 - S_n$) is coupled fiber-optically to a common control unit containing a pump/probe laser, a heating laser, and a series of photodetectors.

inside the cell limited the sensitivity and generated spurious magnetic field signals. It can therefore be advantageous to separate the laser and detector from the vapor cell and couple the light to and from the cell either through optical fibers or through free space. This approach can be especially attractive when large sensor arrays are desirable, so that all sensor heads can be interrogated with the same laser. A sketch of such a magnetometer is shown in Fig. 7.4. Because light from the laser is distributed among many sensor heads, a higher-power laser can be used with better noise characteristics compared to those of VCSELs.

In all of these approaches, care has to be taken that the materials used for the sensors generate no magnetic fields. While this sounds trivial, it must be kept in mind that any conductive material near the sensor volume is a source of magnetic noise through the thermal motion of electrons [18, 19]. Lee and Romalis [20] calculated the noise from high-permeability magnetic shielding and suggested the use of ferrite shields. Furthermore, Griffith *et al.* [21] calculated the noise expected from cell bodies made from high-conductivity silicon to be a few $\mathrm{fT}/\sqrt{\mathrm{Hz}}$. By using silicon materials with lower conductivity, this noise can be reduced to below 1 $\mathrm{fT}/\sqrt{\mathrm{Hz}}$. Furthermore, when the noise in a cell with solid Rb on the cell walls was measured and compared to one with minimal amounts of solid Rb present, a difference of several $\mathrm{fT}/\sqrt{\mathrm{Hz}}$ was found. This suggests that the presence of the Rb on the walls of such a small cell contributes significantly to the sensor noise [21].

7.4 Vapor cells

Until now, all microfabricated optical magnetometers have used micromachined alkali vapor cells. Bulk etching techniques in silicon allow for precisely defined cell geometries with the possibility of thin channels and alkali reservoirs with precision better than 25 μm. Wafer thicknesses ranging from 300 μm to 3 mm have been used and cells with lateral dimensions

Figure 7.5 A variety of microfabricated alkali vapor cells, showing a range of interior dimension from 4 mm to 0.5 mm.

between 100 μm and 5 mm have been demonstrated. The two main etching options that have been explored are the dry method of deep reactive ion etching (DRIE) and the wet etch with potassium hydroxide (KOH) [22]. Some other means of creating cell cavities such as ultrasonic drilling and diamond mechanical drilling have also been successful. A number of microfabricated vapor cells of different sizes are shown in Fig. 7.5.

While silicon microfabrication methods are already well established, the main new challenge in fabricating the vapor cell lies in the filling method. Many different approaches for filling MEMS vapor cells have been attempted for the development of chip-scale atomic clocks [23]. While microfabricated atomic magnetometer cells have slightly different driving requirements compared to those for a clock, many challenges are the same. The most common method of filling cells with alkali atoms takes advantage of anodic bonding [24] of borosilicate glass windows onto the silicon body. This method creates a permanent electrostatic bond between the glass and silicon. An initial bond between the silicon frame and the glass on one side of the cell is usually performed in air. This "preform" is then transferred to a glove box or a vacuum chamber in order to minimize the oxygen or water content in the cell. After deposition of the alkali atoms, the chamber is backfilled with a buffer gas and a second glass wafer is bonded on the other side of the silicon. While the most common buffer gas inside the cells is nitrogen, cells have also been made with neon, argon, helium, hydrogen, and xenon. The actual filling of the cell inside the chamber is often done by creating an atomic beam of alkali atoms that passes through an aperture into the bottom of the cell, where the atoms condense [25]. In a final step, the wafers or wafer chips are diced to form the actual vapor cells. Figure 7.5 shows a variety of such cells.

7.5 Heating and thermal management

Another challenge is maintaining an elevated cell temperature with low electrical power. The short path lengths of the microfabricated cells require that the cell is heated to increase the alkali vapor pressure. For a 1 mm long Rb cell with 1 amagat of nitrogen, a temperature of 150°C is required to obtain an optical density of 5 on resonance. When using electrical heating with resistive films, the disturbance of the atoms from stray magnetic fields should be minimized. First, the heater currents can be modulated at a frequency much higher than the bandwidth of the magnetometer. Alternatively, the currents could be chopped, and measurements taken only when the currents are off. Second, the heater traces can be patterned in such a way that the generated fields are largely canceled. This can be done within the same heater layer and also by use of two heater films with a thin isolating layer in between. This was implemented, for example, by Schwindt *et al.* [14], who used patterned transparent indium–tin-oxide (ITO) traces and a 2 μm thick insulating layer of benzocyclobutene. Mhaskar *et al.* [26] deposited titanium traces with an insulating layer of SU-8.

Magnetic fields induced by heating currents can be eliminated entirely through the use of optical heating. Preusser *et al.* [27] demonstrated a fiber-coupled sensor head that was heated by 915 nm light that was absorbed by the body of the vapor cell. While this was demonstrated in a fiber-coupled system, optical heating allows for the possibility of truly remote sensor heads, where all connections to the central control unit are made with light beams through free space [27].

Finally, in order to minimize the power required for maintaining the cell temperature, good thermal insulation is needed. One option is to suspend the cell in a web of strained polyimide inside a vacuum enclosure, a technology pioneered for chip-scale atomic devices by Mescher *et al.* [28,29]. They were able to stabilize the temperature of a $(1.5\ \text{mm})^3$ vapor cell to 75°C with a total power of less than 10 mW, limited by radiation from the cell surface. Other approaches used silicon [30, 31] or silicon nitride [32] as materials for the suspensions. Finally, alternative methods to vacuum packaging and for radiation shielding have been implemented in slightly larger optical magnetometers. They include packaging with aerogel [33] and patterned gold reflectors [34].

7.6 Performance

The sensitivity of microfabricated atomic magnetometers has improved from the level of $40\ \text{pT}/\sqrt{\text{Hz}}$ in 2004 for an integrated device to $5\ \text{fT}/\sqrt{\text{Hz}}$ in 2010 measured in a table-top experiment. The sensitivity of a variety of magnetometers is plotted in Fig. 7.6. Traces A and B are the measured sensitivity of devices for which the laser, optics, and photodetector are integrated together as shown in Fig. 7.2. Traces C and D are measurements carried out in microfabricated vapor cells, but with table-top optics, which allows a high degree of flexibility in the experimental parameters and hence easier optimization. These last traces therefore give an estimate of what sensitivity is possible in a microfabricated cell.

In the frequency band between 20 Hz and 200 Hz, the best sensitivities measured to date are $\sim 5\ \text{fT}/\sqrt{\text{Hz}}$. This is quite competitive with commercial SQUID-based magnetometers, but of course atomic magnetometers have the considerable advantage that there is no need for cooling of the sensor to cryogenic temperatures. At frequencies below 10 Hz, most of the microfabricated sensors show increasing noise, although careful measurements have been carried out in only a few cases. In some cases, the low-frequency noise has been

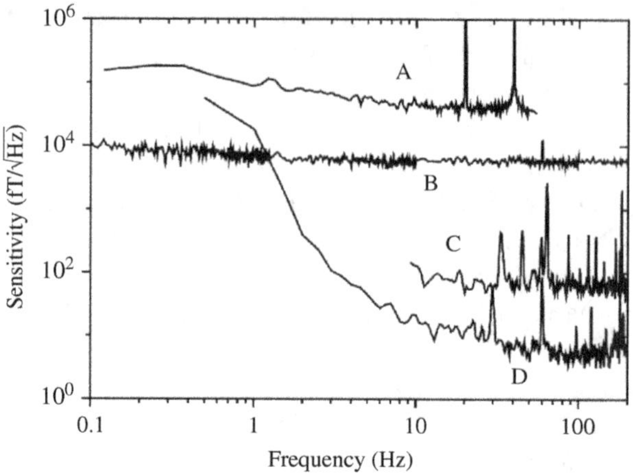

Figure 7.6 Sensitivity of microfabricated atomic magnetometers. (A) Integrated coherent population trapping magnetometer (Schwindt *et al.* [35]), (B) integrated M_x magnetometer (Schwindt *et al.* [14]), (C) SERF magnetometer in microfabricated vapor cell (Shah *et al.* [1]), (D) advanced SERF magnetometer in microfabricated vapor cell (from Griffith *et al.* [21]).

associated with variations of the current flowing through the resistive heaters [26] caused, for example, by time-varying changes in the properties of the heater material. At present, it is not believed that the low-frequency component is a fundamental property of the magnetic sensors.

The bandwidth of microfabricated atomic magnetometers is limited in most cases by the width of the resonance line. Because the resonance width increases as the size of the cell is decreased, highly miniaturized sensors have a natural advantage over their larger counterparts for the measurement of higher-frequency signals: while the larger linewidth (due to more frequent wall collisions) results in worse sensitivity, it simultaneously results in a wider frequency band over which fields can be measured with the same signal-to-noise ratio. The width of this band is important in particular for biomagnetic measurements, for which signals exist from 1 Hz to 100 Hz. Microfabricated atomic magnetometers have demonstrated sensitivities of 6 $\text{pT}/\sqrt{\text{Hz}}$ in the M_x mode with a bandwidth of 1 kHz [14], and 5 $\text{fT}/\sqrt{\text{Hz}}$ in the SERF mode with a bandwidth of 200 Hz [21]. The bandwidth associated with the highest-sensitivity instruments is still largely compatible with many biomagnetic applications.

The dynamic range of microfabricated atomic magnetometers is determined by essentially the same considerations as for larger sensors. When operating in the M_x spin-precession mode, the lowest field that can be measured is determined by the resonance linewidth, which is approximately 500 nT for a cell of interior dimension ~1 mm. Very high magnetic fields can in principle be measured with these sensors, although some instrumentation complications arise at high fields due to the nonlinear Zeeman shift (see Chapter 1) and the large range of drive oscillator frequencies.

In the SERF operating mode, the lowest measurable field is determined by the fundamental sensitivity of the magnetometer. The highest measurable field is determined by the linearity of the resonance slope. For the sensor of Griffith *et al.* [21], for which the resonance width is about 200 Hz, this is about 50 nT. By using field-nulling coils that are actively adjusted to produce zero field at the sensor location, much higher dynamic ranges should be possible.

The heading error of a scalar sensor refers to the change in the sensor reading as the sensor axis is rotated with respect to the field to be measured. Heading errors are generated by a number of effects. Two leading causes are the nonlinear Zeeman shift described above and misalignment of the light propagation axis with the axis of the RF coils [36]. The magnitude of the nonlinear Zeeman shift does not depend on the width of the resonance line and the heading errors due to this effect are therefore expected to be similar in magnitude to those of larger sensors, at least for ^{133}Cs and ^{87}Rb, where the splitting (in Earth's field) is much smaller than the resonance width for both types of magnetometers. Misalignment of the field-coil axis with the light-propagation axis generates a heading error that scales with the resonance width and hence is more important for microfabricated magnetometers than it is for larger sensors.

7.7 Applications of microfabricated magnetometers

As mentioned in the introduction, the main advantages of microfabricated atomic magnetometers over larger sensors are their small size, low power requirements, and potential low fabrication cost. These advantages come at the cost of reduced sensitivity, and hence the applications for which microfabricated atomic magnetometers will be most important are those that have a specific need for microfabricated magnetometers' intrinsic strengths.

Remote detection of magnetic anomalies is one area where low power and low cost may offer considerable advantage. All commercial atomic magnetometers developed to date require several watts of power to run. This essentially prohibits remote, battery-powered operation for extended periods. Microfabricated atomic magnetometers offer the possibility of moderate sensitivity ($\sim 100\ \mathrm{fT}/\sqrt{\mathrm{Hz}}$), while requiring only a few milliwatts of power [see Fig. 7.1(d)]. Li-ion batteries have an energy density of about 0.5 Wh/cm^3. Thus, a 1 cm^3 battery could power a microfabricated atomic magnetometer for a month or more. Hence, remote operation of a microfabricated atomic magnetometer is feasible within the constraints of the determining physics. One could imagine deploying large numbers of these sensors along a perimeter to detect ships, submarines, or vehicles.

A second major application is the detection of magnetic signals generated by the human body. While many parts of the body produce magnetic fields, the two most important are the heart and the brain. These organs generate fields with strengths of about 100 pT outside the chest, and 1 pT outside the head, respectively. To measure these fields, with reasonable signal-to-noise ratio, sensors with sensitivities in the $\mathrm{fT}/\sqrt{\mathrm{Hz}}$ range are needed. Traditionally, measurements of these fields have been limited to SQUID-based magnetic sensors, since only these have had the required sensitivity. However, both large-scale atomic

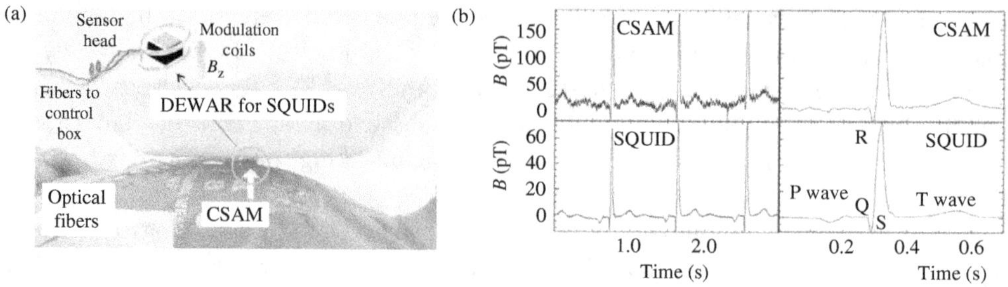

Figure 7.7 (From Knappe *et al.* [40]. Copyright 2010, American Institute of Physics.) Measurement of the magnetic field from a human heart by use of a microfabricated atomic magnetometer. (a) Experimental setup showing a microfabricated atomic magnetometer (CSAM) positioned below a SQUID dewar. (b) Measurements were made simultaneously with a SQUID sensor for comparison.

magnetometers [37–39] and more recently microfabricated atomic magnetometers [40] have been shown to be capable of measuring one or both of these biomagnetic sources. A photograph of an experiment comparing a microfabricated atomic magnetometer to a SQUID-based biomagnetic system is shown in Fig. 7.7(a); signals from a human heart are shown in Fig. 7.7(b).

A third application of microfabricated atomic magnetometers is in low-field nuclear magnetic resonance (NMR). Traditional NMR requires large magnetic fields to (a) polarize the nuclei in the sample and (b) create a nuclear resonance frequency high enough to be detected with high sensitivity by an inductive pick-up coil. Magnetometers with high sensitivity at low frequencies (DC, essentially) are an enabling component for some types of low-field NMR. While SQUID magnetometers have been used [41], there is growing interest in using instead atomic magnetometers [42] due to the lack of need for low-temperature cooling. Microfabricated atomic magnetometers offer an additional advantage in that they could be integrated into portable hand-held instruments for remote imaging [43] or chemical species identification [44].

A basic demonstration of the detection of nuclear magnetization was made by Ledbetter *et al.* [45]. In this experiment, shown schematically in Fig. 7.8(a), hydrogen nuclei (protons) in water were weakly polarized in a permanent magnet; the water then flowed into a microfluidic chip that included a microfabricated vapor cell, inside a shielded environment. The polarized nuclei produced a weak magnetic field that was comparable to the residual magnetic field from the magnetic shields. The orientation of the nuclear polarization was flipped using standard nuclear spin manipulation techniques and the resulting quasi-static change in the magnetic field that occurred as the reoriented atoms flowed into the microfluidic channel was measured, as shown in Fig. 7.8(b). The magnitude of this field change was about 20 pT and was easily detectible by use of the microfabricated atomic magnetometer.

Subsequent experiments have shown that intramolecular J-coupling can be detected at zero magnetic field with microfabricated vapor cells [46], allowing the identification of chemical species without relying on the chemical shift that occurs only at high magnetic

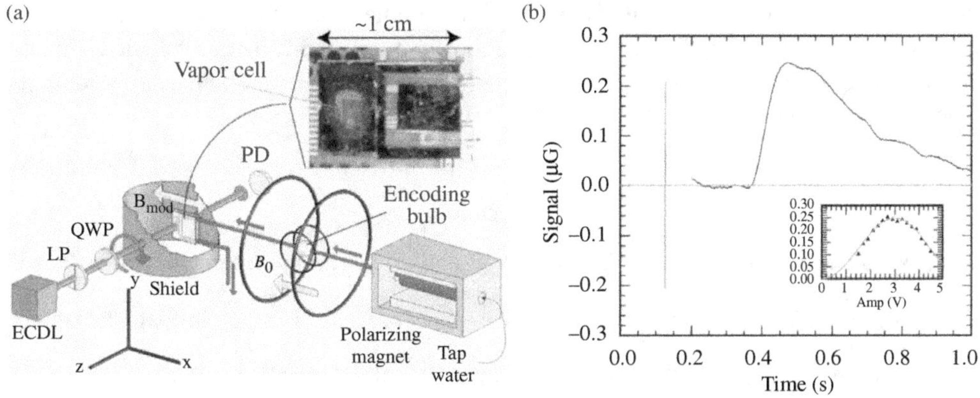

Figure 7.8 (From Ledbetter *et al.* [45].) Measurement of weak nuclear polarization in water by use of a microfabricated atomic magnetometer. (a) Experimental setup showing tap water running through a polarizing magnet and onto the NMR detection chip. (b) Change in the signal observed by the magnetometer after application of a π-pulse in the encoding bulb at the time indicated by the vertical line.

fields. Finally, an enhancement of the nuclear polarization by use of parahydrogen in an experiment with atomic-micromagnetometer detection was demonstrated by Theis *et al.* [47], leading to substantially enhanced magnetic resonance signals.

We anticipate that microfabricated atomic magnetometers will find a number of additional applications, including the measurement of magnetic fields in space and detection of magnetic micro- and nano-particles [40, 48].

7.8 Outlook

This chapter has reviewed recent work to develop highly miniaturized atomic magnetometers. The core technical element is the use of microfabricated alkali vapor cells, which allow the atoms to be confined in volumes below 1 mm^3. These cells can be integrated into sensor heads with volumes as small as 20 mm^3, and can achieve sensitivities as low as 5 fT/$\sqrt{\text{Hz}}$.

We anticipate three future directions for the development of this technology. First, improved sensitivity is always of value, and millimeter-scale sensors are not yet reaching the limits dictated by photon and atom shot noise. Second, we foresee substantial innovative device engineering with a goal of improving the integration of the vapor cells with MEMS, photonic, and fiber-optic components. This is expected to lead to simpler, less expensive and more reliable devices more suited to large-scale manufacturing. Finally, we expect that even smaller magnetic sensors, with cell volumes approaching 10^{-9} cm^3, will be developed that can be operated with extremely low power.

The technology appears highly promising for a broad range of applications. The potential for low-power operation of the sensors suggests that these devices will be very useful for remote detection of magnetic anomalies, where sensors can be deployed in remote locations for long periods, and run on battery power. On the other hand, the potential for low-cost production may substantially benefit applications in biomagnetic instrumentation. Finally, the small size may be important for hand-held nuclear magnetic resonance systems for chemical identification.

References

[1] V. Shah, S. Knappe, P. D. D. Schwindt, and J. Kitching, Nature Photonics **1**, 649 (2007).
[2] *Magnetic Sensors – Emerging Technology Developments,* Frost & Sullivan, Document No. D08C (2006).
[3] S. Knappe, V. Shah, P. D. D. Schwindt, L. Hollberg, J. Kitching, L. A. Liew, and J. Moreland, Appl. Phys. Lett. **85**, 1460 (2004).
[4] D. Budker and M. Romalis, Nature Physics **3**, 227 (2007).
[5] J. C. Allred, R. N. Lyman, T. W. Kornack, and M. V. Romalis, Phys. Rev. Lett. **89**, 130801 (2002).
[6] W. Happer and H. Tang, Phys. Rev. Lett. **31**, 273 (1973).
[7] W. Happer and A. C. Tam, Phys. Rev. A **16**, 1877 (1977).
[8] H. G. Robinson, E. S. Ensberg, and H. G. Dehmelt, Bull. Am. Phy. Soc. **3**, 9 (1958).
[9] M. V. Balabas, T. Karaulanov, M. P. Ledbetter, and D. Budker, Phys. Rev. Lett. **105**, 070801 (2010).
[10] M. V. Balabas, D. Budker, J. Kitching, P. D. D. Schwindt, and J. E. Stalnaker, J. Opt. Soc. Am. B **23**, 1001 (2006).
[11] W. Happer, Rev. Mod. Phys. **44**, 169 (1972).
[12] J. Kitching, S. Knappe, and L. Hollberg, Appl. Phys. Lett. **81**, 553 (2002).
[13] I. M. Savukov, S. J. Seltzer, and M. V. Romalis, J. Magn. Reson. **185**, 214 (2007).
[14] P. D. D. Schwindt, B. Lindseth, S. Knappe, V. Shah, J. Kitching, and L.-A. Liew, Appl. Phys. Lett. **90**, 081102 (2007).
[15] M. A. Perez, U. Nguyen, S. Knappe, E. A. Donley, J. Kitching, and A. M. Shkel, Sens. Act. A **154**, 295 (2009).
[16] X. Song, H. F. Dong, and J. C. Fang, *4th IEEE International Conference on Nano/Micro Engineered and Molecular Systems* 231 (2009).
[17] E. Hodby, E. Donley, and J. Kitching, Appl. Phys. Lett. **91**, 011109 (2007).
[18] J. Clem, IEEE Trans. Magn. **23**, 1093 (1987).
[19] T. Varpula and T. Poutanen, J. Appl. Phys. **55**, 4015 (1984).
[20] S. K. Lee and M. V. Romalis, J. Appl. Phys. **103**, 084904 (2008).
[21] W. C. Griffith, S. Knappe, and J. Kitching, Opt. Exp. **18**, 27167 (2010).
[22] L. A. Liew, S. Knappe, J. Moreland, H. Robinson, L. Hollberg, and J. Kitching, Appl. Phys. Lett. **84**, 2694 (2004).
[23] S. Knappe, in *Comprehensive Microsystems* **3**, edited by Y. Gianchandani, O. Tabata and H. Zappe (Elsevier B. V., Maryland Heights, MO, 2007), p. 571.
[24] G. Wallis and D. I. Pomerantz, J. Appl. Phys. **40**, 3946 (1969).
[25] S. Knappe, P. D. D. Schwindt, V. Shah, L. Hollberg, J. Kitching, L. Liew, and J. Moreland, Opt. Exp. **13**, 1249 (2005).
[26] R. R. Mhaskar, S. Knappe, and J. Kitching, *IEEE International Frequency Control Symposium*, 376 (2010).
[27] J. Preusser, V. Gerginov, S. Knappe, and J. Kitching, *IEEE Sensors Conference 2008*, Leece, Italy, 344–346 (2008). http://dx.doi.org/10.1109/ICSENS.2008.4716451.

[28] R. Lutwak, J. Deng, W. Riley, M. Varghese, J. Leblanc, G. Tepolt, M.Mescher, D. K. Serkland, K. M. Geib, and G. M. Peake, *36th Annual Precise Time and Time Interval (PTTI) Meeting, Washington, DC*, 339–354 (2004).
[29] M. J. Mescher, R. Lutwak, and M. Varghese, *IEEE International Conference on Solid-State Sensors and Actuators*, Seoul, Korea, 311–316 (2005). http://dx.doi.org/10.110.9/SENSOR.2005.1496419
[30] D. W. Youngner, J. F. Detry, and J. D. Zook, United States Patent 6,900,702 (2005).
[31] D. W. Youngner, L. M. Lust, D. R. Carlson, S. T. Lu, L. J. Forner, H. M. Chanhvongsak, and T. D. Stark, *Transducers/Eurosensors* **XXI**, U23 (2007).
[32] M. A. Perez, S. Knappe, and J. Kitching, *IEEE Sensors Conference*, Waikoloa, Hawaii, 2155–2158 (2010). http://dx.doi.org/10.1109/ICSENS.2010.5690546
[33] R. Wyllie, M. Kauer, G. S. Smetana, R. T. Wakai, and T. G. Walker, arXiv:1106.4779v1 (2011).
[34] H. B. Dang, A. C. Maloof, and M. V. Romalis, Appl. Phys. Lett. **97**, 151110 (2010).
[35] P. D. D. Schwindt, S. Knappe, V. Shah, L. Hollberg, J. Kitching, L. A. Liew, and J. Moreland, Appl. Phys. Lett. **85**, 6409 (2004).
[36] A. L. Bloom, Appl. Opt. **1**, 61 (1962).
[37] G. Bison, R. Wynands, and A. Weis, Opt. Exp. **11**, 904 (2003).
[38] H. Xia, A. B. A. Baranga, D. Hoffman, and M. V. Romalis, Appl. Phys. Lett. **89**, 211104 (2006).
[39] C. Johnson, P. D. D. Schwindt, and M. Weisend, Appl. Phys. Lett. **97**, 243703 (2010).
[40] S. Knappe, T. H. Sander, O. Kosch, F. Wiekhorst, J. Kitching, and L. Trahms, Appl. Phys. Lett. **97**, 133703 (2010).
[41] R. McDermott, S. K. Lee, B. ten Haken, A. H. Trabesinger, A. Pines, and J. Clarke, Proc. Nat. Acad. Sci. **101**, 7857 (2004).
[42] V. V. Yashchuk, J. Granwehr, D. F. Kimball, S. M. Rochester, A. H. Trabesinger, J. T. Urban, D. Budker, and A. Pines, Phys. Rev. Lett. **93**, 160801 (2004).
[43] J. A. Seeley, S. I. Han, and A. Pines, J. Magn. Reson. **167**, 282 (2004).
[44] E. E. McDonnell, S. L. Han, C. Hilty, K. L. Pierce, and A. Pines, Anal. Chem. **77**, 8109 (2005).
[45] M. P. Ledbetter, I. M. Savukov, D. Budker, V. Shah, S. Knappe, J. Kitching, D. J. Michalak, S. Xu, and A. Pines, Proc. Nat. Acad. Sci. **105**, 2286 (2008).
[46] M. P. Ledbetter, C. W. Crawford, A. Pines, D. E. Wemmer, S. Knappe, J. Kitching, and D. Budker, J. Magn. Reson. **199**, 25 (2009).
[47] T. Theis, P. Ganssle, G. Kervern, S. Knappe, J. Kitching, M. P. Ledbetter, D. Budker, and A. Pines, Nature Physics **7**, 571 (2011).
[48] D. Maser, S. Pandey, H. Ring, M. P. Ledbetter, S. Knappe, J. Kitching, and D. Budker, Rev. Sci. Instrum. **82**, 086112 (2011).

8

Optical magnetometry with nitrogen-vacancy centers in diamond

V. M. Acosta, D. Budker, P. R. Hemmer, J. R. Maze, and R. L. Walsworth

8.1 Introduction

While atomic magnetometers can measure magnetic fields with exceptional sensitivity and without cryogenics, spin-altering collisions limit the sensitivity of sub-millimeter-scale sensors [1]. In order to probe magnetic fields with nanometer spatial resolution, magnetic measurements using superconducting quantum interference devices (SQUIDs) [2–4] as well as magnetic resonance force microscopes (MRFMs) [5–8] have been performed. However, the spatial resolution of the best SQUID sensors is still not better than a few hundred nanometers [9] and both sensors require cryogenic cooling to achieve high sensitivity, which limits the range of possible applications. A related challenge that cannot be met with existing technology is imaging weak magnetic fields over a wide field of view (millimeter scale and beyond) combined with sub-micron resolution and proximity to the signal source under ambient conditions.

Recently, a new technique has emerged for measuring magnetic fields at the nanometer scale, as well as for wide-field-of-view magnetic field imaging, based on optical detection of electron spin resonances of nitrogen-vacancy (NV) centers in diamond [10–12]. This system offers the possibility to detect magnetic fields with an unprecedented combination of spatial resolution and magnetic sensitivity [8, 12–15] in a wide range of temperatures (from 0 K to well above 300 K), opening up new frontiers in biological [10, 16, 17] and condensed-matter [10, 18, 19] research. Over the last few years, researchers have developed techniques for nanoscale magnetic imaging in bulk diamond [11, 12, 20] and in nanodiamonds [21–23] along with scanning probe techniques [10, 24]. Sensors employing ensembles of NV centers promise even higher sensitivity and the possibility to map out all vector components of the magnetic field [14, 25], as well as wide-field-of-view magnetic field imaging with sub-micron resolution. Pilot NV-ensemble magnetometers and imagers have recently been demonstrated by several groups [15, 18, 19, 26–29].

Optical Magnetometry, ed. Budker, D. and Jackson Kimball, D. F. Published by Cambridge University Press.

The development of these techniques with an eye toward applying them to image novel microscale and nanoscale magnetic phenomena is the focus of this chapter. We will explore the history of optical magnetometry in diamond, the underlying physics which makes nitrogen-vacancy centers such remarkable sensors, and the basic techniques for utilizing their potential.

8.1.1 Comparison with existing technologies

Sensitivity estimates for NV-based magnetometers compare favorably with the currently most sensitive devices, including SQUID sensors and atomic magnetometers. While the ultimate verdict will come from experiments, the prospect of a robust, scalable, solid-state system with a superior operating temperature range (from 0 K to well above room temperature) has many researchers champing at the bit.

Spin-based magnetometers are fundamentally limited by the quantum noise associated with spin projection. Under ideal readout conditions, the minimum detectable magnetic field for a sample of spins with density n in a volume V is given (see Chapters 1 and 2, also Refs. [1, 14]) by

$$\delta B_q \simeq \frac{1}{\gamma}\frac{1}{\sqrt{nVt_mT_2^*}}, \tag{8.1}$$

where $\gamma = 1.761 \times 10^{11}\ \mathrm{s^{-1}T^{-1}}$ is the NV gyromagnetic ratio [30], T_2^* is the electron spin dephasing time, and $t_m \gtrsim T_2^*$ is the measurement time.

Figure 8.1 compares the sensitivity and operating temperatures of the currently most sensitive technologies with two types of diamond sensors being developed, a ~50 μm-scale sensor [15] for low-field NMR detection [31, 32] and general field use, and a 2D surface probe with a spatial resolution of ~10 nm [8, 10, 11, 13, 14, 24] for condensed-matter [19] and biological [16, 17, 27] applications. At centimeter and longer length scales, vapor-based atomic magnetometers are currently the most sensitive devices, but nanoscale spatial resolution is not possible due to thermal motion of the atoms. By employing a high density of a noble gas, vapor-based magnetometry with a spatial resolution of ~10 μm is possible [33], but despite impressive results for mm-scale sensors [31, 34] the projected sensitivity for this length scale is exceeded by the diamond-based sensors presently proposed. Further, for optimal operation, vapor-based cells must be heated to well above room temperature, which increases the power consumption and limits the range of applications.

In the case of SQUID sensors, significantly better spatial resolution is possible, but the operating temperature range is cryogenic, limiting applications. Recently, an Al nano-SQUID was placed on a scanning tip, and the probe of diameter of ~180 nm demonstrated a sensitivity of ~100 $\mathrm{nT}/\sqrt{\mathrm{Hz}}$ [9]. The diamond-based sensors discussed here promise better spatial resolution and comparable magnetic sensitivity without cryogenic operation. Sensors based on giant magnetoresistivity (GMR) [39] and the Hall effect [40, 41] can also be miniaturized to the sub-micron scale and operated at room temperature [39], but their low-frequency sensitivity is typically 1–2 orders of magnitude worse than that of SQUID

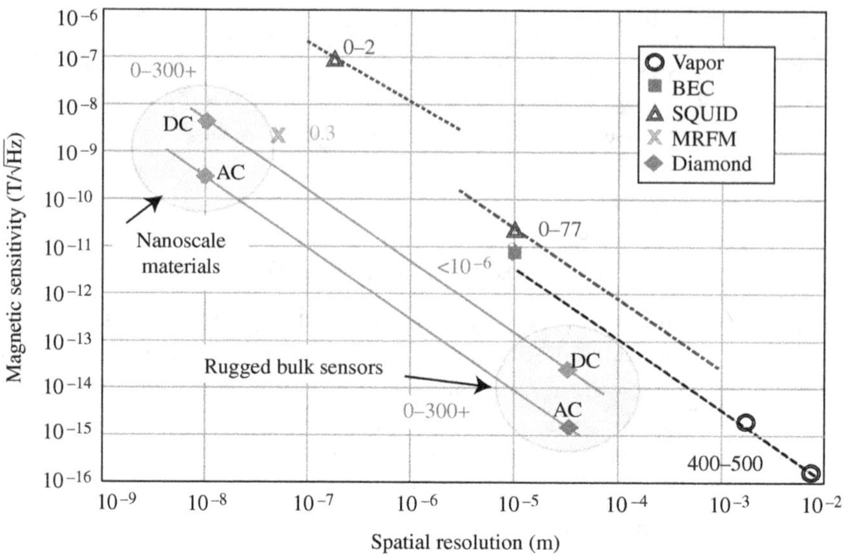

Figure 8.1 Comparison of some of the most sensitive magnetometry technologies. The points for diamond are the projected sensitivities based on Eq. (8.1), and the other points are experimental results for magnetometers based on vapor cells [35,36], BECs [37], low-frequency micro-SQUIDs [38], nano-SQUIDs [9], and narrowband MRFMs [7]. The lines represent the theoretical range and scaling of the spatial resolution for potential sensors based on these technologies (MRFMs and BECs have limited range and complicated scaling and are not shown here). Note that the sensitivity scaling with size for all sensors is $\delta B \propto r_{\text{min}}^{-3/2}$ (see, for example, [1]) except in the case of nano-SQUIDs (without a pickup coil) where the scaling is $\delta B \propto r_{\text{min}}^{-5/4}$ [2]. The numbers are the temperature ranges (in kelvin) for each sensor as reported in the references.

sensors of comparable size. MRFM and Bose–Einstein-condensate (BEC) magnetometers (Chapter 9) also offer impressive spatial resolution and magnetic sensitivity; however, the large magnetic field gradients used in MRFM (approaching 10^6 T/m [5, 42]) and the sub-kelvin-range operating temperatures for both BECs and MRFMs are impractical for many applications.

8.2 Historical background

The use of optically detected magnetic resonance (ODMR, defined here as the detection of spin transitions by optical means) for precision magnetometry is not restricted to gas-phase atoms, though it did indeed begin there [43]. Figure 8.2 shows a brief history of ODMR including recent applications to NV-based magnetometry. Following the first demonstration of ODMR by Fermi and Rasetti in 1925 using an atomic vapor, there was considerable progress made in the 1950s both theoretically [44] and experimentally both for gas-phase atoms [45] and in solids [46, 47]. The grandfather to ODMR, all-microwave electron spin

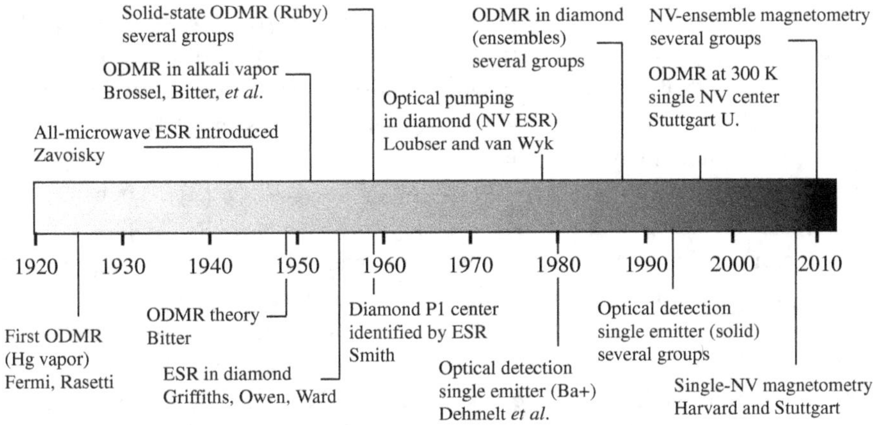

Figure 8.2 Timeline of major developments leading to magnetometry based on ODMR of NV centers in diamond.

resonance (ESR), was developed in the 1940s [48] and soon extended to study defects in diamond [49–51]. While ESR has proven a powerful analytic tool for bulk samples, its reliance on thermal polarization and microwave detection have prevented widespread use in practical magnetometry.[1]

Signal enhancement beyond the thermal polarization limit (optical pumping) of various ESR-active defects in diamond under continuous optical excitation was reviewed by Loubser and van Wyk in 1978 [52]. In that work a particularly luminous "color center," associated with a substitutional nitrogen atom lying nearest-neighbor to a carbon vacancy (the NV center), was identified and the model for the center involving six valence electrons was developed. In the late 1980s, ODMR was observed in high-nitrogen-density, electron-irradiated diamonds [53, 54]. From this and earlier work, it was confirmed that the center was the negatively-charged NV^- center, with ESR transitions occurring in the spin-triplet ground state and optical transitions with an excited state with zero-phonon line at 637 nm [55].

8.2.1 Single-spin optically detected magnetic resonance

Given the ubiquity and importance of optical detection of single quantum systems today, it may come as a surprise that it was only first realized in 1980, when Dehmelt and co-workers observed fluorescence from a single trapped Ba ion [56]. ODMR on a single emitter did not come until 1993 when it was realized by two separate collaborations using cryogenic pentacene molecules embedded in a solid [57, 58]. After this groundbreaking work, the search was on to find a suitable room-temperature system using technologically relevant

[1] See online supporting material (www.cambridge.org/9781107010352).

materials. In 1997 Jörg Wrachtrup and co-workers found such a system when they performed ODMR at 300 K on single NV centers in diamond [59]. In addition to having a paramagnetic ground state and efficient optical pumping mechanism, it was noted that, unlike most color centers, organic molecules, and semiconductor quantum dots, most NV centers in bulk diamond exhibit stable fluorescence that does not bleach or blink. This feature, combined with the low cytotoxicity of the diamond host [60–62] makes NV centers in nanodiamonds ideal candidates for fluorescent markers in biological systems [21, 63–66].

The decade following the initial discovery was marked by major accomplishments in optical and microwave control of the electronic and nuclear degrees of freedom of the NV center and neighboring spins [67–76]. This progress, combined with the realization of extremely long electron-spin coherence times by condensed-matter standards ($T_2 > 1$ ms for isotopically pure diamond [12]) quickly made the NV center a leading candidate for qubits in a solid-state quantum computer [77].

During this time, it was also realized that these properties make NV centers an ideal candidate for optical magnetometry [8, 13, 14]. The first realization of a single-spin magnetometer was simultaneously achieved by two groups (Harvard [11] and Stuttgart [10]). In the last few years several more groups have developed NV-based magnetometers using single spins in bulk diamond [11, 12] and in nanodiamonds on scanning tips [10, 24], as well as large ensembles in micron-scale devices [15, 18, 26, 28]. We review these developments and the underlying physics below.

8.3 NV center physics

The NV center arises when a vacant lattice site lies nearest-neighbor to a substitutional nitrogen atom in the diamond lattice, Fig. 8.3(a). Two types of such centers have been identified, the neutral NV^0 center and the singly charged NV^- center (commonly referred to as the NV center). The latter is of interest to magnetometry because it has a spin-1 ground state [52] with a long spin-coherence time.

The negatively charged NV center is believed to consist of six electrons: the vacant site results in broken bonds or dangling bonds that point toward the vacancy. Three of these electrons come from the three carbons next to the vacancy and the other two electrons come from the nitrogen atom. An extra electron is believed to come from the lattice, probably from other defects such as substitutional nitrogens [78, 79].

8.3.1 Intersystem crossing and optical pumping

One of the crucial features of nitrogen-vacancy centers for practical magnetometry is that they can be optically pumped and interrogated using visible light over a broad range of wavelengths (∼480–640 nm).[2] Figure 8.3(b) shows the level diagram with energy levels and allowed transitions labeled according to a recently proposed model [82, 83, 85]. As discussed in the online supporting material, the center's ground state is a $^3A_2$ spin triplet and optical

[2] See online supporting material (www.cambridge.org/9781107010352).

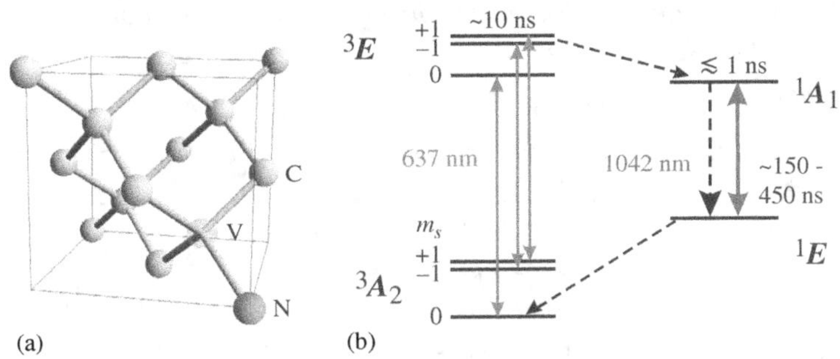

Figure 8.3 (a) Schematic of the NV center in the diamond unit cell. (b) Level diagram for the NV center showing spin-triplet ground and excited states, as well as the singlet system involved in intersystem crossing. Radiative transitions are indicated by solid arrows and nonradiative transitions by dashed arrows. Only those transitions confirmed by experiments are shown. The tentative label of the lower (upper) singlets as ^1E ($^1A_1$) is based on recent theoretical models [80, 81] but other models suggest the reverse ordering [82]. Lifetimes of the excited levels are also shown [83], with the lower singlet lifetime being temperature dependent (~220 ns at room temperature and ~450 ns at 5 K) [83, 84].

transitions connect it with a spin-triplet excited state, ^3E, with the zero-phonon line (ZPL) at ~637 nm. In thermal equilibrium at temperatures above ~1 K, the ground state sublevels are nearly equally populated. After interaction with sufficiently intense light exciting the $^3A_2 \to {}^3E$ transition, the $m_s = \pm 1$ sublevels become depopulated, and eventually 80% or more of the total population accumulates in the $m_s = 0$ ground state [30, 71, 84, 86–88].

This occurs because at least one singlet level lies close in energy to ^3E, and spin–orbit coupling induces triplet–singlet intersystem crossing (ISC) [89]. Experiments have shown that NV centers in the $m_s = \pm 1$ magnetic sublevels have significantly higher probability to undergo ISC [15, 83–85]. NV centers which undergo ISC then decay to another, longer-lived singlet level either nonradiatively or radiatively (ZPL at 1042 nm) [15, 83, 85], after which they cross over to the $^3A_2$ ground state.

By determining the optical-polarization selection rules, experiments [83] have shown that the levels involved in the ISC include ^1E and $^1A_1$.[3] However the ordering of these levels is still under debate [80–82], and the exact decay path responsible for optical pumping remains an open area of research.

Regardless, it is clear that optical excitation results in spin polarization into $m_s = 0$. Furthermore, as NV centers originating in $m_s = \pm 1$ are more likely to decay nonradiatively through the ISC mechanism, NV centers in $m_s = 0$ have a higher probability (per excitation cycle) for fluorescence than those in $m_s = \pm 1$. For dilute samples this fluorescence contrast can be as high as 30%, limited by the difference in branching ratios between $m_s = 0$ and

[3] We note that there exists some evidence that a third level may be involved in the ISC, though the nature of this level is still unknown [83, 84].

Table 8.1. *Coupling coefficients for the ground-state Hamiltonian in Eq. (8.2). We use units where Planck's constant, $h = 1$. ZFS = zero-field splitting.*

Coefficient	Description	Value	Ref.
D	Axial ZFS parameter	2.87 GHz	[92]
E	Transverse ZFS parameter	kHz – MHz	[59,92]
g_s	Electron-spin g-factor	2.003	[30]
$A_{\parallel}$	Axial hyperfine constant	−2.16 MHz	[93]
$A_{\perp}$	Transverse hyperfine constant	−2.7 MHz	[30]
P	Quadrupole splitting parameter	−4.95 MHz	[93,94]
g_I	Nuclear-spin g-factor	0.403	[30]

$m_s = \pm 1$ into and out of the singlet system [10,90]. This spin-selective fluorescence serves as the basis for optical readout in most magnetometry schemes to date [10–12,18,19,24,26,28].

8.3.2 Ground-state level structure and ODMR-based magnetometry

The level structure in the $^3A_2$ ground state is governed by the Hamiltonian $\mathcal{H}_{tot} = \mathcal{H}_S + \mathcal{H}_{SI} + \mathcal{H}_I$, where $\mathcal{H}_S$ refers to the part which affects only the electronic spin ($S = 1$), $\mathcal{H}_{SI}$ is the hyperfine coupling to the nitrogen nucleus ($I = 1$ for ^{14}N),[4] and $\mathcal{H}_I$ is the nuclear portion. These terms can be written as[5]

$$\begin{cases} \mathcal{H}_S &= DS_z^2 + E(S_x^2 - S_y^2) + g_s\mu_B\vec{B}\cdot\vec{S} \\ \mathcal{H}_{SI} &= A_{\parallel}S_zI_z + A_{\perp}(S_xI_x + S_yI_y) \\ \mathcal{H}_I &= PI_z^2 - g_I\mu_N\vec{B}\cdot\vec{I}, \end{cases} \tag{8.2}$$

where μ_B is the Bohr magneton, μ_N is the nuclear magneton, and the other coefficients are defined in Table 8.1, which also lists experimental values.

Figure 8.4(a) displays the fine and hyperfine structure of the ^{3}A$_2$ ground state. The ground state is split by $D \approx 2.87$ GHz due to spin–spin interaction between the unpaired electrons [52,92]. The transverse ZFS, E, is nonzero only in the presence of crystal strain. Here we take $E = 0$, conserving perfect C_{3v} symmetry.[6] In the presence of an axial magnetic field,

[4] In natural abundance, nitrogen is composed of 99.6% ^{14}N, with spin $I = 1$, and 0.4% ^{15}N, with spin $I = 1/2$. We choose to treat only the more naturally abundant ^{14}N nuclei, though there has been some work related to selective implantation and growth of ^{15}NV centers [91].

[5] While, in natural abundance, 98.9% of C atoms are ^{12}C ($I = 0$), some NV centers also have a nearby ^{13}C atom, and consequently a further level structure can exist and has been studied extensively [30,52,79]. We will not consider these special centers here, instead focusing on decoherence and line broadening due to coupling to the bath of randomly located ^{13}C spins in Section 8.3.3.

[6] The effect of nonzero E is discussed in the online supporting material.

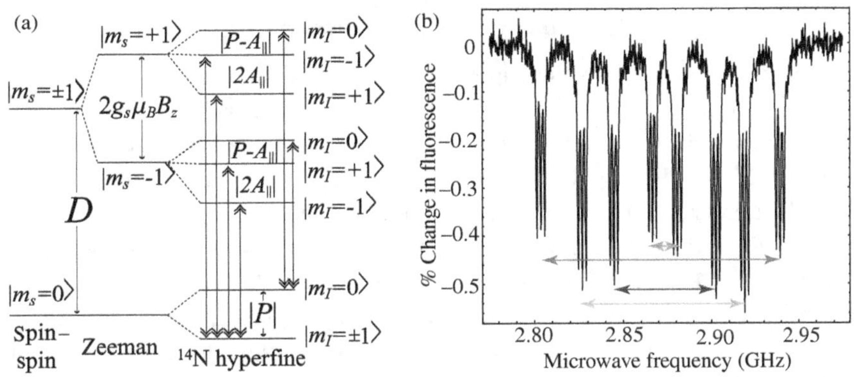

Figure 8.4 (a) Ground-state level structure of the NV center under a small axial magnetic field. Allowed magnetic-dipole transitions are shown with double-headed arrows. The values for the ^{14}N hyperfine splitting can be found in Table 8.1. The nuclear Zeeman effect is not included and E is taken to be zero here. (b) ODMR spectrum (excited at 532 nm) for a low-defect-density CVD diamond [25]. The 24 separate resonances correspond to two different $|\Delta m_S| = 1$ transitions, four different NV orientations, and three different ^{14}N m_I levels. The spectra for each orientation are shifted due to an applied field of ~50 G. Arrows connect resonances of a given orientation.

the $|m_s = \pm 1\rangle$ levels split, each one shifting by $m_S g_S \mu_B B_z$. Additional structure due to coupling to the ^{14}N nucleus is shown with the $m_I = 0$ level separated from $m_I = \pm 1$ by $P \pm A_{\|}$.[7]

The selection rules for magnetic-dipole transitions are $\Delta m_s = \pm 1$ and $\Delta m_I = 0$. These transitions can be driven by an oscillating magnetic field oriented perpendicular to the NV axis. For a single NV center, there are six allowed transitions, as indicated in Fig. 8.4(a). For ensembles, there are four different NV axes and, since only the projection of a small magnetic field on the NV axis affects the transition frequency, this leads to four shifted copies of the single spectrum. For special magnetic-field orientations [i.e., along the (100) crystal orientation], B_z is the same for all four NV orientations, producing only six resonances. However, for most field orientations, B_z is different for each NV orientation, and consequently there are 24 different resonances. This feature makes diamond magnetometers fundamentally vector magnetic sensors, enabling measurement of not only the magnitude but also the direction of an external magnetic field [14, 18, 26, 95, 96].

A fluorescence-detected magnetic-resonance spectrum is shown in Fig. 8.4(b). When the microwave frequency is off resonance, optical transitions are predominately from $|{}^3A_2, m_s = 0\rangle \leftrightarrow |{}^3E, m_s = 0\rangle$ and the fluorescence is at maximum intensity. When the microwaves are tuned to resonance, population is transferred to $m_s = \pm 1$, resulting in non-radiative ISC and consequently diminished fluorescence intensity. This is the basic principle of operation of most diamond-based magnetometers to date [10–12, 15, 18, 19, 24, 26, 28].

[7] We note that due to the negligible spin–orbit coupling in the ground state, the hyperfine splitting arises from magnetic interaction and is not due to electric quadrupole effects as is sometimes the case in alkali atoms.

8.3.3 Interaction with environment

The NV center's spin is largely decoupled from the lattice, and the main sources of decoherence (characterized by T_2 and T_2^* timescales) are from the paramagnetic-impurity spin bath (which dominates for high nitrogen concentration) and interactions with spin-1/2 ^{13}C nuclei [12, 14, 25, 97–99]. Population decay (characteristic timescale T_1) is primarily dominated by three interactions: (1) at room temperature and above it is due to Raman-type interaction with lattice phonons; (2) at lower temperature it is dominated by Orbach-type interaction with local phonons [98, 100, 101]; and (3) at even lower temperature ($\lesssim 100$ K), defect-density-dependent cross-relaxation effects can play a role, and the resulting T_1 can be tuned by more than an order of magnitude using an external magnetic field in some cases [101]. Under most doping conditions and temperatures, $T_1 \gg T_2 \gg T_2^*$, in stark contrast to dense alkali vapor where these timescales are often nearly equal. Below, we give a brief overview of contributions to dephasing and techniques to reach the T_2 and even T_1 limit.

Contributions to T_2^*

A well-known cause of spin dephasing is due to dipolar hyperfine coupling with nearby ^{13}C nuclear spins (natural 1.1% abundance). These nuclear spins are largely unpolarized [102] and produce different effective magnetic fields at the site of each NV center, leading to inhomogeneous broadening of ensemble resonances. Even single NV centers are affected by the ^{13}C spin bath since the unpolarized nuclei precess in the presence of applied or nearby-defect-induced fields, producing a superposition of time-varying magnetic fields with random phase [97]. Based on experiments with single NV centers, this contribution to the spin relaxation rate is $\gamma_C \approx 10^6\ s^{-1}$ [12, 69, 99], and similar results have been reported for ensembles [25, 30]. At very low magnetic fields, $|B_z| \ll E/(g_s\mu_B)$, the NV center is relatively immune to Zeeman shifts, and extremely long values of $T_2^* \approx 40\ \mu s$ have been observed for $m_I = 0$ nuclear spin sublevels in single NV centers [103]. Unfortunately, this regime is not suitable for magnetometry for the same reason T_2^* is long (i.e., insensitivity to magnetic fields).[8] A promising route is to fabricate isotopically pure diamond, where even longer T_2^* times are possible [12, 99].

For dilute samples, the contribution of NV^-–NV^- dipolar interactions to the magnetic-resonance broadening can be approximated by assuming that each NV^- center couples to only the nearest-neighboring NV^- center. For an ensemble, this dipolar coupling leads to a spin-relaxation contribution on the order of $\gamma_{NV} \approx (g_s\mu_B)^2 n_{NV}$, where n_{NV} is the NV^- concentration [14, 25]. For $[NV^-] = 15$ ppm, this corresponds to $\gamma_{NV} \approx 10^6\ s^{-1} \approx \gamma_C$. For ensemble magnetometry using natural-carbon-abundance material, it is worth noting that increasing $[NV^-]$ beyond this level would not improve the low-frequency magnetometer sensitivity, since now $1/T_2^* \propto n_{NV}$ [recall Eq. (8.1)].

[8] However, the NV center can still be a good room-temperature electric-field sensor in this regime [103].

A figure-of-merit for fabricating optimized diamond samples for ensemble work is the nitrogen-to-NV$^-$ conversion efficiency. It is generally assumed that two nitrogen atoms are required to form an NV$^-$ center, one to contribute the actual atom and the other to serve as the donor [25,78,104]. The remaining N^+ is not paramagnetic, but the NV centers and any unconverted nitrogen atoms are. For sufficiently high nitrogen concentration and nitrogen-to-NV$^-$ conversion efficiency of less than 33%, unconverted nitrogen is the main source of NV$^-$ dephasing and dominates over γ_{NV}. Similarly to the estimate of NV$^-$–NV$^-$ interactions, the characteristic timescale for this dephasing is $\gamma_N \approx (g_s\mu_B)^2 n_N$, where we ignore the small difference in g-factors between NV$^-$ and nitrogen [52]. Conversion efficiencies of ~20% [25,105] are now routine for large ensembles, and even higher conversion efficiencies are possible under some conditions [106].

Refocusing the dephasing

Coherent-control techniques can improve the sensitivity for kHz-scale AC fields. As nuclear spins are well protected from the environment, the characteristic correlation times for nuclear spin flips are long compared to T_2 of the NV center. A spin-echo pulse sequence [107] can be employed to remove the effect of these slowly-varying environmental perturbations.

Figure 8.5 shows a timing diagram for such a pulse sequence. NV centers initially in the $m_s = 0$ state are prepared in a superposition via a resonant $\pi/2$ microwave pulse and

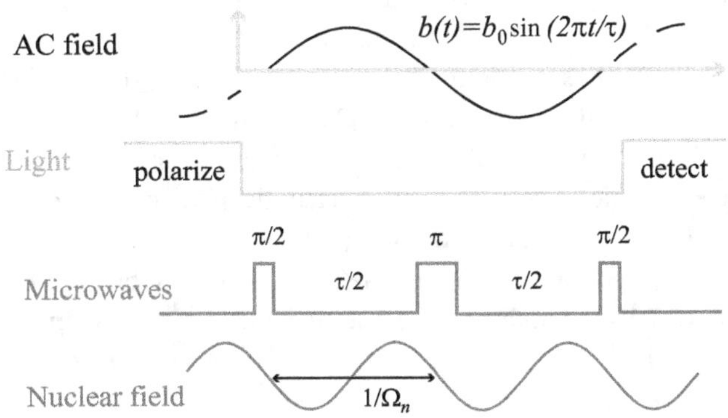

Figure 8.5 Optical and microwave spin-echo pulse sequence used for sensing an AC magnetic field, $b(t)$. NV centers are first polarized into the $m_s = 0$ sublevel. A coherent superposition between the states $m_s = 0$ and $m_s = 1$ (or alternatively $m_s = -1$) is created by applying a microwave $\pi/2$ pulse tuned to this transition. The system freely evolves for a period of time $\tau/2$, followed by a π refocusing pulse. After a second $\tau/2$ evolution period, the electronic spin state is projected onto the $m_s = 0, 1$ (or $m_s = 0, -1$) basis by a final $\pi/2$ pulse, at which point the ground-state polarization is detected optically via spin-dependent fluorescence. The optimal sensitivity is achieved for $\tau \approx T_2$, making this magnetometry scheme ideal for kHz-scale AC fields. The DC magnetic field is adjusted to eliminate the contribution of the randomly phased field produced by ^{13}C nuclear spins (Larmor period, $1/\Omega_n$) by choosing $\tau = 2n/\Omega_n$, for integer n.

evolve over a time $\tau/2$. During this precession window, the nuclear spins also precess, in the presence of a bias magnetic field, with random phase at angular frequency Ω_n. As this precession is not synchronized with the echo sequence, it leads to a time-varying magnetic field which dephases the NV center. However, if the echo sequence is timed such that on each half of the spin echo sequence, the nuclei complete an integer number of Larmor periods, the dephasing due to the oscillating component of the magnetic field produced by the nuclei cancels on each half of the sequence regardless of the phase of the nuclear field. Meanwhile the dephasing due to the constant part of the nuclear fields is refocused by the echo. These two conditions lead to a revival in the NV coherence [108]. Off-axis magnetic fields enhance the Larmor precession of nuclear spin in a position-dependent manner. Therefore, to maximize refocusing, the magnetic field must be aligned along the NV axis [27,68,97].

In essence, the spin-echo technique enables extension of the interrogation time from the limit set by T_2^* up to a value T_2 that is close to the intrinsic spin-coherence time, at the cost of reduced bandwidth and insensitivity to magnetic-field frequencies $\lesssim 1/T_2$. These techniques have been used to improve the sensitivity of diamond magnetometers by more than an order of magnitude to kHz-scale AC fields of known phase [10–12, 27] as well as more modest improvements in sensitivity to randomly fluctuating fields [109, 110]. An important feature of this technique is that it is highly frequency selective, detecting only fields oscillating with frequency in a narrow band around $2\pi/\tau$.

Other pulse sequences can be used to further decouple the NV center from its environment. Examples include Mansfield, Rhim, Ellis, and Vaughan (MREV) sequences to refocus NV–NV coupling [14, 111] and Carr-Purcell-Meiboom-Gill (CPMG) or Uhrig dynamic-decoupling (UDD) pulse sequences to further reduce coupling to all environmental spins [14,109,112–114]. Using these techniques, room-temperature T_2 times of a few milliseconds have been realized using ultrapure diamond [112, 113] and a few hundred microseconds for nitrogen-rich HPHT diamond [80]. Dynamic-decoupling techniques have recently been extended to large ensembles of NV centers, and a collective coherence time reaching $T_2 >$ 2 ms was observed [114, 115]. The ultimate limit is T_1 relaxation due to interaction with the lattice. In most samples, at room temperature, $T_1 \approx 10$ ms, but it is highly temperature dependent [86,98,100,101], and low-temperature devices with coherence times of minutes are theoretically possible.

8.4 Experimental realizations

8.4.1 Near-field scanning probes and single-NV magnetometry

Perhaps the most conceptually simple implementation of a nanoscale magnetometer is using a near-field scanning-probe technique. A nanodiamond with a single NV center is placed at, for example, the end of an atomic-force microscope (AFM) and the resonance frequency (or for AC measurements, the spin echo amplitude) is monitored using the ODMR techniques described in Section 8.3. The probe is scanned laterally just above the sample

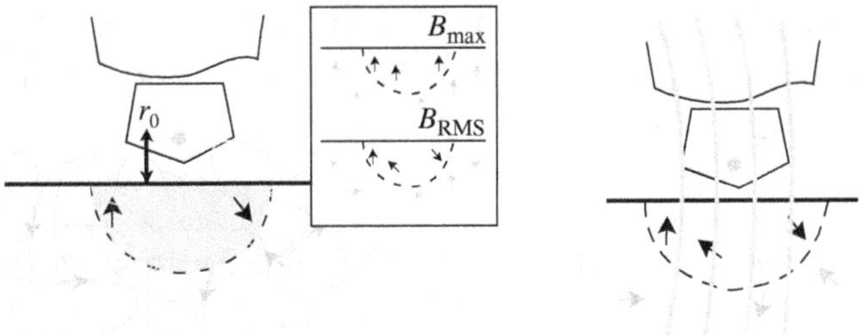

(a) (b)

Figure 8.6 Illustration of high-spatial-resolution magnetometry with a diamond nanocrystal. (a) The dipolar fields from spins in the sample decay rapidly with distance; only those within a distance $\sim r_0$ contribute to the observable signal for the point-like NV center. The inset shows how B_{max} (the magnetic field produced by fully polarized nuclei) and B_{RMS} are related; when few spins are involved, the statistical fluctuations become large. (b) In the presence of a magnetic-field gradient (field lines in grey), only the spins from a small region of the detection volume are precessing at the frequency band center of the detector, enabling even higher spatial resolution. Adapted from Ref. [14].

of interest in order to produce a multidimensional map of the magnetic field produced by spins in the sample. In some situations it may also be possible to attach the sample on the probe and scan over the surface of the diamond. Such a system is called "near-field" because its spatial resolution is limited only by the distance between the NV center and the object of study, not by the diffraction limit set by the wavelength of the fluorescence light.

One of the primary goals of these near-field scanning probes is to detect the signal from a small ensemble of nuclear or electronic spins in the sample of interest. The ultimate limit would be to detect the signal from a single electronic spin [5], or perhaps eventually a single nuclear spin [8, 14]. To see how this may be possible, consider a material with a varying nuclear spin density $n_s(\mathbf{r})$ that is brought in close proximity (distance $\sim r_0$) to the NV center [Fig. 8.6(a)]. Even at cryogenic temperatures, the thermal nuclear spin polarization of the material is small. However, because only a few spins are involved, the distribution of spin configurations leads to an appreciable variance in the spin polarization [116], proportional to $\sqrt{N}$, where $N \approx {r_0}^3 n_s$ is the number of nuclear spins which the sensor is able to detect.

More precisely, this statistical polarization produces a substantial, albeit randomly oriented, magnetic field with projection along the NV axis of [8]:

$$B_{\text{RMS}} \approx \frac{\gamma_n}{2}\left\{\int d^3\mathbf{r}\frac{n_s(\mathbf{r})(1-3\cos\theta(\mathbf{r}))}{|\mathbf{r}|^6}\right\}^{1/2}, \tag{8.3}$$

where γ_n is the nuclear gyromagnetic ratio, $\theta(\mathbf{r})$ is the angle between $\hat{\mathbf{r}}$ and the NV axis at location $\mathbf{r}$, and we have assumed the nuclei are protons (spin $I = 1/2$). In a typical organic molecule the protons are separated by about 1 Å, so $n_s \approx 10^{23}$ cm^{-3}. If the NV center could be positioned $r_0 = 20$ nm from the molecule, the magnetic signal calculated from Eq. (8.3) would be $B_{\text{rms}} \approx 100$ nT, which is equivalent to the field from ~300 proton magnetic moments located 20 nm from the NV center. Such a fluctuating field amplitude is theoretically detectable by a single-NV magnetometer with less than 1 ms of signal averaging (see Fig. 8.1) using an appropriate detection scheme [14, 16, 109, 110].

Further spatial resolution, potentially down to the single-spin level, is possible by employing a large magnetic-field gradient, to encode spatial information spectrally, as is commonly employed in conventional MRI. Using similar techniques to those used in MRFM [42], a magnet near the surface of a substrate can produce gradient fields of order $dB/dr \approx 10^6$ T/m [Fig. 8.6(b)]. The narrow kHz-scale bandwidth of an AC diamond magnetometer ($\sim 1/T_2$) [11], enables it to spectrally distinguish two protons separated by a magnetic-field difference of $\Delta B \approx 0.1$ mT, corresponding to physical separation of $(dB/dr)/\Delta B \approx 0.1$ nm. This implies that individual proton detection may be possible even in organic and biological molecules.

In one of the first demonstrations of the near-field approach performed at Stuttgart [10], a nanodiamond was attached to the tip of an AFM tip [shown schematically in Fig. 8.7(a)], and fluorescence was collected by confocal microscopy using light from a 532 nm laser (coming from below the substrate). A nickel micro-magnet [optical image in Fig. 8.7(b)] was placed on a clear substrate and the near-field diamond nanoscope was scanned in the two lateral dimensions. A microwave field with frequency 2.75 GHz was applied, and the NV^- fluorescence was recorded. The nickel magnet produced a large field gradient such that the microwaves were only on resonance with the NV center's $m_s = 0 \leftrightarrow m_s = -1$ transition in a narrow (~20 nm wide) region near the magnet, producing a thin dark region in fluorescence as NV centers in this region were driven to the darker $m_s = -1$ sublevel [Fig. 8.7(c)]. In the same work, gradient imaging was performed in which the magnet was scanned rather than the diamond, producing resonance rings ~4 nm wide.

At the same time, work was performed at Harvard in AC magnetometry with single NV centers both in bulk as well as in nanodiamonds. There, the principle of AC magnetometry was demonstrated, based on spin-echo sequences (Section 8.3.3). The echo signal as a function of delay time between pump and probe pulses, τ, for a single NV center deep into a bulk crystal is shown in Fig. 8.8(a). An AC magnetic field was then applied, and the time between pumping and probing, τ, was tuned to coincide with one of the revivals in signal due to the nuclear spin bath. Figure 8.8(b) shows the echo signal as a function of AC field amplitude for two magnetic field frequencies, 3.15 and 4.21 kHz. Optimal magnetic sensitivity occurs where the echo signal has maximum slope, and this point can be chosen to coincide with zero AC field by altering the relative microwave pulse phases. Figure 8.8(c) displays the maximum magnetic sensitivity, which peaks at $\sim 30\ \mathrm{nT}/\sqrt{\mathrm{Hz}}$ for this NV center as a function of field frequency.[9]

[9] In work with nanodiamonds, it was seen that T_2 times were only a few μs, and consequently revivals due to the nuclear bath were not observed. Nonetheless a field sensitivity of $\sim 0.5\ \mu\mathrm{T}/\sqrt{\mathrm{Hz}}$ to 380 kHz AC fields was demonstrated for a ~30 nm nanodiamond.

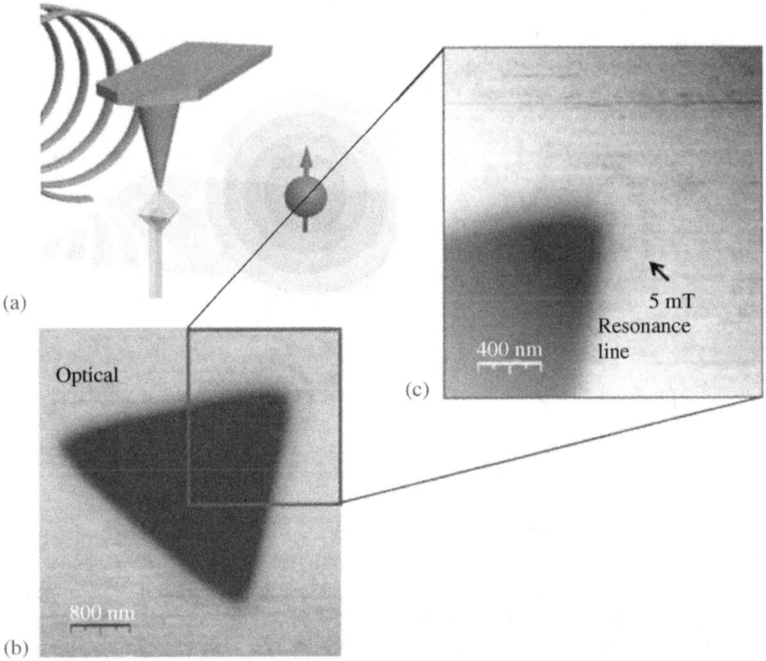

Figure 8.7 (a) Diagram of the magnetic-field-imaging experiment in Ref. [10]. A nanoscale magnetic particle is imaged with a single nitrogen-vacancy defect (within the nanocrystal) fixed at the scanning probe tip. A microscope objective is used to focus light from a 532 nm laser on the NV center, exciting the $^3A_2 \rightarrow {}^3E$ transition on the phonon sideband. The same objective collects fluorescence (i.e., confocal microscopy), which is then spectrally filtered and directed to photodetectors. (b) Fluorescence map in the vicinity of the nickel magnet, recorded using a single NV center on the AFM tip as light source and magnetometer. The triangular shaped magnet is black because laser light is reflected off the magnet and consequently no fluorescence is collected in this region. Inset (c) shows the fluorescence signal from the scanned nitrogen-vacancy center attached to the apex of the AFM tip when resonant microwaves at 2.75 GHz are applied. The arrowed point corresponds to a 5 mT resonance line with the magnetic field tilted by 45° relative to the nitrogen-vacancy axis. Adapted from Ref. [10].

In subsequent work at Stuttgart [12] using isotopically pure diamond, a sensitivity of $\sim$4 nT/$\sqrt{\text{Hz}}$ to AC fields and $\sim$0.5 μT/$\sqrt{\text{Hz}}$ to DC fields was demonstrated for a single NV center deep in the lattice. Recently, a broadband magnetometer using a single NV center in a nanodiamond on an AFM tip was demonstrated using frequency modulation techniques combined with real-time feedback [24]. The sensitivity was only a few μT/$\sqrt{\text{Hz}}$, but a bandwidth from DC to $\sim$0.3 MHz was shown to be possible, which is an important feature for many applications.

In recent work at Harvard, the single-NV AFM approach has been further developed by improved fabrication and control techniques [117], with sensitivities now exceeding those required for imaging and coherent manipulation of single electron spins with imaging resolution $\sim$20 nm and in realistic integration times [118, 119].

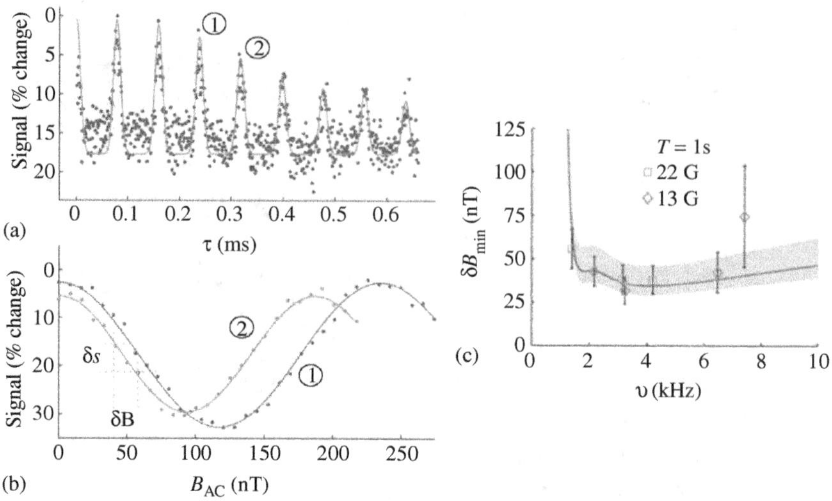

Figure 8.8 (a) Spin-echo measurement for a single NV center more than 1 μm below the diamond surface. The normalized echo signal corresponds to a fractional change in the NV center's fluorescence. Maximal signal corresponds to an average of 0.03 detected photons during the 324-ns photon counting window of a single experimental run. Collapses and revivals are due to interactions with the ^{13}C nuclear spin bath. The revivals occur at half the rate of the Larmor frequency of ^{13}C (here set by $B_{DC} = 22$ G). The spin-echo signal envelope was fitted with an exponential decay function modulated by a strongly interacting pair of nearby ^{13}C (methods are described in Ref. [11]). Magnetometer sensitivity experiments are performed at spin-echo revival peaks to maximize signal. (b) Examples of measured spin-echo signal as a function of B_{AC} for two operating frequencies, 3.15 kHz and 4.21 kHz, corresponding to revivals 1 and 2 indicated in (a). Each displayed point is a result of 7×10^5 averages of spin-echo sequences. The magnetometer is most sensitive to variations in the AC magnetic field amplitude (δB) at the point of maximum slope, with the sensitivity being limited by the uncertainty in the spin-echo signal measurement (δS). The cosine behavior of the signal with respect to AC magnetic field amplitude can be changed to a sine by adjusting the phase of the third microwave pulse by 90°. This change moves the point of maximum magnetometer sensitivity to near zero AC field amplitude. (c) Measured sensitivity of a single NV magnetometer in a bulk diamond sample over a range of frequencies for the external AC magnetic field after averaging for one second. Error bars, standard deviation (s.d.) for a sample size of 30. Also shown is the theoretically predicted sensitivity (solid line), with the shaded region representing uncertainty due to variations in photon collection efficiency. Measurements were carried out at two different DC fields, $B_{DC} = 13$ G and 22 G. Adapted from Ref. [11].

Sensitivity and limitations

The primary limitation in all fluorescence-based magnetometers is fluorescence collection efficiency and limited signal contrast. For sufficiently low measurement contrast, R (relative difference in detected signal depending on spin-projection), the sensitivity using the fluorescence technique can be estimated [14, 15, 25] by modifying Eq. (8.1) as $\delta B_{fl} \approx \delta B_q/(R\sqrt{\eta})$, where η is the detection efficiency and δB_q is the spin-projection-noise-limited minimum magnetic field. Recent single-NV experiments [10–12] yielded typical values of $R \approx 0.2$

and $\eta \approx 0.001$, making the best possible sensitivity, in the absence of excess technical noise or other broadening mechanisms, two to three orders of magnitude worse than the spin-projection-noise limit. For ensembles the contrast is even worse ($R \approx 0.04$) owing to background fluorescence from centers belonging to other orientations or charge states which are out-of-resonance [15, 18, 26, 28]. The contrast is limited by non ideal branching ratios into and out of the dark singlet states [83–85, 90], while η is limited by the field of view of the detection optics and sub-unity quantum efficiency of the detected transition [23, 120].

There has been significant effort in the last few years to increase the collection efficiency by engineering nanoscale waveguides [23, 121, 122], using solid-immersion lenses made of diamond [123–125], or even coupling to metal plasmons [126–128]. While some of these techniques appear promising for future applications, there are still several orders of magnitude room for improvement before reaching fundamental sensitivity limits. Recently, $\eta \approx 0.5$ was realized by surrounding the sides of a diamond chip with photodetectors, thereby collecting the large fraction of fluorescence which is confined between the two parallel diamond chip surfaces by total internal reflection [29].

Another challenge is to obtain long coherence times for NV centers near to the diamond surface. Currently T_2 in nanodiamonds or implanted NV centers within ~10 nm from the surface is just a few microseconds, limiting the fundamental sensitivity. As such NV centers are necessary to realize the impressive spatial resolution promised in Fig. 8.1, a major effort to understand the effect of surface charges and related defects on the NV spin is necessary.

8.4.2 Wide-field array magnetic imaging

One of the features, which makes magnetic sensing with diamond NV centers most promising is the possibility for variable spatial resolution combined with broad measurement bandwidth. So far we have considered single NV magnetometers which offer the best spatial resolution, but ensembles of NV centers can be used to make magnetic field images of much wider fields of view with even greater sensitivity. A wide-field-of-view NV-diamond magnetic field imager employing a shallow NV-enhanced layer on a bulk diamond chip could be used, for example, for real-time, non invasive monitoring of functional activity in cultured neuronal networks, which would provide a new modality for studies of the relationship between microscopic neuronal connectivity and brain circuit function [28].

Figure 8.9 depicts a technique using a wide-field fluorescence microscope to acquire such an image. A green pump laser beam is focused onto a diamond sample with a thin layer of NV centers. The thickness of the layer is chosen to be roughly the desired spatial resolution (typically a few hundred nanometers to a micrometer). The focal point is either a few Rayleigh ranges before (to produce a negative image) or after (positive image) the layer of NV centers and the illuminated region of the NV layer defines the field of view (typically order ~100 μm). The fluorescence is collected from the same objective, spectrally filtered and directed to a charge-coupled device (CCD) for post processing. This proof of principle of

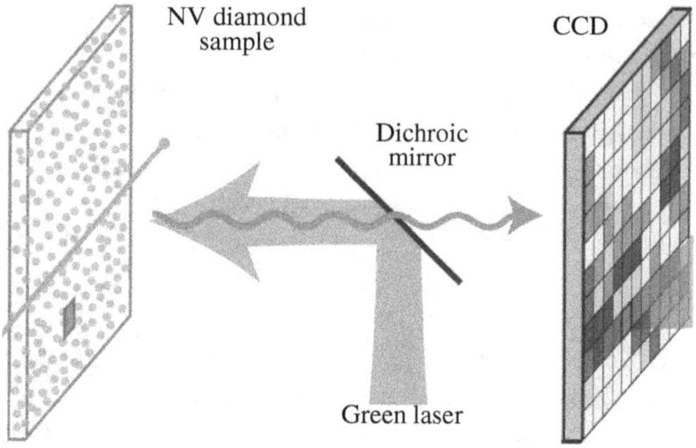

Figure 8.9 Wide-field fluorescence microscope used for magnetic imaging as described in text. The sample under study would be placed in close contact with the thin layer of NV centers. Adapted from Ref. [14].

this scheme was recently demonstrated [26,28]. There a field of view as high as ~140 μm was realized [28] and a spatial resolution of ~0.5 μm, using a high-numerical-aperture oil-immersion objective [26].

8.4.3 NV-ensemble magnetometers

As discussed in Section 8.4.1, the sensitivity of nanoscale fluorescence-based diamond magnetometers is limited due to poor light collection efficiency and diminished measurement contrast. For larger, NV-ensemble devices, the side-collection technique discussed above can provide large improvements in collection efficiency ($\eta \approx 0.5$) [29]. Another avenue may be to use optical rotation [73] or absorption of a probe beam for spin-state detection, as the transmitted light can be easily collected. Unfortunately, the weak spin–orbit coupling in the excited state and the presence of strain in the crystal make the non-spin-conserving transitions necessary for optical rotation negligible, at least for large ensembles at room temperature. However, recently a technique was demonstrated using infrared (IR) absorption on the $^1A_1 \leftrightarrow {}^1E$ transition [15] which circumvents some of the photon shot-noise problems associated with other techniques.

The apparatus is illustrated in Fig. 8.10(a). The principle behind the IR-absorption technique is the following. Under continuous optical pumping, the population of NV centers in the lower, metastable singlet (MS) is detected by monitoring the transmission of the 1042 nm probe beam and this is used to read out the spin polarization of the ensemble. In the absence of resonant microwaves, NV centers are pumped into the $m_s = 0$ ground-state sublevel and there is reduced population in the MS corresponding to maximum probe transmission. Under application of microwaves with frequency $D \pm \gamma B_{z_i}/(2\pi)$, where B_{z_i}

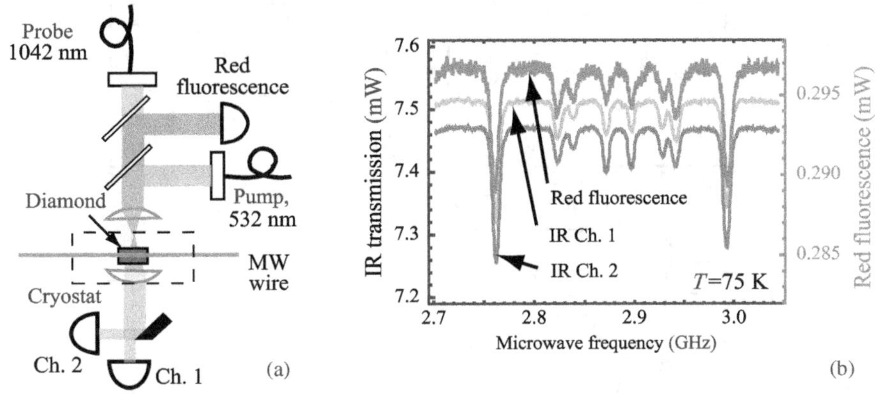

Figure 8.10 (a) IR absorption gradiometer apparatus. The green pump and IR probe beams were focused to a diameter of ~30 μm near the surface of the diamond, and two halves of the transmitted IR beam were detected with separate photodiodes. MW, microwave. (b) Optically detected magnetic resonance at 75 K using the fluorescence method and both halves of the transmitted IR probe. The pump power was 0.8 W and the microwave Rabi frequency was ~1.5 MHz.

is the magnetic field projection along the ith NV orientation, population is transferred to the $m_s = \pm 1$ sublevel, resulting in greater population in the MS and lower probe transmission.

Magnetic-resonance spectra, detected by both fluorescence and IR-transmission, are shown in Fig. 8.10(b). A bias magnetic field of ~4 mT, produced by a permanent magnet, was directed such that each of the four NV orientations had different B_{z_i}, resulting in eight resolved resonances. The contrast of both fluorescence and IR-transmission resonances depends on the change in the MS population, which saturates when the pump rate is $\Omega_p \gtrsim 1/\tau_{\mathrm{MS}}$, where $\tau_{\mathrm{MS}} \approx 0.3$ μs is the MS lifetime [83]. For the IR-transmission resonances, the contrast, R, also depends on the probe's optical depth. At room temperature, for a sample with $[\mathrm{NV}^-] \approx 15$ ppm, $R \approx 0.003$, limited by the weak oscillator strength of the transition [83, 85] and homogeneous broadening of the IR absorption line [83, 85]. The maximum contrast ($R \approx 0.03$) occurs in the 45–75 K temperature range, where the homogeneous and inhomogeneous contributions to the linewidth are approximately equal [129]. This contrast is, coincidentally, nearly the same as the maximum contrast obtained using fluorescence detection, but the signal is much larger due to the higher collected light intensity.

Operation of the device as a magnetometer was accomplished by phase-sensitive detection. An oscillating magnetic field (frequency 40 kHz, amplitude ~0.1 mT) was applied and the resulting photodiode signals were demodulated at the first harmonic using lock-in electronics, Fig. 8.11(a). The measurement bandwidth in this work was limited by the lock-in time constant, but in principle this technique can be used to detect fields with angular frequency approaching the maximum spin polarization rate, $\sim 1/\tau_{\mathrm{MS}}$, without degradation in sensitivity (as demonstrated in Refs. [24, 130]).

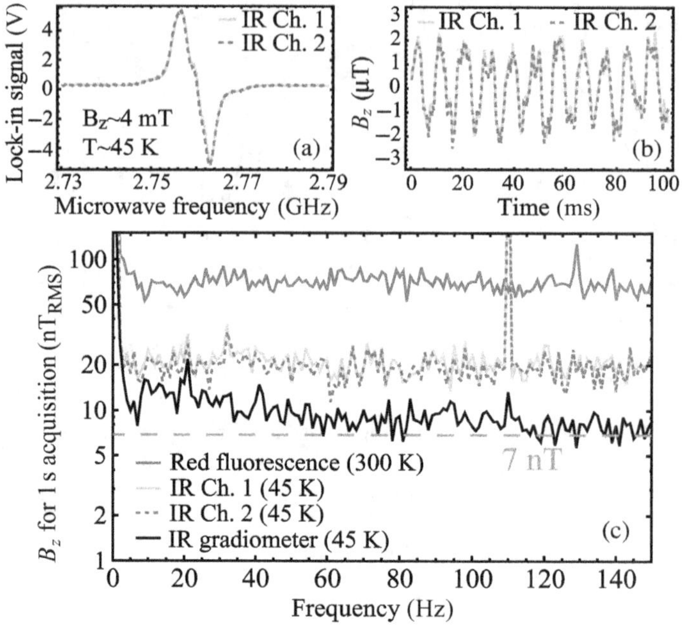

Figure 8.11 (a) Lock-in signal for both IR magnetometer channels. (b) Time-series magnetometer signal, after subtraction of the static bias field, for a 1 μT_{RMS} applied field at 109 Hz. The microwave frequency was tuned to the center of the resonance [zero-crossing in (a)]. (c) Frequency-domain response of the magnetometer output in (b) revealing an IR absorption gradiometer noise floor of 7 nT_{RMS} in 1 s of acquisition.

The Fourier transform of the time-series response for the 109 Hz applied field, Fig. 8.11(c), reveals a noise floor for each channel of $\sim$15 nT_{RMS} in 1 s of acquisition for $\sim$110 Hz frequencies. For comparison, the noise spectrum for a fluorescence-based magnetometer at room temperature is also presented, showing a noise floor of $\sim$60 nT_{RMS} in 1 s of acquisition. Taking the difference of the two IR channels' magnetometer signals gives a noise floor of $\sim$7 nT_{RMS} in 1 s of acquisition for $\sim$110 Hz frequencies. Since these signals correspond to light that has interacted with spatially separate parts of the diamond, the difference signal measures the magnetic field gradient across the beam (effective baseline $\sim$25 μm). The benefit of this gradiometric approach is that in principle technical noise common to both channels, such as laser intensity and ambient field fluctuations, is canceled.

In the absence of technical noise, the IR-absorption technique is limited by photon shot-noise, given by

$$\delta B_p \simeq \frac{1}{\gamma}\frac{\Gamma_{mr}}{R}\sqrt{\frac{E_p}{Pt_m}}, \tag{8.4}$$

where Γ_{mr} is the magnetic-resonance linewidth, E_p is the photon energy, and P is the detected optical power. Analyzing the highest-contrast resonances in Fig. 8.10(b), a photon shot-noise limited sensitivity of $\sim$40 pT for fluorescence collection and $\sim$5 pT for the sum of both IR absorption channels at $t_m = 1$ s can be expected. The latter corresponds to approximately an order of magnitude better sensitivity-per-root-volume than the photon shot-noise limit of recent fluorescence-based ensemble magnetometers [18, 26, 28].

The main drawback of the IR-absorption technique is that the transition is weak, and consequently the first demonstration saw improvements in sensitivity only at cryogenic temperatures. However, it should be possible to extend the technique to room-temperature operation by employing a cavity to increase the optical depth at 1042 nm. A cavity with finesse of $\sim$1000 would be sufficient to optimize the magnetometer response for the sensor presented above. Using two parallel micro-cavities might permit gradiometry with sensitivity approaching the quantum shot-noise limit of $\sim$10 fT in 1 s of acquisition. Such a device would be ideal for low-field NMR detection [32] in, for example, microfluidic devices [31, 130].

8.5 Outlook

By now it should be clear that NV diamond magnetometry has been making rapid progress in the last few years. As discussed in this chapter, NV magnetometry can provide combined magnetic field sensitivity and spatial resolution better than any existing technology, while operating over a wide range of temperatures in a robust, solid-state, optically probed system. NV diamond magnetometers also have excellent properties for both physical and life science applications, including high fluorescence rate; no bleaching or blinking; ability to be fabricated into a wide range of forms (nanocrystals, nanopillars, thin films, AFM tips, bulk chips, etc.); benign chemical properties; and good endocytosis and no cytotoxicity for diamond nanocrystals and other structures used in live-cell imaging.

A remaining challenge is to bridge the significant gap between the theoretically projected and experimentally achieved sensitivities. Progress can be expected from improving the material engineering as well as techniques for photon collection and spin-state readout. Should progress continue its current trajectory, these sensors will become indispensable tools in biology, medical diagnostics, materials science, microfluidics, and defense and security applications. Moreover, NV-based sensors may be able to solve key, unsolved scientific problems in both condensed matter and biological physics. For example, the diamond-AFM technique could be used to image the nanoscale surface structure of antiferromagnetic materials, which consist of a sequence of alternating magnetic moments, resulting in a vanishingly small magnetic field at (typical) distances >10 nm; and of multiferroic materials, which exhibit both magnetic order and ferroelectricity at the same time. In the life sciences, nanoscale NV sensors could provide direct determination of the dynamics of protein domain motion and membrane lipid transport at the single- or few-molecule level, using proteins and lipids that are site-specifically spin-labeled and thus affect proximal NV spins.

References

[1] D. Budker and M. Romalis, Nature Physics **3**, 227 (2007).
[2] J. Clarke and A. I. Braginski, *The SQUID Handbook* **1** (Wiley-VCH, Weinheim, 2004).
[3] M. Jamet, W. Wernsdorfer, C. Thirion, D. Mailly, V. Dupuis, P. Melinon, and A. Perez, Phys. Rev. Lett. **86**, 4676 (2001).
[4] J. P. Cleuziou, W. Wernsdorfer, V. Bouchiat, T. Ondarcuhu, and M. Monthioux, Nature Nanotechnology **1**, 53 (2006).
[5] D. Rugar, R. Budakian, H. J. Mamin, and B. W. Chui, Nature **430**, 329 (2004).
[6] H. J. Mamin, M. Poggio, C. L. Degen, and D. Rugar, Nat. Nano. **2**, 301 (2007).
[7] H. J. Mamin, T. H. Oosterkamp, M. Poggio, C. L. Degen, C. T. Rettner, and D. Rugar, Nano Lett. **9**, 3020 (2009).
[8] C. L. Degen, Appl. Phys. Lett. **92**, 243111 (2008).
[9] A. Finkler, Y. Segev, Y. Myasoedov, M. L. Rappaport, L. Ne'eman, D. Vasyukov, E. Zeldov, M. E. Huber, J. Martin, and A. Yacoby, Nano Letters **10**, 1046 (2010).
[10] G. Balasubramanian *et al.*, Nature **455**, 648 (2008).
[11] J. R. Maze *et al.*, Nature **455**, 644 (2008).
[12] G. Balasubramanian *et al.*, Nature Materials **8**, 383 (2009).
[13] B. M. Chernobrod and G. P. Berman, J. Appl. Phys. **97**, 14903 (2005).
[14] J. M. Taylor, P. Cappellaro, L. Childress, L. Jiang, D. Budker, P. R. Hemmer, A. Yacoby, R. Walsworth, and M. D. Lukin, Nature Physics **4**, 810 (2008).
[15] V. M. Acosta, E. Bauch, A. Jarmola, L. J. Zipp, M. P. Ledbetter, and D. Budker, Appl. Phys. Lett. **97**, 174103 (2010).
[16] L. T. Hall, J. H. Cole, C. D. Hill, and L. C. L. Hollenberg, Phys. Rev. Lett. **103**, 220802 (2009).
[17] L. T. Hall, C. D. Hill, J. H. Cole, B. Stadler, F. Caruso, P. Mulvaney, J. Wrachtrup, and L. C. L. Hollenberg, Proc. Nat. Acad. Sci. **107**, 18777 (2010).
[18] B. J. Maertz, A. P. Wijnheijmer, G. D. Fuchs, M. E. Nowakowski, and D. D. Awschalom, Appl. Phys. Lett. **96**, 92503 (2010).
[19] L. S. Bouchard, E. Bauch, V. M. Acosta, and D. Budker, New J. Phys. **13**, 25017 (2011).
[20] P. C. Maurer *et al.*, Nature Physics **6**, 912 (2010).
[21] B. R. Smith, D. W. Inglis, B. Sandnes, J. R. Rabeau, A. V. Zvyagin, D. Gruber, C. J. Noble, R. Vogel, E. Osawa, and T. Plakhotnik, Small **5**, 1649 (2009).
[22] C. Bradac, T. Gaebel, N. Naidoo, J. R. Rabeau, and A. S. Barnard, Nano Lett. **9**, 3555 (2009).
[23] T. M. Babinec, B. J. M. Hausmann, M. Khan, Y. Zhang, J. R. Maze, P. R. Hemmer, and M. Loncar, Nature Nanotechnology **5**, 195 (2010).
[24] R. S. Schoenfeld and W. Harneit, Phys. Rev. Lett. **106**, 30802 (2011).
[25] V. M. Acosta *et al.*, Phys. Rev. B **80**, 115202 (2009).
[26] S. Steinert, F. Dolde, P. Neumann, A. Aird, B. Naydenov, G. Balasubramanian, F. Jelezko, and J. Wrachtrup, Rev. Sci. Instrum. **81**, 43705 (2010).
[27] P. L. Stanwix, L. M. Pham, J. R. Maze, D. Le Sage, T. K. Yeung, P. Cappellaro, P. R. Hemmer, A. Yacoby, M. D. Lukin, and R. L. Walsworth, Phys. Rev. B **82**, 201201 (2010).

[28] L. M. Pham *et al.*, New J. Phys. **13**, 45021 (2011).
[29] D. Le Sage, L. Pham, N. Bar-Gill, C. Belthangady, M. Lukin, A. Yacoby, and R. Walsworth, Phys. Rev. B **85**, 121202 (2012).
[30] S. Felton, A. M. Edmonds, M. E. Newton, P. M. Martineau, D. Fisher, D. J. Twitchen, and J. M. Baker, Phys. Rev. B **79**, 75203 (2009).
[31] M. P. Ledbetter, I. M. Savukov, D. Budker, V. Shah, S. Knappe, J. Kitching, D. J. Michalak, S. Xu, and A. Pines, Proc. Nat. Acad. Sci. **105**, 2286 (2008).
[32] M. P. Ledbetter, C. W. Crawford, A. Pines, D. E. Wemmer, S. Knappe, J. Kitching, and D. Budker, J. Magn. Reson. **199**, 25 (2009).
[33] E. E. Mikhailov, I. Novikova, M. D. Havey, and F. A. Narducci, Opt. Lett. **34**, 3529 (2009).
[34] V. Shah, S. Knappe, P. D. D. Schwindt, and J. Kitching, Nature Photonics **1**, 649 (2007).
[35] W. C. Griffith, S. Knappe, and J. Kitching, Opt. Express **18**, 27167 (2010).
[36] H. B. Dang, A. C. Maloof, and M. V. Romalis, Appl. Phys. Lett. **97**, 151110 (2010).
[37] M. Vengalattore, J. M. Higbie, S. R. Leslie, J. Guzman, L. E. Sadler, and D. M. Stamper-Kurn, Phys. Rev. Lett. **98** (2007).
[38] J. R. Kirtley, M. B. Ketchen, K. G. Stawiasz, J. Z. Sun, W. J. Gallagher, S. H. Blanton, and S. J. Wind, Appl. Phys. Lett. **66**, 1138 (1995).
[39] D. Wood, K. Ni, D. Schmidt, and A. Cleland, Sensors and Actuators A: Physical **120**, 1 (2005).
[40] G. Boero, M. Demierre, P.-A. Besse, and R. Popovic, Sensors and Actuators A: Physical **106**, 314 (2003).
[41] K. S. Novoselov, S. V. Morozov, S. V. Dubonos, M. Missous, A. O. Volkov, D. A. Christain, and A. K. Geim, J. Appl. Phys. **93**, 10053 (2003).
[42] C. L. Degen, M. Poggio, H. J. Mamin, C. T. Rettner, and D. Rugar, Proc. Nat. Acad. Sci. **106**, 1313 (2009).
[43] E. Fermi and F. Rasetti, Z. Phys. A **33**, 246 (1925).
[44] F. Bitter, Phys. Rev. **76**, 833 (1949).
[45] J. Brossel and F. Bitter, Phys. Rev. **86**, 308 (1952).
[46] S. Geschwind, R. J. Collins, and A. L. Schawlow, Phys. Rev. Lett. **3**, 545 (1959).
[47] I. Wieder, Phys. Rev. Lett. **3**, 468 (1959).
[48] Y. K. Zavoisky, J. Phys. USSR **9**, 245 (1945).
[49] J. H. E. Griffiths, J. Owen, and I. M. Ward, Nature **173**, 439 (1954).
[50] J. H. Griffiths, J. Owen, and I. M. Ward, in *Defects in Crystalline Solids* (The Physical Society, London, 1955), pp. 81–87.
[51] W. V. Smith, P. P. Sorokin, I. L. Gelles, and G. J. Lasher, Phys. Rev. **115**, 1546 (1959).
[52] J. Loubser and J. A. van Wyk, Rep. Prog. Phys. **41**, 1201 (1978).
[53] N. R. S. Reddy, N. B. Manson, and E. R. Krausz, J. Luminesc. **38**, 46 (1987).
[54] E. van Oort, N. B. Manson, and M. Glasbeek, J. Phys. C **21**, 4385 (1988).
[55] G. Davies and M. F. Hamer, Proc. R. Soc. A **348**, 285 (1976).
[56] W. Neuhauser, M. Hohenstatt, P. E. Toschek, and H. Dehmelt, Phys. Rev. A **22**, 1137 (1980).
[57] J. Kohler, J. A. J. M. Disselhorst, M. C. J. M. Donckers, E. J. J. Groenen, J. Schmidt, and W. E. Moerner, Nature **363**, 242 (1993).
[58] J. Wrachtrup, C. von Borczyskowski, J. Bernard, M. Orritt, and R. Brown, Nature **363**, 244 (1993).
[59] A. Gruber, A. Drabenstedt, C. Tietz, L. Fleury, J. Wrachtrup, and C. von Borczyskowski, Science **276**, 2012 (1997).

[60] S.-J. Yu, M.-W. Kang, H.-C. Chang, K.-M. Chen, and Y.-C. Yu, J. Am. Chem. Soc. **127**, 17604 (2005).
[61] A. M. Schrand, H. Huang, C. Carlson, J. J. Schlager, E. Åœsawa, S. M. Hussain, and L. Dai, J. Phys. Chem. B **111**, 2 (2007).
[62] Y.-R. Chang *et al.*, Nature Nanotechnology **3**, 284 (2008).
[63] F. Neugart, A. Zappe, F. Jelezko, C. Tietz, J. P. Boudou, A. Krueger, and J. Wrachtrup, Nano Lett. **7**, 3588 (2007).
[64] O. Faklaris, D. Garrot, V. Joshi, F. Druon, J.-P. Boudou, T. Sauvage, P. Georges, P. A. Curmi, and F. Treussart, Small **4**, 2236 (2008).
[65] J. Tisler *et al.*, ACS Nano **3**, 1959 (2009).
[66] C. Bradac, T. Gaebel, N. Naidoo, M. J. Sellars, J. Twamley, L. J. Brown, A. S. Barnard, T. Plakhotnik, A. V. Zvyagin, and J. R. Rabeau, Nature Nanotechnology **5**, 345 (2010).
[67] F. Jelezko, T. Gaebel, I. Popa, A. Gruber, and J. Wrachtrup, Phys. Rev. Lett. **92**, 76401 (2004).
[68] L. Childress, M. V. G. Dutt, J. M. Taylor, A. S. Zibrov, F. Jelezko, J. Wrachtrup, P. R. Hemmer, and M. D. Lukin, Science **314**, 281 (2006).
[69] M. V. G. Dutt, L. Childress, L. Jiang, E. Togan, J. Maze, F. Jelezko, A. S. Zibrov, P. R. Hemmer, and M. D. Lukin, Science **316**, 1312 (2007).
[70] P. Neumann, N. Mizuochi, F. Rempp, P. Hemmer, H. Watanabe, S. Yamasaki, V. Jacques, T. Gaebel, F. Jelezko, and J. Wrachtrup, Science **320**, 1326 (2008).
[71] P. Neumann *et al.*, Nature Physics **6**, 249 (2010).
[72] G. D. Fuchs, V. V. Dobrovitski, D. M. Toyli, F. J. Heremans, and D. D. Awschalom, Science **326**, 1520 (2009).
[73] B. B. Buckley, G. D. Fuchs, L. C. Bassett, and D. D. Awschalom, Science **330**, 1212 (2010).
[74] E. Togan *et al.*, Nature **466**, 730 (2010).
[75] P. Neumann, J. Beck, M. Steiner, F. Rempp, H. Fedder, P. R. Hemmer, J. A. Wrachtrup, and F. Jelezko, Science **329**, 542 (2010).
[76] G. de Lange, Z. H. Wang, D. Riste, V. V. Dobrovitski, and R. Hanson, Science **330**, 60 (2010).
[77] P. Hemmer and J. Wrachtrup, Science **324**, 473 (2009).
[78] Y. Mita, Phys. Rev. B **53**, 11360 (1996).
[79] A. Gali, M. Fyta, and E. Kaxiras, Phys. Rev. B **77**, 155206 (2008).
[80] P. Delaney, J. C. Greer, and J. A. Larsson, Nano Lett. **10**, 610 (2010).
[81] M. W. Doherty, N. B. Manson, P. Delaney, and L. C. L. Hollenberg, New J. Phys. **13**, 25019 (2011).
[82] Y. Ma, M. Rohlfing, and A. Gali, Phys. Rev. B **81**, 41204 (2010).
[83] V. M. Acosta, A. Jarmola, E. Bauch, and D. Budker, Phys. Rev. B **82**, 201202 (2010).
[84] L. Robledo, L. Childress, H. Bernien, B. Hensen, P. F. A. Alkemade, and R. Hanson, Nature **477**, 574 (2011).
[85] L. G. Rogers, S. Armstrong, M. J. Sellars, and N. B. Manson, New J. Phys. **10**, 103024 (2008).
[86] J. Harrison, M. J. Sellars, and N. B. Manson, Diam. Relat. Mater. **15**, 586 (2006).
[87] G. D. Fuchs, V. V. Dobrovitski, D. M. Toyli, F. J. Heremans, C. D. Weis, T. Schenkel, and D. D. Awschalom, Nature Physics **6**, 668 (2010).
[88] V. M. Acosta *et al.*, Phys. Rev. Lett. **108**, 206401 (2012).
[89] J. Maze, A. Gali, E. Togan, Y. Chu, A. S. Trifonov, E. Kaxiras, and M. D. Lukin, New J. Phys. **13**, 25025 (2011).

[90] N. B. Manson, J. P. Harrison, and M. J. Sellars, Phys. Rev. B **74**, 104303 (2006).
[91] J. R. Rabeau, P. Reichart, G. Tamanyan, D. N. Jamieson, S. Prawer, F. Jelezko, T. Gaebel, I. Popa, M. Domhan, and J. Wrachtrup, Appl. Phys. Lett. **88**, 23113 (2006).
[92] V. M. Acosta, E. Bauch, M. P. Ledbetter, A. Waxman, L. S. Bouchard, and D. Budker, Phys. Rev. Lett. **104**, 70801 (2010).
[93] B. Smeltzer, J. McIntyre, and L. Childress, Phys. Rev. A **80**, 50302 (2009).
[94] M. Steiner, P. Neumann, J. Beck, F. Jelezko, and J. Wrachtrup, Phys. Rev. B **81**, 35205 (2010).
[95] T. P. M. Alegre, C. Santori, G. Medeiros-Ribeiro, and R. G. Beausoleil, Phys. Rev. B **76**, 165205 (2007).
[96] N. D. Lai, D. Zheng, F. Jelezko, F. Treussart, and J.-F. Roch, Appl. Phys. Lett. **95**, 133101 (2009).
[97] J. R. Maze, J. M. Taylor, and M. D. Lukin, Phys. Rev. B **78**, 94303 (2008).
[98] S. Takahashi, R. Hanson, J. van Tol, M. S. Sherwin, and D. D. Awschalom, Phys. Rev. Lett. **101**, 47601 (2008).
[99] N. Mizuochi *et al.*, Phys. Rev. B **80**, 41201 (2009).
[100] D. A. Redman, S. Brown, R. H. Sands, and S. C. Rand, Phys. Rev. Lett. **67**, 3420 (1991).
[101] A. Jarmola, V. M. Acosta, K. Jensen, S. Chemerisov, and D. Budker, Phys. Rev. Lett. **108**, 197601 (2012).
[102] J. P. King, P. J. Coles, and J. A. Reimer, Phys. Rev. B **81**, 73201 (2010).
[103] F. Dolde *et al.*, Nature Physics **7**, 459 (2011).
[104] S. C. Lawson, D. Fisher, D. C. Hunt, and M. E. Newton, J. Phys.: Condens. Matter **10**, 6171 (1998).
[105] I. Aharonovich, C. Santori, B. A. Fairchild, J. Orwa, K. Ganesan, K.-M. C. Fu, R. G. Beausoleil, A. D. Greentree, and S. Prawer, J. Appl. Phys. **106**, 124904 (2009).
[106] B. Naydenov, V. Richter, J. Beck, M. Steiner, P. Neumann, G. Balasubramanian, J. Achard, F. Jelezko, J. Wrachtrup, and R. Kalish, Appl. Phys. Lett. **96**, 163103 (2010).
[107] E. L. Hahn, Phys. Rev. **80**, 580 (1950).
[108] E. Van Oort and M. Glasbeek, Chem. Phys. **143**, 131 (1990).
[109] G. De Lange, D. Riste, V. V. Dobrovitski, and R. Hanson, Phys. Rev. Lett. **106**, 80802 (2011).
[110] A. Laraoui, J. S. Hodges, and C. A. Meriles, Appl. Phys. Lett. **97**, 143103 (2010).
[111] P. Cappellaro and M. D. Lukin, Phys. Rev. A **80**, 32311 (2009).
[112] C. A. Ryan, J. S. Hodges, and D. G. Cory, Phys. Rev. Lett. **105**, 200402 (2010).
[113] B. Naydenov, F. Dolde, L. T. Hall, C. Shin, H. Fedder, L. C. L. Hollenberg, F. Jelezko, and J. Wrachtrup, Phys. Rev. B **83**, 81201 (2011).
[114] N. Bar-Gill, L. M. Pham, C. Belthangady, D. Le Sage, P. Cappellaro, J. R. Maze, M. D. Lukin, A. Yacoby, and R. Walsworth, *Nature Communications* **3**, 858 (2012).
[115] L. M. Pham, N. Bar-Gill, C. Belthangady, D. Le Sage, P. Cappellaro, M. D. Lukin, A. Yacoby, and R. L. Walsworth, Phys. Rev. B **86**, 04521 (2012).
[116] H. J. Mamin, R. Budakian, B. W. Chui, and D. Rugar, Phys. Rev. B **72**, 24413 (2005).
[117] P. Maletinsky, S. Hong, M. S. Grinolds, B. Hausmann, M. D. Lukin, R. L. Walsworth, M. Loncar, and A. Yacoby, Nature Nanotechnology **7**, 320 (2012).
[118] M. S. Grinolds, P. Maletinsky, S. Hong, M. D. Lukin, R. L. Walsworth, and A. Yacoby, Nature Physics **7**, 687 (2011).

[119] S. Hong, M. S. Grinolds, P. Maletinsky, R. L. Walsworth, M. D. Lukin, and A. Yacoby, Nano Lett. **12**, 3920 (2012).
[120] K. Y. Han, S. K. Kim, C. Eggeling, and S. W. Hell, Nano Lett. **10**, 3199 (2010).
[121] C. Santori, P. E. Barclay, K.-M. C. Fu, R. G. Beausoleil, S. Spillane, and M. Fisch, Nanotechnology **21**, 274008 (2010).
[122] A. Faraon, P. E. Barclay, C. Santori, K.-M. C. Fu, and R. G. Beausoleil, Nature Photonics **5**, 301 (2011).
[123] J. P. Hadden, J. P. Harrison, A. C. Stanley-Clarke, L. Marseglia, Y. L. D. Ho, B. R. Patton, J. L. O'Brien, and J. G. Rarity, Appl. Phys. Lett. **97**, 241901 (2010).
[124] P. Siyushev *et al.*, Appl. Phys. Lett. **97**, 241902 (2010).
[125] L. Marseglia *et al.*, Appl. Phys. Lett. **98**, 133107 (2011).
[126] S. Schietinger, M. Barth, T. Aichele, and O. Benson, Nano Lett. **9**, 1694 (2009).
[127] R. Kolesov, B. Grotz, G. Balasubramanian, R. J. Stohr, A. A. L. Nicolet, P. R. Hemmer, F. Jelezko, and J. Wrachtrup, Nature Physics **5**, 470 (2009).
[128] A. Huck, S. Kumar, A. Shakoor, and U. L. Andersen, Phys. Rev. Lett. **106**, 96801 (2011).
[129] K.-M. C. Fu, C. Santori, P. E. Barclay, L. J. Rogers, N. B. Manson, and R. G. Beausoleil, Phys. Rev. Lett. **103**, 256404 (2009).
[130] C. S. Shin *et al.*, arxiv:1201.3152 (2012).

9

Magnetometry with cold atoms

W. Gawlik and J. M. Higbie

9.1 Introduction

The rapid development in recent decades of techniques for producing, trapping, and manipulating cold atoms has, as a side-benefit, made possible new methods of atomic magnetometry. The properties of cold atoms, including long coherence times and excellent spatial localization, are often desirable for high-precision magnetic sensing and allow the techniques of atomic magnetometry to be extended to previously inaccessible regions of parameter space.

Specifically, the appeal of cold atoms for magnetometry lies in the demonstrated potential for high sensitivity at high spatial resolution. Magnetic-field measurements with atoms at finite temperature are generally characterized by motional averaging, in which atoms statistically sample a volume of space determined by the measurement time, the velocity distribution, and, if present, the confinement. For high-spatial-resolution magnetometry, atomic motion must be limited. The average displacement of atoms can be reduced by decreasing the measurement time, but a shorter measurement time is unappealing because it degrades the sensitivity of the measurement. Tighter confinement is an alternative means of reducing atomic motion, and is indeed attractive provided the confinement does not adversely affect the atomic spin coherence. The use of buffer gases in vapor-cell magnetometers is effectively a form of confinement, and indeed allows magnetometry with millimeter-scale spatial resolution. However, buffer gases are not entirely benign for atomic spins, having small but finite spin-destruction collisional cross-sections [1], whose effects would be increasingly deleterious at the high pressures necessary to achieve micrometer-scale resolution. A final alternative is to reduce the velocity spread by cooling the atomic ensemble; indeed, the use of cold atoms permits a significant reduction in motional averaging, and can be achieved with no loss (and potentially an increase) in spin-coherence time. In fact, cold atoms allow

Optical Magnetometry, ed. Budker, D. and Jackson Kimball, D. F. Published by Cambridge University Press.

one to reduce the spread not only in atomic velocities but also in atomic positions, through the use of very tightly confining optical lattices.

9.2 Experimental conditions

9.2.1 Constraints and advantages of using cold atoms for magnetometry

The negligible Doppler broadening of cold-atom optical transitions allows one to address single hyperfine components of these transitions and to eliminate complications caused by unresolved HFS in alkali-atom magnetometers. More importantly, low temperatures and velocities allow significant reduction of the transit relaxation stemming from limited observation time and collisions. Such reduction is significant in the regime where Feshbach resonances occur. For example, Zhang *et al.* [2] demonstrated that a Feshbach resonance can extend the relaxation time of a ground-state coherence over 1 s. Although the standard version of the Feshbach-resonance technique with its strong magnetic field is not practical for magnetometry, its all-optical variant [3] may prove useful in future.

An important advantage of using cold, trapped atoms for magnetometry is high local atomic density. While usual magneto-optical traps (MOTs) do not offer very impressive values (typical diameters of atom clouds on the order of 1 mm, densities of between 10^{10} to 10^{12} atoms/cm^3, and optical densities of about 1), much better atomic localization and higher densities can be reached with optical dipole traps (ODTs). ODTs with single laser beams provide atom clouds of elongated shapes with diameters of about 10 μm and offer optical densities reaching 100 (in the long direction). Crossing two such beams results in a trap where atoms are 3D localized within a volume on the order of $(10\ \mu\text{m})^3$. Yet tighter localization may be achieved with optical lattices.

Experiments with cold atoms are subject to many constraints. The most essential is the need to use a complicated apparatus which is inflexible and, in general, is specialized to a single atomic species. Other limitations are associated with high sensitivity of cold atoms to certain external perturbations. In particular, cold atoms are easily expelled from a trap by collisions with hot atoms and by radiation pressure from the imaging or probing beams. Below, we show that this is a serious problem for studies of nonlinear near-resonant interactions. Moreover, for observables associated with atomic spins, a good control of magnetic fields is necessary, particularly when MOT fields are switched off.

9.2.2 Cold samples of atoms above quantum degeneracy

In this section we focus on atoms with temperatures above quantum degeneracy which we call cold *thermal atoms*. For magnetometric applications cold thermal atoms are available in MOTs, ODTs, and/or in a free expansion after their release from a trap.

A MOT needs a special arrangement of laser beams and inhomogeneous magnetic field. Although the net magnetic field should be zero at the center of an ideal MOT, the finite size of a trapped atom cloud and imperfect balancing of counterpropagating beams result

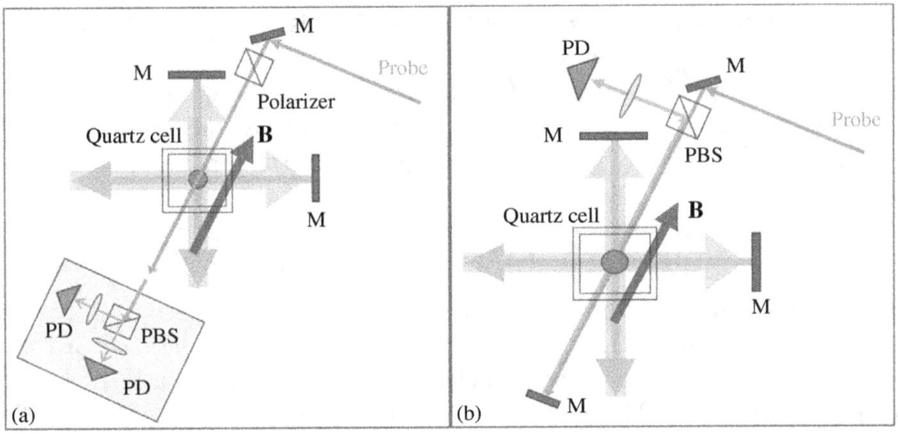

Figure 9.1 The setups for measurements of the Faraday rotation with cold atoms in a MOT. The wide arrows schematically indicate the retroreflected MOT beams. (a) The setup of the experiment with a single probe-beam passage and a balanced polarimeter. (b) Arrangement with retroreflection of a probe beam for reduction of the light pressure effects. Photodetector PD records only light which changed polarization by double passage across the cold sample. M labels mirror; PBS, polarizing beam splitter; PD, photodetector. Direction of the magnetic field **B** necessary for the observation of the Faraday rotation is indicated.

in atomic perturbation by nonzero magnetic field. In Ref. [4], Brzozowski *et al.* applied Raman spectroscopy for diagnostics of an operating MOT and analyzed inhomogeneous broadening caused by a finite atom cloud size and the magnetic field gradient. Spectroscopic experiments with operating MOTs have been described also in Refs. [5] and [6].

Figure 9.1 depicts a typical experimental setup used for magneto-optic experiments with a MOT. In the arrangement of Fig. 9.1(a), the polarization rotation of a linearly polarized probe beam traversing the cold-atom sample is measured by a two-channel polarimeter consisting of a polarizing beam splitter (PBS) and photodiodes (PDs). In Fig. 9.1(b), the probe beam is reflected and traverses the sample twice. A PD measures the intensity of the polarization component of the probe beam which is orthogonal to the initial polarization after two passes across the sample. If the medium's dichroism is negligible, this intensity is proportional to the square of the rotation angle θ^2. Thanks to the double passage, not only is the rotation doubled, but more importantly, the effect of the light pressure is significantly reduced. This is essential in the case of experiments with nonlinear Faraday rotation (Section 9.4) which necessitate appropriately strong probe beams.

High-precision magnetometric applications require switching off the MOT. The measurements can be performed either on an unperturbed cloud of atoms falling freely under gravity, or after transferring atoms to another kind of a trap, usually an ODT. The first way is relatively simple and often used but has an important drawback: the measurement time is limited by atom number losses to no more than 20 ms or even shorter for narrow probe beams. The second technique alleviates that problem, yet is more demanding and not free from losses during the atom transfer.

In addition to relaxation and atom loss, in the case of nonlinear magneto-optics with cold atoms, the relevant limitation becomes the decoherence resulting from spontaneous emission. It can be reduced by sufficient detuning of the ODT lasers.

9.3 Linear Faraday rotation with trapped atoms

Faraday rotation with cold atoms can be described in a standard way, i.e., as a measure of an optical anisotropy expressed as a difference of refractive indices n_+ and n_- for the σ^+ and σ^- components of a linearly polarized probe beam. Rotation per unit length is $(\omega/2c)\mathrm{Re}[n_+ - n_-]$, hence the net rotation is obtained by integration along the probe beam,

$$\theta = \frac{\omega}{2c}\int_0^L \mathrm{Re}[n_+ - n_-]dz\,, \tag{9.1}$$

where ω is the light frequency, L the length of the medium, and c the speed of light.

Using the standard approach (see, for example, Ref. [7]), one can easily relate the medium's refractive indices with its dipole transition moments and density-matrix elements

$$n_\pm - 1 \propto \sum_{e,g} \mathrm{Re}\left[d_{eg}^\pm \rho_{eg}^\pm\right], \tag{9.2}$$

with $d_{eg}^\pm$ being the matrix elements of the dipole moment associated with the $\sigma^\pm$-polarized light-beam component, and $\rho_{eg}^\pm$ the related density-matrix elements. The summation goes over all ground and excited-state sublevels $|g\rangle$ and $|e\rangle$, linked by allowed transitions. In the stationary regime, $\rho_{eg}^\pm$ can be expressed as

$$\rho_{eg}^\pm = \frac{1}{\delta_{eg} - i\Gamma/2}\sum_{e'g'}(\Omega_{eg'}\rho_{g'g} - \rho_{ee'}\Omega_{e'g})\,, \tag{9.3}$$

where $\delta_{\alpha\beta}$ and $\Omega_{\alpha\beta}$ denote, respectively, the light detuning and Rabi frequency for the $|\alpha\rangle \leftrightarrow |\beta\rangle$ transition, and $\Gamma/2$ is the relaxation rate of the optical coherence. The polarization index $\pm$ is related to the magnetic quantum numbers of states $|e\rangle$, $|g\rangle$ by standard selection rules for polarized light. Relations (9.2) and (9.3) indicate that optical coherences, and consequently also the refractive indices and rotation angle, depend on the density-matrix elements $\rho_{g'g}$ and $\rho_{ee'}$ which represent populations of, and coherences between, Zeeman sublevels of the ground and excited states.

For low light intensity, in the linear regime when optical pumping and coherences are negligible, atomic media are in thermal equilibrium, hence their density matrices are diagonal and contain only equilibrium atomic populations. In such cases, the refractive indices are expressed as sums over all components of the given transition of the standard dispersive functions associated with individual magnetic sublevels of the ground and excited states, weighted by equilibrium populations and corresponding transition probabilities.

For a three-level Λ system with $F = 1$ and $F' = 0$ and small Zeeman splitting ($\omega_L \ll \Gamma$), the rotation angle can be expressed as

$$\theta \propto \frac{\omega_L(\rho_{--} + \rho_{++} + 2\mathrm{Re}[\rho_{-+}]) + (\rho_{--} - \rho_{++})\delta - \Gamma \mathrm{Im}[\rho_{-+}]}{\delta^2 + (\Gamma/2)^2}, \tag{9.4}$$

where ω_L denotes the Larmor frequency ($\omega_L = g\mu_B B$) and δ is the probe-light detuning from the $F = 1 \leftrightarrow F' = 0$ transition resonance (g is the Landé factor and μ_B the Bohr magneton). With no optical pumping, $\rho_{--} = \rho_{++}$, and there is no coherence in the system, $\rho_{-+} = 0$; the only nonvanishing contribution to the rotation is the linear Faraday effect represented by the first term of Eq. (9.4). This kind of rotation associated with the presence of a magnetic field is called the *diamagnetic rotation* and allows determination of magnetic fields by measurement of polarization rotation of a probe light beam.

When optical pumping creates some population imbalance between the $|-\rangle$ and $|+\rangle$ sublevels, the second term in Eq. (9.4) reflects the so called *paramagnetic rotation* which occurs even without external magnetic field [8]. The two kinds of rotation possess different symmetries as a function of the light frequency and magnetic field which can be used to disentangle complicated mixed signals. For example, as seen from Eq. (9.4), the paramagnetic contribution cancels when $\delta = 0$, even if optical pumping creates a population difference. Still, in such cases, diamagnetic Faraday rotation is affected by optical pumping which results in different amplitudes of the $n_\pm(\delta)$ curves and needs to be taken into account in detailed analysis of the Faraday effect.

The rotation signal for the diamagnetic effect reaches a maximum at $\delta = 0$. Solid-state or dense gas samples are opaque on and near resonance and are not amenable to studies of rotation. However, in case of dilute atomic vapors, resonance absorption is weak and does not hinder measuring the magneto-optical rotation near the line center. One can then benefit from a resonant increase of dispersion and rotation, which becomes orders of magnitude larger than in standard solid-state Faraday materials [9]. For $\delta = 0$, the rotation can be expressed by a simple formula

$$\theta(B) \propto N \frac{\omega_L}{(\omega_L)^2 + (\Gamma/2)^2}, \tag{9.5}$$

where N stands for atomic density.

As can be seen, the rotation is a linear function of the magnetic field when the Zeeman splitting is much smaller than the atomic linewidth, $\omega_L \ll \Gamma/2$. This has obvious application for magnetometry since the measurement of an angle can be directly converted to measurement of a magnetic field. For hot vapors, averaging over the atomic velocities extends considerably the range ΔB where $\theta(B)$ is linear to $\Delta_D/(\mu_B g)$, where Δ_D is the Doppler width. For cold atoms, however, ΔB is determined by the much smaller natural linewidth Γ.

As stated above, maximum rotation per given magnetic field is obtained for the resonant case, $\delta \approx 0$. However, resonant light considerably perturbs cold atoms by heating the atomic cloud, which limits the interaction time. For nonresonant probe beams this effect is less serious but in this case circular dichroism may create ellipticity in the transmitted probe beam. Consequently, the rotation signals need to be carefully disentangled from the ellipticity.

First studies of the linear Faraday rotation with cold atoms were the experiments with ^{7}Li atoms of Franke-Arnold *et al.* [10] and with ^{85}Rb atoms of Labeyrie *et al.* [11] in MOTs. Franke-Arnold *et al.* reported very strong resonant enhancement of the rotation signal and noted its potential for magnetometric measurements. An important feature of the observed rotation signals was the time dependence of the observed signals exhibiting strong loss of atoms from the trap. To alleviate the effect of atom loss, the measurement was performed in a short time window after switching off the MOT. Despite these attempts, the losses were considerable because of hyperfine optical pumping which caused deformation of the measured $\theta(B)$ dependences. Only for a short time (6 μs) was this dependence as predicted for linear Faraday effect with a constant number of atoms.

In the work of Labeyrie *et al.* [11], linear Faraday rotation with cold Rb atoms was studied in a high-density MOT. The probe beam was spectrally purified by a Fabry–Perot interferometer and its power was reduced to 0.1 μW, which corresponded to a saturation parameter on the order of 10^{-3}. The weak probe allowed accurate studies of linear rotation. By capturing a large number of atoms (3×10^9), high optical density (OD ≈ 24) was reached which created problems with proper analysis of the results, and required taking into account the laser spectrum, lensing effect, etc. Working in a high-density regime, Labeyrie *et al.* found it necessary to use higher magnetic fields to reduce resonance absorption and open a transmission window by Zeeman tuning. At 8 G, they reached 150° rotation, which corresponds to a Verdet constant of about 5×10^5 rad (TM).

A nice demonstration of an interplay between the diamagnetic and paramagnetic effects in cold atoms has been given by Choi *et al.* [12] who investigated how preparation of the sample by selective population of a single Zeeman sublevel affects the magneto-optic response of a cold cesium sample. They paid considerable attention to elimination of any possible optical nonlinearity caused by the probe beam: its intensity was reduced to the nW/mm^2 level and measurement time limited to 400 μs. Two special cases were considered. One with the sample prepared by optical pumping to $m_F = 0$, the second with the $m_F = +3$ sublevel of the $^2S_{1/2}$, $F = 3$ ground state. The two cases gave very different results. For $m_F = 0$, the Faraday-rotation angle was a familiar symmetric function of the probe-beam detuning, and the dichroism (ellipticity) was an asymmetric one. For $m_F = 3$ the opposite was true. This demonstrated different behavior of the sample when not initially oriented (all population in $m_F = 0$) and when possessing initial orientation or magnetization (all population in $m_F = 3$). The former case is associated with the diamagnetic Faraday effect with the rotation caused exclusively by the magnetic field, the latter with the paramagnetic rotation, when the rotation is produced by the initial orientation even without an external field.

There are two important cases in which paramagnetic rotation becomes significant. One is when a nonzero magnetic field applied to an oriented system generates Larmor precession, the frequency of which is directly related to the strength of the magnetic field. The potential of using paramagnetic rotation in cold atomic samples for magnetometric measurements has been demonstrated by Isayama *et al.* in an experiment with ^{85}Rb atoms cooled to about 10 μK [13]. After switching off the trap fields, the sample was prepared by optical pumping with a pulse of circularly polarized light and probed by a weak probe beam detuned by about 3 GHz from the resonant transition. Stray magnetic fields were reduced by a magnetic shield around the MOT chamber and compensating coils. The setup was able to measure applied fields of about 2 mG with 180 μG precision.

The second important case is when the rotation measurements are used for diagnostics of quantum observables (in particular nondestructive diagnostics). For such measurements, a nonresonant light beam senses birefringence of the sample prepared by a specific orientation of magnetic moments without substantially perturbing the sample by light absorption and spontaneous emission. The success of the method is based on the direct relation between the rotation angle θ of the polarization of a weak probe beam and the angular momentum (spin) projection F_z along a given direction

$$\theta = -\frac{\sigma \rho l}{12} \frac{\Gamma}{\delta} \frac{\langle \widehat{F}_z \rangle}{F}, \tag{9.6}$$

where the product $\sigma \rho l$ is the optical depth on resonance ($\sigma = 3\lambda^2/2\pi$ being the resonant photon scattering cross-section, ρ the atomic number density, and l the optical path length through the sample). This relation allows quantum-state diagnostics by the rotation measurement. Such a possibility has been demonstrated with atoms in an optical lattice [14] and with various kinds of traps [15, 16]. Many of these applications implement techniques such as spin squeezing for reducing spin-measurement noise below the standard shot-noise limit. More details on these developments can be found in Chapters 2 and 3 of this book.

Atom traps make it possible to perform magnetometry with a good spatial resolution. This has been demonstrated by Terraciano *et al.* in their experiments with optical dipole traps [17, 18]. Blue detuning of 25 GHz enabled trapping of 10^7 ^{85}Rb atoms and measurement of magnetic fields with a 100 μG precision in a 500 μm volume.

9.4 Nonlinear Faraday rotation

9.4.1 Low-field, DC magnetometry

The nonlinear Faraday effect (NFE) has been widely studied in alkali-metal atomic vapor cells [7, 19]. The nonlinearity is caused by a Zeeman polarization, which encompasses both population imbalance and coherences of Zeeman sublevels that can be created by an appropriately strong light beam. According to Eq. (9.4), existence of nonzero off-diagonal density-matrix elements yields an extra contribution to the overall rotation. Studies of NFE can be performed with either a single beam which acts as pump and probe at the same time, or with two beams, where the two roles are separated.

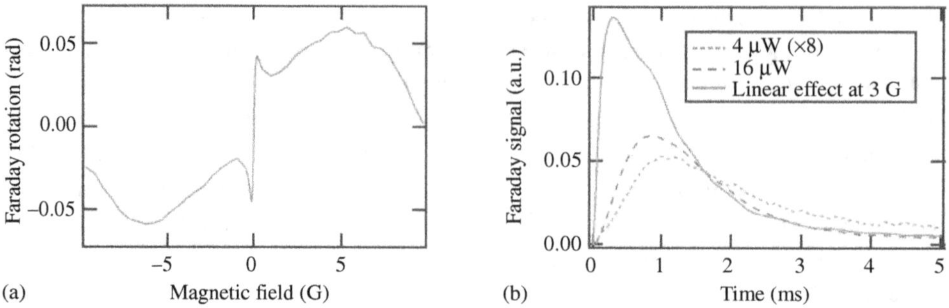

Figure 9.2 (a) Nonlinear Faraday rotation in a cold-atom sample recorded at 2 ms probing time. The wide structure is the linear and the narrow central resonance is the nonlinear Faraday effect, power-broadened by the 64 μW laser beam. (b) Time dependence of the squared rotation recorded for two laser powers: 4 μW (magnified ×8) and 16 μW at a magnetic field of 45 mG compared to the linear rotation at 3 G. For more details see Ref. [20].

Cold-atom experiments studying the nonlinear Faraday effect are generally more demanding than those studying linear Faraday rotation. Not only are they more technically challenging, they are also more prone to various systematic errors, e.g., those caused by expelling atoms from the trap by light pressure. This becomes a serious problem when the light beam needs to be appropriately strong to create the nonlinearity.

A study of NFE with a cold-atom sample of ^{85}Rb released from MOT has been reported in Ref. [20]. In this experiment the atoms were captured and cooled in a standard MOT. Subsequently, the cooling (trapping) laser and the MOT quadrupole magnetic field were switched off and a magnetic field of a certain value was applied along the probe-beam direction (the Faraday field). Finally, the linearly polarized probe beam was switched on and the rotation of its polarization plane was measured. After typically 10 ms, the probe beam and Faraday field were switched off, and the trap fields were switched on to recapture the expanding atomic cloud. The sequence was repeated for each Faraday-field value. A typical resonance observed in the experiment is shown in Fig. 9.2(a). It consists of a broad structure of about 10 G width associated with the linear Faraday effect and a narrow central feature, which is a signature of the nonlinear Faraday rotation. The central resonance of a 20 mG width corresponds to the presence of $\Delta m = 2$ coherences in the $F = 3$ ground state of ^{85}Rb.

9.4.2 Coherence evolution

By using a special time sequence of the light beams and Faraday magnetic field, the time evolution of the superposition state can be studied. Figure 9.2(b) depicts typical time dependences obtained for two different light intensities at a magnetic field corresponding to the maximum nonlinear contribution, compared with the time dependence of a linear Faraday rotation at 3 G. While the linear signal rises very fast, the NFE signal needs some time to develop. As seen, the time becomes shorter for higher light intensity. Such experiments allow diagnostics of superposition states created in a cold-atom sample. The narrow width of the feature (some mG with the described setup) shows the potential of NFE with cold atoms for precision magnetometry with interesting features: sub-mG sensitivity, large dynamic range (zero-field to several gauss), and sub-mm spatial resolution in magnetic-field mapping.

9.4.3 High-field, amplitude-modulated magneto-optical rotation

The ground-state coherences dephase when Larmor precession becomes faster than the coherence relaxation time, hence direct observation of stationary NFE signals is limited to a narrow range of magnetic fields around $B = 0$. For observation of NFE in a wider range it is necessary to use frequency [21] or amplitude [22] modulation of light (see Chapter 6). In this arrangement, strobed pumping creates modulated Zeeman coherence and phase sensitive detection is used to extract the magneto-optical rotation amplitude. Under the conditions of the experiments described in Refs. [21] and [22], in addition to the zero-field resonance, two other resonances appear in the demodulated rotation signal when the modulation frequency Ω_m matches $\pm$ twice the Larmor precession frequency in a given magnetic field. These high-field resonances result from optical pumping synchronous with the Larmor precession. The width of these resonances can be as narrow as the zero-field resonance. Figure 9.3 shows a NFE signal with two amplitude-modulated magneto-optical rotation resonances at ± 3 G that are the evidence of driving $|\Delta m| = 2$ coherences at nonzero magnetic fields. The position of these resonances is insensitive to the light intensity, which allows for precision magnetometry of nonzero magnetic fields. In Ref. [20] it was demonstrated that the modulation technique allows measurement of fields up to ± 9 G.

9.4.4 Paramagnetic nonlinear rotation

Many experiments on paramagnetic rotation have been performed with nonresonant light which is associated with light shifts. Smith *et al.* [14] demonstrated that the tensor component of the light shift adds a nonlinear term to the spin Hamiltonian, whose magnitude is dependent on the angle between the laser polarization and the magnetic field. The nonlinearity vanishes when the relative angle $\theta = \arctan\sqrt{2} \approx 54°$, but is maximized for $\theta = 0$. This dependence has been experimentally verified in Ref. [18]. The nonlinear Hamiltonian is useful for nondestructive preparation and measurement of a quantum state [14, 23]. In particular, in Ref. [23] it has been demonstrated that the nonlinear Hamiltonian allows reaching the so-called "super-Heisenberg" limit of interferometric sensitivity. In that limit, the sensitivity scales as $N^{-(k-1/2)}$ with N the number of atoms which exhibit a k-particle interaction. This idea has been applied by Napolitano *et al.* in an experiment [24] in which super-Heisenberg scaling was demonstrated with cold ^{87}Rb contained in an ODT.

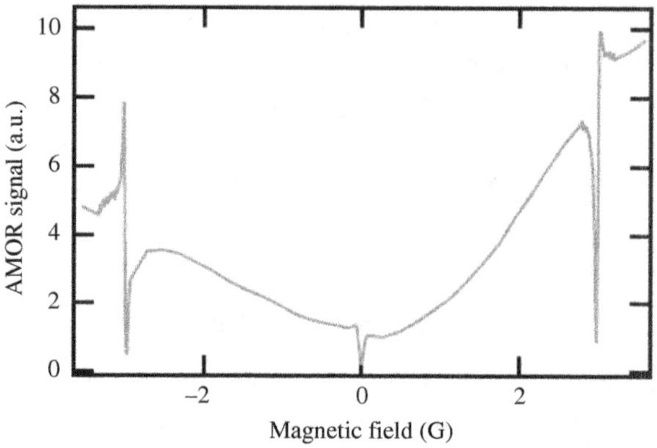

Figure 9.3 Nonlinear Faraday rotation with amplitude-modulated light (AMOR) in a cold-atom sample. The plot shows the square of the rotation angle. The side resonances occur when the modulation frequency equals twice the Larmor frequency in a given magnetic field. From Ref. [20]

9.5 Magnetometry with ultra-cold atoms

9.5.1 Overview of ultra-cold atomic magnetometry methods

Measurements via density modulations

One method of sensing magnetic fields with cold atoms, pioneered by Wildermuth *et al.* [25,26], involves exploiting the mechanical force on an alkali atom due to an inhomogeneous magnetic field. At temperatures in the microkelvin range and below, atomic momentum is sufficiently small that this magnetic force can appreciably modify atomic motion. The most familiar example of this magnetic force is the magnetic trap, which has played an important role in the production and study of cold-atom samples [27,28]. The modification of the trapping potential produced by an applied inhomogeneous field can thus be used to measure the applied field. Atoms in a weak-field-seeking state, for instance, will tend to collect in regions where the applied field is smallest; by measuring the atomic density, Wildermuth *et al.* determined the variations in the applied field. Although this method works for both thermal and quantum-degenerate atoms, the atomic density mirrors spatial variations of the magnetic field more accurately at lower temperature and most accurately with a Bose-condensed sample. Below the Bose-condensation transition temperature, the ability of the atomic density to adapt to the magnetic potential-energy landscape is limited not by temperature but by the atoms' quantum kinetic energy and collisional interaction energy. For condensates in the Thomas–Fermi regime [29], the condensate's density profile is determined by the chemical potential μ, with variations in the confining potential inducing perturbations on the density. The precision with which density can be ascertained then becomes an effective limit on the magnetic sensitivity. Under optimal conditions, the density measurement is limited by atom shot noise. Wildermuth *et al.*, using a ^{87}Rb Bose–Einstein Condensate (BEC) in a highly elongated magnetic trap have demonstrated a sensitivity of

$10\,\mu G$ (1 nT) at a distance of $3\,\mu m$ and used the magnetic field measurement to infer the source current distribution in a microfabricated wire. The authors suggest that a factor of 100 improvement in sensitivity is achievable by making use of a Feshbach resonance to tune the interaction strength near to zero, though this requires operation at a very specific bias magnetic field, which is sometimes undesirable, for example in measurements on magnetizable samples. The use of a magnetic trap with finite bias field in this technique implies that the measurement is primarily sensitive only to a single vector component of the magnetic field; for certain applications this is useful, while for others a direct measurement of the magnitude of the field is most helpful. The spatial resolution in this method is limited to around one healing length ($\xi \equiv \sqrt{\hbar^2/2m\mu}$) by the kinetic-energy cost of creating small features in the condensate density [29]. In other words, the condensate density can be said to low-pass filter the potential with a characteristic filter scale of ξ. Thus lowering the chemical potential to improve sensitivity also has the effect of reducing spatial resolution.

Spinor-condensate magnetometer

A spinor Bose condensate is a condensate with an atomic-spin degree of freedom. In practice, this is typically realized by optical trapping of cold atoms using a far-detuned linearly polarized laser beam. The optical dipole potential produced by such a laser beam is to a good approximation independent of the spin state, allowing the spin to evolve without direct influence from the motional confinement.

With a spinor condensate, one can also measure the effects of the magnetic field on the internal rather than the external atomic state, that is to say on the atomic spin, rather than the atom's motion. In principle, this can be achieved making use of either the longitudinal or transverse spin. The longitudinal spin, i.e., the component parallel to a uniform imposed guiding magnetic field, would respond to adiabatically applied magnetic perturbations by tilting in the direction of the net local vector field, much as iron filings align themselves with field lines near a permanent magnet. Alternatively, atoms prepared with spins transverse to the applied magnetic field will precess at the Larmor frequency determined by the magnitude of the local magnetic field. In either case, spin-sensitive detection can subsequently be used to extract information about the spatial dependence of the magnetic field. The possible magnetic-field sensitivity in the transverse-spin case, however, is much greater, as it takes advantage of the long spin coherence time for a direct measurement of phase or frequency.

In the simplest version of such a measurement, a circularly polarized off-resonant probe laser beam is incident on a transversely magnetized condensate in a single-well optical trap. Because of the difference in dipole matrix elements for atoms magnetized parallel and antiparallel to the probe, regions with parallel magnetization will phase shift the light much more than those with antiparallel magnetization. This spatially dependent phase shift can then be imaged by the phase-contrast technique [30].

In general, the atomic sample acts as a spin-state-dependent birefringent and phase-shifting medium for polarized imaging light. Depending on the configuration of the detector, both magnetization (corresponding to $\Delta m = 1$ coherences and synonymous to the atomic orientation polarization moment, see Section 1.3) and nematicity (corresponding to $\Delta m = 2$

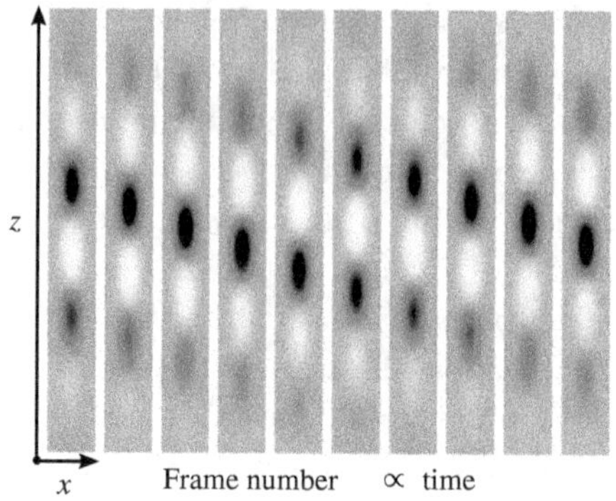

Figure 9.4 Illustration of BEC magnetization imaging (numerical calculation). The condensate has been allowed to evolve in a magnetic gradient prior to the imaging sequence. The frames in the image are taken at regularly spaced times to sample the Larmor precession of atomic spins. A given pixel of the image oscillates in intensity as the magnetization at that location precesses into and out of the imaging axis.

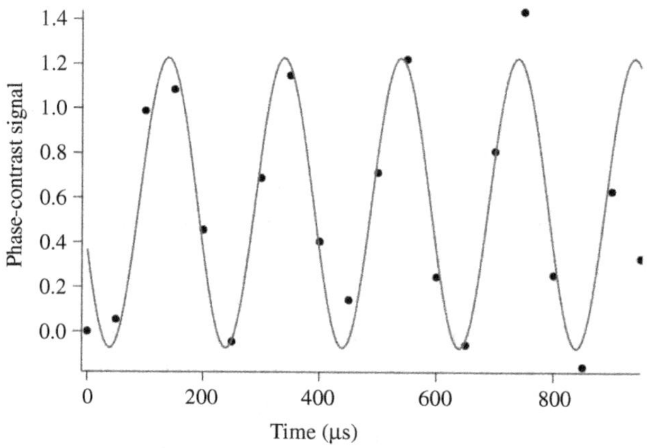

Figure 9.5 Oscillation of magnetization (phase-contrast) signal at a single camera pixel as a function of time. Regions of high and low signal correspond to magnetization parallel and antiparallel to the imaging axis as the atomic spins precess. The accumulated phase gives a high-precision measurement of the magnetic field at this location. (With permission of D. M. Stamper-Kurn.)

coherences and synonymous with atomic alignment, see Section 1.3) can be measured. Detection of the spatially varying phase of Larmor precession has been achieved by taking a rapid sequence of non destructive images of the atomic gas [31]. In such a sequence of images, a given pixel of the image will oscillate between high and low intensity, allowing the local phase of oscillation to be extracted as illustrated in Figs. 9.4 and 9.5. Nondestructive imaging is not an absolute requirement; indeed in principle the phase of Larmor precession in a constant field could be compared to that of a local oscillator by taking a single well-timed image. In practice, however, the degree of magnetic-field stability required is not attainable in most experiments to date, and the Larmor-precession signal in one portion of the condensate is instead phase-referenced to that in another portion, yielding a gradiometric measurement. Such a measurement allows operation in the presence of fluctuating magnetic fields, provided these fields have low spatial frequencies. Because in this case the phase of precession must be extracted from the image signal itself without prior knowledge of its value, it is convenient to sample the Larmor precession multiple times on the same experimental run, which can only be done if the imaging method is nondestructive.

The method of spinor-condensate magnetometry has been demonstrated experimentally by Vengalattore *et al.* [32] using an optically stimulated magnetic field with a Gaussian profile of RMS width 24 μm and 1.66 μG height. The obtained single-shot uncertainty was 0.9 pT in an 11 μm-square region of the condensate. For the actual experimental duty cycle, this corresponds to a sensitivity of 8.3 pT/$\sqrt{\text{Hz}}$, which could be improved to around 0.5 pT/$\sqrt{\text{Hz}}$ by making use of improvements in the experimental duty cycle similar to those demonstrated by Barrett *et al.* [33] and Kinoshita *et al.* [34]. This sensitivity exceeds that of current scanning SQUID magnetometers by an order of magnitude. The technique, moreover, represents a measurement of the magnetic field simultaneously throughout the field of view, resulting in immunity to temporal drifts.

Optical-lattice magnetometry

A variant of the foregoing method makes use of atoms trapped in an optical lattice. This confinement geometry offers a number of advantages over a single-well potential. First, the tight confinement ensures that, to good approximation, the measurement reliably records the magnetic field at a point, rather than a spatio-temporal average. Second, the tight confinement of the lattice can prevent atoms from recoiling after an atom–photon interaction [35, 36]. Interaction of subsequent photons with the atomic density or polarization modulation formed by recoiling atoms can result in superradiant Rayleigh or Raman scattering of the detuned probe beam. This deleterious effect may be obviated by the tight confinement of the lattice. Third, localization of single atoms in sufficiently deep lattice sites eliminates collisional effects, including spin-changing interactions between pairs of atoms (which, as discussed in Section 9.5.3, can lead to instability of certain spinor states [37, 38]) and three-body recombination (which results in atom number loss from the trap). Because an optical lattice shares with other optical traps the possibility of nondepolarizing confinement, a longer lifetime due to reduced three-body recombination translates directly into a longer spin coherence time, potentially in the tens of seconds (for good vacuum).

Realizing these advantages, however, requires a somewhat smaller atomic density relative to a single-well trap. For a ^{87}Rb sample in an optical lattice of wavelength 1024 nm, for instance, a single atom per lattice site implies a density of around 7.5×10^{12} cm^3, approximately an order of magnitude lower than in a typical single-well condensate. Although this lower density reduces sensitivity for a fixed high spatial resolution, it potentially also allows a larger measurement area or field of view (assuming a fixed atom number). The features enumerated here, including long coherence time and isolation from collisional perturbations, make this method an intriguing and very promising future direction for cold-atom magnetometry.

9.5.2 Figures of merit

We have stated that the key advantage of ultra-cold atomic magnetometry is its high spatial resolution. In order to make this claim quantitative, it is useful to develop a figure of merit which takes into account both magnetic sensitivity and spatial resolution. A key quality of a magnetometer is its sensitivity to small changes in the magnetic field. Although the magnetic field can be different at each point in space, all real magnetometers report a spatial average over some finite region. For instance, in an evacuated and antirelaxation-coated vapor cell, an atom may statistically sample the entire volume of the cell between pumping and probing, so that the magnetic field measured is the average field in the interior of the vapor cell. Alternatively, in a vapor cell at high buffer-gas pressure, an atom diffuses only a short distance during the spin coherence time, and the detected magnetic field will be an average over the region of overlap of the pumping and probing laser beams. In the latter case, a tradeoff exists between spatial resolution, which is best for a small detection volume, and sensitivity, which is best for a larger volume containing a larger number of atoms. In this case, it is useful to make use of a figure of merit which captures both spatial resolution and magnetic sensitivity, as introduced by Allred *et al.* [39]. Indeed, for application to magnetic microscopy, where a spatial map of magnetic field strength is desired, such a figure of merit is necessary to make meaningful comparison between sensing techniques. The standard quantum limit of magnetometric sensing uncertainty is given in Chapter 1 by Eq. (1.1).

The measurement time T can be related to the magnetic-field detection bandwidth; for instance, if the instrument reports the average value of the magnetic field in a time T, then the time domain version of this relation

$$\bar{B}(t) = \frac{1}{T}\int_t^{t+T} dt' B(t') \tag{9.7}$$

can be converted to the Fourier-space relation

$$\widetilde{\bar{B}}(\nu) = e^{-i\pi\nu T}\left(\frac{\sin\pi\nu T}{\pi\nu T}\right)\widetilde{B}(\nu), \tag{9.8}$$

where $\widetilde{B}$ denotes the Fourier transform of $B(t)$ and similarly for $\widetilde{\bar{B}}$. The equivalent bandwidth $\Delta\nu$ (i.e., the bandwidth of a "box" transfer function which would transmit the same mean-square magnitude of white noise) can be obtained from the square magnitude of this transfer function by integration:

$$\Delta\nu = \int_0^\infty d\nu\left(\frac{\sin\pi\nu T}{\pi\nu T}\right)^2 = \frac{1}{2T}. \tag{9.9}$$

For different averaging schemes, the precise relation between T and $\Delta\nu$ will differ by an order-unity numerical factor, as illustrated by a low-pass filter of time constant T, whose transfer function $(1+2\pi i\nu T)^{-1}$ yields the relation $\Delta\nu = 1/4T$.

In a similar way, we can relate the spatial length scale of a measurement to its spatial bandwidth. In a map of magnetic-field strength, the measured signal can be binned into pixels, either in hardware or in software, and the spatial averaging inherent in this binning is closely analogous to the temporal averaging outlined above, giving a similar correspondence between the spatial scale in each dimension $L_{x,y,z}$ and the associated spatial bandwidth $\Delta\nu_{x,y,z}$:

$$\Delta\nu_{x,y,z} = \frac{1}{2L_{x,y,z}}. \tag{9.10}$$

We can thus generalize the usual figure of merit, i.e., the magnetic field uncertainty per unit temporal bandwidth or the square root of the magnetic-field power spectral density, to the magnetic-field uncertainty per unit spatiotemporal bandwidth:

$$\mathcal{F} = \frac{\Delta B}{\sqrt{\prod_i \Delta\nu_i}}, \tag{9.11}$$

where the index i in the product runs over both time and spatial dimensions of the measurement. The quantity $\mathcal{F}$ may aptly be termed the "local" sensitivity.

As a concrete instance, we consider mapping the magnetic field in a two-dimensional plane, with simple averaging in time and in the x–y dimensions. For fixed density, the number of atoms N is given by $N = nL_xL_yL_z$, where n is the number density, L_x and L_y are the averaging or bin sizes in the measurement (x–y) plane, and L_z is the depth along the imaging axis. From Eq. (1.1), the field uncertainty for averaging time T is

$$\Delta B = \frac{\hbar}{g\mu_B\sqrt{nL_xL_yL_z\,\tau T}}, \tag{9.12}$$

and we have also for an averaging filter, $\Delta\nu = 1/2T$, $\Delta\nu_{x,y} = 1/2L_{x,y}$, so that the figure of merit (9.11) becomes

$$\mathcal{F}_{\mathrm{2d}} = \frac{\hbar\sqrt{8}}{g\mu_B\sqrt{nL_z\,\tau}}. \tag{9.13}$$

Similarly we can define the figure of merit appropriate to sensing magnetic fields in a three-dimensional region of space, or volumetric local sensitivity:

$$\mathcal{F}_{\mathrm{3d}} = \frac{4\hbar}{g\mu_B\sqrt{n\tau}}. \tag{9.14}$$

We note that the units of $\mathcal{F}_{\mathrm{2d}}$ are (in the cgs system) $\mathrm{G\,cm}/\sqrt{\mathrm{Hz}}$, while the units of $\mathcal{F}_{\mathrm{3d}}$ are $\mathrm{G\,cm^{3/2}}/\sqrt{\mathrm{Hz}}$.

Using these figures of merit, we can compare ultra-cold-atom magnetometry to various other techniques. The demonstrated local sensitivity of spinor condensate magnetometry is

$0.17\,\mathrm{pG\,cm^{3/2}}/\sqrt{\mathrm{Hz}}$ (for three-dimensional measurements) and $10\,\mathrm{pG\,cm}/\sqrt{\mathrm{Hz}}$ (for two-dimensional measurements) with a resolution below $10\,\mu\mathrm{m}$. By way of comparison, the spin-exchange-relaxation-free (SERF) magnetometry technique [39,40], with its extremely high sensitivity, has a demonstrated volumetric local sensitivity of $3\,\mathrm{pG\,cm^{3/2}}/\sqrt{\mathrm{Hz}}$ at length scales of a few millimeters, and a projected local sensitivity (limited by alkali–alkali spin-destruction collisions) of $0.1\,\mathrm{pG\,cm^{3/2}}/\sqrt{\mathrm{Hz}}$. The scanning superconducting quantum interference device (SQUID) microscopy technique [41] operates at a typical sensitivity of $0.7\,\mathrm{nG\,cm}/\sqrt{\mathrm{Hz}}$ at a resolution of around $10\,\mu\mathrm{m}$. These numbers illustrate the exceptionally high sensitivity of spinor magnetometry at several-micron length scales.

9.5.3 Details of spinor magnetometry

Spinor physics

There are many good references on the physics of spinor condensates [42–45]. Here we briefly review those features most important to magnetometry with spin-1 atoms. The energy of a condensate with (three-component) wave function Ψ is given by

$$H = \int d^3\mathbf{r} \left\{ \frac{\hbar^2}{2m} |\nabla\Psi|^2 + \Psi^\dagger U(\mathbf{r})\Psi + \frac{c_0}{2}(\Psi^\dagger\Psi)^2 + \frac{c_{\mathrm{spin}}}{2}(\Psi^\dagger \mathbf{F}\Psi)^2 \right\}, \tag{9.15}$$

where the terms in the integrand are, in order, the kinetic and potential energies, the spin-independent collisional interaction energy, and the spin-dependent collisional interaction energy, and $\mathbf{F}$ denotes the vector of dimensionless spin-1 angular momentum operators. The coupling constants c_0 and c_{spin} are given by $c_0 = \frac{4\pi\hbar^2}{m}(a_0 + 2a_2)/3$ and $c_{\mathrm{spin}} = \frac{4\pi\hbar^2}{m}(a_2 - a_0)/3$, where a_2 and a_0 are the scattering lengths for the total spin-2 and total spin-0 scattering channels. Although the coefficient c_{spin} is much smaller than the coefficient c_0 for typical species (in particular, both ^{87}Rb and ^{23}Na), it is crucial in determining the nature of the spinor ground state. Specifically, if $c_{\mathrm{spin}} > 0$, then states which minimize the square magnitude of the expectation value $\langle \mathbf{F} \rangle$ will be energetically favored by the condensate. Such a condensate is called *polar* or *antiferromagnetic*, and is exemplified by sodium. The mean-field spinor ground-state wave function of such a condensate has spin projection $m_F = 0$ along some axis. In the alternative case that $c_{\mathrm{spin}} < 0$, states with maximum square magnitude of the expectation value $\langle \mathbf{F} \rangle$ are energetically favorable. A condensate of this sort is referred to as *ferromagnetic*, and ^{87}Rb is a paradigmatic example of a ferromagnetic condensate. The ground state of a ferromagnetic spin-1 condensate has all spins in the $m_F = +1$ state along an arbitrary axis.

In order to make spatially resolved measurements of magnetic field, the condensate cannot remain in its spinor ground state, however, but must be subjected to spatially inhomogeneous evolution. Thus it is important to understand not merely the ground state, but also the nature of the collective excitations from the ground state. One can view the condensate as the vacuum state from which collective excitations and quasiparticles may be created and into which they may disappear. Collisional interactions modify the properties of these elementary excitations from those of free particles, as worked out by Bogoliubov [46] and, for the case of spinor condensates, by Ho [42] and by Ohmi and Machida [43]. The dispersion relation of these elementary excitations depends both on the spinor wave function of the underlying condensate and on the spin state of the excitations themselves. For noninteracting particles in a homogeneous gas, the dispersion relation is given by $\hbar\omega = \epsilon_k$ where we define $\epsilon_k = \hbar^2 k^2/2m$. For a homogeneous ferromagnetic spin-1 condensate in the $m_F = 1$ spin state, Ho and Ohmi *et al.* determined the dispersion relations for each of the possible spin states. For collective excitations with spin projection $m_F = 0$ the dispersion relation is that of a free particle, $\hbar\omega = \epsilon_k$; for excitations with $m = -1$, the dispersion relation possesses a gap proportional to the spin interaction energy $\hbar\omega = \epsilon_k + 2c_{\text{spin}}n$; and for $m_F = +1$ excitations are of the "regular" Bogoliubov phonon type, with dispersion relation $\hbar\omega = \sqrt{\epsilon_k^2 + 2c_{\text{spin}}n\epsilon_k}$. (Here n is the condensate number density.) In the case of an antiferromagnetic condensate of $m_F = 0$ atoms, these authors determined that there exist excitations with spin character $m_F = \pm 1$, whose dispersion relation is Bogoliubov-like but with the spin-dependent coupling constant determining the cross-over between linear and quadratic dependence, $\hbar\omega = \sqrt{\epsilon_k^2 + 2c_{\text{spin}}\epsilon_k n}$, and excitations with spin character $m_F = 0$, which, being in the same spin state as the condensate, have a standard Bogoliubov dispersion relation, $\hbar\omega = \sqrt{\epsilon_k^2 + 2c_0\epsilon_k n}$.

One important point which can be seen from these results is that if one prepares a condensate in the "wrong" spin state, e.g. a ferromagnetic condensate, with $c_{\text{spin}} < 0$ in an antiferromagnetic ($m_F = 0$) state, the dispersion relation $\hbar\omega = \sqrt{\epsilon_k^2 + 2c_{\text{spin}}\epsilon_k n}$ can yield an imaginary frequency, indicating that this state is unstable and will decay to other spin states, as is in fact seen experimentally [37, 38].

Spatial resolution

Since each atom in the condensate is delocalized within a spatial extent determined by the condensate wave function, one might erroneously assume that no spatial information about magnetic or other perturbations on a smaller length scale could be gleaned. In fact, the wave function itself registers perturbations on small scales; in other words, although each atom is delocalized within the volume of the trapped cloud, the spin and spatial degrees of freedom can become correlated if the perturbation is spatially inhomogeneous, and the wave function encodes these correlations. Thus measurements performed on the ensemble can yield complex patterns of atomic spin. The spatial resolution with which the magnetic field can be determined from measurements of these spin textures is limited both by diffraction of the imaging light and by the response of the spinor condensate to the applied field.

The response of the condensate spin state to an applied field is, to first approximation, simply to precess around it. However, for long times or small features, the spinor dynamics must also be considered. The time evolution of the condensate wave function neglecting atom–atom correlations is given by the Gross–Pitaevskii equation,

$$i\hbar\frac{\partial\Psi}{\partial t}=\frac{\hbar^2}{2m}\nabla^2\Psi+U(\mathbf{r})\Psi+c_0(\Psi^\dagger\Psi)\Psi+c_{\text{spin}}(\Psi^\dagger\mathbf{F}\Psi)\cdot\mathbf{F}\Psi. \tag{9.16}$$

To analyze the spinor dynamics, we can write the condensate wave function as

$$\Psi(\mathbf{r},t)=\psi_g(\mathbf{r})\chi(\mathbf{r},t)e^{-i\omega_g t}, \tag{9.17}$$

where ψ_g is the scalar ground state with energy $\hbar\omega_g$, and $\chi(\mathbf{r},t)$ is the spinor wave function at subsequent times. Although this factorization may be performed without loss of generality, it is most useful when the condensate density profile remains nearly constant, and the dynamics are primarily confined to the spin degrees of freedom. The Gross-Pitaevskii equation, rewritten in these terms, becomes

$$i\hbar\frac{\partial\chi}{\partial t}=-\frac{\hbar^2}{2m}\nabla^2\chi-\frac{\hbar^2}{m}\nabla\log\psi_g\cdot\nabla\chi+c_{\text{spin}}|\psi_g|^2(\chi^\dagger\mathbf{F}\chi)\cdot\mathbf{F}\chi+g_F\mu_B\mathbf{B}(\mathbf{r})\cdot\mathbf{F}\chi. \tag{9.18}$$

The last term is typically dominant, so that to first approximation, the spinor wave function of atoms in this regime, exposed to the perturbing magnetic field $\mathbf{B}(\mathbf{r})$ is simply

$$\chi(\mathbf{r},t)=e^{-ig_F\mu_B\mathbf{B}(\mathbf{r})\cdot\mathbf{F}t/\hbar}\chi(\mathbf{r},0), \tag{9.19}$$

where $\mathbf{F}$ are the dimensionless spin matrices, g_F is the atomic g-factor, μ_B is the Bohr magneton, and t is the time. In other words, the atoms remain in the spatial ground state, but their spins are rotated by the local magnetic field. This approximation is valid for short times or weakly inhomogeneous fields, but for longer times or very short distances, the quantum kinetic energy term $-\frac{\hbar^2}{2m}\nabla^2\chi$ contributes to dispersion of small spinor features, as illustrated in Fig. 9.6.

The response of a condensate to an applied perturbation can also be understood with reference to its elementary excitations. These collective excitations can be created by an applied inhomogeneous field. Indeed, if we consider a condensate in the $m_F=+1$ state, subjected to an applied inhomogeneous transverse field $B_x(\mathbf{r})$, then the Zeeman energy, given by

$$H_{\text{Zeeman}}=g_F\mu_B\int d^3\mathbf{r}\,\Psi^\dagger(\mathbf{r})B_x(\mathbf{r})F_x\Psi(\mathbf{r}), \tag{9.20}$$

can be written to leading order in the condensate order parameter as the Fourier-space sum,

$$H_{\text{Zeeman}}\approx g_F\mu_B\sqrt{\frac{N}{2}}\sum_{\mathbf{k}}\widetilde{B}_x(\mathbf{k})a_{0,\mathbf{k}}+\widetilde{B}_x^*(\mathbf{k})a_{0,\mathbf{k}}^\dagger, \tag{9.21}$$

where $a_{m,\mathbf{k}}$ is the annihilation operator for the quasiparticle with spin projection m and angular wave vector $\mathbf{k}$. The above expression makes explicit that the spatial Fourier component of the magnetic field with wave vector $\mathbf{k}$ directly creates collective excitations with the

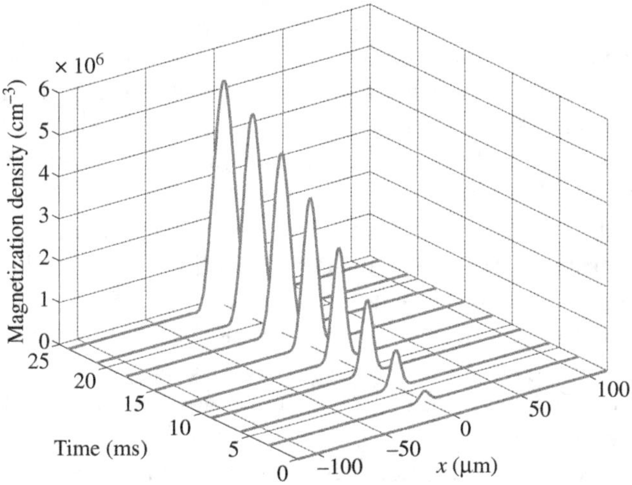

Figure 9.6 Dispersion of transverse magnetization (simulation result). A ferromagnetic condensate is exposed to a small Gaussian-profile region of magnetic field transverse to its magnetization axis, resulting in local rotation of the magnetization into the third, perpendicular direction. In addition to growth of magnetization (due to Larmor precession of the magnetization), the Gaussian packet spreads as a result of its quantum kinetic energy of confinement.

same wave vector. These continually created collective excitations then evolve according to the modified dispersion relation determined by the collisional interaction with the condensate, as specified earlier. As the excitations propagate outward from the region of their creation, they contribute to a blurring of the magnetization pattern and lower the spatial resolution.

In contrast to a single-well spinor-condensate magnetometer, where spinor dynamics imposes a tradeoff between resolution and evolution time, an optical-lattice magnetometer possesses atoms anchored to their respective lattice sites. As a result, such dispersion or quantum diffusion cannot occur. Moreover, in an optical lattice the atom is localized within a small region around the minimum of the lattice potential. Thus, although the spatial coverage is poor at length scales smaller than the lattice spacing, and although achieving lattice-resolved imaging resolution remains experimentally challenging [47–49], the potential fundamental resolution of the method is in the range of 50 nm.

9.5.4 Comparison with thermal-atom magnetometry

It is interesting to ask to what extent the excellent sensitivity achieved by spinor condensate magnetometry is a result of using a quantum-degenerate atomic sample, and to what extent similar results could be obtained from a nondegenerate sample.

To first approximation, the atomic velocity distribution is not relevant to the measurement of Larmor frequency, because the velocity and spin degrees of freedom are largely decoupled, and magnetometric performance is governed by coherence of the atoms' spin state, not their spatial state. Indeed, even in room-temperature gases, in which large numbers of velocity states are incoherently populated, it is straightforward to create atomic spin states that are out of thermal equilibrium, and, for example, at near zero effective temperature. However, as noted above, for spatially resolved measurements, the proper figure of merit is sensitivity to the magnetic field in a particular region of space, and for this local sensitivity the motional atomic degrees of freedom are significant.

In an optical lattice, it is not required that the atoms be Bose-condensed to achieve high magnetic sensitivity, merely that their temperature be low enough for strong lattice confinement. This example suggests that quantum degeneracy plays no fundamental role in cold-atom magnetometry.

Certain differences between quantum-degenerate and nondegenerate samples arise simply because the former typically exhibit a higher number density. This difference in density is not fundamental, for a thermal gas can in principle always be prepared at the same density as a condensate, provided one is free to adjust the temperature. Nevertheless, the condensate's higher density has several important consequences. Most obviously, a high number density yields a higher local sensitivity, other factors being equal.

Moreover, classically, a high density of atoms produces a short mean free path and reduced diffusion in a given length of time. Thus, in a thermal gas there are several qualitatively different regimes. At low density, collisions are rare, and atoms follow (classically) time reversible trajectories. On average such atoms will explore the entire energetically available trap volume, though individual trajectories may be restricted to lower-dimensional submanifolds, as in the case of commensurate harmonic trap frequencies. At intermediate densities, randomizing collisions become more frequent, so that trajectories are no longer time reversible but still tend to explore the available volume. Finally, at high densities, collisions are so frequent that an atom can diffuse only a short distance in its spin coherence time. In both the intermediate and high-density regimes, the spin dynamics and spin coherence are increasingly modified by spin-exchange collisions. Moreover, at high densities collisional losses such as molecular recombination become significant.

With a quantum-degenerate sample, one is not justified in speaking of well-defined atomic trajectories; nevertheless, certain phenomena are analogous to their classical counterparts in a thermal gas. At low condensate density, each atom occupies a single-particle wave function which is unmodified by the presence of other atoms and determined purely by the physics of a single atom in its trapping potential. At higher densities, the condensate wave function is modified by the collisional mean-field interaction, but remains well described by a single wave function. For the usual (stable) case where these interactions are repulsive, the atoms occupy a region that is wider in position space and correspondingly narrower in momentum space. This velocity narrowing is beneficial for spatially resolved magnetometry.

9.5.5 Applications

Ultra-cold-atom magnetometry is applicable to a wide range of fundamental scientific and practical problems. Spinor-condensate magnetometers possess a number of important advantages, some common to other atomic magnetometers, such as noncryogenic operation, "automatic" calibration in terms of fundamental constants, and reduced low-frequency flicker noise, and some due to the specific nature of the ultra-cold-atom system, including high local sensitivity, ultra-high-vacuum compatibility, simultaneous scan-free measurement over the entire field of view, and wideband sensor transparency, permitting concurrent implementation of magnetic-field imaging and optical imaging.

In vacuo applications

Because the production of ultra-cold atoms and the exploitation of their long coherence time require ultra-high vacuum conditions, the most straightforward applications are those involving vacuum-compatible measurement samples. A potential fundamental-physics application is comagnetometry for electron electric dipole-moment searches, particularly those which themselves make use of cold atoms [50], as discussed in Chapter 18. Application to condensed-matter systems is promising, including such candidate systems as vortex structures and dynamics in type II superconductors, mapping of microfractures and grain boundaries in ferrous materials, and detection of solid-state impurity spin states.

Aigner *et al.* have recently employed cold-atom magnetometry to study long-range correlations in an electric current through disordered conductors [51]. Potential measurements of technological interest include nondestructive evaluation of microchips, including measurement of leakage currents and current irregularities in integrated circuits.

Atmospheric-pressure samples

Although the long coherence time of ultra-cold atoms requires ultra-high vacuum, it is possible to envision using a trapped atomic gas in vacuum to measure fields produced by a sample outside the vacuum system. In order to maintain the technique's high spatial resolution, however, the field-producing and field-detecting samples would need to be within $1-10\,\mu\text{m}$ of each other. Though technologically challenging, the feasibility of manufacturing ultra-thin vacuum windows capable of sustaining atmospheric pressure and of maintaining ultra-cold atoms near a room-temperature dielectric surface without deleterious effects [52] has been well demonstrated. Samples requiring such a vacuum membrane include biological cells, particularly those known to conduct current, such as neurons and heart cells, and high-outgassing materials.

References

[1] W. Franzen, Phys. Rev. **115**, 850 (1959).
[2] R. Zhang, S. R. Garner, and L. V. Hau, Phys. Rev. Lett. **103**, 233602 (2009).
[3] M. Theis, G. Thalhammer, K. Winkler, M. Hellwig, G. Ruff, R. Grimm, and J. Denschlag, Phys. Rev. Lett. **93**, 123001 (2004).

[4] T. Brzozowski, M. Brzozowska, J. Zachorowski, M. Zawada, and W. Gawlik, Phys. Rev. A **71**, 013401 (2005).
[5] J. Nash and F. Narducci, J. Mod. Opt. **50**, 2667 (2003).
[6] H. Ling-Xiang and W. Yu-Zhu, Chinese Phys. **13**, 754 (2004).
[7] W. Gawlik and S. Pustelny, in *Horizons in World Physics: New Trends in Quantum Coherence and Nonlinear Optics* **263**, edited by R. Drampyan (Nova Science Publishers, New York, 2009), p. 45.
[8] J. Manuel and C. Cohen-Tannoudji, Compt. Rend. Acad. Sci. (Paris) **257**, 413 (1963).
[9] W. Gawlik, J. Kowalski, R. Neumann, H. Wiegemann, and K. Winkler, J. Phys. B **12**, 3873 (1979).
[10] S. Franke-Arnold, M. Arndt, and A. Zeilinger, J. Phys. B **34**, 2527 (2001).
[11] G. Labeyrie, C. Miniatura, and R. Kaiser, Phys. Rev. A **64**, 033402 (2001).
[12] J. M. Choi, J. M. Kim, J. H. Lee, Q. H. Park, and D. Cho, Phys. Rev. A **71**, 043409 (2005).
[13] T. Isayama, Y. Takahashi, N. Tanaka, K. Toyoda, K. Ishikawa, and T. Yabuzaki, Phys. Rev. A **59**, 4836 (1999).
[14] G. A. Smith, S. Chaudhury, and P. S. Jessen, J. Opt. B **5**, 323 (2004).
[15] M. Takeuchi, T. Takano, S. Ichihara, Y. Takasu, M. Kumakura, T. Yabuzaki, and Y. Takahashi, Appl. Phys. B **83**, 107 (2006).
[16] M. Kubasik, M. Koschorreck, M. Napolitano, S. R. de Echaniz, H. Crepaz, J. Eschner, E. S. Polzik, and M. W. Mitchell, Phys. Rev. A **79**, 043815 (2009).
[17] M. L. Terraciano, S. E. Olson, M. Bashkansky, Z. Dutton, and F. K. Fatemi, Phys. Rev. A **76**, 053421 (2007).
[18] M. Terraciano, M. Bashkansky, and F. K. Fatemi, Phys. Rev. A **77**, 063417 (2008).
[19] D. Budker, W. Gawlik, D. F. Kimball, S. M. Rochester, V. V. Yashchuk, and A. Weiss, Rev. Mod. Phys. **74**, 1153 (2002).
[20] A. Wojciechowski, E. Corsini, J. Zachorowski, and W. Gawlik, Phys. Rev. A **81**, 053420 (2010).
[21] D. Budker, D. F. Kimball, V. V. Yashchuk, and M. Zolotorev, Phys. Rev. A **65**, 055403 (2002).
[22] W. Gawlik, L. Krzemień, S. Pustelny, D. Sangla, J. Zachorowski, M. Graf, A. O. Sushkov, and D. Budker, Appl. Phys. Lett. **88**, 131108 (2006).
[23] M. Napolitano and M. Mitchell, New J. Phys. **12**, 093016 (2010).
[24] M. Napolitano, M. Koschorreck, B. Dubost, N. Behbood, R. J. Sewell, and M. W. Michell, Mitchell, Nature **471**, 486 (2011).
[25] S. Wildermuth, S. Hofferberth, I. Lesanovsky, E. Haller, L. Andersson, S. Groth, I. Bar-Joseph, P. Krüger, and J. Schmiedmayer, Nature **435**, 440 (2005).
[26] S. Wildermuth, S. Hofferberth, I. Lesanovsky, S. Groth, P. Krüger, J. Schmiedmayer, and I. Bar-Joseph, Appl. Phys. Lett. **88**, 264103 (2006) [arXiv:cond-mat/0512520].
[27] A. L. Migdall, J. V. Prodan, W. D. Phillips, T. H. Bergeman, and H. J. Metcalf, Phys. Rev. Lett. **54**, 2596 (1985).
[28] H. F. Hess, G. P. Kochanski, J. M. Doyle, N. Masuhara, D. Kleppner, and T. J. Greytak, Phys. Rev. Lett. **59**, 672 (1987).
[29] F. Dalfovo, S. Giorgini, L. Pitaevskii, and S. Stringari, Rev. Mod. Phys. **71**, 463 (1999).
[30] F. Zernike, Monthly Not. R. Astronom. Soc. **94**, 0377 (1933).
[31] J. Higbie, L. Sadler, S. Inouye, A. Chikkatur, S. Leslie, K. Moore, V. Savalli, and D. Stamper-Kurn, Phys. Rev. Lett. **95**, 50401 (2005).
[32] M. Vengalattore, J. Higbie, S. Leslie, J. Guzman, L. Sadler, and D. Stamper-Kurn, Phys. Rev. Lett. **98**, 200801 (2007).

[33] M. D. Barrett, J. A. Sauer, and M. S. Chapman, Phys. Rev. Lett. **87**, 010404 (2001).
[34] T. Kinoshita, T. Wenger, and D. Weiss, Phys. Rev. A **71**, 11602 (2005).
[35] R. Dicke, Phys. Rev. **89**, 472 (1953).
[36] P. Jessen and I. Deutsch, Adv. Atomic Mol. Opt. Phys. **37**, 95 (1996).
[37] L. Sadler, J. Higbie, S. Leslie, M. Vengalattore, and D. Stamper-Kurn, Nature **443**, 312 (2006).
[38] W. Zhang, D. Zhou, M. Chang, M. Chapman, and L. You, Phys. Rev. Lett. **95**, 180403 (2005).
[39] J. Allred, R. Lyman, T. Kornack, and M. Romalis, Phys. Rev. Lett. **89**, 130801 (2002).
[40] I. Kominis, T. Kornack, J. Allred, and M. Romalis, Nature **422**, 596 (2003).
[41] J. R. Kirtley and J. P. Wikswo, Annu. Rev. Mater. Sci. **29**, 117 (1999).
[42] T.-L. Ho, Phys. Rev. Lett. **81**, 742 (1998).
[43] T. Ohmi and K. Machida, J. Phys. Soc. Jpn **67**, 1822 (1998).
[44] C. Pethick and H. Smith, *Bose-Einstein Condensation in Dilute Gases* (Cambridge University Press, Cambridge, 2002).
[45] M. Ueda and Y. Kawaguchi, arXiv:1001.2072 (2010).
[46] N. Bogoliubov, J. Phys. **11**, 23 (1947).
[47] K. Nelson, X. Li, and D. Weiss, Nature Physics **3**, 556 (2007).
[48] W. Bakr, J. Gillen, A. Peng, S. Fölling, and M. Greiner, Nature **462**, 74 (2009).
[49] J. Sherson, C. Weitenberg, M. Endres, M. Cheneau, I. Bloch, and S. Kuhr, Nature **467**, 68 (2010).
[50] C. Chin, V. Leiber, V. Vuletić, A. Kerman, and S. Chu, Phys. Rev. A **63**, 33401 (2001).
[51] S. Aigner, L. Pietra, Y. Japha, O. Entin-Wohlman, T. David, R. Salem, R. Folman, and J. Schmiedmayer, Science **319**, 1226 (2008).
[52] D. Harber, J. McGuirk, J. Obrecht, and E. Cornell, J. Low Temp. Phys. **133**, 229 (2003).

10
Helium magnetometers

R. E. Slocum, D. D. McGregor, and A. W. Brown

10.1 Introduction

Optically pumped helium (He) magnetometers have provided magnetic field data for military, aeromagnetic survey, space exploration and geophysical laboratory applications for over five decades. The characteristics of He magnetometers that have made them instruments of choice for these varied applications include high sensitivity, high accuracy, simplicity of the resonance line, small heading errors due to light shifts, temperature independence of resonance cells, linear relationship between the magnetic field and the resonance frequency, excellent stability for gradiometer operation and robustness for field and space use. Scalar He magnetometers can easily be configured for omnidirectional operation with no moving parts to provide full sensitivity on all headings relative to the magnetic field direction.

Helium magnetometers have two types of optical pumping radiation sources. All He magnetometers manufactured from 1960 to 1990 utilized an RF electrodeless discharge He-4 lamp as an optical pumping source of 1083 nm resonance radiation which is composed of three closely spaced He-4 resonance lines D0, D1, and D2. In the 1980s, the development efforts for a single-line pump source for both He magnetometers and basic research on He isotopes resulted in both high-efficiency semiconductor lasers and optical fiber lasers at 1083 nm. Laser-pumped He magnetometers are characterized by sensitivities up to two orders of magnitude better than lamp-pumped He magnetometers and are more accurate, smaller, and very stable for use in magnetic gradiometers. L. D. Schearer provided a comprehensive review of the beginning science of He-4 magnetometers [1] and a review of the first 25 years of progress in optically pumped He magnetometers [2]. In 2000, Slocum and Smith provided an overview of developments over the first four decades [3]. McGregor provided a complete analysis of the physics of laser pumped He-4 magnetometers in 1987 [4], and this analysis was expanded and updated in 2010 by Plante *et al.* [5].

Optical Magnetometry, ed. Budker, D. and Jackson Kimball, D. F. Published by Cambridge University Press.

This chapter provides an overview of the past 50 years of He magnetometers. Helium space magnetometers are discussed separately in Chapter 15. The principles of operation for the most common scalar He magnetometers and vector space magnetometers are presented in Section 10.2. Key elements of a He magnetometer include the He isotope, optical pumping source, optical pumping process, and production and observation of paramagnetic and parametric resonances. The resonance element in these instruments is the isotope He-4 excited by a weak RF discharge. The main optical pumping sources used since the 1990s are laser sources resulting in dramatic improvements in sensitivity, accuracy, and miniaturization. Methods for producing resonance effects are paramagnetic resonance for scalar magnetometers and parametric resonance for vector magnetometers which are observed by optical methods.

Helium magnetometers began with the first optical pumping of the metastable level of He by Colegrove and Franken [6] at the University of Michigan in 1958. The Franken patent [7] was purchased by Texas Instruments (TI) in 1959, leading to the first successful optically pumped scalar He magnetometer at TI in 1960 [8], where more than 1500 scalar He-4 magnetometers were manufactured for use in US Navy aircraft for airborne Magnetic Anomaly Detection (MAD) of submarines. Slocum and Reilly at TI developed the first space vector He magnetometer in 1961 [9], the first in a series of space magnetometers built by E. J. Smith's magnetometer team at the Jet Propulsion Laboratory and flown on a set of seven pioneering planetary and Earth-orbital missions described in Chapter 15.

All He-4 magnetometers manufactured in the 20th century were lamp-pumped instruments. The rapid development of laser-pumped He magnetometers over the last two decades was led by a group of French laboratories and by Polatomic, Inc., in the USA. The French group began collaborative research and development efforts with L. D. Schearer of the University of Missouri-Rolla on laser pump sources for He isotopes [10]. The collaborative team continues to develop significant applications of laser-pumped He isotopes in their home laboratories. Collaborators included M. Leduc and F. Laloe of the Laboratoire Kastler Brossel, Ecole Normale Superieure, Paris (ENS). J. Hamel represented the Laboratory for Spectrocopie Atomic ISMRA at Caen (ISMRA-Caen). In 1996, LETI-CEA Advanced Technologies of Grenoble, France, was selected to design and build a laser pumped He-4 magnetometer for the SWARM satellites scheduled for launch in 2013. The SWARM instrument is discussed in Chapter 15. In the USA, Polatomic was encouraged in 1988 by the US government to engage in research and development of laser-pumped He magnetometers. Polatomic pioneered the practical application of 1083 nm semiconductor distributed Bragg reflector (DBR) laser pump sources for high-performance He magnetometers. US Navy sponsorship of laser He magnetometer development was shifted in 1988 to Polatomic for development of the next generation laser-pumped AN/ASQ-233 He-4 system for US Navy anti-submarine aircraft.

10.2 Helium magnetometer principles of operation

Comprehensive discussions of operating principles for scalar and vector He-4 magnetometers are presented here. Six elements characterizing each He magnetometer are the (1) He

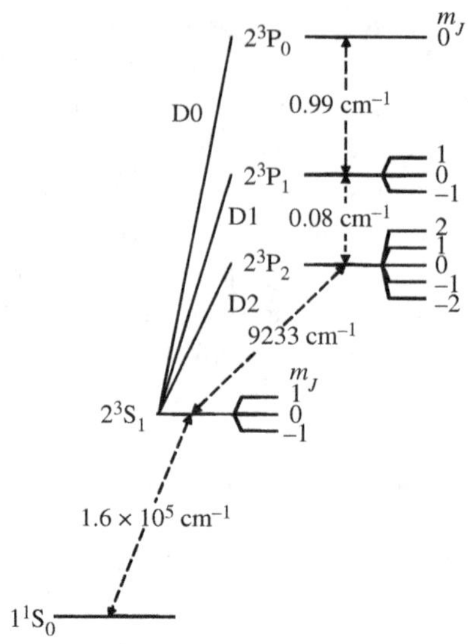

Figure 10.1 Energy levels for optical pumping of metastable He-4.

isotope, (2) optical pumping source, (3) optical pumping process, (4) method for observing optically pumped He, (5) generation of magnetic resonance signals, and (6) sensor and electronic components.

10.2.1 Helium resonance element

A glass cell is filled with the He-4 isotope which has a nucleus containing two protons and two neutrons with no nuclear spin. The resonance element consists of He-4 atoms excited to the 2^3S_1 metastable energy level (Fig. 10.1). A weak RF discharge produces a population of metastable atoms, since electromagnetic transition to the ground state is forbidden. In practice, metastable He atoms are produced by electron collisions with ground-state atoms in a weak RF electrodeless discharge maintained in glass cells ranging in size from 1 cm^3 to 60 cm^3.

When a magnetic field $\mathbf{B}_0$ is present, Zeeman splitting separates the He-4 metastable level into magnetic states $m = 0, \pm 1$ (Fig. 10.2). The three magnetic states are populated by metastable He-4 atoms with spin = 1 where the magnetic moments of atoms in each spin state are oriented parallel ($m = -1$), oriented antiparallel ($m = +1$), or aligned perpendicular ($m = 0$) to the ambient magnetic field. In the absence of optical pumping, the magnetic spin states are in thermal equilibrium and are equally populated.

10.2.2 Helium optical pumping radiation sources

The optical pumping source provides a collimated beam of polarized 1083 nm radiation which is directed through the resonance cell. An RF electrodeless discharge He-4

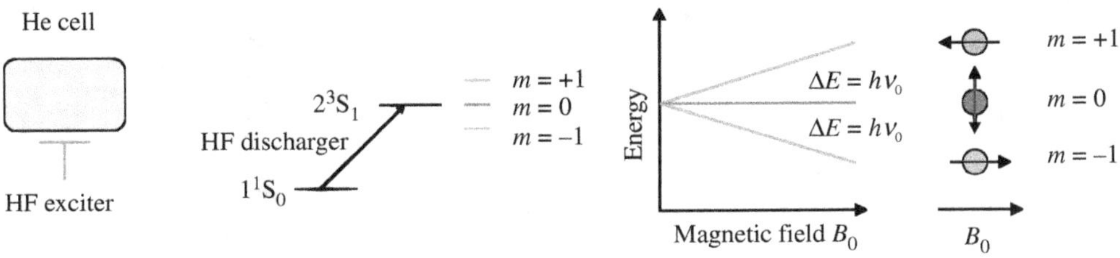

Figure 10.2 Helium-4 energy level diagram with Zeeman splitting.

lamp with an excitation frequency in the 20–40 MHz range was the exclusive choice for magnetometers prior to 1990. Helium-4 lamps are low-noise sources and emit resonance radiation made up of three spectral lines (D0, D1, and D2) corresponding respectively to the three optical transitions between the 2^3S_1 level and the three P-levels (Fig. 10.1).

Optical pumping varies the populations in the three magnetic states by depopulating the $m = 0$ state or depopulating either the $m = +1$ or $m = -1$ state. The pumping efficiency is low for lamp radiation since the D0 and D1 lines pump counter to the D2 line [1,4,11]. It can be seen from the calculated absorption coefficients that a significant improvement in optical pumping efficiency can be achieved by single-line pumping using the D0 line [4]. A single-line pumping source at the wavelength of the 1083 nm D0 absorption line became technically feasible with the advent of 1083 nm lasers in the 1980s and proved extremely important for He magnetometry. McGregor's 1987 prediction [4] of significant sensitivity improvements from laser pumping was confirmed when LNA laser pumping of metastable He for space magnetometers showed a factor of 45 increase in resonance signal amplitude [12]. Although the lasers of choice for He pumped magnetometers initially appeared to be diode-laser-pumped LNA and neodymium-doped optical fibers, in 1988 Polatomic and JPL teamed with General Optronics to demonstrate the first practical 1083 nm semiconductor laser using a composition-tuned InGaAsP chip. InGaAs was later found to be a more stable material at 1083 nm, which was at the long-wavelength limit. SDL had developed the InGaAs lasers for 980 nm and 1020 nm and were successful in extending stable narrow line performance out to 1083 nm. The InGaAs DBR diode laser has become the pump-source of choice for a number of advanced space and military He magnetometers. A critical factor for design of high-sensitivity magnetometers is low-noise laser drivers and line-locking loops for the InGaAs DBR lasers [13]. An alternate laser pump source for some He magnetometers is the diode-pumped fiber laser.

The polarization state of the pumping beam is typically circular, linear, or unpolarized. In Colegrove and Franken's original He-4 optical pumping experiment [6], the lamp radiation was unpolarized and the pumping efficiency was very low. The majority of laser and lamp pump sources employ circularly polarized pumping beams [7], although Slocum in 1988 showed that linearly polarized laser pumping can produce resonance lines with strength equal to that produced by the circularly polarized source.

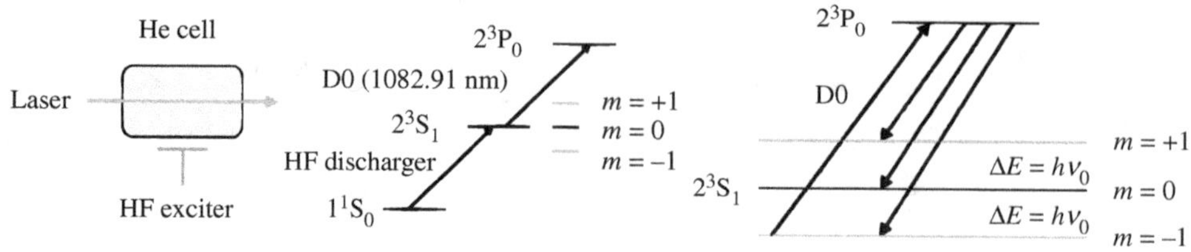

Figure 10.3 Single-line pumping in magnetic spin states of the He-4 2^3S_1 metastable level with spin 1.

10.2.3 Optical pumping of metastable helium

A simple explanation of optical pumping of He-4 will be illustrated with single-line laser pumping of the three magnetic states of the 2^3S_1 level (Fig. 10.3). Discussion of lamp optical pumping in He-4 can be found in the early references [1,4,11]. The three Zeeman levels $m = +1, 0, -1$ in the presence of ambient field $\mathbf{B}_0$ have separation energy $\Delta E = h\nu_0$ between adjoining spin states where $\nu_0 = [\gamma_e/(2\pi)]B_0$ and $\gamma_e/(2\pi) = 28.02495266$ Hz/nT. In the Earth's field, the $m = +1, 0, -1$ sublevels are equally populated in an active RF discharge and approach thermal equilibrium since they are populated at equal rates by electron excitation of ground-state atoms. The metastable lifetimes are in the range of 10^{-3} to 10^{-4} s. Optical pumping drives the metastable population to a nonequilibrium distribution over the spin states.

When circularly polarized pumping radiation at the D0 wavelength (1082.91 nm) is directed along the $\mathbf{B}_0$ direction, the normalized spin state absorption coefficients are 16:0:0 for m equal to +1:0:-1 respectively. Absorption of atoms in the $m = -1$ state lifts them to the 2^3P_0 level from where they decay at equal rates to all three 2^3S_1 states. This optical pumping process rapidly depletes the $m = -1$ spin state and transfers the atoms to the $m = 0$ and $m = +1$ states. Since atoms in the 2^3S_1 level have a magnetic moment, a net magnetic moment is built up antiparallel to the ambient magnetic field direction. If the helicity of the circularly polarized radiation is reversed, atoms in the $m = +1$ state have a relative absorption coefficient of 16 while atoms in the other two Zeeman sublevels produce zero absorption. Optical pumping now depopulates the $m = +1$ state and the optical pumping process transfers the 2^3S_1 atoms to the $m = 0$ and $m = -1$ states producing a net magnetic moment parallel to the ambient field. When linearly polarized light is substituted for circularly polarized light, the relative absorption coefficients are 0:16:0 for m equal to +1:0:-1, respectively, so atoms from the $m = 0$ state are transferred to the other two states. The resonance signal amplitude produced when the linearly polarized beam is directed along an axis perpendicular to $\mathbf{B}_0$ with the direction of linear polarization along $\mathbf{B}_0$ is equal to the amplitude produced for a circularly polarized beam parallel to $\mathbf{B}_0$.

Discharge effects

The effectiveness of He optical pumping depends on the pumping beam intensity and metastable atoms' relaxation effects in the RF discharge [1,4,11]. The metastable He atoms are created by an RF discharge driven by a voltage from an RF oscillator applied to external electrodes on the He cell. The drive frequency is typically in the range 20–50 MHz. In the discharge, free electrons collide with the ground-state He atoms and produce 2^3S_1 metastable He atoms. The more important collisional atomic processes in the He discharge are

(1)	$He + e \rightarrow He^* + e$	(metastable excitation)
(2)	$He^* + e \rightarrow He + e$	(metastable de-excitation)
(3)	$He^* + e \rightarrow He^+ + e + e$	(ionization by electron)
(4)	$He^* + He^* \rightarrow He^+ + He + e$	(ionization by metastable atom)
(5)	$He^* + He + He \rightarrow He_2^* + He$	(formation of excited He molecule)
(6)	$He^+ + He + He \rightarrow He_2^+ + He$	(formation of ionized He molecule)
(7)	$He^* + e \rightarrow He(2^3P) + e$	(excitation of He 2^3P atom)

Here "He" represents a ground-state (1^1S_0) He atom, "e" denotes an electron, "He^*" is a metastable (2^3S_1) He-4 atom, "He^+" is a He-4 ion, "He_2^*" is an excited He-4 diatomic molecule, "He_2^+" is an ionized He-4 diatomic molecule, and "$He(2^3P)$" denotes a He-4 atom in one of the 2^3P levels (Fig 10.1). Electrons are excited by the RF discharge and create metastable atoms through process (1). The metastable atoms in turn are involved in the creation of free electrons through processes (3) and (4). The amplitude of the RF field for the discharge is typically adjusted so that the total absorption of 1083 nm radiation in the He cell is in the range between 10% and 20%. The discharge adjustment alters the magnetic resonance by trading narrow linewidth (dim discharge) for an increase in metastable atoms (bright discharge).

The rate at which the polarized metastable atoms relax to an unpolarized state has a large effect on magnetometer sensitivity since the metastable relaxation rate influences both the magnetic resonance amplitude and linewidth. Effects that contribute to the metastable relaxation rate include diffusion of metastable atoms to the cell wall (diffusional broadening), collisions of metastable atoms with other particles in the discharge (collisional broadening), interaction of the metastable atoms with the pumping light (light broadening), and relaxation of the metastable polarization due to motion in a magnetic gradient (magnetic gradient broadening). In magnetometer applications the relaxation process is usually dominated by collisional broadening and light broadening. Processes (2), (3), (4), (5), and (7) contribute to collisional broadening. Since metastable atoms are de-excited by contact with the wall surface rather than diffusion into the glass, wall coating technology has not proved effective for He-4 cells.

Light shifts

Optical pumping with circularly polarized light produces two types of light shift of the magnetic resonance frequency of He-4 magnetometers. The first shift is the *virtual* light shift of the magnetic states by off-resonant circularly polarized radiation. A small deviation of the laser wavelength from the absorption line center results in a significant change in

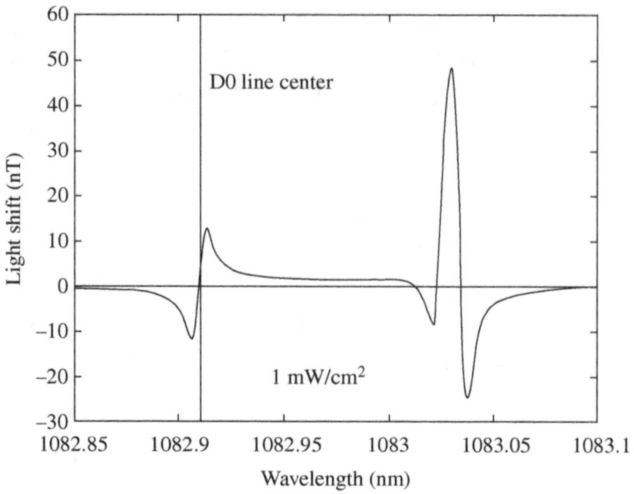

Figure 10.4 Virtual light shift for He-4 metastable atoms interacting with circularly polarized radiation.

resonant frequency. A plot of the theoretical virtual light shift of the D0 states as a function of pumping wavelength for circularly polarized radiation is shown in Fig. 10.4 for a typical radiation intensity. The second shift is the *real* light shift caused by coherence transfer from optically excited states, and is a factor only at higher gas pressures where P-level mixing occurs. Both light-induced resonance frequency perturbations were predicted and analyzed by Barrat and Cohen-Tannoudji [14, 15]. The light shifts in He-4 were first investigated by Schearer [16] and resulted in a dual-beam approach to null the light-induced heading errors in airborne magnetometers. The dependence of the resonance frequency shift on the angle between the light beam and the field was investigated in He-4 in order to reduce heading errors in He-4 magnetometers [11, 17]. A typical value for the real and virtual shifts of the lamp-pumped He-4 resonance frequency was found to be 0.1 nT for the real shift and ±0.4 nT for the virtual shift. The coherence transfer shift is not present in He-4 magnetic resonance if laser radiation at the D0 wavelength is employed to pump between the 2^3S_1 level and the 2^3P_0 level (Fig. 10.1). If the pumping light is perfectly linearly polarized, the virtual light shift and real light shift are completely eliminated, which is ideal for high-accuracy applications. The Absolute Scalar Magnetometer designed by LETI-CEA has reported accuracy better than 45 pT with linearly polarized radiation [18].

10.2.4 Observation of optically pumped helium

For conventional He-4 magnetometers, the optical pumping beam radiation is monitored with an IR photodetector sensitive at 1083 nm (Fig. 10.5). Transmission of the pumping beam through the sample monitors the optically pumped state of the resonance sample. The

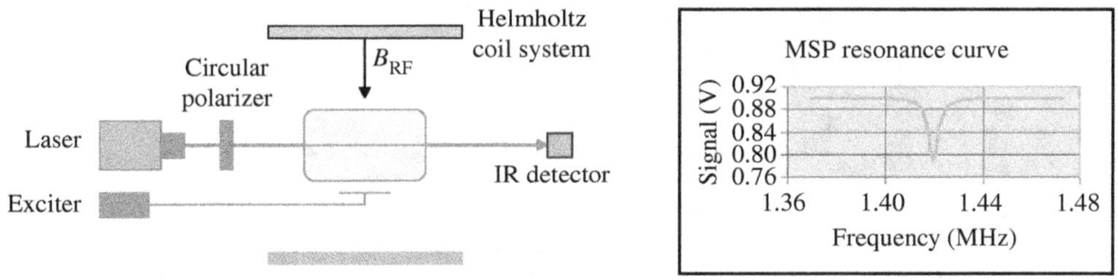

Figure 10.5 (a) Monitoring the state of optical pumping in a He cell and (b) standard paramagnetic resonance (MSP) curve (right).

resonance element functions as a magnetically tuned variable-density optical filter. When the sample is optically pumped, the high absorption state is depopulated and the sample has maximum transparency. When magnetic resonance alters the spin state populations or destroys the polarization, absorption increases as the high-absorption magnetic states are repopulated. The absorption and optical pumping efficiency varies with the angle between the pumping beam direction and the ambient magnetic field direction. The transmitted pumping beam is monitored to observe changes in the sample caused by paramagnetic resonance and parametric resonance conditions in the cell. A servo loop utilizes an error signal produced by high-frequency modulation of the laser about the D0 line center and observed on the photodetector to lock the laser wavelength to the D0 line.

10.2.5 Observation of magnetic resonance signals in optically pumped helium

The optically monitored field-dependent magnetic resonance signals are used to observe and track either the scalar value or vector components of the ambient magnetic field. Paramagnetic resonances produced by magnetically-driven spin precession are used in scalar He magnetometers. Parametric resonances are observed for vector component magnetometers.

Paramagnetic resonance: magnetically-driven spin precession (MSP) scalar mode

Conventional scalar He magnetometers utilize magnetically-driven spin precession (M_z MSP) to induce magnetic resonance transitions between the three magnetic states shown in Fig. 10.3. For MSP magnetic resonance, the spins of the atoms are forced to precess in phase by a small magnetic field $\mathbf{B}_1$ oscillating at or near the resonance frequency. The optical pumping is destroyed by magnetic resonance in the three Zeeman-split states of the 2^3S_1 He atoms. In the M_z mode (see Chapter 4 for general discussion of M_z magnetometers), an RF oscillator applies a swept resonance-frequency voltage to coils around the He cell (Fig. 10.5) to drive the $\mathbf{B}_1$ field. Sweeping the oscillator frequency through the Larmor frequency produces a magnetic resonance observed on the transmission through the cell with maximum absorption occurring at ν_0 (Fig. 10.6). In the M_z mode, a frequency modulation between the resonance curve inflection points generates an error signal used to lock the oscillator to ν_0.

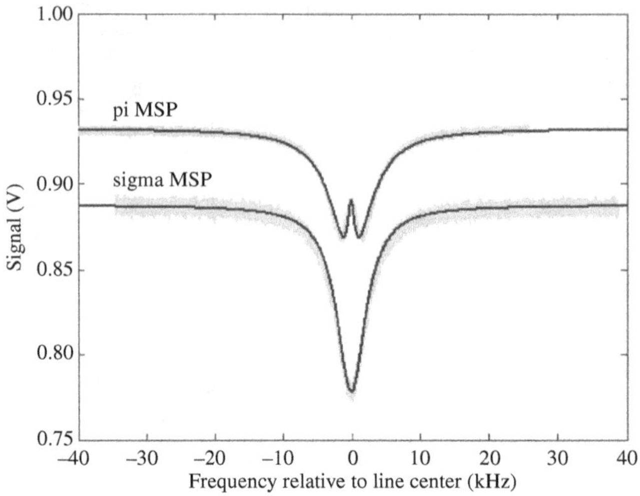

Figure 10.6 Magnetic resonance curves with linearly polarized pumping light (upper curve) and circularly polarized pumping light (lower curve). In both curves the gray bands are the experimental data, and the black curves are theoretical fits to the experimental data.

Pi-pumping magnetic resonance

Pi-pumping designates optical pumping of He-4 using linearly polarized radiation. In investigations of pi-pumping at Polatomic, McGregor compared magnetic resonance curves for circularly polarized and linearly polarized pumping radiation [19]. Cheron *et al.* also compared theoretical and experimental data for linearly polarized pumping of He-4 [20]. In Fig. 10.6, the resonance curve obtained using linearly polarized light is displayed as the upper curve (labeled pi MSP). The lower curve (labeled sigma MSP) is for circularly polarized light with the same cell and operating conditions. The double inverted peak in the pi resonance curve is a characteristic feature predicted by theory and observed experimentally when linearly polarized pumping light is used.

The resonance curve amplitudes for the linearly polarized pi-MSP magnetic resonance and the circularly polarized sigma-MSP magnetic resonance behave differently with respect to orientation of the pumping beam direction to the ambient magnetic field. The theoretical variations of the two types of magnetic resonances are shown in Fig. 10.7 as a function of the angle θ. For sigma-MSP magnetic resonance θ represents the angle between the laser light propagation direction and the direction of the ambient magnetic field. For pi-MSP magnetic resonance, θ is the angle between the laser polarization plane (plane in which the laser radiation electric field oscillates) and the direction of the ambient magnetic field. The pi-MSP resonance amplitude is more sensitive to direction than is the sigma-MSP resonance amplitude. From the signal amplitude curves of Fig. 10.7, one can show that a MSP omnidirectional sensor using linearly polarized light requires six cells compared to three cells for a sensor using circularly polarized light. Because of the orientation dependence of the pi-MSP

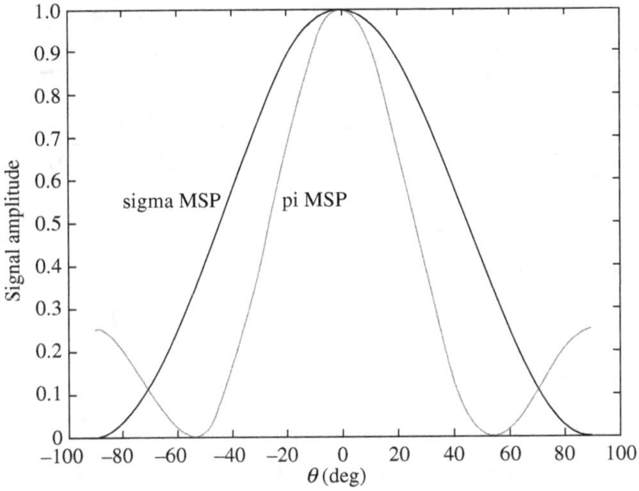

Figure 10.7 Signal amplitude variation with orientation for pumping with linearly polarized light (pi MSP) and with circularly polarized light (sigma MSP).

magnetic resonance, linearly polarized light pumping is especially attractive in stationary magnetometer applications where omnidirectionality is not required. The electronics for a magnetometer using linearly polarized light is nearly identical to a magnetometer using circularly polarized light. The major difference between the two types of magnetometer is the need for the pi-MSP magnetometer to operate in the region between the two minima. The electronics involved in the startup procedure must find the center frequency of the resonance curve and accommodate the fact that the sign of the curvature of the resonance curve changes when one passes from the region outside of either minimum to the interior region between the two minima.

Parametric resonance: bias field nulling (BFN) vector mode

Lamp-pumped vector He magnetometers were developed jointly by TI and JPL and utilize the bias field nulling (BFN) technique that was first demonstrated in a space magnetometer by Slocum and Reilly in 1963 [9]. The BFN technique has been used extensively over a 50-year period in vector He magnetometers developed for space applications (Chapter 15). The BFN technique is an application of parametric resonances, which are described in detail by Cohen-Tannoudji [21, 22]. The BFN technique relies on the zeroth-order parametric resonance and observes optically pumped samples under the condition of zero magnetic field. Slocum also utilized higher-order parametric resonances in vector magnetometers for Earth field applications [23, 24]. A self-calibrating laser-pumped vector magnetometer employing the BFN technique has recently been developed at Polatomic.

A phenomenological description of the vector magnetometer using the BFN mode includes a cell of metastable He atoms that are optically pumped and monitored by a

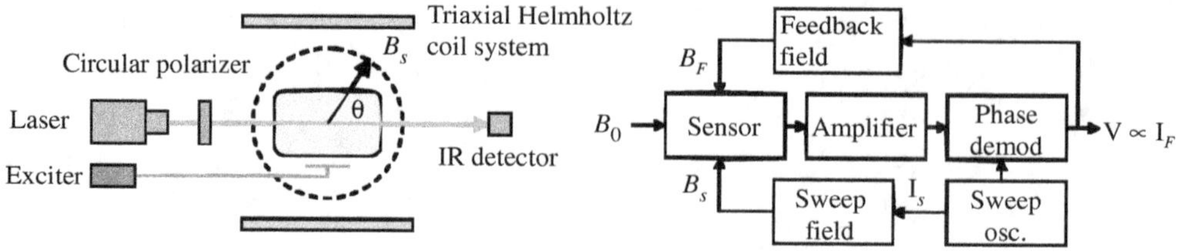

Figure 10.8 Bias Field Nulling (BFN) method for the parametric resonance in the *xz* plane with conceptual diagram of the detection scheme.

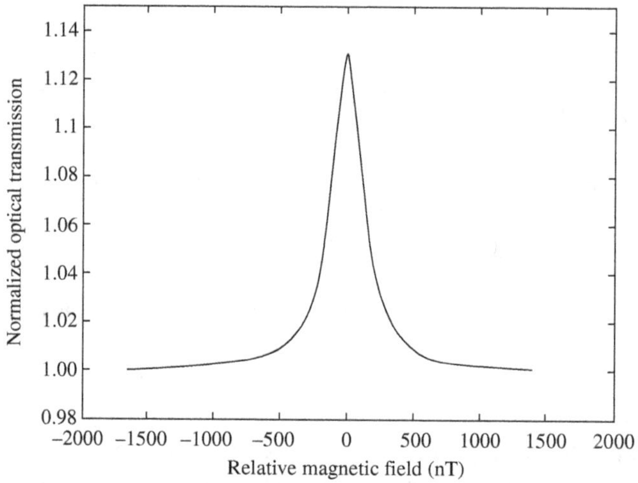

Figure 10.9 The 0th-order parametric resonance is indicated by the transmitted light intensity near zero magnetic field on a magnetic axis perpendicular to the optical axis. These data were obtained using circularly polarized laser light and a 4 cm^3 He-4 cell. When the magnetic field is swept along the optical axis through zero field, no change in transmission is observed, demonstrating that optical pumping occurs at zero field and is identical to the optically pumped state with the field parallel to the optical axis.

circularly polarized beam of 1083 nm radiation (Fig. 10.8). In the BFN vector mode, the magnetic field is measured by sensing changes in the absorption that occur in the optically pumped He subjected to a rotating low-frequency magnetic field in a plane containing the pumping beam direction while the magnetic field in the cell is actively reduced to near zero. The parametric resonance curve shown in Fig. 10.9 is observed when the magnetic field in the He cell is varied around the zero point along a single axis. This BFN method uses three sets of orthogonal coils to null the ambient magnetic field components along the beam direction and two perpendicular directions. The bias field current in each coil required to null the field along its axis is proportional to the vector component of the magnetic field in the axial direction and is determined from the calibration constant of the coil.

A beam of circularly polarized 1083 nm radiation pumping the metastable He-4 atoms in the absorption cell is analyzed to produce an error signal for controlling the nulling fields of the coils. The orthogonal coils generate the nulling fields along with a low-frequency rotating magnetic field applied to the He cell to vary the optical pumping efficiency. The optical pumping efficiency is proportional to $\cos^2\theta$, where θ is the angle between the residual magnetic field and the pumping beam direction. The rotating magnetic field is applied in a plane containing the pumping radiation direction (z axis) and a perpendicular axis (x or y). The rotating field is produced by sinusoidal currents in the z coil and the x (or y) coil of the triaxial coil system that surrounds the He cell. One sinusoidal current is applied to the coil that is aligned with the optical axis (z axis), and the other sinusoid is applied to an orthogonal coil with a 90° phase difference to produce a rotating field in that plane. When no ambient magnetic field is present, the transmission of the pumping light through the He cell is modulated at the second harmonic 2ω of the frequency ω of the rotating field, which is typically less than $2\pi \times 500$ Hz. When a varying ambient magnetic field is encountered, the absorption is modulated at the second harmonic 2ω along with an additional component at ω, the fundamental rotation frequency of the applied field. The signal at frequency ω contains two phase-sensitive components proportional respectively to the residual field components along the z and x (or y) axes. The two ω components are synchronously detected and generate error signals for controlling the coil currents to null the ambient field components along the z and x (or y) axes. Since the rotating field is applied in a plane containing the z axis and either the x or y axis, the sweep field plane is sequentially commutated between the two orthogonal planes (xz and yz) on alternate cycles. In this way two of the three vector components of the ambient magnetic field are determined on each sweep cycle. The coil currents in each coil axis generated by the error signals are proportional to the corresponding magnetic field components and are superimposed on the sweep field modulations which have a frequency higher than the vector magnetometer detection band.

A second method for obtaining vector measurements with a He magnetometer utilizes sinusoidal field perturbations [25]. A triaxial coil set applies small modulations to the He cell that is used for the scalar measurements. The vector components are extracted by demodulating the optical signal at the three distinct modulation frequencies used for each respective axis. Since this technique measures the projection of the magnetic field onto each vector axis using only scalar measurements, the vector measurement noise levels are effectively amplified by the ratio of scalar field magnitude and the perturbation field amplitude (typically a factor of $\sim 10^3$) compared to the scalar noise level. This significantly limits the sensitivity of the vector measurements. The bandwidth of the vector measurements is also limited to a small fraction of the scalar bandwidth. While these performance features limit the scientific utility of this approach, this technique does achieve simultaneous scalar/vector measurements in a single He magnetometer. Both vector measurement techniques require calibration of the vector measurements [26]. Requirements for all vector He magnetometers include precise knowledge of the gain, zero-point offset, and mechanical alignment (orthogonality) for each axis to achieve optimum accuracy. This is accomplished by using complementary scalar measurements from the same instrument or a separate nearby scalar magnetometer. By subjecting the vector and scalar magnetometers to a variety of field orientations via maneuvers or external coils, the correction factors that minimize the difference between the scalar value and vector magnitude can be determined. Calibration of the vector component measurements can reduce accuracy errors to the level of 1 nT or better.

10.3 Conclusions

Although He optical magnetometers are often less well known to commercial magnetometer customers than alkali vapor magnetometers, lamp-pumped He-4 magnetometers have played significant roles in military and geophysical airborne magnetometry for more than 50 years. Innovations in He magnetometers have occurred at a rapid pace over the last 10 years following the advent of 1083 nm laser pump sources. Single-line laser pumping resulted in an improvement in sensitivity of more than two orders of magnitude over lamp-pumped magnetometers without sacrifices in portability or stability. Robust laser-pumped airborne scalar magnetometers with unequaled sensitivity and simplicity for military surveillance and geophysical surveys with sampling rates up to 1200 Hz are now proven technology. Robust and stable field units have been demonstrated approaching 40 $\mathrm{fT}/\sqrt{\mathrm{Hz}}$ sensitivity with excellent accuracy improved by the absence of light shifts. Vector magnetometers having high stability and accuracy are opening the door to high resolution vector and tensor gradiometers without the need for cryogenic cooling required by SQUID magnetometers. Traditionally, He magnetometers have almost exclusively been available to US Government sponsors, but the door is open to "dual-use" projects for nonmilitary geophysical and laboratory applications. If you have a magnetic measurement requirement, it is now possible to include laser-pumped He magnetometers on your shopping list. Recent advances in He magnetometry make it possible to consider all available magnetometer types in order to select the instrument best suited for the task at hand. The capabilities of He magnetometers have made He-4 airborne magnetometers the exclusive choice of the US Navy for more than 40 years. The following characteristics of the Multi-Mode Magnetic Detection System (3MDS) sensor developed by Polatomic for the US Navy illustrate many of the performance capabilities of He magnetometers:

- Laser pumping He-4 for highest sensitivity – Laser pumping of He-4 achieves the highest sensitivity (0.3 $\mathrm{pT}/\sqrt{\mathrm{Hz}}$) of any commercial sensor available for aeromagnetic surveys and military surveillance.
- Omnidirectional sensing – The He-4 sensor is configured to have full sensitivity on all headings anywhere in the world with no dead zones. Magnetometers with dead zones for certain headings relative to the magnetic field require mechanical alignment that varies with survey location on the Earth's surface.
- Resonance cell stability – He-4 cells are extremely stable because of the inert nature of the noble gas and lack of hyperfine structure. Excitation by an RF discharge provides

excellent performance when two He magnetometers are used as a gradiometer. Magnetometers often have stability problems in a gradiometer when the atomic density varies with temperature and a heating element is used for thermal control.

- High sample rate – The 3MDS sample rate of 430 Hz permits detection of extremely low frequency (ELF) radiation in surveillance applications and detection of electromagnetic radiation emitted by hydrocarbon deposits.
- Locked-oscillator resonance loop – A locked-oscillator design has flat response over the entire 215 Hz measurement band, whereas in the self-oscillator designs instruments become noisier as the detection frequency increases.
- Small heading errors – He-4 has very small light shifts in the detection signal resulting in low heading errors defined as false signals observed when the sensor changes orientation.
- Integrated flight system – The 3MDS is a turn key unit that includes all flight system elements including GPS, accelerometers, vector magnetometers, and platform noise compensation computer that are integrated into a "towed bird" unit.
- Flight platform noise compensation – A state-of-the-art magnetic noise compensation system developed for the US Navy is included in each 3MDS unit.

The most cost-effective magnetometer is the one that can actually do the job for the lowest price. The additional cost of a custom-designed He magnetometer could ultimately be worthwhile if it is the most effective solution for a particular measurement or surveillance problem. The increasing use of laser-pumped He magnetometers and the decreasing price of 1083 nm lasers is making these instruments more affordable relative to other types of magnetometers for both commercial and military customers.

References

[1] L. D. Schearer, in *Advances in Quantum Electronics* (ed. J. R. Singer) (Columbia University Press, New York, 1961), p. 239.
[2] L. D. Schearer, Ann. Phys. Fr. **10**, 845 (1985).
[3] R. E. Slocum and E. J. Smith, Contrib. Geophys. Geodesy **31**, 99 (2001).
[4] D. D. McGregor, Rev. Sci. Instrum. **58**, 1067 (1987).
[5] M. K. Plante, D. L. MacFarlane, D. D. McGregor, R. E. Slocum, W. M. Sampson, A. W. Brown, Phys. Rev. A **82**, 013837 (2010).
[6] F. D. Colegrove and P. A. Franken, Phys. Rev. **119**, 680 (1960).
[7] P. A. Franken, United States Patent 3,122,702 (February 25, 1964).
[8] A. R. Keyser, J. A. Rice, L. D. Schearer, J. Geophys. Res. **66**, 4163 (1961).
[9] R. E. Slocum and F. N. Reilly, IEEE Trans. Nucl. Sci. **NS-10**, 165 (1963).
[10] L. Schearer, M. Leduc, F. Laloe, and J. M. Hamel, United States Patent 4,806,864 (February 21, 1989).
[11] R. E. Slocum, *Orientation Dependent Resonance and Non-resonance Effects in Optically Pumped Triplet Helium* (Ph.D. Dissertation, University of Texas at Austin, 1969).
[12] R. E. Slocum, L. D. Schearer, P. Tin, and R. Marquedant, J. Appl. Phys. **64**, 6615 (1988).
[13] R. E. Slocum, United States Patent 5,036,278 (July 30, 1991).

[14] J. P. Barrat and C. Cohen-Tannoudji, J. Phys. Radium **22**, 443 (1961).
[15] C. Cohen-Tannoudji, Compt. Rend. **252**, 394 (1961).
[16] L. D. Schearer, Phys. Rev. **127**, 512 (1962).
[17] R. E. Slocum, Rev. Phys. Appl. (Paris) **5**, 109 (1970).
[18] T. Jager, J-M. Léger, F. Bertrand, I. Fratter and J.-C. Lalaurie, IEEE Sensors **2010**, 1564 (2010).
[19] R. Slocum and D. McGregor, *Advanced Optically-driven Spin Precession Magnetometer for ASW*, Phase II Final Technical Report, Naval Air Warfare Command STTR Contract No. N68335-06-C-0041 (March 2008).
[20] B. Cheron, H. Gilles, J. Hamel, O. Moreau, and E. Noel, J. Phys. II (France) **6**, 175 (1996).
[21] A. Kastler, Phys. Today **20**, 34 (1967).
[22] C. Cohen-Tannoudji, J. Dupont-Roc, S. Haroche, and F. Laloe, Phys. Rev. Appl. (Paris) **5**, 95 (1970).
[23] R. E. Slocum, Phys. Rev. Lett. **29**, 1642 (1972).
[24] R. E. Slocum and B. I. Marton, IEEE Trans. Magn. **MAG-9**, 221 (1973).
[25] O. Gravand, A.Khokhlov, J.L. Le Mouel, and J.M. Leger, Earth Planets Space **53**, 949 (2001).
[26] J. M. G. Merayo, P. Brauer, F. Primdahl, J. R. Petersen, and O. V. Nielsen, Meas. Sci. Technol. **11**, 120 (2000).

11

Surface coatings for atomic magnetometry

S. J. Seltzer, M.-A. Bouchiat, and M. V. Balabas

11.1 Introduction and history

Paraffin films and other surface coatings have played a decisive role in the emergence and development of optical magnetometry. When alkali atoms in the vapor phase collide with the bare surface of a glass container, they disappear inside the glass and are replaced in the vapor phase by another atom with random spin orientation. With a mean free path of the dimensions of the cell (typically on the order of 1 to several cm), the collision frequency is much too high, $10^4\ s^{-1}$, to maintain the substantial spin polarization required for practical applications. In order to prevent this detrimental effect, vapor cells include either an inert buffer gas [1–3] or an antirelaxation surface coating [4]. In the presence of a noble gas at a pressure from 10^{-2} to a few atmospheres, the alkali atoms diffuse very slowly from the center of the cell to the glass walls, and their orientation is only very slightly affected by gas collisions. However, there are several advantages to the use of a surface coating instead of buffer gas. If the static magnetic field is not homogeneous, then resonance lines suffer from inhomogeneous broadening in the presence of the gas [5–7]. In addition, the optical pumping process is perturbed by the buffer gas [8, 9]: (i) it is more efficient at the center of the cell than near the uncoated walls, so that the atomic orientation is inhomogeneous inside the cell; (ii) the pump beam absorption line is broadened, and its profile varies with the distance from the entrance window. These effects are unfavorable for the production of alignment in the ground state.

In the 1950s, Ramsey proposed containment of atoms in a paraffin-coated cell [10], and in 1958 Dehmelt, Robinson, and Ensberg observed that the orientation of alkali atoms was indeed preserved during collisions if the glass walls were coated with paraffin [4]. Shortly afterward, silane coatings [11, 12] were also successfully demonstrated to preserve alkali polarization. In the absence of buffer gas, alkali atoms can move freely throughout the cell, abating the problems described earlier. After several decades of experimentation, it is now

Optical Magnetometry, ed. Budker, D. and Jackson Kimball, D. F. Published by Cambridge University Press.

known how to produce paraffin-coated surfaces that permit an alkali atom to experience tens of thousands of bounces, or even more, without losing its orientation, allowing for very long spin coherence times to be maintained.

Soon after the discovery of paraffin coatings, their success was tarnished by significant variation of observed quality among published results. This was due partly to difficulties encountered during the coating fabrication, leading to irreproducible results, and partly also to the necessity of detecting a well-defined observable quantity: in a given coated cell, relaxation of different observables may be dominated by mechanisms of different origins. Considerable effort was rapidly engaged with the aim of obtaining reproducible results in the presence of coatings. In 1960, Bouchiat and Brossel started a detailed study of the relaxation of Rb atoms on paraffin-coated walls [13, 14] and clarified most of the facets of the complex problem of alkali atom relaxation in wall-coated cells. A summary of their results given in Section 11.2 forms the basis of our understanding of antirelaxation coatings. Practical applications of wall coatings for magnetometry were pioneered at ENS in Paris under the direction of Kastler and Brossel with high sensitivity to very weak magnetic fields first demonstrated[1] in 1969 [15]. A deuterated paraffin-coated cell provided a zero-field level crossing line (Hanle effect) with a width of only 1.4 μG and allowed, for instance, the measurement of a static magnetic field of about 1 nG generated by free precession of optically oriented ^{3}He nuclei at low density ($\sim 10^{13}$ cm^{-3}) [16].

In the decades since, coatings have become an indispensable part of many high-sensitivity alkali vapor magnetometers, as discussed in other chapters of this book. In Section 11.3 we detail a typical procedure for the fabrication of paraffin-coated cells. Indeed, paraffin is currently ubiquitous in magnetometry cells operated at temperatures below its melting point (which depends on the precise material but is typically near 60°C, with measurements performed up to 87°C using deuterated polyethylene [13]), and recent developments allow the use of coatings at even higher temperatures. Two very recent developments will further extend the utility of antirelaxation coatings and their importance for sensitive magnetometry. The first development is the recent discovery by Balabas and co-workers that alkene coating materials, which contain at least one unsaturated C=C double bond per molecule (as opposed to paraffins, which are nominally alkanes with only saturated C–C single bonds), are very effective at preserving ground-state alkali polarization during collisions [17]. This result was very surprising, as it runs counter to decades of conventional wisdom that highly polarizable materials should be more relaxing. However, experimentation with alkenes was suggested by the results of Raman spectroscopy of an effective paraffin material (the Russian "pwMB" wax described in Section 11.3) inside a potassium cell just after it was prepared [18], which revealed the presence of C=C double bonds, thus showing that such bonds are not overly depolarizing. The double bond signature was absent in spectra taken in the same cell one year later, indicating that potassium atoms modified the coating material, possibly making it more "inert." X-ray photoelectron spectra of the same material also indicated the presence of double bonds [19]. After initial passivation upon first exposure to alkali

[1] See supplemental material.

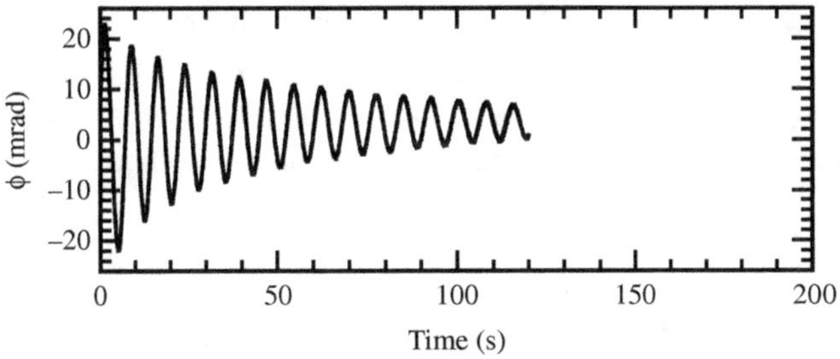

Figure 11.1 Measurement of magnetometer optical rotation signal, proportional to transverse alkali spin polarization in an alkene-coated cell, showing bi-exponential decay times of 13 s and 77 s. (Adapted from Ref. [20].)

vapor, alkene materials may present a more passive surface to the alkali vapor than saturated paraffins. Relaxation times as long as $T_2 = 77$ s (shown in Fig. 11.1) have been demonstrated for rubidium spins using alkene coatings with about 20 carbon atoms per molecule [20], equivalent to approximately one million bounces, an improvement of more than one order of magnitude over any other known coating material. These results are currently difficult to reproduce, with relaxation times shorter than 20 s being more typical in most cells, which is still very impressive. The precise nature of the physical and chemical interactions between alkali atoms and the alkene surface, both during and after passivation, remains to be studied.

The other recent major advance in coating technology is the demonstration of effective coating materials at high temperatures, permitting magnetometry with dense alkali vapors (see, for example, Chapter 5). Seltzer and Romalis first showed that octadecyltrichlorosilane [OTS, $CH_3(CH_2)_{17}SiCl_3$] preserves spin polarization for both potassium and rubidium at temperatures above 160°C, with slow degradation of the coating observed at 170°C [21]. The quality of the cells produced was highly variable, with the best cell allowing more than 2000 bounces, although several hundred bounces was more typical. It was found to be necessary to include at least some quenching gas to prevent radiation trapping, thus partially restricting the motion of atoms in the cell, although a hybrid optical-pumping technique using multiple alkali species may partially mitigate this problem [22]. Balabas and co-workers have also shown that longer-chain alkenes with roughly 30 carbons per molecule show promise as a high-temperature coating material, capable of operating at temperatures in excess of 100°C. The antirelaxation properties of the coating are not as good as those demonstrated by alkenes with 20 carbon atoms, and typical lifetimes in cm-sized cells are on the order of 10 ms, comparable to those observed using OTS (M. V. Balabas *et al.*, personal communication, 2011).

These developments provide hope for further discoveries in the near future. Magnetometers and similar devices would greatly benefit from surface coatings with enhanced effectiveness and stability. The production of coated cells is currently very time intensive and laborious, and the results can vary from one laboratory to another, so new methods need to be developed for mass production of cells. In addition, new coating materials able to withstand temperatures above 300°C would allow coating of microfabricated vapor cells [23] (see Chapter 7). A better understanding of atom–surface interactions will be necessary to design and engineer novel materials for these purposes; recent approaches to the study of these interactions that incorporate surface science methodology will be presented in Section 11.5.

11.2 Wall relaxation mechanisms

In this section, we give a general description of the relaxation process on paraffin surfaces, based on a statistical representation of the physical mechanisms occurring on a microscopic scale. This is the result of a 5-year experimental study by Bouchiat and Brossel, which advanced by trial and error to arrive at a detailed interpretation within a theoretical framework [13, 14]. The work involved preparation and measurement of a notably large number of spherical Pyrex bulbs (most of them 6 cm in diameter) coated with various types of paraffins and filled with either ^{87}Rb or ^{85}Rb vapor, *always without additional gas.*

11.2.1 Origin and time dependence of the disorienting interaction

In the absence of gas, an alkali atom flies in a straight line from wall to wall in an average time τ_v, determined by the dimensions of the cell and the thermal velocity of the atom, during which its evolution involves only the Zeeman and hyperfine interactions. As shown in Fig. 11.2, the relaxation rate of the spin orientation (the observable $\langle S_z \rangle$) exhibits a temperature dependence typical of physical adsorption in the range 20–60°C. This is a clear indication that the atom does not scatter elastically off the surface, but rather gets adsorbed for a certain time interval when it strikes the wall [24], on average τ_s. During this dwell time it is acted upon by the perturbation $\mathcal{H}_1(t)$, which is a random function of time with correlation time τ_c. There is no correlation between perturbations occurring at two different sites, so that $\tau_c \lesssim \tau_s$. The fraction of time that an atom spends adsorbed on the surface is therefore $\tau_s/(\tau_s + \tau_v)$. Above 60°C an additional process occurs, linked to the

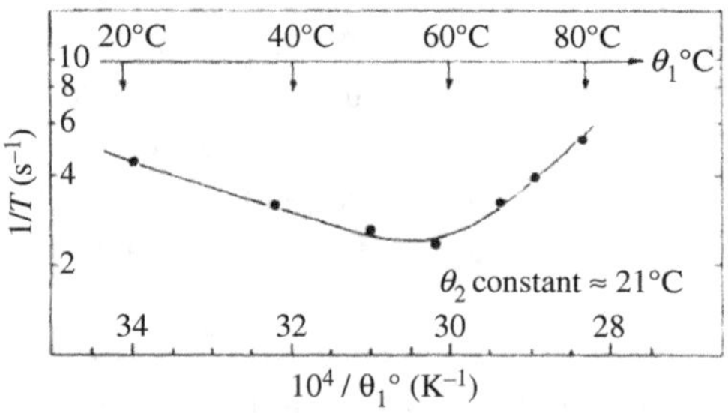

Figure 11.2 Wall relaxation rate variation for the observable $\langle S_z \rangle$ in a ^{87}Rb cell coated with hydrogenated polyethylene, when the wall temperature, θ_1, is varied while the reservoir temperature $\theta_2 < \theta_1$ and the vapor pressure remain fixed. For a quantitative interpretation see the supplemental material. (Figure from Ref. [14].)

fact that a fraction of the atoms disappear inside the coating. In low magnetic fields other main characteristics are apparent:

1. Since an oriented atom experiences many collisions before being disoriented, collisions are *weak*. The motional narrowing condition [25] applies: $\langle|\mathcal{H}_1(t)|^2\rangle\tau_c^2 \ll 1$.
2. One finds that the relaxation time scales as the dimension of the cell. Such an effect indicates that relaxation does take place on the walls and that $\tau_s \ll \tau_v$.
3. Relaxation has been studied for several saturated paraffin chains $(CH_2)_n$ with widely different n values. Relaxation times depend only weakly on n: they are slightly longer $\simeq 30\%$ for light paraffins eicosane and dotriacontane ($n = 20$ and 32, respectively) than for a polyethylene sample (saturated chains with $n \gtrsim 100$, melting at $\sim 120°C$). Very similar results are also obtained with silane coatings. Therefore, the main characteristics of relaxation seem to be determined by the CH_2 group common to all these coatings.
4. Strikingly, on deuterated polyethylene, termed $(CD_2)_n$, relaxation times are found to be $\sim$5 times longer than on ordinary polyethylene $(CH_2)_n$, reaching 1 s for ^{87}Rb and 2 s for ^{85}Rb in 6 cm-diameter spherical cells [26]. The nuclear magnetic moment of ^{12}C is zero, while it is smaller for deuterons than for protons. *Thus, at least in part, the disorientation process involves the dipole–dipole (dd) interaction between the nuclear spin* **K** *of the proton (or deuteron) and the electron spin* **S**. Theoretically, if this were the only interaction between the atom and the wall, then the relaxation rates would scale as $K(K+1)\gamma_K^2$, where γ_K is the gyromagnetic ratio of the spin **K**, so that one would expect an improvement by a factor $\frac{3}{8}\gamma_H^2/\gamma_D^2 = \frac{3}{2}\mu_H^2/\mu_D^2 = 16$ with deuterated paraffins. (This is actually reduced to 12.8, due to the 1.7% H impurity in the deuterated sample employed.) As one observes an improvement only by a factor of 5, there must be a *second disorienting interaction*.

The best explanation seems to be provided by the spin–orbit interaction postulated by Bernheim to explain the observed relaxation of alkali atoms in the presence of a buffer gas, through binary collisions with noble gas atoms [27, 28]. In this case, the spin–orbit interaction is deformed when the electron clouds overlap, and furthermore the electron's orbital motion is coupled to the nuclear rotation. To second order, both interactions give rise to a disorienting interaction which can be formally written $\gamma\mathbf{S}\cdot\mathbf{N}$ [29, 30], where **N** represents the relative angular momentum of the colliding pair. A similar interaction is present during atom–wall collisions, with the role of $\gamma\mathbf{N}$ played by an effective random magnetic field $\mathbf{H}_2(t)$ that acts on **S** at the same time as the field $\mathbf{H}_1(t)$ generated by the protons (or deuterons) of the paraffin. However, the field $\mathbf{H}_2(t)$ involves the velocity of the Rb atom trapped in the surface potential, and it exhibits a strikingly different time dependence, as seen in Fig. 11.3. Theoretically, this is accounted for by introducing the correlation function [25], $\langle\mathbf{H}_j(t)\mathbf{H}_j(t+\tau)\rangle = H_1^2\exp(-\tau/\tau_{c1}) + H_2^2\exp(-\tau/\tau_{c2})$ of the j component of the total field, involving the RMS magnitudes and correlation times of both fields.

11.2.2 Methods of investigation

For arriving at a quantitative interpretation, it is necessary to keep under rigorous control many effects which may obscure the wall collisions. The observed physical quantity is

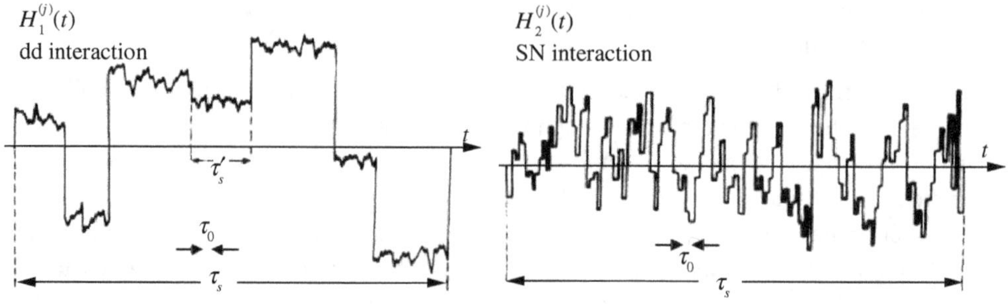

Figure 11.3 Time variations of the j component of the random fields experienced by a Rb atom adsorbed on the wall. $\mathbf{H}_1(t)$ is due to the dipole–dipole interactions and $\mathbf{H}_2(t)$ to the spin–orbit interactions. Vertical scales are arbitrary. Here $\tau_0 \sim 10^{-12}$ s is the period of thermal vibration of the adsorbed atom in the surface potential, $\tau_s = \tau_0 \exp(E_a/kT)$ represents the mean dwell time on the wall, and E_a is the average adsorption energy $\gtrsim kT$. Since E_a varies on the atomic scale with an average site-to-site amplitude $\Delta E_a \lesssim E_a$, the atom can migrate over the surface, jumping between neighboring sites whenever its kinetic energy exceeds ΔE_a. The dwell time at a given site is $\tau_s' = \tau_0 \exp(\Delta E_a/kT)$. There is no correlation between the fields $\mathbf{H}_1(t)$ at two neighboring sites, so τ_s' represents the correlation time τ_{c1} of this field. By contrast, the effective field $\mathbf{H}_2(t)$ involves the relative velocities of the adsorbed atom and atoms of the wall. Under the effect of thermal fluctuations, changes in amplitude *and sign* occur on the timescale of atomic vibrations. $\mathbf{H}_2(t)$ is expected to have a short correlation time $\tau_{c2} \simeq 10^{-12}$ s. (Reproduced from Ref. [14].)

determined only if one knows with great precision the optical conditions of detection [31, 32]. It is in addition crucial to avoid contamination of the walls by the alkali or by traces of vapor or foreign gas, and to control the effect of exchange collisions (see supplemental material). The reservoir effect also has to be properly taken into account: an atom which evaporates from the surface of the small reservoir of metal reaches the cell body after many bounces on the wall, but it may eventually return to the metal surface by the reverse process. Before being lost to the reservoir, an atom survives inside the bulb for a limited time given by the average time of the above sequence, $\bar{t}$, which varies with the cell geometry. This is the longest relaxation time one may measure in the cell for any observable. This effect was predicted by calculations and confirmed by measurements on specially designed cells, as described in the supplemental material.

Theoretically, the problem of relaxation of alkali atoms induced by *weak* collisions on uniformly coated walls, assuming a disorienting interaction of a magnetic type, has been solved [33] by applying the theory of relaxation in gases and liquids [25]. For clarity, we first consider the case of a single correlation time, and we define the rate which determines the evolution of all the observable quantities:

$$C = \frac{2}{3}(\tau_s/(\tau_s + \tau_v))\, \gamma_S^2\, H^2 \tau_c. \tag{11.1}$$

This parameter can be interpreted as the Larmor frequency shift $\gamma_S H$ induced by the random field, reduced by the motional narrowing factor $\gamma_S H \tau_c$ and by the probability for one atom to be located on the wall at any given time. Moreover, it is known that a time-dependent interaction induces transitions between atomic levels with energy separation $\hbar\omega$ only if there is a nonzero component at the frequency ω in its Fourier spectrum, $\propto j(\omega) = 1/(1+\omega^2\tau_c^2)$ for an exponential correlation function. For a given observable, the frequency(ies) involved depends on the energy of the transitions accompanying the decay. For instance, if we consider the observable $\langle \mathbf{S}\cdot\mathbf{I}\rangle$, the difference of populations between the two hyperfine (HF) states involves only the HF splitting frequency ΔW, then the relaxation rate is simply

$$\frac{1}{T_H} = C\,j(\Delta W)\,.$$

The observable $\langle I_z\rangle$ also has a single relaxation rate

$$\frac{1}{T_n} = C[j(\omega_F)+j(\Delta W)]/(2I+1)^2\,,$$

but it involves the two eigenfrequencies, HF and Zeeman, $\omega_F = \omega_S/(2I+1)$, of the alkali atom in low DC magnetic fields. The appearance of the ground-state multiplicity $2I+1$ in this expression can be understood by noting its connection with the number of elementary interactions required to achieve a complete decay.

The relaxation of $\langle S_z\rangle$ is characterized by two rates,

$$\frac{1}{T_e} = C\left\{\left[j(\omega_F)-j(\Delta W)\right]/(2I+1)^2+j(\Delta W)\right\}$$

and $1/T_n$, the rate associated with $\langle I_z\rangle$. Both rates are nearly equal for $\Delta W\tau_c \gg 1$, but differ by the large factor $(2I+1)^2/2$ in the opposite limit of short correlation times. This provides a preliminary interpretation of the measurements. For CH_2 coatings the relaxation curve of ^{87}Rb atoms does not clearly deviate from an exponential, while for ^{85}Rb atoms with a larger nuclear spin and smaller HF splitting the two rates are distinguishable, indicating that the correlation time of the dominant dd interaction is long. By contrast, the two rates appear conspicuously in the relaxation curves of $\langle S_z\rangle$ on CD_2 coated walls, indicating that in this latter case the spin–orbit interaction with a very short correlation time prevails.

By varying the DC magnetic field, and thus the atomic Zeeman frequency, one expects to explore the whole frequency spectrum $j(\omega)$ and determine τ_c. For instance, in CH_2-coated cells one should see a rapid decrease of T_n when $\omega_F\tau_c$ becomes of the order of 1. However, things are less simple when the interesting frequency range coincides with the field domain where hyperfine decoupling occurs [33]. First, theory shows that the relaxation of $\langle S_z\rangle$ involves several rates because of HF mixing. In semilog coordinates the decay curve is not a straight line, and one defines a "pseudo" time constant as the inverse of the slope of the tangent to this curve at the point where the ordinate is half-maximum. The same convention is used to analyze the experimental data discussed in the next subsection. The

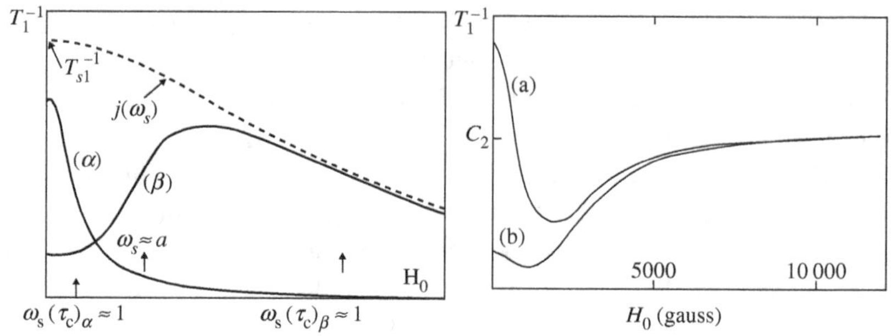

Figure 11.4 Theoretical variations of T_1^{-1} versus the DC magnetic field H_0 (arbitrary vertical scale). T_1 represents the "pseudo" time constant characteristic of $\langle S_z \rangle$ decay for ^{87}Rb, starting from $\langle S_z \rangle = \frac{1}{2}$. *Left*: with only one interaction of magnetic type [Curves (α) and (β) correspond respectively to the case of a long, $\Delta W \tau_c \gg 1$, and a short correlation time $\Delta W \tau_c \ll 1$. The dashed curve represents the field variations, T_{S1}^{-1}, for a spin **S** uncoupled from **I**.]. *Right*: with two uncorrelated interactions of magnetic type present, $\tau_{c1} = 3 \times 10^{-10}$ s, $\tau_{c2} = 10^{-12}$ s, $C_1/C_2 = 16$ for curve (a) and 1 for curve (b). The value of C_2, common to both curves, determines T_1^{-1} in fields above 5000 G. (Figure from Ref. [14].)

result depends on the initial conditions and is different for $\langle S_z \rangle_{t=0} = +\frac{1}{2}$ or $-\frac{1}{2}$. Second, for complete decoupling the electronic spin behaves as if it were isolated, and therefore its relaxation loses the large lengthening factor $(2I+1)^2/2$. This gives rise, by contrast, to a rate increase for $\omega_S \simeq \Delta W$. Figure 11.4 presents the theoretical predictions for long and short correlation times as well as for the more complicated case of practical interest here, where there are two interactions present with very different correlation times.

If two *uncorrelated* interactions are simultaneously present, they can be treated independently, so that the global relaxation rate is simply the sum of the individual ones. Thus, the wall relaxation rates can be described as the sum of the rates describing the dipole–dipole and the spin–orbit interactions, considered as if each one was acting alone.

11.2.3 Quantitative interpretation

Measurements have been performed separately on the two isotopes, ^{87}Rb and ^{85}Rb, for the two types of polyethylene coatings, $(CH_2)_n$ and $(CD_2)_n$, in order to obtain the relaxation rates of $\langle S_z \rangle$ and $\langle \mathbf{S} \cdot \mathbf{I} \rangle$. Each cell provides three time constants, except $(CH_2)_n$ cells which provide two (T_H and a "pseudo" time constant for $\langle S_z \rangle$). With this set of data, *corrected for the reservoir effect*, one has 11 independent determinations available to search for six independent parameters: C_1^H, C_2^H, C_1^D, C_2^D, τ_{c1}, and τ_{c2}, i.e., the characteristic rates [Eq. (11.1)] of the two interactions for both coatings and their correlation times. The results of this global fit are submitted to cross-checks, since expected relations among the parameters are not *a priori* assumed.

The best fit [13] of theory to experiment is obtained with correlation times $\tau_{c1} = 4 \times 10^{-10}$ s, $\tau_{c2} \ll 10^{-11}$ s, and rates $C_1^H = 62\ \mathrm{s}^{-1}$, $C_1^D = 5\ \mathrm{s}^{-1} \simeq C_1^H/12.8$, $C_2^H \simeq C_2^D = 2.2\ \mathrm{s}^{-1}$

(in 6 cm-diameter bulbs). The agreement is satisfactory, taking into account the dispersion (less than 15%) among the rates measured in cells prepared under nominally identical conditions. Using Eq. (11.1) and the above values of C_1^H and τ_{c1}, we obtain a determination of the quantity $(\tau_s/\tau_v)(\gamma_s H_1^H)^2$. On the other hand, the origin of the $\mathbf{H}_1$ field has been clearly identified as the nuclear dipole field seen by a Rb atom adsorbed on the coating, hence its magnitude is constrained by physical arguments. Taking for the Rb–proton average distance a value characteristic of physical adsorption $r_{SK} = 2.5 \pm 0.2$ Å, one obtains the average proton field magnitude $4\pi \times 10^{-7}\mu_p/r_{SK}^3$ in the range 7–14 G. The average dwell time of Rb atoms on the wall is then extracted by reporting this field magnitude in the expression of C_1^H. Thus we obtain a value of the dwell time, τ_s, lying inside the range (0.4–2.0)$\times 10^{-9}$ s (compatible with the condition $\tau_s \gtrsim \tau_{c1}$, with both parameters determined independently). We have to underline that a dwell time as long as 10^{-6} s, claimed by the authors of Ref. [34] as described in Section 11.5, is in conflict with the ensemble of results summarized here: to account for the measured rate C_1^H, such a long dwell time would imply a proton field ~30 times smaller (a fraction of 1 G), which implies an average adsorption distance exceeding 7 Å that is difficult to explain.

The magnetic field dependence of the "pseudo" time constant T_1 of $\langle S_z \rangle$ was measured for ^{87}Rb and ^{85}Rb up to 5000 G [13,35] and compared with the theoretical curve computed numerically for the set of the parameters determined in low fields. The results confirm the existence of two types of interaction. In particular, the fast decrease of $1/T_1$ observed around 400 G is pronounced for $(CH_2)_n$ coatings and much reduced for $(CD_2)_n$ materials. This is another proof for the existence of a correlation time of a few 10^{-10} s, associated with the field of the coating nuclear spins, much reduced by passing from $(CH_2)_n$ to $(CD_2)_n$. The results also confirm that the second interaction has a very short correlation time, $\tau_{c2} \ll 10^{-11}$ s, and that its magnitude is nearly the same for both coatings. Supposing $\tau_{c2} \approx 10^{-12}$ s, we get $H_2^H = H_2^D \approx 50$ G. Although this field is stronger than H_1^H, its very short correlation time makes its wall relaxation effect much smaller.

Notwithstanding all the information collected from relaxation studies, independent determination of the surface interaction parameters, in particular a more direct determination of the dwell time, would be welcome. There are possibilities based on modern surface science methods applied to the study of alkali-surface interactions (Section 11.5). Instead of the statistical global description obtained in terms of averaged parameters, they can provide several local probes of the surface at a microscopic scale. Combining both complementary approaches to study the same individual samples looks like a promising way to make further progress.

11.3 Coating preparation

Two coating preparation techniques are described here. One method is based on condensation of the coating material onto the "cold" inner surface of the cell: the material evaporates from a heater placed in the center of the cell volume, while the cell is vacuum pumped.

The other technique is based on condensation of a saturated vapor of the material onto the "hot" inner surface of the cell during slow uniform cooling of the cell volume. The latter technique is used more often at present because it allows for control of the coating preparation temperature.

First, let us estimate the temperature one needs in order to prepare a coating with a thickness of $L = 50$ nm, provided that the coating has uniform thickness in a 10 cm-diameter cell. The paraffin density is about $\rho \approx 0.8$ g/cm^3, and the number of paraffin molecules in the coating is

$$N_c = \frac{4\pi\rho R^2 L}{(14n+2)m}, \tag{11.2}$$

where R is the cell radius, n is the number of carbon atoms per molecule, and $m = 1.66 \times 10^{-24}$ g is the atomic mass unit. For a spherical cell, if all molecules condense on the inner wall after the cell cools down, then the paraffin vapor density must be

$$n_c = \frac{N_c}{V_c} = \frac{3\rho L}{(14n+2)mR}. \tag{11.3}$$

We have $n_c = 1/(14n+2) \times 1.4 \times 10^{18}$ cm^{-3} for our parameters. The vapor concentration can be calculated from pressure using the Antoine equation:

$$\log_{10} P = A - B/(C+T), \tag{11.4}$$

where P is in torr, and T is the temperature in kelvin. A, B, and C are the Antoine constants, which can be found in Ref. [36]. Thus, the temperature can be found as

$$T = B[A - \log_{10}(n_c k) - \log_{10} T]^{-1} - C, \tag{11.5}$$

where we have used the ideal-gas equation $p = nkT$. For tetracontane $C_{40}H_{82}$ ($A = 6.5962$, $B = 2735.8$, $C = -199.05$), $n_c = 2.57 \times 10^{15}$ cm^{-3}, and the temperature is 566 K.

One first needs to select the initial material for coating. M. V. Balabas has used the Russian polyethylene wax ТУ N 38302116-81 (consisting of the remains of low-pressure polyethylene production) for most of the cells he has prepared in the last 30 or so years; the resulting material is occasionally referred to as "pwMB" wax in order to distinguish it from other paraffins. In addition, it was recently found that a very effective alkene-based material can be produced by using Chevron Phillips alpha olefin fraction C20-24 (CAS Number 93924-10-8) [20]. In order to remove components with high saturated vapor pressure, the material is distilled at a temperature of 200–220°C using a two-step distillation process using a glass distiller shown in Fig. 11.5. The distiller consists of a container for the initial material, diffusers to prevent direct travel of molecules without collisions with the walls, a container for the distilled fraction, an oven, and a thermocouple for active temperature control. As the first step, light fractions are distilled at about 200°C for the Russian polyethylene wax and at about 80°C for the alkene material (the Russian and alkene materials melt at 60°C

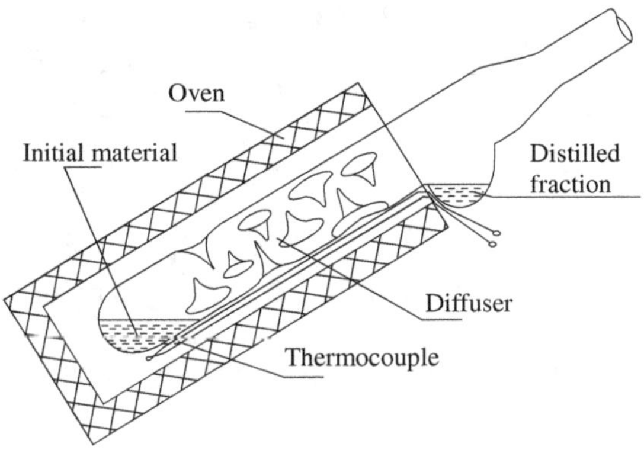

Figure 11.5 Design of the glass distiller used in the preparation of coating materials.

and 30°C, respectively). For the Russian wax, the material remains are then removed and put again into a similar distiller and again distilled at 220°C, with the resulting distilled fraction being used for coating preparation. For the alkene material, the remains of just a single distillation are used for coating. The residual pressure is maintained around 10^{-5} torr during the distillation procedure, and each distillation step lasts for 8 hours.

Thorough washing of the cell is vital before it can be filled. Balabas uses a 10% solution of hydrochloric acid to clean the inner surface of the cell. The surface is washed for several minutes with a volume of solution equal to about one-quarter of the cell volume. After that, the surface is washed with the same amount of distilled water a few times (usually four times, but a set of cells was successfully prepared with a single wash). A 4% solution of hydrofluoric acid was also tested, and no significant variation of the coating quality was observed.

Special glassware must be prepared for the filling procedure, shown in Fig. 11.6. It consists of the cell (1), a container for the coating material (7), and a tube with a breakable joint (5). The cell consists of the main body (1), a sidearm or stem (3) as a metal container, and a diaphragm in between with a small aperture of less than 1 mm diameter (2). The bottleneck constrictions (4), (8), (9), and (11) are to facilitate vacuum sealing. A magnetic hammer (10) is used to break the joints (5) and (12) under vacuum, when the cell and the glass container with metal are connected to the vacuum system for filling the cell stem with alkali metal. The part containing the alkali metal (12) is not attached to the rest of the glassware until the cell is ready to be filled with metal.

The first step is placement of a magnetic needle with a drop of the coating material on its sharp end into position (14), which will not be heated. After that, the whole construction (1)–(10) is sealed on to the vacuum system (15), and the part shown in the dashed box is heated to 420°C under vacuum pumping. When the residual pressure in the construction drops to 10^{-5} torr, the construction is cooled down to room temperature.

The next step is to move the needle with the coating material into the container (7) at position (16), followed by mild heating of the needle to melt the material. The liquid

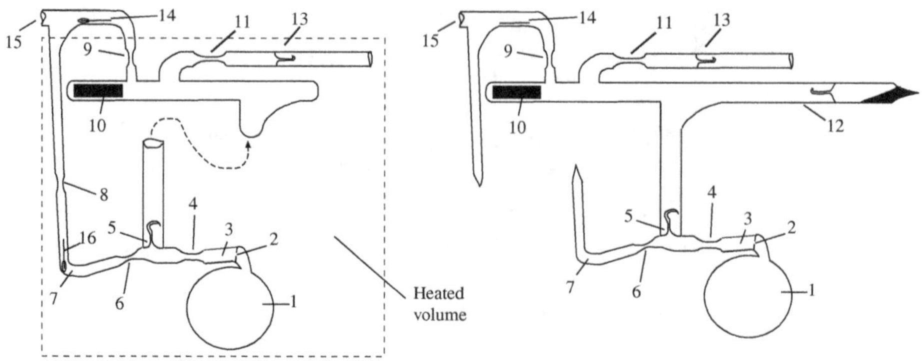

Figure 11.6 Glassware for coating preparation, (left) during paraffin deposition and (right) during metal distillation. See text for the key.

material moves from the needle onto the glass surface of the container, and the needle with the remains of the material is removed from the container. After that, the part of the glass construction including parts from (1) to (7) is sealed off through the bottleneck (8) from the vacuum system under vacuum pumping.

At this point, the cell is ready to be coated. To perform the coating, the detached construction is placed into an oven with intense air circulation provided to even out the temperature inside the oven, and the oven is heated up to the desired temperature. The construction is kept at this temperature for 1 hour, and then the heaters of the oven are turned off. During the next few hours the oven is cooled down to room temperature. Thus, when the construction is removed from the oven, the coating is prepared, and the cell is ready to be filled with metal.

The next step is to wrap the cell volume (1) up to the diaphragm (2) in thermal insulation (ceramic tape) to protect the coating against possible damage by the flame during the metal-distillation procedure. The stem (2) must be free from the thermal insulation. The last step of the cell preparation is metal distillation into the stem. First, the construction (1)–(7) with folded cell and the container with metal (12) are sealed on to the vacuum system as shown by the dashed lines in Fig. 11.6. The metal is prepared separately through a reaction of an alkali salt. During vacuum pumping, the parts of the construction are heated with the soft flame of a glassblowing torch for degassing. The part of the (1)–(7)-construction from stem (3) to bottleneck (6) is heated to remove the coating. After that, the container (7) with the coating material remains is sealed off from the construction. When the residual pressure drops below 10^{-5} torr, the breakable joints are broken with the hammer, and the metal is distilled from the container (12) into the stem (3) with a soft flame from the torch.

The last operation is to seal off the cell from the vacuum system through the bottleneck (4). Once the cell is sealed off, the insulation is removed, and the cell is placed into an oven for curing. The remains of the alkali metal are distilled into container (13), and the whole glass construction is sealed off from the vacuum system through bottleneck (9) under vacuum.

It is generally observed that a "fresh" coated cell does not show any resonant light absorption because the density of the alkali metal vapor is very low inside the coated volume. The cell must be cured by uniformly heating to the temperature appropriate for the coating material (for example, 60–70°C for pwMB wax) for at least several hours to obtain a pressure that is comparable to, but generally lower than, the alkali saturated vapor pressure. It is still not known in detail what happens to the coating during the curing process, but the result is that the coating becomes saturated with the alkali atoms physically and/or chemically. The pressure remains lower than the saturated vapor pressure even after many years, indicating that there is a continuing, irreversible flow of alkali atoms into the coating. The alkali polarization lifetime increases as the cell cures, and experimentally recorded lifetime values can eventually exceed 1 second for paraffins and 70 seconds for alkenes. The kinetics of the curing process are discussed in detail in the supplementary material.

11.4 Light-induced atomic desorption (LIAD)

Surface coatings also have the potential to benefit future alkali–metal magnetometers by providing a method of achieving increased atomic density through the light-induced atomic desorption (LIAD) effect [37, 38]. During the curing phase, organic coatings used in alkali cells absorb a large number of alkali atoms, and afterward the coatings continue to absorb atoms indefinitely at a slower rate. Upon illumination by nonresonant light, many coatings eject atoms into the interior of the cell, enabling the alkali density in the cell to surpass the saturated vapor pressure at the cell temperature. In some cases, the density can increase by more than an order of magnitude.

The LIAD effect allows both temporary and sustained increase in alkali density without heating the cell, which can be advantageous for applications where elevated cell temperatures are undesirable because of potential damage either to components of the cell itself, such as the coating, or to objects located near the cell. Thus, magnetometers may operate at room temperature with a higher vapor density than would otherwise be possible. The wall relaxation rate in paraffin cells has been observed to depend on cell temperature, so the use of LIAD instead of heating to increase density can result in longer relaxation times and therefore better magnetometer sensitivity [39]. In addition, the density can be changed more quickly using LIAD than by varying the cell temperature because the latter must overcome thermal inertia, and it can be dynamically controlled using active feedback techniques (see the supplementary material) [40, 41].

LIAD was initially discovered by Gozzini *et al.* in 1993 [37], after observing unexpected fluorescence by sodium atoms at room temperature in a polydimethylsiloxane (PDMS, $CH_3[Si(CH_3)_2O]_nSi(CH_3)_3$) coated cell upon irradiance by D1 or D2 laser light, although the effect was found to be stronger for light at shorter wavelengths (and not dependent on the atomic resonance structure). LIAD has since been demonstrated in siloxane coatings (containing molecules with Si-O units) with other alkali elements [38, 42–44], in paraffin

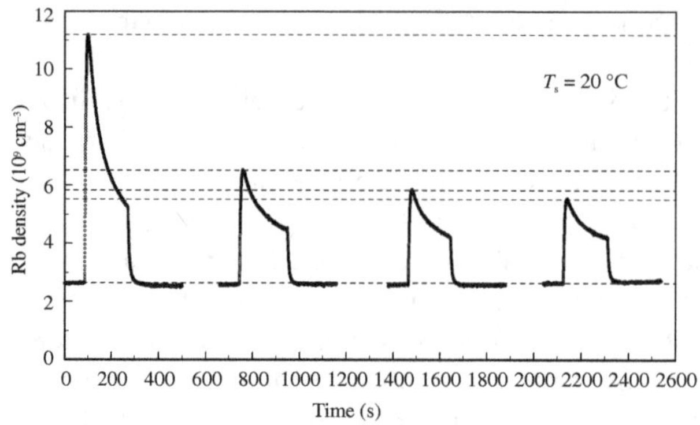

Figure 11.7 Measured Rb vapor density upon repeated exposures to desorbing light in a paraffin-coated cell, showing rapid increase at the start of exposure followed by slow decay, as well as an overall decrease in density for successive exposures. (Adapted from Ref. [45].)

[45, 46], and in octadecyltrichlorosilane [47, 48]. Light-induced desorption of alkali atoms from bare glass [49] and from glass coated with a superfluid ^{4}He film [50, 51] has been observed, as well as from other uncoated materials including quartz [52] and stainless steel [53]. While the photodesorption rate from bare glass is smaller than from glass coated with a siloxane, paraffin, or silane film, it can be increased by using nanoporous glass with a large total surface area [54].

The dynamics of LIAD are illustrated in Fig. 11.7. The initial alkali vapor density in the cell n_0 is typically slightly less than the saturated vapor density. After the desorbing light turns on, the density increases rapidly to some maximum value $n_{\max}$ as atoms are ejected from the coating. The density then decreases slowly while the desorbing light remains on because the flux of atoms out of the coating cannot fully compensate for losses due to condensation, reabsorption, and motion of atoms into the stem of the cell. After the light turns off, desorption ceases and the density decays back to its initial value n_0; the density often undershoots n_0 and requires time to recover. Subsequent applications of desorbing light result in smaller values of $n_{\max}$, with a waiting time on the order of one to several hours required to obtain the same maximum density for the same experimental conditions.

The physical and/or chemical mechanisms behind LIAD are not yet understood, although several phenomenological models have been developed, which are discussed briefly in the supplementary material. In addition to inducing the ejection of alkali atoms from the surface of the coating, application of light evidently enhances diffusion of alkali atoms from the bulk of the coating toward the surface [42, 55]. There is a threshold frequency required for light to stimulate desorption, which suggests that some minimum photon energy is necessary to remove the atoms from potential energy wells within the bulk of the coating. The nature of these energy wells is unknown and may vary between materials, but one possibility is a reversible chemisorption process in the bulk [19, 37, 56]. Experimental observations lead to an estimated potential on the order of 1 eV, consistent with the measured threshold wavelengths [45,55]. The need to replenish the reservoir of atoms within the coating explains the observed decline in vapor density once it reaches its initial maximum at the beginning of illumination, as well as the long waiting time between LIAD pulses required to maintain the same yield. The probability of atomic adsorption during each collision with the coated surface is measured to be on the order of 10^{-6} [13, 14, 45, 57], so atoms are ejected much more quickly during LIAD than they can be replaced. Continued research into the actual mechanisms of LIAD and the dependence of desorption efficiency on coating composition and history [19], in addition to new methods of dynamic control of the vapor density, should allow the integration of LIAD in future magnetometry devices for achievement of nonthermal alkali vapor density.

11.5 Recent characterization methods

Recent advances in the performance and practicality of alkali-metal magnetometers have led to renewed interest in studying effective surface coatings, in the interest of discovering or engineering superior new coating materials for various experimental conditions, such as the elevated temperatures required for sensitive operation of SERF magnetometers [58] or construction of microfabricated vapor cells [23]. Understanding the properties of both the coatings themselves and their interactions with alkali atoms is essential to this goal. Several recent studies have applied the techniques of surface science to the study of antirelaxation coatings, while others have employed novel methods for the study of the alkali–surface interface or the quality of individual coatings.

For example, atomic force microscopy (AFM) imaging reveals the topography of the coated surface. AFM images of octadecyltrichlorosilane (OTS) surfaces exposed to rubidium reveal different clustering behavior of rubidium on the surface, depending on whether the OTS coating is a self-assembled monolayer or a multilayer, as shown in Fig. 11.8. For monolayer films, the rubidium atoms form numerous small "islands" located at the

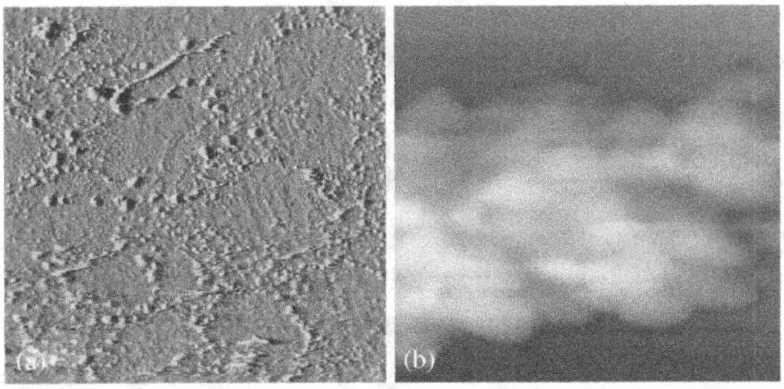

Figure 11.8 5 μm×5 μm AFM height images of OTS films after exposure to rubidium vapor; lighter colors represent greater surface height and indicate the presence of rubidium clusters. (a) Image of a monolayer OTS coating, showing small rubidium clusters at domain boundaries. (b) Image of a large rubidium cluster in a multilayer OTS film. (Adapted with permission from Ref. [59]. Copyright 2009, American Institute of Physics.)

grain boundaries of the film with size on the order of hundreds of nanometers, while for multilayers without apparent grain boundaries the rubidium atoms form a small number of widely spaced, significantly larger islands with size on the order of several micrometers; the overall surface area covered by such clusters is much greater for the monolayers, which may explain the shorter alkali polarization lifetimes observed with monolayers compared to multilayer OTS coatings, since the regions covered with clusters likely act as defects that relax alkali atoms upon collision [59]. Minimizing cluster formation, for instance by reducing the presence of grain boundaries, could therefore lead to improved polarization lifetimes. Similar clusters are also observed on paraffin coatings exposed to rubidium [19]. In addition, AFM images of different paraffin materials reveal that the surface does not necessarily need to have a crystalline structure in order to preserve alkali polarization, which is consistent with observations that a paraffin coating can remain effective even when partially or fully melted [19].

X-ray photoelectron spectroscopy (XPS) has been used to study the bonding of alkali atoms to the molecules of coated surfaces. XP spectra reveal that alkali atoms react with and bind to both monolayer and multilayer OTS films, with alkali atoms located throughout the thickness of the multilayer film [59], although the alkali atoms remain electrically isolated from the carbon atoms in the coating (this is evident in the different extent of charging experienced by the rubidium and carbon atoms during the XPS measurement) [60]. An XPS study of paraffin surfaces shows no evidence that alkali atoms react with tetracontane, but they do react and bind to carbon atoms in the custom-made paraffin material produced from the Russian wax as described in Section 11.3, referred to as "pwMB" wax. Unlike tetracontane, the pwMB wax was shown to contain unsaturated C=C double bonds by near-edge x-ray absorption fine structure (NEXAFS) spectroscopy, and it is possible that alkali atoms attack these unsaturated bonds to form alkali–carbon bonds within the coating [19]. Another possibility is that the high polarizability of the double bonds makes the adsorption energy especially large, and, hence, the dwell time at their location anomalously long. Alkali atoms bound chemically to a coating material might serve as a reservoir for the LIAD effect, and a comparison of the LIAD yields of several different paraffin and alkene materials does suggest some weak correlation between the presence of unsaturated bonds in a material with the observation of LIAD from that material [19].

A more specialized technique for the study of the alkali–coating interface is the use of evanescent waves, which interact only with atoms located within about 1 micrometer from the surface [61]. Pump and/or probe beams incident on the cell wall at angles greater than the critical angle experience total internal reflection at the interface between the (coated) glass and the cell interior; the effective length of the cell can be changed by inclusion of a movable prism to act as the cell wall, in order to vary the fraction of time that the atoms spend near the surface. Surface-interaction parameters for rubidium measured using this method include dwell time during a collision on the order of microseconds on Surfasil (a commercially available coating material whose main ingredient is dichlorooctamethyltetrasiloxane) and tenths of a microsecond on OTS depending on temperature; the probability of depolarization during collision on the order of 10^{-2} on Surfasil and 10^{-3} on OTS (the latter value being consistent with lifetime measurements for multilayers [21]); and average phase shift during collision less than 0.2 mrad for both materials [62–64]. For saturated paraffin, this technique measured a dwell time on the order of microseconds [34]. These measurements lead to a determination of the high-temperature dwell time τ_0 (see Section 11.2) on the order of several nanoseconds, about three orders of magnitude longer than the usually assumed value $\sim$1 ps based on the period of vibration within an adsorption site; the authors of Ref. [63] speculate that this implies each atom samples 10^3 adsorption sites during each stay within the coating. As discussed earlier, this appears to contradict the conclusion of Section 11.2, possibly due to the presence of nitrogen buffer gas, which may change the nature of the surface; multimolecular adsorption is a complex problem considered in Ref. [24], Chapter V.

While these methods are very useful for studying the physical and chemical properties of the coatings, the technique with perhaps the most immediate benefit for magnetometer development is measuring the number of polarization-preserving collisions allowed by different coating materials. One approach is to manufacture a large number of cells and observe their properties, such as in a recent study of more than 250 paraffin-coated cells in which most cells had a longitudinal linewidth below 4 Hz, limited mainly by loss of atoms to the stem, as shown in Fig. 11.9 [65]. However, manufacture of cells is time consuming and labor intensive, and it is more efficient to coat flat substrates if the purpose is only to study the coatings rather than to produce usable cells. Coated quartz plates have been assembled into cubic cells for subsequent measurement of Zeeman and hyperfine linewidths

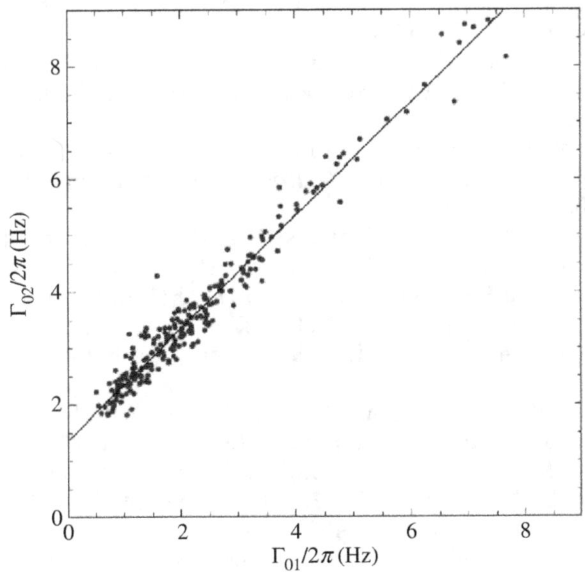

Figure 11.9 Comparison of the "intrinsic" lifetimes (in the limit of zero pump beam power) $\Gamma_{01} = 1/T_1$ and $\Gamma_{02} = 1/T_2$ of 241 identical paraffin-coated Cs vapor cells (approximately 28 mm ID spheres). (Reprinted with permission from Ref. [65]. Copyright 2009, American Institute of Physics.)

and hyperfine frequency shifts [66]. In addition, coated glass and silicon plates have been inserted into a prototype "reusable" alkali vapor cell, which in principle may allow for easy replacement of coated samples for rapid testing [67]. Both of these techniques have been demonstrated as effective for studying the relaxation properties of coatings and have the potential to allow comparisons between different materials or deposition techniques under variable ambient conditions. The use of flat substrates also permits easy study of the samples using surface-science methods both before and after characterization. Thus, various surface coatings can be tested with the goal of discovering the physical and chemical requirements of quality antirelaxation coatings, leading to novel coating technology in order to enhance the performance of future optical magnetometers.

Acknowledgments

Scott Seltzer thanks Alexander Pines for his generous support.

References

[1] J. P. Barrat, J. Brossel, and A. Kastler, C. R. Acad. Sci. **239**, 1196 (1954).
[2] J. Brossel, J. Margerie, and A. Kastler, C. R. Acad. Sci. **241**, 865 (1955).
[3] P. L. Bender, Ph.D. thesis, Princeton University (1956).
[4] H. G. Robinson, E. S. Ensberg, and H. G. Dehmelt, Bull. Am. Phys. Soc. **3**, 9 (1958).
[5] S. F. Watanabe and H. G. Robinson, J. Phys. B **10**, 931 (1977).
[6] G. D. Cates, S. R. Schaefer, and W. Happer, Phys. Rev. A **37**, 2877 (1988).
[7] S. Pustelny, D. F. Jackson Kimball, S. M. Rochester, V. V. Yashchuk, and D. Budker, Phys. Rev. A **74**, 063406 (2006).
[8] F. Masnou-Seeuws and M.-A. Bouchiat, J. Phys. (Paris), **28** 406 (1967).
[9] S. Knappe and H. G. Robinson, New J. Phys. **12**, 065021 (2010).
[10] N. F. Ramsey, Rev. Sci. Instrum. **28**, 57 (1957).
[11] M. A. Bouchiat, T. R. Carver, and C. M. Varnum, Phys. Rev. Lett. **5**, 373 (1960).
[12] C. O. Alley, *Advances in Quantum Electronics* (Columbia University Press, New York, 1961), p. 120.
[13] M.-A. Bouchiat, Ph.D. thesis, Université de Paris (1964), available in French at http://tel.ccsd.cnrs.fr/tel-00011797.
[14] M.-A. Bouchiat and J. Brossel, Phys. Rev. **147**, 41 (1966).
[15] J. DuPont-Roc, S. Haroche, and C. Cohen-Tannoudji, Phys. Lett. A **28**, 638 (1969).
[16] C. Cohen-Tannoudji, J. DuPont-Roc, S. Haroche, and F. Laloë, Phys. Rev. Lett. **22** 758 (1969).
[17] M. V. Balabas, K. Jensen, W. Wasilewski, H. Krauter, L. S. Madsen, J. H. Müller, T. Fernholz, and E. S. Polzik, Opt. Express **18**, 5825 (2010).
[18] M. V. Balabas, Ph.D. thesis, S. I. Vavilov State Optical Institute (1995).
[19] S. J. Seltzer, D. J. Michalak, M. H. Donaldson, M. V. Balabas, S. K. Barber, S. L. Bernasek, M.-A. Bouchiat, A. Hexemer, A. M. Hibberd, D. F. Jackson Kimball, C. Jaye, T. Karaulanov, F. A. Narducci, S. A. Rangwala, H. G. Robinson, A. K. Shmakov, D. L. Voronov, V. V. Yashchuk, A. Pines, and D. Budker, J. Chem. Phys. **133**, 144703 (2010).
[20] M. V. Balabas, T. Karaulanov, M. P. Ledbetter, and D. Budker, Phys. Rev. Lett. **105**, 070801 (2010).

[21] S. J. Seltzer and M. V. Romalis, J. Appl. Phys. **106**, 114905 (2009).
[22] M. V. Romalis, Phys. Rev. Lett. **105**, 243001 (2010).
[23] L.-A. Liew, S. Knappe, J. Moreland, H. Robinson, L. Hollberg, and J. Kitching, Appl. Phys. Lett. **84**, 2694 (2004).
[24] J. H. De Boer, *The Dynamical Character of Adsorption* (Oxford University Press, New York, 1963).
[25] A. Abragam, *Principles of Nuclear Magnetism* (Oxford University Press, New York, 1961), Chap. 8.
[26] M.-A. Bouchiat and J. Brossel, C. R. Acad. Sci. **254**, 3828 (1962).
[27] R. A. Bernheim, J. Chem. Phys. **36**, 135 (1962).
[28] R. J. McNeal, J. Chem. Phys. **37**, 2726 (1962).
[29] R. J. McNeal, J. Chem. Phys. **40**, 1089 (1964).
[30] R. M. Herman, Phys. Rev. **136A**, 1576 (1964) and **137A**, 1062 (1965).
[31] M.-A. Bouchiat, J. Physique (Paris), **26**, 415 (1965).
[32] M.-A. Bouchiat and F. Grossetête, J. Physique (Paris), **27**, 353 (1966).
[33] M.-A. Bouchiat, J. Physique (Paris) **24**, 371 and 611 (1963).
[34] E. Ulanski and Z. Wu, Appl. Phys. Lett. **98** 201115 (2011).
[35] J. Vidal-Couret, M.-A. Bouchiat, J. Nasser, and J. Brossel, C. R. Acad. Sci. **260**, 1904 (1965).
[36] J. Dykyj, J. Svoboda, R. C. Wilhoit, M. Frenkel, and K. R. Hall, "2 Hydrocarbons, C27 to C100," in *SpringerMaterials – The Landolt-Börnstein Database*, ed. K. R. Hall (1999). Available online at http://dx.doi.org/10.1007/10680373_8.
[37] A. Gozzini, F. Mango, J. H. Xu, G. Alzetta, F. Maccarrone, and R. A. Bernheim, Nuovo Cimento Soc. Ital. Fis. D **15**, 709 (1993).
[38] M. Meucci, E. Mariotti, P. Bicchi, C. Marinelli, and L. Moi, Europhys. Lett. **25**, 639 (1994).
[39] M. T. Graf, D. F. Kimball, S. M. Rochester, K. Kerner, C. Wong, D. Budker, E. B. Alexandrov, M. V. Balabas, and V. V. Yashchuk, Phys. Rev. A **72**, 023401 (2005).
[40] T. Karaulanov, M. T. Graf, D. English, S. M. Rochester, Y. J. Rosen, K. Tsigutkin, D. Budker, E. B. Alexandrov, M. V. Balabas, D. F. Jackson Kimball, F. A. Narducci, S. Pustelny, and V. V. Yashchuk, Phys. Rev. A **79**, 012902 (2009).
[41] A. Burchianti, A. Bogi, C. Marinelli, E. Mariotti, and L. Moi, J. Mod. Opt. **57**, 1305 (2010).
[42] C. Marinelli, K. A. Nasyrov, S. Bocci, B. Pieragnoli, A. Burchianti, V. Biancalana, E. Mariotti, S. N. Atutov and L. Moi, Eur. Phys. J. D **13**, 231 (2001).
[43] S. Gozzini and A. Lucchesini, Eur. Phys. J. D **28**, 157 (2004).
[44] C. Marinelli, A. Burchianti, A. Bogi, F. della Valle, G. Bevilacqua, E. Mariotti, S. Veronesi, and L. Moi, Eur. Phys. J. D **37**, 319 (2006).
[45] E. B. Alexandrov, M. V. Balabas, D. Budker, D. English, D. F. Kimball, C.-H. Li, and V. V. Yashchuk, Phys. Rev. A **66**, 042903 (2002); Phys. Rev. A **70**, 049902(E) (2004).
[46] S. Gozzini, A. Lucchesini, L. Marmugi, and G. Postorino, Eur. Phys. J. D **47**, 1 (2008).
[47] A. Cappello, C. de Mauro, A. Bogi, A. Burchianti, S. Di Renzone, A. Khanbekyan, C. Marinelli, E. Mariotti, L. Tomassetti, and L. Moi, J. Chem. Phys. **127**, 044706 (2007).
[48] S. J. Seltzer, Ph.D. thesis, Princeton University (2008).
[49] S. Du, M. B. Squires, Y. Imai, L Czaia, R. A. Saravanan, V. Bright, J. Reichel, T. W. Hänsch, and D. Z. Anderson, Phys. Rev. A **70**, 053606 (2004).

[50] A. Hatakeyama, K. Oe, K. Ota, S. Hara, J. Arai, T. Yabuzaki, and A. R. Young, Phys. Rev. Lett. **84**, 1407 (2000).
[51] A. Hatakeyama, K. Enomoto, N. Sugimoto, and T. Yabuzaki, Phys. Rev. A **65**, 022904 (2002).
[52] J. Brewer and H. G. Rubahn, Chem. Phys. **303**, 1 (2004).
[53] B. P. Anderson and M. A. Kasevich, Phys. Rev. A **63**, 023404 (2001).
[54] A. Burchianti, C. Marinelli, A. Bogi, J. Brewer, K. Rubahn, H. G. Rubahn, F. Della Valle, E. Mariotti, V. Biancalana, S. Veronesi, and L. Moi, Europhys. Lett. **67**, 983 (2004).
[55] S. N. Atutov, V. Biancalana, P. Bicchi, C. Marinelli, E. Mariotti, M. Meucci, A. Nagel, K. A. Nasyrov, S. Rachini, and L. Moi, Phys. Rev. A **60**, 4693 (1999).
[56] J. H. Xu, A. Gozzini, F. Mango, G. Alzetta, and R. A. Bernheim, Phys. Rev. A **54**, 3146 (1996).
[57] V. Liberman and R. J. Knize, Phys. Rev. A **34**, 5115 (1986).
[58] I. K. Kominis, T. W. Kornack, J. C. Allred, and M. V. Romalis, Nature **422**, 596 (2003).
[59] D. M. Rampulla, N. Oncel, E. Abelev, Y. W. Yi, S. Knappe, and S. L. Bernasek, Appl. Phys. Lett. **94**, 041116 (2009).
[60] S. Suzer, E. Abelev, and S. L. Bernasek, Appl. Surf. Sci. **256**, 1296 (2009).
[61] K. F. Zhao and Z. Wu, Appl. Phys. Lett. **89**, 261113 (2006).
[62] K. F. Zhao, M. Schaden, and Z. Wu, Phys. Rev. A **78**, 034901 (2008).
[63] K. F. Zhao, M. Schaden, and Z. Wu, Phys. Rev. Lett. **103**, 073201 (2009).
[64] K. F. Zhao, M. Schaden, and Z. Wu, Phys. Rev. A **81**, 042903 (2010).
[65] N. Castagna, G. Bison, G. Di Domenico, A. Hofer, P. Knowles, C. Macchione, H. Saudan, and A. Weis, Appl. Phys. B **96**, 763 (2009).
[66] Y. W. Yi, H. G. Robinson, S. Knappe, J. E. Maclennan, C. D. Jones, C. Zhu, N. A. Clark, and J. Kitching, J. Appl. Phys. **104**, 023534 (2008).
[67] S. J. Seltzer, D. M. Rampulla, S. Rivillon-Amy, Y. J. Chabal, S. L. Bernasek, and M. V. Romalis, J. Appl. Phys. **104**, 103116 (2008).

12
Magnetic shielding

V. V. Yashchuk, S.-K. Lee, and E. Paperno

12.1 Introduction

The ability to obtain well-characterized, stable magnetic field conditions independent of the Earth's magnetic field ($\sim$0.3–0.6 G) and environmental perturbations is the key to many urgent fundamental and applied investigations using high-precision magnetometry.

State-of-the-art magnetometers operate with a sensitivity on the level of 1–10 $\times$ 10^{-12} G/$\sqrt{\text{Hz}}$ (see Chapters 4–10), whereas the amplitude spectrum of geomagnetic noise (see, for example, Ref. [1]) varies from about 0.2 G/$\sqrt{\text{Hz}}$ at 10^{-5} Hz (about a day observational period) to 2×10^{-8} G/$\sqrt{\text{Hz}}$ at 1 Hz (about a second observational period) with an approximately inverse-power-law (fractal-type) frequency dependence, scaled as $f^{-1.4}$ [1]. In order to provide the required stability of the magnetic field, different techniques of magnetic shielding and stabilization have been brought into practice.

Despite the considerable amount of literature on experiments vitally dependent on the efficacy of magnetic shielding, the details of the design of shielding systems and their performance are scarce and, until now, scattered among numerous publications.

In this chapter, we review the physical principles of magnetic shielding and discuss design approaches, performance characterization, and the use of shielding systems. We begin in Section 12.2 with a discussion of fundamentals and basic approaches to designing and optimally using a ferromagnetic shield. Peculiarities of ferrite-based and superconducting shields are considered in Sections 12.3 and 12.4, respectively.

12.2 Ferromagnetic shielding

Depending on the frequency of external magnetic fields, ferromagnetic shielding occurs through different physical mechanisms. For static and low-frequency fields, the most

Optical Magnetometry, ed. Budker, D. and Jackson Kimball, D. F. Published by Cambridge University Press.

important mechanism is flux shunting due to a high magnetic permeability, μ, of the material (see, for example, reviews [2–6]).

At higher frequencies, the skin effect starts to play an essential role, becoming predominant at frequencies $\geq$ 10–50 Hz for rather thick (thickness $t \geq 1$ mm) shielding shells made of high-permeability ($\mu \geq 10^5$) materials. Practically, a ferromagnetic shield optimized for shielding static fields automatically provides efficient shielding of oscillating fields. Therefore, we concentrate the discussion on the case of static and low-frequency fields. Shielding in the intermediate-frequency regime where both mechanisms are equally important is discussed, for example, by Hoburg [7].

12.2.1 Simplified estimation of ferromagnetic shielding efficiency for a static magnetic field

To shield a space, the external magnetic flux is directed around it via a high-permeability ferromagnetic material [Fig. 12.1(a)]. The efficiency of a shield can be characterized by the shielding ratio

$$T \equiv \frac{B_{in}}{B_0}, \tag{12.1}$$

where B_0 is the homogeneous magnetic field existing before introducing the shield, and B_{in} is the field induced inside the shield due to B_0.

We should note here that in the literature there are different definitions of the shielding efficiency (ratio, factor). Here, we use the terminology of Ref. [3] where the shielding efficiency (ratio) is defined as in Eq. (12.1).

According to Rowland's law (Ohm's law for magnetic circuits) [8, 9], for static fields the shielding efficiency depends on the relation of the effective resistances (reluctances) of the ferromagnetic "yoke," $R_f = L_f/\mu S_f$, and that of the inner space, $R_s = L_s/S_s$ [see Fig. 12.1(a)], where L_i is the characteristic length and S_i is the cross-section of the corresponding magnetic circuit (see, for example, Refs. [6, 10, 11]). These magnetic reluctances constitute a divider for the magnetic flux $\Phi_0 = B_0 S_0$ dividing it into Φ_{in} and Φ_f:

$$\frac{\Phi_{in}}{\Phi_f} = \frac{R_f}{R_s}. \tag{12.2}$$

The resulting shielding ratio for a single ferromagnetic shell shown in Fig. 12.1(a) can be estimated supposing $L_f \approx L_s \approx X$, $S_0 \approx S_s$, $S_s \propto X^2$, and $S_f \propto tX$:

$$T_a = \frac{\Phi_{in} S_0}{\Phi_0 S_s} \approx \frac{\Phi_{in}}{\Phi_f + \Phi_{in}} \approx \frac{R_f/R_s}{1 + R_f/R_s} \propto \frac{\frac{X}{\mu t}}{1 + \frac{X}{\mu t}} \approx \frac{X}{\mu t}, \tag{12.3}$$

where X is the characteristic size of the shell, such that $\mu t \gg X$.

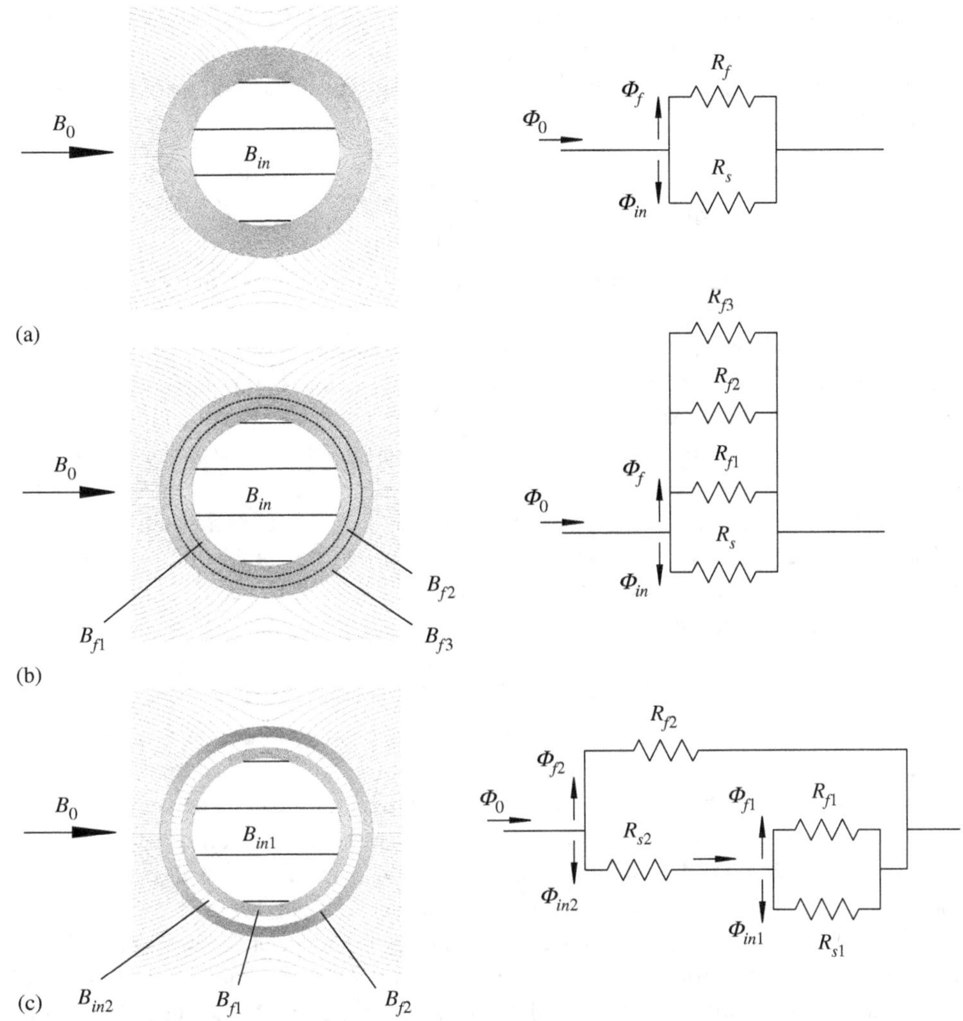

Figure 12.1 Magnetic circuit approach to estimating the performance of a multilayer shield. The terms are defined in the text.

12.2.2 Multilayer ferromagnetic shielding

According to Eq. (12.3), a way to improve shielding efficiency is to increase the shell thickness t as illustrated in Fig. 12.1(b), where the thick cylindrical shield is divided into three equal thickness shells. Being perfectly coupled, these three shells can be thought of as a circuit of reluctances connected in parallel and conducting nearly the same flux, which gives

$$T_b \propto \frac{X}{\mu(t_1 + t_2 + t_3)}, \tag{12.4}$$

supposing $X \gg t_i$.

However, a much more efficient way to improve shielding efficiency is to use multilayer shielding as shown in Fig. 12.1(c). In this case, the equivalent magnetic circuit consists of two dividers providing an overall shielding ratio proportional to the product of the shielding ratios of the separate shells:

$$T_c = \frac{\Phi_{in1} S_0}{\Phi_0 S_{s1}} \approx \frac{\Phi_{in}}{\Phi_{in} + \Phi_{f2}} \frac{\Phi_{in1}}{\Phi_{in1} + \Phi_{f1}} \propto \frac{X_1}{\mu_1 t_1} \frac{X_2}{\mu_2 t_2} . \tag{12.5}$$

Here, for simplicity, we use the assumptions that $S_0 \approx S_{s1} \approx S_{s2}$ and that the gap between the layers is on the order of X_2.

Efficiency of the multilayer design can be illustrated using an example of the particular shielding geometry depicted in Fig. 12.1. For this case and assuming $\mu = 60$, numerical simulation using ANSYS Maxwell 13 software [12] shows that the shielding efficiency obtained with a thick single shell [Fig. 12.1(a)] is the same as with two thinner nested shells with a significantly lower total weight [Fig. 12.1(c)]. For a higher μ, shielding with two separated layers appears to be even more efficient than that of the single shell.

Thus, introducing air gaps between the shells can help maintain the same shielding efficiency despite removing a significant part of the shielding material. Simply put: it is worthwhile shielding the shielding shells.

Effect of shell shape

The advantage of using multilayer shielding was demonstrated both experimentally and theoretically as far back as the end of the nineteenth century [13–16]. In the course of the twentieth century, a number of publications were devoted to the development of an analytical approach to derive simple formulae for the shielding efficiency for different geometries: cubic and cylindrical shields of finite length [2, 7, 17–20] and spherical and spheroidal shields [21–24]. A systematic review of early works as well as a theoretical treatment and comparison with practical realizations can be found in the work of Sumner *et al.* [2]. The analysis of various shielding geometries and configurations is presented in detail in a book by Rikitake [3]. A comparison of the analytical expressions from Refs. [2] and [18] for the axial-field shielding with the results of numerical calculations is given in Ref. [19], where analytical formulae for better approximation of the shielding efficiency for a multilayer cylindrical shield are also presented. Four-layer cylindrical shields with hemispherical and conical end caps are described in Ref. [25].

The dependence of the shielding efficiency on the shapes and sizes of shells of a multilayer shield can be summarized by a simplified formula:

$$T_{\text{tot}} \equiv \frac{B_{in}}{B_0} \approx T_n \times \prod_{i=1}^{n-1} T_i \left[1 - \left(\frac{X_{i+1}}{X_i} \right)^k \right]^{-1} ; \; T_i \approx \frac{X_i}{\mu_i t_i} . \tag{12.6}$$

In Eq. (12.6), T_i is the shielding efficiency of a separate ith shell, X_i ($X_i > X_{i+1}$) is the layer's radius or length (depending on the relative orientation of the magnetic field and the layer),

t_i and $\mu_i \gg X_i/t_i$ are the thickness and magnetic permeability, and n is the number of shells. Good approximations of the powers k are $k = 3$ for a spherical shield, $k = 2$ and $k = 1$ for the transverse and axial shielding ratios of a cylindrical shield with flat lids, respectively.

Equation (12.6) shows that spherical shells provide the best shielding (for shields of comparable dimensions). Moreover, spherical shields provide a uniform field within the shielded area. (Note that infinitely long cylindrical shields also possess this property.)

Unlike spherical shields, the shielding efficiency of cylindrical shields depends on the direction of the applied field. The axial shielding factor of closed cylindrical shields decreases with the shield length (i.e., the shielding improves as the length-to-diameter ratio increases). For a single-shell shield with equal length and diameter, it is 90% compared to the transverse shielding factor. Increasing the shield length decreases the axial shielding factor in approximately inverse proportion, for example, down to about 46% and 21% for the length increase by a factor of 2 and 4, respectively [19]. To obtain substantial axial shielding with open-ended cylindrical shields, their lengths should significantly exceed their diameters [19, 26]. It is important also to note that open-ended cylindrical shields with optimized length [19, 26] can provide better axial shielding, both in terms of the shielding ratio and of the residual field homogeneity, than their closed counterparts with flat caps. A simplified explanation is that shielding caps "attract" an extra magnetic flux that leaks inside the shield. Note also that in the limit of infinitely long cylinders, either open or closed, axial shielding vanishes, approaching 1.

Thus a spherical shield is preferable if one thinks of the equal shielding efficiency in all directions, the overall size, and cost of the material. The axial shielding efficiency of closed axial cylindrical shields can be improved by using conical caps (see Section 12.2.5).

Optimal shell separation

As shown above, a multilayer shield with thin shells and relatively wide gaps between them can be as effective as a much heavier and more expensive thick, single-layer shield. To minimize the total size of a multilayer shield, the air gaps between its shells should be optimized.

For two coaxial infinitely long cylinders with the same thickness t, relative permeability μ, and outer diameters D_1 and D_2, such that $D_1 > D_2$, the transverse shielding ratio for $T_i = D_i/(\mu t) \ll 1$ is approximately given by [27] [compare with Eq. (12.6)]:

$$T_{\text{tot}} = T_2^2 \frac{D_1}{D_2} \left(1 - \frac{D_2^2}{D_1^2}\right)^{-1}. \tag{12.7}$$

Straightforward differentiation of Eq. (12.7) suggests an optimal ratio of diameters of $D_1/D_2 = \sqrt{3}$. Assuming $D_2 \gg t$, the optimal gap, Δ_{opt}, for the double-shell infinitely long cylindrical shield under consideration is $\Delta_{\text{opt}} \approx 0.73 D_2$. Similar consideration for two concentric spherical shells gives the optimal ratio of the diameters of $D_1/D_2 = \sqrt[3]{4} \approx 1.59$ that corresponds to the optimal gap of $\Delta_{\text{opt}} \approx 0.59 D_2$.

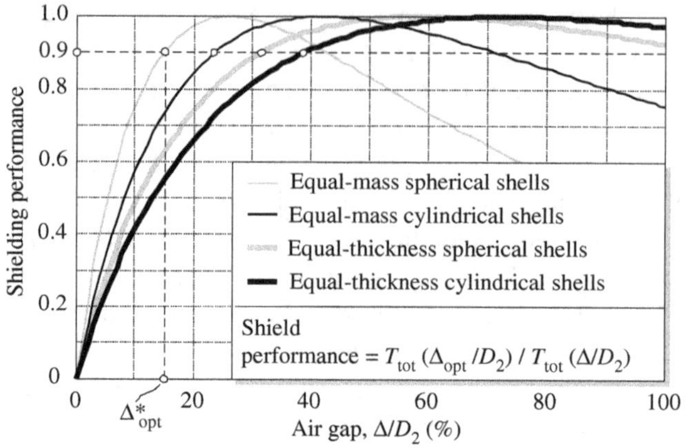

Figure 12.2 The effect of the relative air gap Δ/D_2 on the shielding performance of double-shell spherical and infinitely long cylindrical ferromagnetic shields. Equal-mass shields provide the most effective shielding per unit mass, but most practical shields are made using constant-thickness material. (Adapted with permission from Ref. [28]. Copyright 2005, American Institute of Physics.)

Figure 12.2 shows the shielding performance of double-shell spherical and infinitely long cylindrical ferromagnetic shields calculated assuming either equal mass or equal thickness of the shells [28]. For all the cases depicted in Fig. 12.2, the shielding performance, defined as the ratio of the shielding efficiencies for optimized and unoptimized size of the gap, rises quickly at first on increasing the gap and, after reaching the maximum at the optimal gap, gradually decreases.

Note that in Fig. 12.2, a 90% performance is reached at a gap significantly smaller than Δ_{opt}. For example, an equal-mass double-shell spherical shield would have 90% performance at $\Delta \approx 0.15D_2$; for a similar shield combined of long cylinders, $\Delta \approx 0.24D_2$. In a more practical case, where the shell thickness is kept constant, a double-shell spherical and a cylindrical shields with 90% performance would have $\Delta \approx 0.32D_2$ and $\Delta \approx 0.39D_2$, respectively. These gap values are significantly smaller than the exact optimal values found above.

Therefore from the point of view of shield cost and size, it is reasonable to define the optimal shell separation, Δ^*_{opt}, as the smallest gap that provides 90% of the maximum shielding performance. Separating shells with the Δ^*_{opt} gap rather than a gap that provides best shielding allows one to make the multilayer shield much more compact, without significantly sacrificing its shielding performance.

The optimal shell separation Δ^*_{opt} can be found analytically only for the simplest shielding assemblies, such as spherical and infinitely long cylindrical shields [27]. Multishell shields that are most widely used in practice have relatively short, closed cylindrical shells. Their shielding performance is typically limited by the axial shielding, so if effective shielding in all directions is desired, it is sufficient to optimize the shell separation focusing on the axial shielding efficiency.

Reference [28] shows that in contrast to spherical and infinitely long cylindrical shields, for equal-thickness closed axial cylindrical shields the optimal separation is surprisingly small. Depending on the shield geometry, Δ^*_{opt} varies between 5% and 17% of the inner shell diameter.

For practical purposes, the axial shielding with two closed cylindrical shells and relatively narrow air gaps can be estimated by using empirical formula [28]:

$$T_{\text{tot}} \approx 1.92\, T_2^2\, \alpha\, (1 + 0.1/\Delta_a + 0.06 \times \alpha^2/\Delta_r)\,, \tag{12.8}$$

where Δ_a and Δ_r are axial and radial air gaps relative to the inner shell diameter, D_2, and $\alpha = L_2/D_2$ is the aspect ratio of the inner shell.

Effect of openings

The design of a shield should usually provide access via multiple openings into the shielded area, for example, for laser beams, connecting wires of the inner coils, etc. Unfortunately, the openings also allow magnetic fields to penetrate inside the shield.

The magnetic field penetrating through an open end of a cylindrical shield decreases exponentially (Refs. [3, 18], for example):

$$B_{in} \approx B_0 e^{-\beta x/D}\,, \tag{12.9}$$

where B_{in} is either a transverse or longitudinal fringing field at the distance x from the shield end; the factor β equals about 7.0 and 4.5 for transverse and longitudinal fields, respectively; and D is the inner diameter of the shield. According to Eq. (12.9), at a distance of one diameter from the shield end, transverse and longitudinal fields are attenuated by factors of 10^3 and 10^2, respectively.

An opening in a shield with a diameter d much smaller than the characteristic sizes of the shield can be thought of as a removed magnetic dipole; so that the field from the opening should drop as x^{-3}. More accurate estimation of the components of an axial magnetic field penetrating through partial openings in the caps of a cylindrical shield can be obtained with a numerically verified empirical expression given in Ref. [29].

In order to reduce the effect of openings, one can widen the air gaps between the shielding shells [30]. The penetration of magnetic field can also be suppressed with cylindrical ferromagnetic collars mounted over the openings in the shells [31]; see also Section 12.2.5. Practically, with an aspect ratio for a collar of $l/d \approx 2{-}3$, the penetrating field becomes negligible.

Note that for alternating fields, openings can even enhance the shielding factor [32]. This is because the field penetrating through the shield walls has a phase shift with respect to the outside field; whereas the field coming in through the openings is in phase with the outside field (low frequencies). The phase difference of the fields depends on the shield design. At some frequency, it can become close to π, providing a resonance enhancement of the shielding factor. In Ref. [33], due to use of the phase-shift mechanism, an order of magnitude enhancement of the shielding factor at 22 Hz has been achieved.

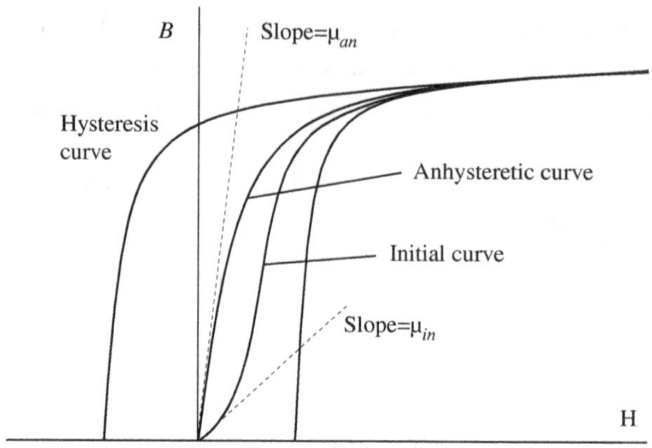

Figure 12.3 Hysteresis magnetization curve of a ferromagnetic material and various definitions of permeability.

12.2.3 Optimization of permeability: annealing, degaussing, shaking, tapping

Shielding efficiency strongly depends on the magnetic permeability of the shielding material [Eqs. (12.3)–(12.6)], which itself depends on the applied field and exhibits hysteretic behavior (see Fig. 12.3).

The best shielding is obtained well away from magnetic saturation of the shielding material. As shown in Fig. 12.1(c), magnetic flux density in multilayer shields is much higher in the outermost shell than in inner ones. Therefore, the outermost shell should be checked for saturation at the outset. For typical ferromagnetic shielding materials, such as mumetal [34,35], the maximum magnetic flux density should be below 2×10^3 G. In the Earth's field this implies requirements for the shield geometry of $D/t \leq 4000$ for transverse shielding and $L^3/tD^2 \leq 1.6 \times 10^7$ for axial shielding in the case of $2 \leq L/D \leq 40$ [2].

The magnetic permeability of the shielding material is also significantly affected by the treatment history of the material, its residual imperfections and stresses, as well as by handling and mounting.

Annealing

Imperfections and stresses in a shielding material, which obstruct the free motion of the magnetic domain boundaries, are significantly reduced by applying an appropriate annealing procedure. With high nickel content soft-ferromagnetic alloys, such as mumetal [34, 35], permalloy [36], and CO-NETIC [37], widely used for magnetic shielding, annealing in a dry hydrogen atmosphere is strongly recommended as the final step in the fabrication process.

Hydrogen annealing provides multiple effects. The annealing removes impurities, such as carbon and sulfur, and alters the material's crystal structure by removing stresses and aligning the grains (see, for example, Refs. [38, 39] and references therein). As a result, the material's permeability and shielding performance are significantly improved [40–42].

Martin and Snowdon [42] have investigated the magnetic shielding effect of CO-NETIC foil of 0.1 mm thickness over a broad range of temperatures, from 4.2 K to 293 K. It was shown that annealed shields were an order of magnitude more effective at all temperatures.

If annealing is performed at zero magnetic field, the magnetization properties of the material are described by the initial magnetization curve [43–45] (Fig. 12.3). The slope of the curve at low magnetic fields gives an initial permeability, μ_{in}. For high-permeability materials, μ_{in} is typically smaller by an order of magnitude than the maximum permeability, $\mu_{\max}$. For example, for a perfectly annealed CO-NETIC AA alloy the initial permeability is specified to be $\mu_{in} = 30\,000$, whereas $\mu_{\max} = 450\,000$ [46].

Degaussing

With a large shielding ratio, the main factor determining the residual magnetic field inside the shield is the residual magnetization of the innermost shielding layer. This residual magnetization is subject to change due to the applied magnetic field (a result of hysteresis), and also due to handling, mounting, and loading of the annealed shield. Disassembling-assembling a shield [31] can increase the residual magnetization of the innermost layer, leading to an increase of the residual magnetic field in the shielded volume up to 300 μG, compared to the typical value after degaussing of 10 to 50 μG.

In order to increase the effective permeability and decrease the residual magnetic field inside the shield, a special procedure known as demagnetizing (degaussing) is used. Degaussing consists in applying to the shielding material a slowly decaying alternating magnetic field. The maximum value of the applied field should produce a field in the material enough for its complete saturation. Reference [47] recommends the maximum demagnetizing field to be five times the material coercivity H_c, which is specified to be $H_c = 15$ mOe for CO-NETIC AA [46]. After such treatment, the magnetization properties of the ferromagnetic material are described with an anhysteretic magnetization curve (Fig. 12.3) and corresponding anhysteretic (ideal) permeability μ_{an} (see, for example, Refs. [38, 43, 45]), which can significantly exceed μ_{in} [48].

It should be emphasized that demagnetizing improves the shielding performance only for purely static fields by generating the magnetization of the shield that compensates an applied field and results in a lower field inside the shield. For shielding of the varying part of the external field after demagnetizing, μ_{in} is still appropriate. The relation between the demagnetizing field and anhysteretic magnetization is the subject of investigation in Ref. [49]. Demagnetizing is usually used to bring the shield into ideal conditions at its installation and periodically during its operation. However, the advantage of the large anhysteretic permeability can also be used for shielding of slowly varying fields if a demagnetizing (*shaking*) oscillating field is steadily applied. This type of operation can yield much better shielding efficiency [50, 51]; see also the dedicated subsection below.

Usually demagnetizing is accomplished by winding a special coil around a shielding shell (in solenoidal and/or toroidal configuration) and supplying the coil with AC current at an industrial frequency of 50 or 60 Hz [48, 52–54]. The demagnetizing current is gradually reduced from a large (material-saturating) value to zero, for example, by means of a variable autotransformer, as was done in Ref. [53]. For a smooth change of the current, a current-control system based on an induction variometer (variocoupler) was developed in Ref. [54]. It ensures a smooth decrease in the demagnetization current by a factor of 10^3 without sudden variations, which arise as the autotransformer runner jumps from one turn of its coil to another.

Another demagnetization arrangement is used in Refs. [17, 55–57]. A current is passed through the center of the shields using either an existing vacuum can as a conductor [17, 56, 57] or a dedicated wire on the axis of the shields [55]. Due to ineffective single-turn-coil geometry, the demagnetizing current has to be large, up to 3 kA at 50 Hz, requiring a rather complicated current control system [57].

Note that demagnetization of a ferromagnetic material can be achieved by applying a static magnetic field, oppositely directed with respect to the residual magnetization. After the applied magnetic field is removed, the magnetization is restored only partially. By "overshooting" the applied field, it is possible to achieve nearly complete demagnetization. Static-field demagnetization of a ferromagnetic shield was used by Zolotorev [58] in order to locally zero the residual magnetization over the "bad" spots of a shield.

Mechanical shaking and tapping

Understanding the effects of mechanical treatment of a ferromagnetic shield is not straightforward. The handling and mounting of an annealed shield, as well as the loading stresses, shock, vibration, and applied magnetic fields are known to cause degradation of the permeability and the residual magnetization. For example, in Refs. [25, 47] a decrease of the permeability by a factor of 2 and even more was observed. In contrast to this observation, shaking and vibrating a ferromagnetic shield can also result in partial demagnetization, similar to the effect of degaussing with an oscillating magnetic field [47]. Accordingly, a number of researchers have observed that gentle tapping on the shield can help the degaussing process to lower the residual magnetization of the shield [59]. The same mechanical factors can also increase the residual magnetic field noise in the shielded volume.

In soft ferromagnetic materials placed in a low magnetic field with a strength near that of the coercive field, the magnetization process is mainly due to domain wall motion [38]. Interaction of the motion with the disorder present in the material due to impurities, lattice dislocations, residual stresses, etc., leads to sudden changes (jumps) of the magnetization. These jumps are known as the Barkhausen effect [60]. The Barkhausen effect plays a fundamental role in the demagnetization process (see, for example, Refs. [61–63]). In the application to ferromagnetic shielding practice, it is important to note that temperature variation, plastic and dynamic deformations, and ultrasonic vibration in soft ferromagnetic materials (such as permalloy and mumetal) strongly affect the Barkhausen effect [62, 64, 65].

Shaking

Increasing effective permeability of a ferromagnetic shield material by shaking (that is by applying an alternating magnetic field directed to circulate along the shielding layer) was first implemented in Refs. [50] and [51]. With a large enough amplitude of the shaking field and at a frequency several times higher than the frequency of the varying outside magnetic field, the average magnetization of the shielding material follows the anhysteretic magnetization curve (Fig. 12.3). The resulting effective permeability appears to be significantly larger than the initial permeability of the shielding material. For effective shaking, the RMS amplitude of the applied field should be on the order of the coercivity, that is about 0.02–0.05 Oe for magnetically soft materials (see, e.g., Refs. [38, 48, 51, 52]).

The improvement of the shielding efficiency due to shaking depends on shielding design, arrangement, and environmental conditions. In Ref. [52] with a single-layer rectangular shield biased by the Earth's magnetic field, shaking at 50 Hz increases the shielding efficiency by a factor of 5. An increase by a factor of 8 has been reported for a two-layer shield [51].

While implementing shaking, it is necessary to take care of possible leakage of the shaking field into the shielded volume [52, 66–69]. The problem of the leakage field can be overcome if shaking is applied only to the outer shells; then the innermost layer shields the leakage field, as implemented, e.g., in Refs. [54, 66, 70]. However, this approach leads to a low efficiency of shaking. For the case of a cubic magnetically shielded room with three nested cubic shells [66], shaking applied to two outer shells improves the shielding only by a factor of 2.

An interesting approach to avoid leakage of the shaking field was used in Ref. [57]. In this set-up, a five-layer shield suppresses the magnetic field inside a vacuum system placed inside the shield and consisting of a hollow cylindrical vacuum chamber with axial input and output tubes. The vacuum system is used as a conductor for an alternating current (1–3 A) that induces a shaking field in the innermost shell. The returned current is passed through the outside shield case. In this arrangement, shaking improves transverse field shielding by a factor of approximately 10 and the longitudinal one by a factor of ~7, without leakage of the shaking field.

12.2.4 Magnetic-field noise in ferromagnetic shielding

Even with perfect shielding of external magnetic fields, modern precision measurements can be limited by magnetic field noise at the level of $1-10 \times 10^{-11}$ G/$\sqrt{\text{Hz}}$ caused by thermal fluctuation of magnetic domains (magnetization fluctuation, see Section 12.3.3) and thermal agitation of conduction electrons (Johnson–Nyquist thermal currents) in the innermost layer of the magnetic shield itself [33, 71–73]. Superconducting magnetic shields (see Section 12.4) do not generate magnetic noise, but thermal radiation shields, required for their use with room-temperature samples, typically generate magnetic noise at a $1-3 \times 10^{-11}$ G/$\sqrt{\text{Hz}}$ level [74]. For comparison, the thermal magnetic noise from a human body is much lower, approximately 10^{-12} G/$\sqrt{\text{Hz}}$ [75].

A comprehensive analysis of the magnetic field noise from high-permeability magnetic shields with different geometry is presented in Ref. [73] based on the fluctuation-dissipation theorem. A back-of-the-envelope consideration of magnetic field fluctuations due to

Johnson–Nyquist thermal currents and a thorough discussion of the practical consequences for high-precision measurements can be found in Ref. [76].

For a large, thin metal sheet of resistivity ρ at finite temperature T, the RMS magnitude of the magnetic induction $\langle B^2 \rangle$ at a distance a from the sheet is given (in quasistatic approximation) by (see, for example, Ref. [76])

$$\langle B^2 \rangle \propto \frac{4k_B T \Delta f \delta}{c^2 \rho a^2}, \tag{12.10}$$

where k_B is the Boltzmann constant, c is the speed of light in vacuum, Δf is the measurement bandwidth, and δ is the sheet thickness. Equation (12.10) directly follows from the Nyquist theorem for voltage noise across a resistor (see, for example, Ref. [77]).

The magnetic field noise [Eq. (12.10)] found in the quasistatic approximation is independent of frequency (white noise). However, at high frequencies, currents induced in the metal by a changing thermal magnetic field and the magnetic field induced by these currents are not negligible. The induced currents tend to significantly reduce magnetic field fluctuations (self-shielding effect) at frequencies above the cut-off frequency f^* given by [76]

$$f^* \propto \frac{c^2 \rho}{2\pi \delta a}. \tag{12.11}$$

Magnetic field noise inside a shield can be significantly diminished by carefully designing the shell structure of the shield. According to Eq. (12.10), the thickness of the innermost layer should be as small as possible. However, it would lead to a proportional increase of the cut-off frequency, and, therefore, to a relative gain of magnetic noise at high frequencies. An increase of distance a (and, accordingly, an innermost layer size) is beneficial for the decrease of both the noise level and the cut-off frequency. Another effective way is to use an innermost layer made of a material with low electrical conductivity, such as ferrite (see Section 12.3).

When designing a shield, the reflection effect should also be taken into account. As pointed out in Refs. [71] and [78] a mumetal plate behind a good conductor plate amplifies the thermal magnetic noise. The noise of the conductor plate is suppressed if the order of the plates is changed.

Note that the detected noise level depends on the magnetic field sensor and the method of detection. The noise detected by a finite-size sensor decreases, relative to that measured by an infinitely small sensor, with increasing the distance to a conducting plate. Use of gradiometric measurements also enables reduction of the detected magnetic field noise by an order of magnitude [71].

12.2.5 Examples of ferromagnetic shielding systems

The Yashchuk et al. shielding system

The configuration of a four-layer ferromagnetic shielding system developed at Berkeley [31] for investigation of nonlinear magneto-optical effects with alkali vapor cells [79] is shown in Fig. 12.4.

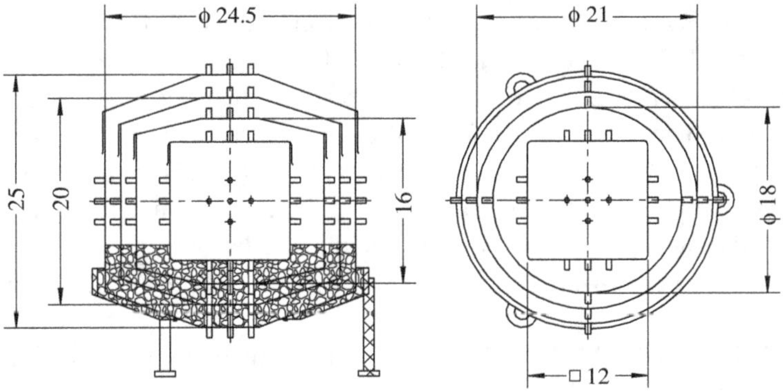

Figure 12.4 Geometrical configuration of the four-layer magnetic shield at Berkeley [31]. Dimensions in inches.

The design shown in Fig. 12.4 employs for the three outer layers an approximation to a spherical shape, simpler to manufacture than a true sphere. The innermost shield is in the shape of a cube with rounded edges. This allows compensation of the residual magnetic field and its gradients as well as application of relatively homogeneous fields with a system of nested 3D coils of cubic shape. The field homogeneity is increased by image currents, due to the boundary conditions at the interface of the high-permeability material, which make the short coils into effectively infinite solenoidal windings.

The layers are spaced with polyurethane foam to reduce mechanical stress. All CO-NETIC parts (0.4 mm thick) were annealed in a hydrogen atmosphere upon manufacturing. For demagnetization, a degaussing AC current with the maximum amplitude of 2 A is applied to the inner coils. The shield is designed to allow optical access through a number of 1/2-inch-diameter holes. Each of the holes has an associated 1-inch stem preserving the shielding properties.

Note that variation of the laboratory magnetic field due to motion of an elevator (about 10 m from the laboratory) is up to 5 mG. Variation due to the Sun's magnetic activity, measured during a magnetic field storm, reaches 30 mG. Fields of such values are monitored with a fluxgate magnetometer. The signal from a fluxgate magnetometer is used to partially compensate and stabilize the laboratory's magnetic field via a feedback-loop system consisting of a 3D set of relatively large outside coils and a current amplifier.

The careful design of the Berkeley magnetic shielding system allows for an overall shielding efficiency of $\sim 10^{-6}$ (roughly the same in all directions) and a long-term magnetic-field stability inside the innermost shield at the level of 0.1 μG. Five-layer versions of this basic design have demonstrated passive shielding ratios of better than 10^{-7} [80].

Magnetically shielded rooms

An ultimate approach to magnetic shielding for biomedical measurements is the employment of magnetically shielded rooms (MSRs) accommodating both a subject and the measurement equipment. An MSR is used to create a wide measuring space with a residual magnetic noise at a level of several $fT/\sqrt{Hz}$ in a magnetically and vibrationally hostile city environment.

A typical MSR [47, 81–89] consists of three to eight nested layers made of high-permeability ferromagnetic plates a few millimeters thick. At least one electromagnetic shielding layer, which is important at frequencies above 1 Hz, is also included. This layer is usually made of about 10-mm-thick aluminum plates. Electric and magnetic continuities of the conductive and magnetic layers are maintained by overlay strips. The junctions in the aluminum layer can be electroplated with silver or gold to further improve the conductivity [81].

Commercially available MSRs [82] with three layers (permalloy, aluminum, permalloy) have a 12 m^2 floor space, about 0.5 m wall thickness, weigh 6.8 tonnes, and provide a shielding field attenuation of about 40, 60, and 100 dB at 0.01, 1, and 100 Hz, respectively.

An extraordinary MSR is described in Ref. [83]. It has inside and outside sizes of 2.9 and 6 m cubes and comprises seven layers of mumetal, with 27 mm total thickness, and an aluminum layer. The MSR provides field attenuation of about 97, 153, and 166 dB at 0.01, 1, and 5 Hz, respectively.

12.3 Ferrite shields

Ferrites are metal oxides with ferrimagnetic order. Ferrites with low coercivity, or soft ferrites, are widely used for high-permeability, low-loss magnetic components in radiofrequency, microwave, and power electronics. The primary advantage of ferrites in these applications is low electromagnetic power loss due to low electrical conductivity. At 100 kHz, for example, a typical manganese–zinc ferrite has an electrical conductivity 6–8 orders of magnitude lower than that of silicon–iron, which leads to reduction in eddy-current losses by the same factor.

Given the intimate relationship between eddy-current loss and Johnson–Nyquist noise, it is no surprise that ferrites make excellent candidates for low-noise magnetic components in compact magnetometers. Room-temperature magnetic shields and a flux concentrator made of soft ferrites have been demonstrated [33,90,91]. As atomic magnetometers become increasingly more sensitive, the possibility of creating a subfemtotesla magnetic-field environment with a nonmetallic magnetic shield has gained much interest. In this section we survey properties of ferrites as they pertain to the design and construction of a magnetic shield for table-top experiments.

12.3.1 Permeability

Two important classes of soft ferrites are manganese–zinc (MnZn) and nickel–zinc (NiZn) ferrites. The former has higher permeability, whereas the latter has better high-frequency loss properties. The initial permeability of MnZn ferrites ranges from several hundred to several thousand. This is smaller than the permeability of mumetal. In order to obtain a high shielding factor, therefore, it is preferable to use ferrite in the innermost layer of a multilayer magnetic shield, in which the outer layers are made of conventional magnetic metal. Thermal magnetic noise of the latter is then shielded by the ferrite inner layer.

The permeability of ferrites varies significantly with temperature. Two temperatures of interest are the compensation temperature, T_{comp}, and the Curie temperature, T_{curie}. At $T = T_{\text{comp}}$ the magnetocrystalline anisotropy, which is the energy barrier for domain rotation, is minimized and permeability thereby increased. Ferrimagnetic to paramagnetic phase transition occurs at T_{curie}, with accompanying rise and collapse of magnetic susceptibility. Many commercial-brand soft ferrites are engineered in chemical composition to allow maximum permeability or minimum temperature coefficient around room temperature. Variation of permeability by as much as 50% is common in MnZn ferrites in the temperature range of 0 to 100°C. For an extensive review of temperature dependence of permeability and other properties of soft ferrites see Ref. [92].

12.3.2 Fabrication and the effect of an air gap

Like any ceramic materials, ferrites are brittle and cannot be readily formed into sheets, tapes, and other low-dimensional or complex shapes. This makes construction of a large magnetic shield (such as MSR) entirely out of ferrite very costly and perhaps impractical. At Princeton a cylindrical enclosure has been constructed [33] with the inner volume measuring 10 cm (height) × 10 cm (diameter), out of three—one for the sidewall and two for endcaps—MnZn ferrite pieces. The sidewall piece, a hollow cylinder with 1 cm wall thickness, was formed by machining from a solid cylindrical block of ferrite, fabricated by powder mixing and sintering [93].

As ferrite shields are assembled from multiple parts, low-permeability (air or glue) gaps are introduced due to surface roughness and unevenness. According to magnetic-circuit theory (Section 12.2.1), an air gap inserted in a high-permeability magnetic medium behaves like a low-conductivity gap inserted into a current-carrying wire. If the gap length is X_g and the length of the path through the ferrite is X, the total magnetic resistance is

$$R_m = \sum \frac{X}{\mu S} = \frac{X_g}{1 \cdot S_g} + \frac{X}{\mu_{\text{ferr}} \cdot S} \approx \frac{1}{\mu_{\text{ferr}} S}\left(\mu_{\text{ferr}} X_g + X\right), \qquad (12.12)$$

where μ_{ferr} is the relative permeability of the ferrite, and we assume that the cross-sectional areas of the flux path in the air gap (S_g) and in the ferrite (S) are approximately the same. We see that the gap effectively increases the magnetic length of a flux path by $\mu_{\text{ferr}} X_g$. Therefore, for an air gap not to affect the shielding factor significantly, the gap length should be small compared to the original path length in the ferrite divided by μ_{ferr}. For $X = 10$ cm, $\mu_{\text{ferr}} = 5000$, X_g should be much less than 20 μm. Modern grinding and lapping techniques allow preparation of a sizable ferrite block with surface flatness within a few microns.

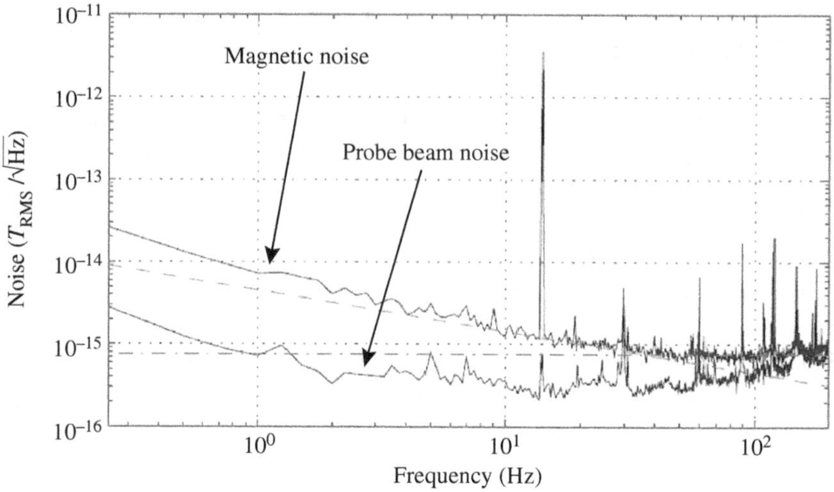

Figure 12.5 Magnetic-field noise from the Princeton ferrite shield (adapted with permission from [33]. Copyright 2007, American Institute of Physics). The dashed line represents noise of $4.4\,f^{-1/2}$ fT where f is in Hz. In comparison, white noise from a mumetal shield of the same size was estimated to be 18 fT/$\sqrt{\text{Hz}}$.

12.3.3 Thermal noise

Thermal fluctuations of magnetic domains produce magnetic-field noise inside a ferrite shield with power spectral density rising as $1/f$ with decreasing frequency (Fig. 12.5). Quantitative evaluation of magnetic-field noise emanating from a linear passive medium can most easily be done by invoking the generalized Nyquist relationship, also known as the fluctuation-dissipation theorem. In this approach, the problem of calculating equilibrium magnetic-field noise is conveniently replaced by that of an energy-loss calculation in a driven medium. A number of commercial finite-element analysis packages (e.g., Ref. [12]) allow calculation of electromagnetic power loss in an arbitrarily shaped magnetic shield in three dimensions. The shield is harmonically driven by a point dipole or a small test coil located where magnetic-field noise is to be calculated. The power loss is converted to the magnetic-field noise by

$$S_B(f) = \frac{2k_BTP(f)}{\pi^2 p^2 f^2}\,. \tag{12.13}$$

Here, $S_B(f)$ is the power spectral density of the magnetic-field noise at frequency f, k_BT is the thermal energy at temperature T, p is the amplitude of the driving dipole moment, and $P(f)$ is the power loss spectrum.

For metallic shields, $P(f)$ is almost completely dominated by the eddy current loss, $P(f) \propto f^2$, which gives rise to a frequency-independent (white) magnetic-field noise attributable to fluctuating electric currents. On the other hand, energy loss associated with magnetic viscosity, or hysteresis loss,[1] exhibits a frequency spectrum $P(f) \propto f$. The resulting $1/f$ noise has been observed in inductor cores, ferrite magnetic shields [33], and flux concentrators [90].

Analytical calculation of the magnetic-field noise in a cylindrically symmetric shield [73] has shown that as the frequency decreases, the $1/f$ noise surpasses the white noise when the skin depth of the shield becomes comparable to the thickness of the shield divided by the square root of the magnetic loss tangent, $\tan\delta_{\text{loss}}$. For a 1-mm-thick mumetal, the crossover frequency is estimated to be about 0.1 Hz, whereas for a 1-cm-thick MnZn ferrite, it is on the order of 1 kHz.

For a cylindrically symmetric geometry, the magnetic-field noise due to eddy-current loss is given by

$$\sqrt{S_B(f)} = \frac{G\sqrt{k_B T \sigma t}}{ca}, \tag{12.14}$$

where G is a dimensionless geometrical factor of order unity, a is the linear dimension of the shield, and t and σ are its thickness and conductivity, respectively. The corresponding formula for the $1/f$ noise is

$$\sqrt{S'_B(f)} = \frac{\sqrt{kT\tan\delta_{\text{loss}}/\mu_r}}{ca\sqrt{\omega t}}G', \tag{12.15}$$

where the magnetic loss factor $\tan\delta_{\text{loss}}/\mu_r$ has replaced the electrical conductivity, and the thickness term appears in the denominator. The latter is valid as long as t is sufficiently large to ensure that the ferrite enclosure has a shielding factor much greater than unity.

Equation (12.14) indicates that the white noise of MnZn ferrite with its low conductivity should be $\sim$3 orders of magnitude smaller than that of a metallic ferromagnetic shield, even if the latter were an order of magnitude thinner (for example, 1 mm vs. 1 cm). Despite the rise of the magnetic-domain noise at low frequencies, ferrite shields can still show 1–2 orders of magnitude less noise than a ferromagnetic shield in the frequency range of 10–100 Hz (Fig. 12.5).

Commercial-brand soft ferrites are typically optimized for loss at radio frequencies and loss factors below 1 kHz are poorly documented. Laboratory tests on several MnZn ferrite samples from different manufacturers indicate that the lowest loss factor in the zero-frequency limit is in the range of $\tan\delta_{\text{loss}}/\mu_r = 10^{-7}$–$10^{-6}$. It may be possible that customized process optimization could further bring down the low-frequency loss factor while maintaining a high permeability.

12.4 Superconducting shields

The use of a superconducting magnetic shield in atomic magnetometry was reported as early as 1976 [94]. In SQUID (superconducting quantum interference device) magnetometry, on

[1] We refer to magnetic energy loss in the limit of vanishing driving field, which in the ferrite literature is distinguished from high-power hysteresis loss and is often called "residual loss."

the other hand, it is common to have a SQUID chip inside a superconducting niobium shield and couple magnetic field signals originating from the outside of the shield to the sensor via a superconducting flux transformer. Because of the high cost of large-scale cryogenic cooling, superconducting shields are often designed to provide shielding over a relatively small volume, say, of 1 liter or less. The resulting proximity of the shield to the magnetometer generally raises two questions—noise and back-action. In the case of a superconducting shield, the Johnson–Nyquist noise current from unpaired electrons [94] is many orders of magnitude lower than the Johnson–Nyquist noise in a metallic shield. The low temperature of the shield also means that any thermal noise from a given loss mechanism is suppressed accordingly. Often a bigger concern with a superconducting shield in terms of noise is trapped magnetic flux in the imperfections of the material or holes in the shield, which can drift over time and is hard to control over repeated cool-down cycles. The issue of back-action of the shield is discussed at the end of this section.

12.4.1 Principles

Magnetic field repulsion or perfect diamagnetism is a defining property of a superconductor. Magnetostatics involving a simply connected superconducting material can be modeled by assigning the latter a vanishing relative permeability, $\mu_r = 0$. In the magnetic circuit analogy, such material represents a magnetic insulator, $X/(\mu t) = \infty$. For a multiply connected superconducting piece, such as an enclosure with holes, proper modeling requires additional consideration, namely flux conservation in a superconducting ring [95]. Flux conservation makes the magnetic field distribution inside a multiply connected shield dependent on the history of the shield. Figure 12.6 shows magnetic field lines around a hollow cylinder to which an external magnetic field was applied before (a) and after (b) the superconducting transition occurred. Note that in case (a), the field inside the cylinder is even slightly increased due to the superconductor (flux focusing). In order to create a zero-field shielded region, therefore, it is important to cool down the superconducting shield itself in a (temporary) zero-field environment. A conventional ferromagnetic shield is often used for such a purpose.

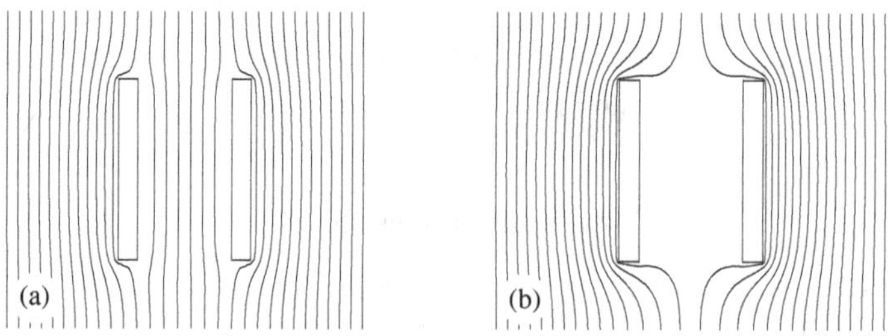

Figure 12.6 Magnetic field lines around an open superconducting cylinder. (a) Field was applied before the cylinder became superconducting. (b) Field was applied after the superconducting transition.

Zero-field cool-down is less critical for shielding against a time-dependent magnetic field. However, it is still beneficial because flux trapped in the shield can cause field drift inside the shield due to temperature changes, thermal activation, and mechanical vibration. Removal of trapped flux after cool-down is considered practically difficult, although a method was discussed in Ref. [96].

Completely closed superconducting enclosures have an infinitely high shielding efficiency. The effect of a hole is comparable to that in a high-permeability shield. The magnetic field inside an open cylindrical superconducting shield falls off with distance x from the opening as Eq. (12.9), with $\beta = 7.66$ and 3.68 for axial and transverse applied fields, respectively [97].

12.4.2 Materials and fabrication

A small-volume superconducting magnetic shield operating at the liquid-helium temperature can readily be assembled from machined lead blocks and rolled lead sheets. The lead parts can be joined together with conventional Pb-Sn solder, which becomes superconducting at around 7 K. For example, a closed-bottom, open-top lead cylinder with 7.5 cm diameter and 40 cm height can be constructed entirely by soldering along the seam of the sidewall and around the bottom cap. A shielding efficiency in excess of 10 000 can be achieved in a shield made in this fashion.

The shielding efficiency can be further enhanced by nesting multiple layers of superconducting shields. As an early example, a three-layer, open-top cylindrical lead shield with inside dimensions of 51 cm (length) and 12 cm (diameter) was reported to have achieved a shielding efficiency of 10^6, limited by trapped magnetic field in the walls of the shield [98]. In this example, each lead layer had a hemispherical bottom formed by mold casting and cold pressing. The hemispherical shape helped expel the Earth's magnetic field effectively from the central axis of the shield radially to the outside of it as the shield was immersed in liquid helium in an upright position.

A lighter and more mechanically reliable shield can be constructed from Nb and Nb alloys. A 1-cm-diameter Nb–Zr alloy tube often houses a SQUID and other superconducting components in SQUID magnetometry. Both Nb and Nb alloys are machinable with conventional shop tools. Nb parts can also be welded together in an argon atmosphere. Due to the high cost of material and custom fabrication, solid monolithic Nb shields have been limited primarily to compact-sensor applications, encompassing a volume of no more than several cubic centimeters. An interesting recent development is electrolytic coating of Nb and Nb alloys on a normal conducting substrate. A spherical superconducting shield was fabricated this way to achieve an isotropic shielding efficiency of 10^7 [99].

Magnetic shields have also been made from high-transition-temperature superconducting materials for operation at temperatures substantially higher than 4 K. A BSCCO (bismuth strontium calcium copper oxide) superconducting shield has been demonstrated for biomagnetic measurements using SQUIDs [100]. The shield was cooled by circulating cold helium gas and was large enough to accommodate a human subject for magnetoencephalography

(MEG) studies. Magnetic-field noise was reported to be reduced by a factor of 100 at 1 Hz compared to a magnetically shielded room commonly used in MEG studies.

12.4.3 Image field

Magnetic shields of both field-absorbing and field-repelling types alter the field distribution in space and therefore can affect the magnetometer measurement. For an atomic magnetometer that bases the measurement on the precession frequency of a relatively concentrated magnetic moment in space, the image field from magnetic shields can affect the measurement through a shift in the precession frequency. This may be compared with the back-action of magnetic shields in coil-based magnetometry, in which the presence of the shield alters the self-inductance of the coil and thereby changes its sensitivity calibration.

Suppose that a magnetic moment $\mathbf{p} = (p_x, p_y, p_z)$ is placed at a position $\mathbf{r} = (0, 0, z)$ and an infinitely large, high-permeability plate occupies a space defined by $-h < z < 0$. The boundary condition for the magnetic field in $z > 0$ is $B_x = B_y = 0$ on $z = 0$. The magnetic field at $\mathbf{r}$ in the upper half space can be calculated by noting that the boundary condition is satisfied by an image dipole $\mathbf{q} = (-p_x, -p_y, p_z)$ placed at $(0,0,-z)$. The field at $\mathbf{r}$ is

$$\frac{1}{(2z)^3}(p_x, p_y, 2p_z)\,. \tag{12.16}$$

The vector in the parenthesis can be decomposed as

$$(p_x, p_y, 2p_z) = (p_x, p_y, p_z) + (0, 0, p_z)\,, \tag{12.17}$$

in which the first term represents a field parallel to the dipole, and thus does not affect its precession, and the second term represents an effective static field which introduces a frequency shift of

$$\Delta\omega = \gamma \frac{p_z}{8z^3} \tag{12.18}$$

for a dipole precessing about the z axis.

A similar analysis can be followed for the reaction field from a superconducting plate. Here the boundary condition, $B_z = 0$ on the xy plane, is satisfied by an image dipole $\mathbf{q} = (p_x, p_y, -p_z)$ at $(0,0,-z)$. The field at $\mathbf{r}$ is

$$\frac{1}{(2z)^3}(-p_x, -p_y, -2p_z)\,. \tag{12.19}$$

and the frequency shift is given by the negative of Eq. (12.18).

For an atomic magnetometer with cell magnetization M, the frequency shift scales as $M\,(R/z)^3$, where R is the linear dimension of the cell and z is the distance to the shield's wall. For a cell which is nominally spherical but possesses a small degree of nonsphericity, the residual dipolar field of the cell resulting from the nonsphericity can be comparable to the reaction field from a nearby magnetic shield. For example, a nominally spherical cell that is elongated along the z axis by 0.1% and is uniformly magnetized along the same axis would have a dipolar self-field that is 80% of the reaction field from a high permeability plate located at a distance five times the radius of the cell.

The image method can be extended to find the reaction field of magnetic shields made of multiple plates.

The dipolar reaction field inside a cubical space defined by six planes at $x = \pm a/2$, $y = \pm a/2$, $z = \pm a/2$ can be calculated by assigning image dipoles at $\mathbf{r}_{mnl} = (am, an, al)$ according to

$$\mathbf{p}_{mnl} = \left(p_x(-1)^{n+l}, p_y(-1)^{l+m}, p_z(-1)^{m+n}\right) \tag{12.20}$$

for a high-permeability shield and

$$\mathbf{p}_{mnl} = \left(p_x(-1)^{m}, p_y(-1)^{n}, p_z(-1)^{l}\right) \tag{12.21}$$

for a superconducting shield. Here m, n, l are integers from $-\infty$ to $+\infty$ and $\mathbf{p} = (p_x, p_y, p_z)$ is the source dipole at the origin. The reaction field at the origin is parallel to $\mathbf{p}$ and is given by

$$\mathbf{B}_{\text{shield, high}\,\mu} = \frac{1}{a^3}\mathbf{p}\cdot 5.354 \tag{12.22}$$

and

$$\mathbf{B}_{\text{shield, supercon}} = \frac{1}{a^3}\mathbf{p}\cdot(-9.698). \tag{12.23}$$

Square and spherical shields are two examples in which the reaction field on the dipole at the center is parallel to the dipole itself, and, thus, exerts no torque on the dipole.

The dipolar reaction field of an infinitely long, cylindrical superconducting shield is calculated in Ref. [101].

References

[1] C. G. Constable, S. C. Constable, in *Frontiers and Challenges in Geophysics*, edited by R. S. J. Sparks and C. J. Hawkesworth (American Geophysical Union, 2004), pp. 147–160.
[2] T. J. Sumner, J. M. Pendlebury, and K. F. Smith, J. Phys. D **20**, 1095 (1987).
[3] T. Rikitake, *Magnetic and Electromagnetic Shielding* (TERRAPUB and D. Reidel, Tokyo, Dordrecht, 1987).
[4] I. B. Khriplovich and S. K. Lamoreaux, *CP Violation Without Strangeness: Electric Dipole Moments of Particles, Atoms, and Molecules* (Springer-Verlag, Berlin, New York, 1997).
[5] D. Budker, D. F. Kimball, S. M. Rochester, and V. V. Yashchuk, Phys. Rev. A **65**, 033401 (2002).
[6] K. L. Kaiser, *Electromagnetic Shielding* (Taylor and Francis, Boca Raton, 2006).
[7] J. F. Hoburg, IEEE Trans. Electromag. Compat. **37**, 574 (1995).
[8] H. A. Rowland, Phil. Magn. **46**, 140(1873).
[9] A. Morley and E. Hughes, *Principles of Electricity*, 5th ed. revised by W. Bolton (Longman, London, 2007).

[10] E. Paperno and I. Sasada, IEEE Trans. Magn. (Japan) **24**, 40 (2000).
[11] I. Sasada, E. Paperno, and P. Ripka, in P. Ripka, ed., *Magnetic Sensors and Magnetometers* (Artech House, Boston and London, 2001).
[12] ANSYS, Inc., *ANSYS Maxwell* (http://www.ansoft.com/products/em/maxwell/).
[13] A. W. Ruker, Phil. Magn. **37**, 95 (1894).
[14] H. E. J. G. du Bois, Ann. **63**, 348 (1897).
[15] H. E. J. G. du Bois, Ann. Physik **65**, 1 (1898).
[16] A. P. Wills, Phys. Rev. **9**, 193 (1899).
[17] D. Kleppner, H. C. Berg, S. B. Crampton, N. F. Ramsey, R. F. C. Vessot, H. E. Peters, and J. Vanier, Phys. Rev. **138**, 972 (1965).
[18] A. J. Mager, IEEE Trans. Magn. **6**, 67 (1970).
[19] E. Paperno, H. Koide, and I. Sasada, J. Appl. Phys. **87**, 5959 (2000).
[20] E. Paperno, H. Koide, and I. Sasada, IEEE Trans. Magn. **37**, 2881 (2001).
[21] T. Rikitake, J. Geomag. Geoelect. **44**, 637 (1992).
[22] T. Rikitake, J. Geomag. Geoelect. **44**, 919 (1992).
[23] T. Rikitake, J. Geomag. Geoelect. **45**, 589 (1993).
[24] T. Rikitake, J. Geomag. Geoelect. **47**, 639 (1995).
[25] D. U. Gubser, S. A. Wolf, and J. E. Cox, Rev. Sci. Instrum. **50**, 751 (1979).
[26] E. Paperno, I. Sasada, and H. Naka, IEEE Trans. Magn. **35**, 3943 (1999).
[27] A. K. Thomas, IEEE Trans. Electromagn. Compat. **10**, 142 (1968).
[28] E. Paperno, S. Peliwal, M. V. Romalis, and A. Plotkin, J. Appl. Phys. **97**, 10Q104 (2005).
[29] L. H. Chang, Ch. Wang, G. H. Luo, M. C. Lin, J. Magn. Magn. Mater. **239**, 591 (2002).
[30] E. Paperno, M. Romalis, and Y. Noam, IEEE Trans. Magn. **40**, 2170 (2004).
[31] V. V. Yashchuk, D. Budker, and M. Zolotorev, AIP Conf. Proc. **457**, 177 (1999).
[32] A. J. Mager, IEEE Trans. Magn. **6**, 67 (1970).
[33] T. W. Kornack, S. J. Smullin, S.-K. Lee, and M. V. Romalis, App. Phys. Lett. **90**, 223501 (2007).
[34] Magnetic Shield Co., *MuMetal* (http://www.magnetic-shield.com/mumetal.html).
[35] Amuneal Manufacturing, Co., *Magnetic Shielding Materials* (http://www.amuneal.com/magnetic-shielding/theory-design/magnetic-shielding-materials).
[36] E-Song America, Inc., *Permalloy – Magnetic Field Shielding* (http://www.esongamerica.com/resources/Permalloy.pdf).
[37] Magnetic Shield Co., *Sheet and Foil Material Selection Guide* (http://www.magnetic-shield.com/literature.html).
[38] R. M. Bozorth, *Ferromagnetism* (IEEE Press, Piscataway, NJ, 1993).
[39] K. Gupta, K. K. Raina, and S. K. Sinha, J. Alloys Comp. **429**, 357 (2007).
[40] D. J. Snee, J. Appl. Phys. **38**, 1172 (1967).
[41] W. D. Kehr, J. Appl. Phys. **41**, 1857 (1970).
[42] D. L. Martin and R. L. Snowdon, Rev. Sci. Instrum. **46**, 523 (1975).
[43] R. Boll, ed., *Soft Magnetic Materials: Fundamentals, Alloys, Properties, Products, Applications, The Vacuumschmelze Handbook* (Heyden, London, 1979).
[44] H. Warlimont, in *Magnetic Hysteresis in Novel Magnetic Materials*, edited by G. C. Hadjpanayis (Kluwer Academic Publishers, Dordrecht, Boston, 1997), pp. 685–709.
[45] G. Bertotti, *Hysteresis in Magnetism: for Physicists, Materials Scientists, and Engineers* (Academic Press, San Diego, 1998).

[46] Magnetic Shield Co., *Stress Annealed CO-NETIC AA Alloy* (http://www.magnetic-shield.com/products/conetic.html).
[47] E. Baum and J. Bork, J. Magn. Magn. Mater. **101**, 69 (1991).
[48] S. M. Freake and T. L. Thorp, Rev. Sci. Instrum. **42**, 1411 (1971).
[49] H. J. de Wit, J. App. Phys. **81**, 1838 (1997).
[50] T. Spooner, Phys. Rev. **25**, 527 (1925).
[51] D. Cohen, Appl. Phys. Lett. **10**, 67 (1967).
[52] V. O. Kelha, R. Peltonen, and B. Rantala, IEEE Trans. Magn. **16**, 575 (1980).
[53] E. D. Commins, S. B. Ross, D. DeMille, and B. C. Regan, Phys. Rev. A **50**, 2960 (1994).
[54] Y. V. Borisov, S. N. Ivanov, V. M. Lobashev, and Y. V. Sobolev, Nucl. Instrum. Methods A **357**, 115 (1995).
[55] R. Kunski and J. Vanier, J. Phys. E **15**, 1207 (1982).
[56] A. N. Kozlov, Y. V. Nikitenkio, and Y. V. Taran, Nucl. Instrum. Methods **192**, 379 (1982).
[57] Y. V. Taran, Instrum. Exper. Techn. **27**, 1487 (1984).
[58] M. S. Zolotorev (private communications).
[59] E. B. Alexandrov, M. S. Zolotorev, S. K. Lamoreaux, and V. V. Yashchuk (private communications).
[60] H. Barkhausen, Z. Phys. **20**, 401 (1919).
[61] E. P. T. Tyndall, Phys. Rev. **24**, 439 (1924).
[62] V. M. Rudyak, Sov. Phys. Uspekhi **13**, 461 (1971).
[63] G. Durin, S. Zapperi, J. Appl. Phys. **85**, 5196 (1999).
[64] F. Forster, H. Wetzel, Zs. Metallkunde **33**, 115 (1941).
[65] H. Markert, Phys. Sta. Sol. **20**, K67–70 (1967).
[66] V. O. Kelha, J. M. Pukki, R. S. Peltonen, A. J. Penttinen, R. J. Ilmoniemi, and J. J. Heino, IEEE Trans. Magn. **18**, 260 (1982).
[67] I. Sasada, S. Kubo, and K. Harada, J. Appl. Phys. **64**, 5696 (1988).
[68] I. Sasada and Y. Ohnaka, IEEE Trans. J. Magn. Japan **9**, 33 (1994).
[69] I. Sasada, E. Paperno, and H. Koide, J. Appl. Phys. **87**, 5962 (2000).
[70] I. S. Altarev, Y. V. Borisov, N. V. Borovikova, A. I. Egorov, S. N. Ivanov, E. A. Kolomensky, M. S. Lasakov, V. M. Lobashev, V. A. Nazarenko, A. N. Pirozhkov, A. P. Serebrov, Yu. V. Sobolev, and E. V. Shulgina, Phys. Atom. Nucl. **59**, 1152 (1996).
[71] J. Nenonen, J. Montonen, and T. Katila, Rev. Sci. Instrum. **67**, 2397 (1996).
[72] I. K. Kominis, T. W. Kornack, J. C. Allred, and M. V. Romalis, Nature **422**, 596 (2003).
[73] S.-K. Lee and M. V. Romalis, J. Appl. Phys. **103**, 084904 (2008).
[74] L. Vanthull, R. A. Simpkins, and J. T. Harding, Phys. Lett. **24A**, 736 (1967).
[75] T. Varpula and T. Poutanen, J. Appl. Phys. **55**, 4015 (1984).
[76] D. Budker, D. F. Kimball, and D. P. DeMille, *Atomic Physics: An Exploration through Problems and Solutions*, 2nd ed. (Oxford University Press, New York, 2008), Problem 8.16, pp. 403–406.
[77] C. Kittel and H. Kroemer, *Thermal Physics* (W. H. Freeman, San Francisco, 1980).
[78] R. Maniewski, I. Derka, T. Katila, T. Ryhanen, and M. Seppanen, Med. Biol. Eng. Comput. **23**, 9 (1985).
[79] D. Budker, V. Yashchuk, and M. Zolotorev, Phys. Rev. Lett. **81**, 5788 (1998).
[80] S. Xu, S. M. Rochester, V. V. Yashchuk, M. H. Donaldson, and D. Budker, Rev. Sci. Instrum. **77**, 083106 (2006).

[81] D. Cohen, U. Schlapfer, S. Ahlfors, M. Hamalainen, and E. Halgren, in H. Nowak, ed., *Biomag 2002. Proceedings of the 13th International Conference on Biomagnetism* (VDE Verlag, Berlin, 2003).
[82] NEC TOKIN, Inc., *Product Catalog* (http://www.nec-tokin.com/english/product/pdf_dl/magnetic_shielding_chamber.pdf).
[83] J. Bork, H.-D. Hahlbohm, R. Klein, A. Schnabel, in *Biomag 2000. Proceedings of the 12th International Conference on Biomagnetism*, J. Nenonen, ed., (VDE Verlag, Berlin, 2001) pp. 970–973.
[84] A. Mager, Naturwissenschaften **69**(8), 383–8 (1982).
[85] A. Mager, in *Biomagnetism*, S. N. Eme, H. D. Hahlbohm, and H. Lubbing, eds. (Walter de Gruyter, Berlin, 1981), pp. 51–78.
[86] S. Erne, H. D. Hahlbohm, H. Scheer, and Z. Trontelj, in *Biomagnetism*, S. N. Eme, H. D. Hahlbohm, and H. Lubbing eds. (Walter de Gruyter, Berlin, 1981), pp. 79–87.
[87] V. O. Kelha, in *Biomagnetism*, S. N. Eme, H. D. Hahlbohm, and H. Lubbing, eds. (Walter de Gruyter, Berlin, 1981), pp. 33–50.
[88] G. Kajiwara, K. Harakawa, and H. Ogata, IEEE Trans. Magn. **32**, 2582 (1996).
[89] K. Harakawa, G. Kajiwara, K. Kazami, H. Ogata, and H. Kado, IEEE Trans. Magn. **32**, 5226 (1996).
[90] W. C. Griffith, R. Jimenes-Martinez, V. Shah, S. Knappe, J. Kitching, Appl. Phys. Lett. **94**, 023502 (2009).
[91] H. B. Dang, A. C. Maloof, M. V. Romalis, Appl. Phys. Lett. **97**, 151110 (2010).
[92] E. C. Snelling, *Soft Ferrites: Properties and Applications*, 2nd ed. (Butterworths, London, 1988), Chapter 3.
[93] Ceramic Magnetics, Inc., *Mn-Zn Ferrites* (http://www.cmi-ferrite.com/).
[94] B. A. Andrianov, L. V. Sidorkina., Zh. Tekh. Fiz. **46**, 2007 (1976) [Sov. Phys. Tech. Phys. **21**, 1176 (1976)].
[95] C. Cordier, S. Flament, and C. Dubuc, IEEE Trans. Appl. Supercond. **9**, 4702 (1999).
[96] J. Clem, M. M. Fang, S. L. Miller, J. E. Ostenson, Z. X. Zhao, and D. K. Finnemore, Appl. Phys. Lett. **47**, 1224 (1985).
[97] J. Claycomb, and J. H. Miller, Jr., Rev. Sci. Instrum. **70**, 4562 (1999).
[98] S. I. Bondarenko, V. I. Sheremet, S. S. Vinogradov, and V. V. Ryabovol, Sov. Phys. Tech. Phys. **20**, 73 (1975).
[99] V. N. Kolosov, and A. A. Shevyrev, Inorg. Mater. **46**, 28 (2010).
[100] H. Ohta, T. Matsui, and T. Uchikawa, Neurol. Clin. Neurophysiol. **2004**, 58 (2004).
[101] H. J. M. ter Brake, *SQUID Magnetometers*, Ph.D. dissertation (University of Twente, Netherlands, 1986).

Part II

Applications

13

Remote detection magnetometry

S. M. Rochester, J. M. Higbie, B. Patton, D. Budker, R. Holzlöhner, and D. Bonaccini Calia

13.1 Introduction

Shortly after the inception of atomic magnetometry, alkali-vapor magnetometers were being used to measure the Earth's magnetic field to unprecedented precision. During the same era, Bell and Bloom first demonstrated all-optical atomic magnetometry through synchronous optical pumping [1] (see Chapters 1 and 6). In this approach, optical-pumping light is frequency- or amplitude-modulated at harmonics of the Larmor frequency ω_L to generate a precessing spin polarization within an alkali vapor at finite magnetic field [2, 3]. Although this technique received considerable attention from the atomic physics community for its applicability to optical pumping experiments, Earth's-field alkali-vapor atomic magnetometers continued to rely on radiofrequency (RF) field excitation for several decades (see Chapter 4). Upon the advent of diode lasers addressing alkali and metastable helium transitions, synchronously pumped magnetometers experienced a revival beginning in the late 1980s. In recent years, such magnetometers have found applications in nuclear magnetic resonance detection [4] (see also Chapter 14), quantum control experiments [5], and chip-scale devices intended for spacecraft use [6] (see also Chapters 7 and 15).

All-optical magnetometers possess several advantages over devices that employ RF coils. RF-driven magnetometers can suffer from cross-talk if two sensors are placed in close proximity, since the AC magnetic field driving resonance in one vapor cell can adversely affect the other. All-optical magnetometers are free from such interference. When operated in self-oscillating mode [7], RF-driven magnetometers require an added $\pm 90°$ electronic phase shift in the feedback loop to counter the intrinsic phase shift between the RF field and the probe-beam modulation. In an all-optical magnetometer this same phase shift can be achieved simply by varying the relative orientations of the pump and probe beam polarizations [8]. Most importantly, all-optical magnetometers require no physical connection between the driving electronics and the alkali-vapor cell. This allows completely remote

Optical Magnetometry, ed. Budker, D. and Jackson Kimball, D. F. Published by Cambridge University Press.

interrogation of the magnetic resonance in a faraway atomic sample. In this chapter, we describe two applications of this principle. In Section 13.2, we discuss an experimental demonstration of remotely interrogated all-optical magnetometry using a Rb vapor cell. In Section 13.3, we describe a proposed technique for employing the technology of laser guide stars to perform magnetometry using sodium in the mesosphere.

13.2 A remotely interrogated all-optical ^{87}Rb magnetometer

An all-optical remote detection magnetometer has been reported in Ref. [9]. A schematic of the magnetometer is shown in Fig. 13.1. The unshielded sensor was similar to that described in Ref. [8] in that the pump and probe beams were derived from a single laser whose frequency was stabilized by a dichroic atomic vapor laser lock (DAVLL) [10]. The atomic sample consisted of an antirelaxation-coated [11] alkali-vapor cell containing enriched ^{87}Rb and no buffer gas; the longitudinal spin relaxation time of atoms within the cell was 1.2 s. The laser beams were carried from an optics and electronics rack to a launcher assembly via polarization-maintaining optical fibers; the pump beam amplitude was modulated with a fiberized Mach–Zehnder electro-optic modulator (EOM). At the launcher the collimated output beams were linearly polarized and aimed at a sensor head placed 10 meters away. This assembly contained the ^{87}Rb cell within an enclosure heated to 34.5°C by a 1.7 kHz alternating current flowing through counter-wrapped heating wires. These wires comprised the only physical contact between the experimental apparatus and the atomic sample. In principle such heating is not necessary, but it was employed here to boost the optical-rotation signal above electronic interference from AM radio stations.

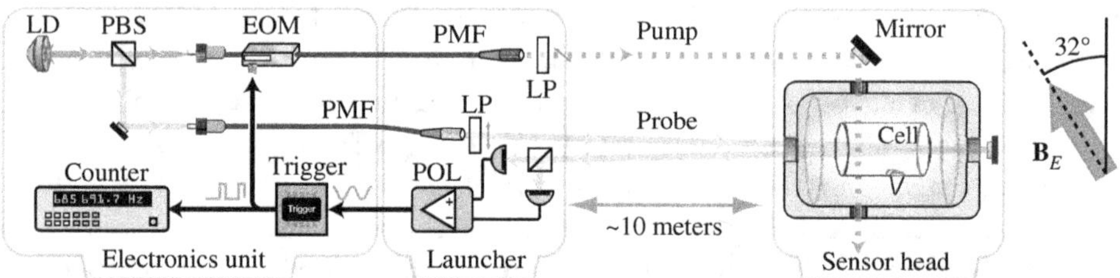

Figure 13.1 Schematic of the remote detection magnetometer. The laser diode (LD) beam was split with a polarizing beamsplitter (PBS) and coupled into two polarization-maintaining fibers (PMF). The pump beam amplitude was modulated by a fiberized EOM. At the launcher both beams were collimated, sent through linear polarizers (LP), and aimed at the sensor head. The pump beam was polarized horizontally, creating atomic alignment perpendicular to the ambient field. The probe beam reflected off a mirror aimed at a balanced polarimeter (POL) within the launcher. In self-oscillating mode (depicted), the polarimeter output was conditioned to drive the EOM directly. In driven-oscillation mode, the EOM was driven by a swept frequency source and the polarimeter output demodulated with a lock-in amplifier. The Earth's field $\mathbf{B}_E$ was independently measured to be 32° from the vertical in the direction depicted. (Reprinted with permission from Ref. [9]. Copyright 2012, American Institute of Physics.)

The probe beam traveled horizontally through the optical cell in a double-pass configuration, reflecting off a mirror behind the cell and propagating back toward the launcher. There a balanced polarimeter split the probe beam into orthogonal polarizations which were projected onto two photodiodes, allowing optical rotation to be measured. Synchronous optical pumping at $2\omega_L$ created atomic alignment[1] [12] within the ^{87}Rb vapor which produced time-varying optical rotation of the probe polarization at ω_L and $2\omega_L$. The ω_L harmonic arises when the field is tilted away from the direction of the probe beam propagation vector [13]. In the current experiment the ambient geomagnetic field pointed 32° from the vertical, leading to an optical rotation component at ω_L which was several times larger than the $2\omega_L$ harmonic.

The Zeeman shifts of the alkali ground-state sublevels (total electron spin $J = 1/2$) at a magnetic field B can be calculated from the Breit–Rabi equation [14]:

$$E(F, m_F) = -\frac{\mathcal{A}_{\text{hfs}}}{4} - g_I \mu_B m_F B \pm \frac{\mathcal{A}_{\text{hfs}}(I+\frac{1}{2})}{2}\sqrt{1 + \frac{4m_F x}{2I+1} + x^2}\,, \tag{13.1}$$

where E is the energy of the ground-state sublevel with quantum numbers F and m_F ($F = I \pm \frac{1}{2}$ being the total angular momentum of the ground state and I the nuclear spin), $\mathcal{A}_{\text{hfs}}$ is the hyperfine structure constant, and the perturbation parameter x is given by

$$x \equiv \frac{(g_J + g_I)\mu_B B}{\mathcal{A}_{\text{hfs}}(I+\frac{1}{2})}\,. \tag{13.2}$$

Here g_J and g_I are the electron and nuclear g-factors, respectively.[2] At low magnetic fields, a linear approximation to Eq. (13.1) predicts a single resonance at ω_L for all transitions with $\Delta m_F = 1$ and another resonance at $2\omega_L$ for all $\Delta m_F = 2$ transitions. At Earth's field (B_E), these resonances split into sets of resolved transitions due to higher-order corrections. In the work of Ref. [9] the laser was tuned to address the $F = 2$ ground-state hyperfine manifold of ^{87}Rb, yielding four resonances with $\Delta m_F = 1$ and three with $\Delta m_F = 2$. The magnetometer is nominally designed to probe $\Delta m_F = 2$ resonances in order to reduce systematic errors [15], but in this study it could also be operated near ω_L due to the field configuration.

In driven-oscillation mode, a function generator was used to drive the EOM and the probe beam optical rotation was detected with a lock-in amplifier. The modulation frequency was swept around $2\omega_L$ in order to map out the $\Delta m_F = 2$ magnetic-resonance curve, which was recorded on an oscilloscope. An example data set is shown in Fig. 13.2. For these data, the pump beam power was set to 260 μW peak with a 50% duty cycle and the probe beam was 55 μW continuous. Three resonances separated by the ~70 Hz nonlinear Zeeman splitting are clearly visible (see Chapter 1, Sec. 1.1.2, and Chapter 6, Sec. 6.5).

[1] Optical pumping with circularly polarized light modulated at ω_L creates atomic *orientation*, a net magnetization within the vapor. Linearly polarized light modulated at $2\omega_L$ generates *alignment*, a preferred axis of polarization which can be viewed as a twofold-symmetric spin distribution. The former represents a coherence between states with $\Delta m_F = 1$; the latter a coherence between states with $\Delta m_F = 2$ (where the quantization axis is along $\mathbf{B}_E$). See Chapters 1 and 6 for more details.

[2] Here we employ the sign convention of g_I such that the nuclear magnetic moment $\boldsymbol{\mu}_N$ is given by: $\boldsymbol{\mu}_N = g_I \mu_B \mathbf{I}$.

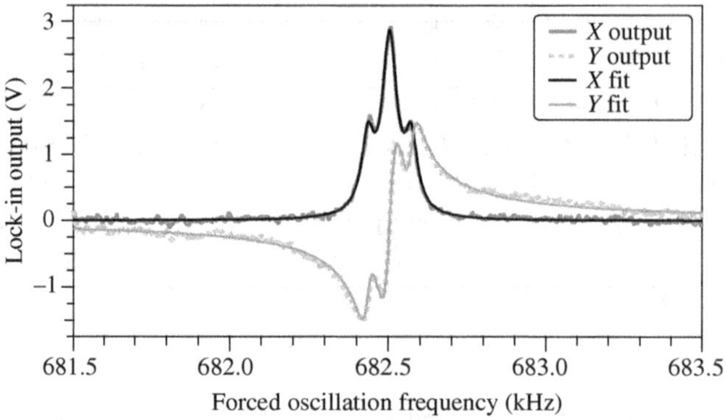

Figure 13.2 Driven-oscillation data recorded with the remote magnetometer. A spurious background has been subtracted from the data, which were then fitted to the spectrum predicted by Eq. (13.1). The data and the fit have been re-phased to portray purely absorptive and dispersive quadratures. (Reprinted with permission from Ref. [9]. Copyright 2012, American Institute of Physics.)

Due to electrical interference, a phase-coherent signal was picked up by the lock-in amplifier; this produced a frequency-dependent offset even when the pump and probe beams were blocked. This spurious baseline was subtracted from the data in Fig. 13.2 and the data were fitted to the three-Lorentzian magnetic-resonance spectrum predicted by Eq. (13.1). According to this fit, the central ($m_F = -1 \rightarrow m_F = +1$) magnetic resonance occurs at 682 504.318 $\pm$ 0.050 Hz (1σ uncertainty). Converting the best-fit frequency uncertainty into a field uncertainty yields a magnetic sensitivity of 3.5 pT. The lock-in time constant was 10 ms and the frequency sweep rate was 200 Hz/s, such that most of the spectrum was recorded within a span of $\sim$2 seconds. For the fitting procedure, the data were averaged in 10 ms bins in order to reduce correlations in point-to-point noise which would erroneously reduce the estimated frequency uncertainty. As a cross-check of this sensitivity figure, many sets of simulated data were generated with random noise which was statistically equivalent to the off-resonant noise measured in the experiment. Repeated least-squares fitting of this simulated data yielded a root-mean-square scatter of 0.046 Hz in best-fit frequency (equivalent to 3.3 pT) when all other fitting parameters were held fixed. The off-resonant noise indicates that if the EOM driving frequency were set to the zero-crossing of the dispersive trace shown in Fig. 13.2 and a steady-state experiment performed, fluctuations of $\gtrsim$ 9.6 pT could be detected with a 1 Hz noise bandwidth.

This magnetometric sensitivity was achieved in spite of several sub-optimal experimental conditions: high pump and probe beam powers, excessive pump duty cycle, analog data transmission, and sub-optimal field orientation. In a reference sensor consisting of identical components interrogated non-remotely [16], optimization of the pump power and duty cycle yielded an optical rotation signal 14 times larger than that shown in Fig. 13.2. This implies that optimization of the pump beam characteristics could immediately yield sub-pT sensitivity in the remote scheme. Moreover, these signals were recorded in an unshielded

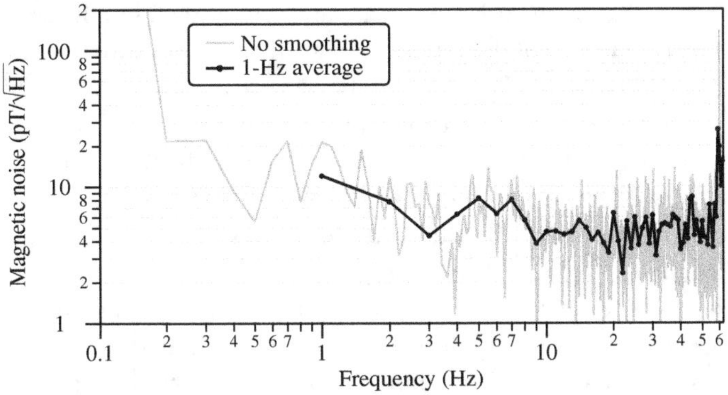

Figure 13.3 Power spectral density (PSD) of the gradiometer signal. The beat frequency of the sensors was calculated as a function of time, converted into a field difference, and Fourier-transformed to yield the magnetic noise as a function of frequency. The thin gray trace is the Fourier transform; the thicker black trace is the same data smoothed into 1 Hz bins. The mean noise floor between 1 Hz and 50 Hz is 5.3 pT/$\sqrt{\text{Hz}}$. Ambient 60-Hz magnetic-field noise can be clearly seen. (Reprinted with permission from Ref. [9]. Copyright 2012, American Institute of Physics.)

environment and subject to fluctuations in the ambient field which were often larger than 10 pT/s. Most likely the sensitivity demonstrated here is limited by genuine field fluctuations. Improved sensitivities can therefore be expected in future experiments, particularly if a gradiometric scheme is employed.

In self-oscillating mode the polarimeter output was conditioned by a triggering circuit to drive the EOM directly, generating a positive feedback loop and causing the system to oscillate spontaneously at the magnetic-resonance frequency. A passive band-pass filter of width 10 kHz centered around ω_L was included in the loop to reduce broadband noise fed into the triggering circuit. (Oscillation at $2\omega_L$ was also possible, but less robust due to AM radio interference and smaller signal amplitude.) The probe beam power was 50 μW leaving the launcher; the pump beam power was 10 μW time-averaged with a low (10–20%) duty cycle. To quantify the magnetometer's performance we mixed down its self-oscillation signal with that of the reference sensor using the lock-in amplifier, with the reference sensor acting as the external frequency reference and the remote sensor as the signal input. Helmholtz coils near the test sensor generated a field offset seen by the two magnetometers. This gradient was tuned to generate a self-oscillation beat frequency of ~275 Hz and the lock-in time constant set to 1 ms. The output of the lock-in amplifier was recorded with a data acquisition card and saved to a computer.

Figure 13.3 shows an analysis of the magnetic-field noise observed by the sensors. The output of the lock-in was digitally filtered with a 125 Hz band-pass filter about the intermediate frequency, then fitted with a running sine wave in 8 ms segments to calculate the beat frequency as a function of time. This frequency was then converted into a fluctuation about Earth's field using Eq. (13.1), assuming that both magnetometers were oscillating on the same $\Delta m_F = 1$ resonance.[3] A Fourier transform of the field difference yielded the power spectral density (PSD) of the reported magnetic noise. The average noise floor from 1 Hz to 50 Hz was 5.3 pT/$\sqrt{\text{Hz}}$. It is not clear how much of the PSD noise floor arises from sensor noise and how much can be attributed to current noise in the power supply driving the Helmholtz coils or fluctuations in the ambient field gradient. A principal advantage of the self-oscillating scheme is its high bandwidth – AC magnetic fields of frequency $\gtrsim 1$ kHz and magnitude 1 nT have been detected with high SNR using a version of the reference sensor described here.

Future embodiments of this magnetometer will use a Maxwell's fish-eye lens retroreflector, based upon a graded-index sphere, in place of the probe mirror in the sensor head. Initial polarimetry tests using a 633 nm laser and a digital polarimeter indicated that the retroreflector preserves linear polarization of a probe beam to within a few degrees of ellipticity (the measurement error of the polarimeter). Use of this optical element will obviate the need for alignment of the sensor, allowing fully remote interrogation of the magnetic field. In future tests the free-space baseline of the magnetometer will be increased, with the expectation that this technique can be extended to distances of several hundred meters before atmospheric seeing becomes a significant noise source [17]. Beyond this distance scale, adaptive optics techniques may become necessary to mitigate the effects of atmospheric turbulence and retain magnetometric sensitivity.

A sensitive remote magnetometer capable of being interrogated over several kilometers of free space would be highly desirable in several applications, including ordnance detection, perimeter monitoring, and geophysical surveys. Inexpensive manufacturing of the cell/retroreflector package would allow many such sensors to be widely distributed and interrogated by a single optical setup. Further research in remote magnetometry will also contribute to recently proposed efforts to measure the Earth's magnetic field using mesospheric sodium atoms and laser guide-star technology, described in the next section.

13.3 Magnetometry with mesospheric sodium

Measurements of geomagnetic fields are an important tool for peering into the Earth's interior, with measurements at differing spatial scales giving information about sources at corresponding depths. Mapping of fields on the few-meter scale can locate buried ferromagnetic objects (e.g., unexploded ordnance or abandoned vessels containing toxic waste), while maps of magnetic fields on the kilometer scale are used to locate geological formations promising for mineral or oil extraction. On the largest scale, the Earth's dipole field gives information about the geodynamo at depths of several thousand kilometers. Magnetic field variations at intermediate length scales, in the range of several tens to several hundreds of kilometers, likewise offer a window into important scientific phenomena, including the behavior of the outer mantle, the solar-quiet dynamo in the ionosphere [18], and ionic

[3] The *relative* field fluctuation measurement does not have any substantial dependence upon which $\Delta m_F = 1$ resonance is chosen; this remains true even if different resonances are assumed for the two sensors.

currents as probes of ocean circulation [19], a major actor in models of climate change. To avoid contamination from local perturbations, measurements of such slowly varying components of the magnetic field must typically be made at a significant height above the Earth's surface (e.g., measurements of components with a spatial-variation scale of 100 km require an altitude of ~100 km) and with high sensitivity (around 1 nT). Though magnetic mapping at high altitude has been realized with satellite-borne magnetic sensors [20–22], the great expense of multisatellite missions places significant limitations on their deployment and use. Here we discuss a high-sensitivity ground-based method of measuring magnetic fields from sources near Earth's surface with 100-km spatial resolution.[4] The method exploits the naturally occurring atomic sodium layer in the mesosphere and the significant technological infrastructure developed for astronomical laser guide stars (LGS). This method promises to enable creation of geomagnetic observatories and of regional or global sensor arrays for continuous mapping and monitoring of geomagnetic fields without interference from ground-based sources.

The envisaged measurement is a form of atomic magnetometry, adapted to the conditions of the mesosphere. The principle is to measure spin precession of sodium atoms by spin-polarizing them, allowing them to evolve coherently in the magnetic field, and determining the post-evolution spin state. Spin-polarization of mesospheric sodium is achieved by optical pumping, as proposed in the seminal paper on sodium LGS by Happer *et al.* [23]. In the simplest realization, the pumping laser beam is circularly polarized and is launched from a telescope at an angle nearly perpendicular to the local magnetic field, as shown in Fig. 13.4. The magnetic field causes transverse polarization to precess around the field at the Larmor frequency. To avoid "smearing" of atomic polarization by this precession, the optical-pumping rate is modulated near the Larmor frequency, as first demonstrated by Bell and Bloom [1] (see Chapters 1 and 6). When the modulation and Larmor frequencies coincide, a resonance results, and a substantial degree of atomic polarization is obtained. The atomic polarization in turn modifies the sodium fluorescence, which is detected by a ground-based telescope, allowing the sodium atoms to serve as a remote sensor of the magnitude of the magnetic field in the mesosphere. Specifically, the resonance manifests itself as a sharp increase in the returned fluorescence for the D2 line of sodium, or a decrease for the D1 line, as a function of the modulation frequency. By stationing lasers and detectors on a few-hundred-kilometer grid, a simultaneous map of magnetic fields may be obtained; alternatively, the laser and detector can be mounted on a relocatable stable platform such as a ship or truck to facilitate magnetic surveying. In contrast to ground-based and satellite-based measurements, the platform is not required to be non-magnetic or magnetically quiet.

[4] We note that resolution, as used here, refers to the ability to distinguish fields from separate sources. Sources on the Earth's surface produce fields in the mesosphere which vary only over 100-km distances; closer sources, however, such as those in the ionosphere, could be detected in the mesosphere with a spatial resolution substantially better than 100 km.

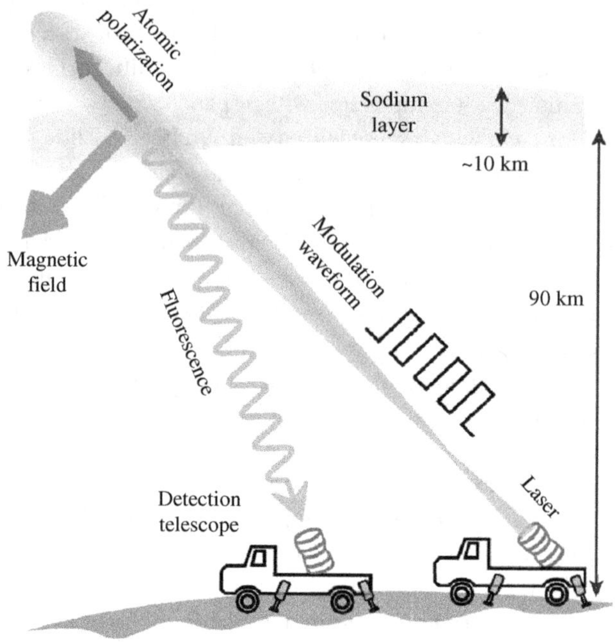

Figure 13.4 Fluorescent detection of magneto-optical resonance of mesospheric sodium. (Diagram not to scale.) Circularly polarized laser light at 589 nm, modulated near the Larmor frequency, pumps atoms in the mesosphere. The resulting spin polarization (pictured as instantaneously oriented along the laser beam propagation direction) precesses around the local magnetic field. Fluorescence collected by a detection telescope exhibits a resonant dependence on the modulation frequency.

The magnetometric sensitivity of this technique is governed by the number of atoms involved in the measurement, the coherence time of the atomic spins, and detector solid angle. A detailed discussion of mesospheric properties relevant to LGS has recently been given in Ref. [24]. Briefly, the sodium layer has an altitude of ~90 km, thickness of ~10 km, temperature of 180 K, and number density of $3 \times 10^9\,\mathrm{m}^{-3}$. This density is low by vapor-cell standards, but the interaction volume and atom number can be large, limited chiefly by available laser power. The coherence time is set primarily by collisions with other atmospheric molecules and secondarily by atom loss from the region being probed (e.g., due to diffusion or wind). A velocity-changing collision occurring after an atom is pumped typically removes the atom from the subset of velocity classes which are near-resonant with the laser light; as a consequence, these collisions result in an effective decay of spin polarization of the atoms that interact with the light. Moreover, spin-exchange collisions of sodium atoms with unpolarized paramagnetic species, predominantly O_2, result in a randomization of the electron spin, and therefore also lead to decay of sodium polarization. To our knowledge, the spin-exchange cross-section of oxygen with sodium has not been measured. However, its magnitude can be estimated from other known spin-exchange cross-sections, leading to an expected spin-damping time on the order of 250 μs. The detector solid angle for a 1-m^2 telescope, assuming isotropic emission, is 8×10^{-13} sr.

The atom number can be maximized either by increasing the interaction volume or by including more velocity classes. Since the most probable atomic velocity component along the laser beam propagation direction is zero, broadening the laser spectrum to include velocity classes away from zero velocity offers diminishing returns as spectral widths approaching the Doppler linewidth (~1 GHz) are reached. By contrast, defocusing the laser beam to illuminate a larger region in space suffers no limitation other than the amount of laser power that can be supplied.

The expected relaxation rate, however, can be sharply reduced by broadening the laser linewidth, because velocity-changing collisions become less effective at removing polarized atoms from resonant velocity classes as the laser linewidth approaches the Doppler linewidth. The optimal strategy, then, involves both broadening the laser spectrum to approximately one Doppler linewidth and defocusing the laser beam to the extent permitted by the total laser power. One expects on intuitive grounds that the optimum laser intensity resonant with a single velocity class should be such that the characteristic rate of optical pumping $\Gamma_p \equiv \gamma_0 I/2I_{\text{sat}}$ (where I is the laser intensity and $I_{\text{sat}} \approx 60\,\text{W/m}^2$ is the saturation intensity of the sodium cycling transition) is on the same order as the decay rate of atomic polarization.

Because the velocity-changing collision rate is significant on the spin-relaxation timescale but small compared to the natural lifetime of the sodium excited state, the mesospheric conditions represent a new and little-explored regime of atomic physics and optical pumping, requiring the development of new theoretical techniques [25] and numerical approaches. In Ref. [26], a detailed numerical ground-state density-matrix analysis of spin precession and optical pumping on the D1 and D2 transitions of sodium has been presented. A number of velocity classes (typically sixteen) were included in the model, spanning the Doppler distribution. A circularly polarized pump laser beam was assumed, oriented at right angles to a magnetic field of 0.5 G, with a spin-exchange collision time of 250 μs and a velocity-changing collision time of 50 μs. The laser spectrum was taken to be Lorentzian with a half-width of 400 MHz, which was found to be near-optimal. Since light of this spectral width does not resolve the sodium excited-state hyperfine structure, the hyperfine effect was neglected (the use of a repumping laser to prevent depopulation to the other hyperfine ground state was assumed). Since the optical intensity resonant with any velocity group is substantially lower than the saturation intensity, the ground-state method is expected to be accurate; comparisons were also made using the full (ground and excited-state) optical Bloch equations for selected parameters.

Sample spectra for the laser tuned to the D1 and the D2 lines are shown in Fig. 13.5. The photon shot-noise-limited magnetometric sensitivity, neglecting technical photometric noise, can be calculated from the width and peak height of these spectra. A constant launched

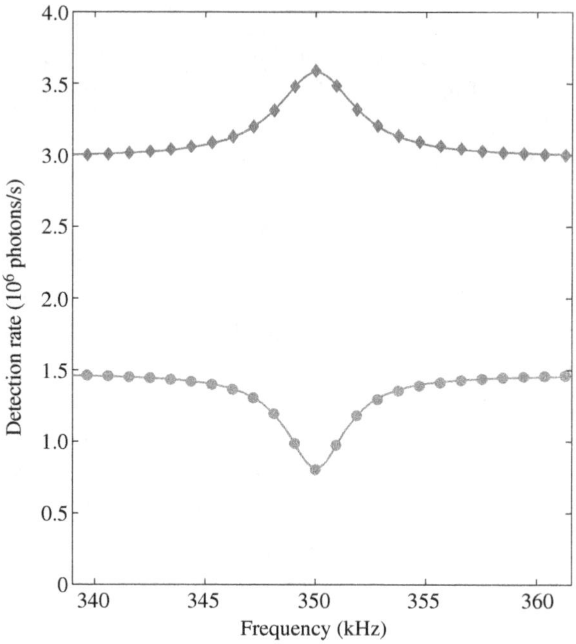

Figure 13.5 Calculated magneto-optical resonance profiles for mesospheric sodium. The resonances shown correspond to the D2 (upper curve, diamonds) and D1 (lower curve, circles) sodium lines. Symbols are the results of numerical calculations, and solid lines are Lorentzian fits to these results. Calculations are for an intensity $I = 28\,\mathrm{W/m^2}$, a laser linewidth of 400 MHz, and a modulation duty cycle of 20% with a detector collection area of $1\,\mathrm{m^2}$.

pump laser power of 20 W is maintained by varying the laser spot size on the mesosphere. Contour plots of sensitivity versus duty cycle and intensity are shown in Fig. 13.6.

The optimum sensitivity, occurring on the D1 transition at a duty cycle of 20% and peak pump intensity of $32\,\mathrm{W/m^2}$, is $0.44\,\mathrm{nT}/\sqrt{\mathrm{Hz}}$ (i.e., if magnetic fluctuations within a bandwidth $\Delta\nu$ are measured, the root-mean-square noise floor of the measurement will be $0.44\,\mathrm{nT}\sqrt{\Delta\nu/1\,\mathrm{Hz}}$). This sensitivity is within approximately one order of magnitude of the limit set by quantum spin-projection noise, the difference being due to hyperfine structure and Doppler broadening. The optimum on the D1 line offers superior magnetometric sensitivity in part because the D1 resonances are dark; i.e., they have reduced fluorescence, with correspondingly reduced photon shot noise and broadening. If technical rather than fundamental noise sources dominate, then the D2 resonance may be preferable for its larger signal size. The D2 line is routinely used by existing LGS lasers, while the nearby D1 line should be readily attained by current lasers with minimal modifications.

The calculated magnetometric sensitivity, which is limited by currently available laser power and may therefore be expected to improve with advances in laser technology, is useful for the envisaged geophysical applications, which require measurement of fields,

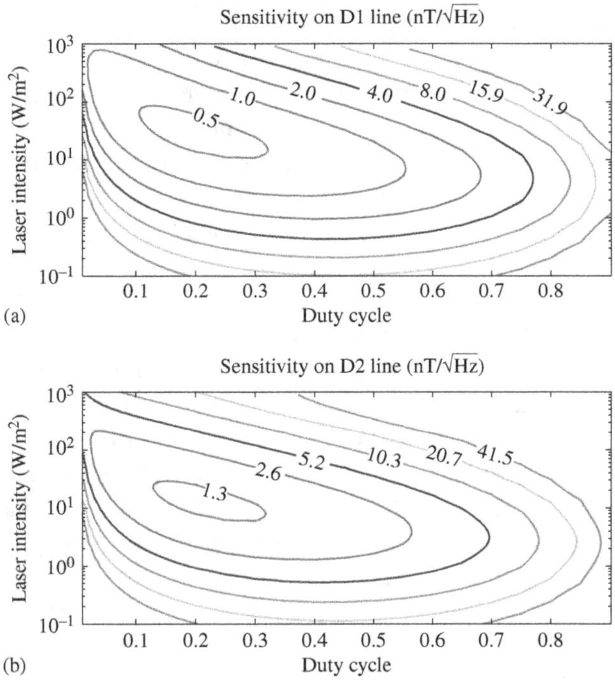

Figure 13.6 Contour plot of calculated magnetometer sensitivity as a function of pumping duty cycle and intensity. Sensitivity is calculated using a laser linewidth of 400 MHz, a detector area of 1 m^2, a spin-exchange collision time of 250 μs, and a velocity-changing collision time of 50 μs. Contours are logarithmically spaced at intervals of one octave.

e.g., in the 1–10 nT range for ocean circulation [19] and the tens of nanotesla range for the solar-quiet dynamo [18]; moreover, the dynamic range of the measurement is not subject to any simple physical limit, as the resonance technique works well and with similar sensitivity at any magnetic-field strength. Moreover, the sensitivity of this technique could be further enhanced by as much as five orders of magnitude by reflecting a laser from a rocket- or satellite-borne retroreflector, instead of being limited to the small fraction of fluorescence emitted toward the detecting telescope. Though lacking the appealing simplicity of an entirely ground-based apparatus, such a hybrid approach has the potential of extremely high sensitivity, while avoiding the challenging magnetic cleanliness requirements of spacecraft-borne sensors.

Note that at the cost of increasing the relaxation rate, the excited-state hyperfine structure can be spectrally resolved by using a narrow-linewidth laser, potentially allowing detection of magneto-optical resonances involving higher polarization multipoles such as alignment, which is prepared by pumping with linearly polarized light [27] (see Chapters 1 and 6). This capability has the additional practical implication that magnetic fields in the mesosphere could be efficiently sensed in regions (e.g., near the poles) where the Earth's magnetic field is near-vertical. In such locations it is more practical for the pump laser beam to be parallel to the field than perpendicular, a geometry suitable for magnetometers based on atomic alignment but not for those employing atomic orientation.

Although the operating laser beam intensity is not high by laser standards, the possibility of accidental illumination of aircraft and satellites must be taken into account. The safety of magnetometric measurements, however, is not substantially different from present-day LGS observations, and can be assured by similar protocols for shutting off the pump laser to accommodate occasional fly-overs, either prescheduled or monitored in real time.

Because the width of the resonance at the optimum sensitivity is around 5 kHz, effects that are important in other atomic magnetometers, such as the quadratic Zeeman shift, which splits the resonance by ~150 Hz, and the natural magnetic inhomogeneity [28], which broadens the resonance by ~700 Hz, are relatively minor. Temporal variations of the magnetic field are, in principle, merely part of the measured signal, and not an instrumental limitation. However, large enough fluctuations could make it difficult to track the resonance frequency. A likely upper bound for the magnetic fluctuations on timescales of 1 s to 100 s can be taken to be a typical observed value at the Earth's surface of around 1 nT. As this is again substantially smaller than the resonance linewidth, one can expect that except during magnetic storms, it should not be difficult to keep the laser modulation frequency on resonance. Variations of the height and density of the sodium layer itself are an additional practical concern.

A real measurement will require reducing sensitivity to such variations through comparison of on-resonant and off-resonant signals, e.g., by dithering the modulation frequency or by employing spatially separated pump beams with different modulation frequencies. The calculation further assumes that optical shot noise from the detected fluorescence is the dominant noise source; if daytime operation is desired, the spectral intensity of scattered background light will be comparable to or somewhat larger than the fluorescence signal. This can be mitigated by detecting the fluorescence synchronously with the pump-laser modulation, though at the cost of increased technical complexity.

A further deviation from the idealized calculation comes from turbulence in the lower atmosphere, which results in different, random phase shifts of the laser beam in different patches of space. In the typical LGS application, the far-field diffraction pattern in the mesosphere from these phase patches consists of filaments whose individual lateral size is set by the numerical aperture of the laser launch telescope, but whose collective extent is governed [29] by the patch size (~0.1 m). Fluctuation of these filaments in time results in undesirable changes in pump-laser intensity, motion of the illuminated column, and variation of the returned fluorescence. Although these effects will prevent fine-tuning of laser intensity, such variations are not anticipated to strongly affect the sensitivity obtainable with the proposed technique.

Acknowledgments

The authors of Section 13.2 would like to thank collaborators Charles Stevens and Joseph Tringe (LLNL) for initiating this project and providing the retroreflector, as well as Mikhail

Balabas for the antirelaxation-coated ^{87}Rb cells. We also thank Mark Prouty, Ron Royal, and Lynn Edwards at Geometrics, Inc., for experimental assistance and use of magnetometric facilities. We acknowledge support from Victoria Franques through the Department of Energy's National Nuclear Security Agency (NNSA-NA-22)NA 22, Office of Nonproliferation Research and Development. This work was also supported in part by the Navy (contract N68335-06-C-0042), by the DoE Office of Nuclear Science (Award DE-FG02-08ER84989), and by NSF (ARRA 855552).

The authors of Section 13.3 acknowledge stimulating discussions with Peter Milonni, William Happer, Michael Purucker, and Stuart Bale. This work is supported by the NGA NURI program.

References

[1] W. E. Bell and A. L. Bloom, Phys. Rev. Lett. **6**, 280 (1961).
[2] W. Gawlik, L. Krzemien, S. Pustelny, D. Sangla, J. Zachorowski, M. Graf, A. O. Sushkov, and D. Budker, Appl. Phys. Lett. **88**, 3 (2006).
[3] V. Acosta, M. P. Ledbetter, S. M. Rochester, D. Budker, D. F. Jackson Kimball, D. C. Hovde, W. Gawlik, S. Pustelny, J. Zachorowski, and V. V. Yashchuk, Phys. Rev. A **73**, 053404 (2006).
[4] G. Bevilacqua, V. Biancalana, Y. Dancheva, and L. Moi, J. Magn. Reson. **201**, 222 (2009).
[5] S. Pustelny, M. Koczwara, L. Cincio, and W. Gawlik, Phys. Rev. A **83**, 043832 (2011).
[6] H. Korth, K. Strohbehn, F. Tejada, A. Andreou, S. McVeigh, J. Kitching, and S. Knappe, Johns Hopkins APL Technical Digest **28**, 248 (2010).
[7] A. L. Bloom, Appl. Opt. **1**, 61 (1962).
[8] J. M. Higbie, E. Corsini, and D. Budker, Rev. Sci. Instrum. **77**, 113106 (2006).
[9] B. Patton, O. O. Versolato, D. C. Hovde, E. Corsini, J. M. Higbie, and D. Budker, Appl. Phys. Lett. **101**, 083502 (2012).
[10] V. V. Yashchuk, D. Budker, and J. R. Davis, Rev. Sci. Inst. **71**, 341 (2000).
[11] M. V. Balabas, T. Karaulanov, M. P. Ledbetter, and D. Budker, Phys. Rev. Lett. **105**, 070801 (2010).
[12] V. V. Yashchuk, D. Budker, W. Gawlik, D. F. Kimball, Y. P. Malakyan, and S. M. Rochester, Phys. Rev. Lett. **90**, 253001 (2003).
[13] S. Pustelny, W. Gawlik, S. M. Rochester, D. F. Jackson Kimball, V. V. Yashchuk, and D. Budker, Phys. Rev. A **74**, 063420 (2006).
[14] M. Auzinsh, D. Budker, and S. Rochester, *Optically Polarized Atoms: Understanding Light–Atom Interactions* (Oxford University Press, Oxford, 2010).
[15] C. Houde, B. Patton, O. O. Versolato, E. Corsini, S. Rochester, and D. Budker, in Proc. SPIE 8046. *Unattended Ground, Sea and Air Sensor Technologies and Applications XIII*, edited by E. M. Carapezza, 80460Q (2011).
[16] E. Corsini, Ph.D. thesis, University of California at Berkeley (2012).
[17] A. L. Buck, Appl. Opt. **6**, 703 (1967).
[18] W. H. Campbell, Pure Appl. Geophys. **131**, 315 (1989).
[19] R. H. Tyler, S. Maus, and H. Luhr, Science **299**, 239 (2003).
[20] E. Friis-Christensen, H. Lühr, and G. Hulot, Earth, Planets, Space **58**, 351 (2006).
[21] J. A. Slavin, G. Le, R. J. Strangeway, Y. Wang, S. A. Boardsen, M. B. Moldwin, and H. E. Spence, Geophys. Res. Lett. **35**, 2107 (2008).

[22] M. Purucker, T. Sabaka, G. Le, J. A. Slavin, R. J. Strangeway, and C. Busby, Geophys. Res. Lett. **34**, 24306 (2007).
[23] W. Happer, G. J. MacDonald, C. E. Max, and F. J. Dyson, J. Opt. Soc. Am. A **11**, 263 (1994).
[24] R. Holzlöhner, S. M. Rochester, D. Bonaccini Calia, D. Budker, J. M. Higbie, and W. Hackenberg, Astron. Astrophys. **510**, A20 (2010).
[25] S. W. Morgan and W. Happer, Phys. Rev. A **81**, 042703 (2010).
[26] J. M. Higbie, S. M. Rochester, B. Patton, R. Holzlöhner, D. Bonaccini Calia, and D. Budker, Proc. Nat. Acad. Sci. **108**, 3522 (2011).
[27] M. Auzinsh, D. Budker, and S. M. Rochester, Phys. Rev. A **80**, 053406 (2009).
[28] *National Geophysical Data Center*, http://www.ngdc.noaa.gov/geomagmodels/IGRFWMM.jsp.
[29] D. L. Fried, J. Opt. Soc. Am. **56**, 1372 (1966).

14
Detection of nuclear magnetic resonance with atomic magnetometers

M. P. Ledbetter, I. Savukov, S. J. Seltzer, and D. Budker

14.1 Introduction

Nuclear magnetic resonance (NMR) is a powerful analytical tool for elucidation of molecular form and function, finding application in disciplines including medicine (magnetic resonance imaging), materials science, chemistry, biology, and tests of fundamental symmetries [1–6]. Conventional NMR relies on a Faraday pickup coil to detect nuclear spin precession. The voltage induced in a pickup coil is proportional to the rate of change of the magnetic flux through the coil. Hence, for a given nuclear spin polarization, the signal increases linearly with the Larmor precession frequency of the nuclear spins. Since the thermal nuclear spin polarization is also linear in the field strength, the overall signal is roughly proportional to B^2, motivating the development of stronger and stronger magnetic fields. Additionally, an important piece of information in NMR is the so-called chemical shift, which effectively modifies the gyromagnetic ratios of the nuclear spins depending on their chemical environment. This produces different precession frequencies for identical nuclei on different sites of a molecule, and the separation in precession frequencies is linear in the magnetic field. For these reasons, tremendous expense has been spent on the development of stronger magnets. Typical spectrometers feature 9.4 T superconducting magnets, corresponding to 400 MHz proton precession frequencies, and state-of-the-art NMR facilities may feature 24 T magnets, corresponding to 1 GHz proton precession frequency. While the performance of such machines is impressive, there are a number of drawbacks: superconducting magnets are immobile and expensive (roughly \$500 000 for a 9.4 T magnet and console) and require a constant supply of liquid helium. An important parameter in NMR is the linewidth, which determines the number of lines that may be resolved in a complex spectrum. In conventional NMR, the linewidth is typically limited by residual field gradients, despite great effort to achieve high field homogeneity, at the part-per-billion level. In the context of magnetic resonance imaging (MRI), high magnetic fields are problematic in that they

Optical Magnetometry, ed. Budker, D. and Jackson Kimball, D. F. Published by Cambridge University Press.

interfere with pacemakers, and metallic implants often cause severe distortion in high-field MRI. Even more serious than image distortion, patients with ferromagnetic shrapnel are at risk of further bodily harm due to strong mechanical forces associated with field gradients.

These drawbacks provide motivation for investigation of NMR in low magnetic fields. An additional advantage of low-field NMR is the fact that it is possible to achieve extremely homogeneous magnetic fields on an absolute scale, yielding narrow lines and accurate determination of coupling parameters. The problem of low nuclear spin polarization in low field can be overcome by prepolarization in a much larger magnetic field than is used during the measurement process, or by using any number of hyperpolarization techniques such as dynamic nuclear polarization [7], spin-exchange optical pumping [8], or parahydrogen-induced polarization [9–12]. For instance, Earth-field magnetic resonance spectroscopy [13, 14] and imaging [15] have been conducted using prepolarization and inductive pickup coils. Due to the low (2 kHz) proton precession frequency afforded by the Earth's field, these experiments suffer from low sensitivity, necessitating longer signal averaging and much larger sample volumes than is common in high-field applications.

There are therefore compelling reasons to conduct low-field NMR experiments using detectors that are better suited to the low-frequency regime. For example, the sensitivity of superconducting quantum interference devices (SQUIDs) [16] is independent of the detection frequency, making them currently the most common alternative to coils for NMR in fields below 1 T [17, 18]. SQUIDs have been used at low field to measure both chemical shifts [19] and scalar J couplings [20], as well as to perform simple imaging experiments [21, 22]. Perhaps the most vivid demonstration of the power of low-field NMR to open up new applications is the use of SQUIDs to detect magnetic resonance signals [23] and to image [24] samples inside metallic enclosures, which would be impossible at stronger fields due to the greatly reduced skin depth at high frequencies. Recently, mixed sensors comprising a superconducting flux transformer coupled to a giant magnetoresistive (GMR) sensor [25] have also been demonstrated for detection of nuclear magnetic resonance [26,27] and nuclear quadrupole resonance [28] at low fields.

The downside of SQUIDs and mixed-GMR sensors is that they contain superconducting components and thus require cooling to cryogenic temperatures. The need for a constant supply of cryogenic liquids limits the portability and significantly increases the operating costs of these devices, making them impractical for many magnetic resonance applications.

In contrast, optical magnetometers have no such need and can be engineered to perform for extended lengths of time with minimal upkeep, making them an ideal choice for portable magnetic resonance devices. The resonance frequency of the magnetometer can be tuned by applying a magnetic field, enabling detection of magnetic fields from DC to radiofrequency (RF). In the case of an RF magnetometer, the sensitivity is nearly independent of frequency. A theoretical treatment [29], the results of which are presented in Fig. 14.1, indicates that a potassium magnetometer has better sensitivity than a surface coil occupying the same $16\ \text{cm}^3$ volume for frequencies below $\approx$50 MHz, assuming both devices are operating close to their fundamental sensitivity limits. This cross-over frequency increases as the devices become smaller, making the comparison especially favorable for miniature magnetometers.

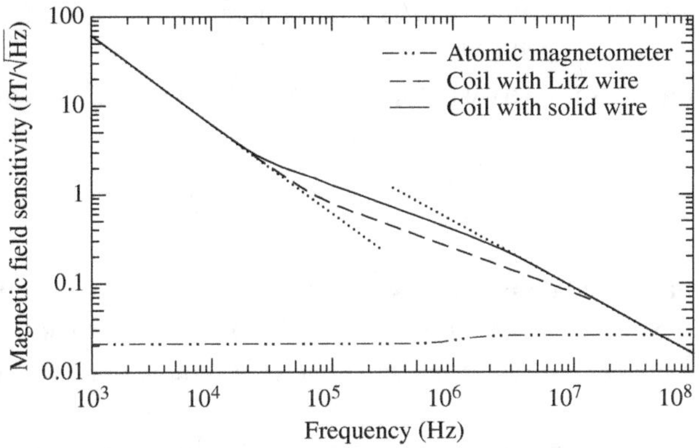

Figure 14.1 Sensitivity as a function of frequency for a surface pickup coil with volume 16 cm^3, compared to a potassium radiofrequency magnetometer occupying the same volume. Each device is assumed to operate at its ideal sensitivity limit, and the number of turns and diameter of wire for the coil are optimized at each frequency. (From Ref. [29].)

14.2 The NMR Hamiltonian

A detailed discussion of the varied phenomena in nuclear magnetic resonance can be found in excellent books such as those by Slichter [1] and Ernst, Bodenhausen, and Wokaun [2]. A complete discussion is beyond the scope of this chapter, so we will confine ourselves to a very brief review. In NMR, there are a variety of contributions to the nuclear spin Hamiltonian. The Zeeman term, describing the interaction of the nuclear spins with an applied magnetic field, is given by

$$H_Z = \sum_i \hbar\gamma_i(1-\sigma_i)\mathbf{I}_i \cdot \mathbf{B}. \tag{14.1}$$

Here γ_i is the gyromagnetic ratio of the ith spin, with spin $\mathbf{I}_i$, $\mathbf{B}$ is the magnetic field, and σ_i is the chemical shift. In conventional NMR, the Zeeman term usually dominates all others by many orders of magnitude. Typical proton chemical shifts are on the order of 10 ppm.

Nuclear quadrupole interactions for spins with $I > 1/2$ arise due to the interaction of the nuclear electric quadrupole moment with a molecule's internal electric field gradient,

$$H_Q = \frac{eQ}{6I(2I-1)} \sum_{\alpha,\beta} \frac{\partial^2 V}{\partial x_\alpha \partial x_\beta} \left[\frac{3}{2}(I_\alpha I_\beta + I_\beta I_\alpha) - \delta_{\alpha\beta} I^2 \right]. \tag{14.2}$$

Here e is the charge of the proton, Q is the quadrupole moment of the nucleus, and V is the electrostatic potential associated with the local electronic structure. The energy scale associated with nuclear quadrupole resonance (NQR) can be as high as hundreds of MHz. For ^{14}N, with $I = 1$, of particular relevance to detection of explosives [30], typical NQR frequencies are on the order of several MHz. In liquid or gas phase NMR, molecular tumbling rapidly averages to zero the effects of H_Q on coherent evolution. Residual quadrupole couplings may, however, contribute to relaxation.

Dipole–dipole interactions between the ith and jth spins have the form

$$H_D = \frac{\hbar\gamma_i\gamma_j}{r_{ij}^3}\left[\mathbf{I}_i \cdot \mathbf{I}_j - 3(\mathbf{I}_i \cdot \hat{\mathbf{r}}_{ij})(\mathbf{I}_j \cdot \hat{\mathbf{r}}_{ij})\right]. \tag{14.3}$$

Here r_{ij} is the distance separating the two nuclei and $\hbar\gamma_i\gamma_j/r_{ij}^3$ are on the order of kHz. Similar to quadrupole interactions, coherent evolution due to dipole–dipole interactions often vanishes in liquids, although it is responsible for relaxation and nuclear Overhauser effects [1].

Finally, the smallest contributions to the NMR Hamiltonian are scalar spin–spin couplings, of the form

$$H_J = \hbar\sum_{ij} J_{ij}\mathbf{I}_i \cdot \mathbf{I}_j. \tag{14.4}$$

Such couplings are known as J couplings and are the result of a second-order hyperfine interaction due to overlap of the electronic wave function with two nuclei. For ^{13}C–^{1}H connected by a single bond, these couplings are usually on the order of 150–200 Hz. The size of the coupling decreases rapidly with the number of intervening bonds and is typically not observable through more than five bonds. J couplings involving heavy nuclei can be substantially larger. NMR spectra exhibiting such couplings can be used to obtain important information about molecular spin topology as well as bond strength, angle, and torsion.

14.3 Challenges associated with detection of NMR using atomic magnetometers

One of the primary difficulties associated with using atomic magnetometers for detection of NMR is that the gyromagnetic ratio of alkali spins differs from those of nuclear spins by two to three orders of magnitude. For the case of ^{87}Rb, with nuclear spin $I = 3/2$, $\gamma_{\text{alkali}}/2\pi = g_s\mu_B/(2I+1)/h \approx 700$ kHz/G, while the gyromagnetic ratio of protons is $\gamma/2\pi = 4.257$ kHz/G. The response of the magnetometer to weak oscillating signals, such as those generated by precessing nuclei, typically has a Lorentzian profile about the alkali Larmor precession frequency. Depending on the magnetometer configuration, this may be in the range of 0.01–200 Hz. Hence, if the alkali spins are immersed in the same magnetic field (greater than several mG) as the NMR sample, the signal due to nuclear spins precessing around the magnetic field will fall far outside the magnetometer's bandwidth. There are several ways around this difficulty.

Remote detection: Another method for dealing with this issue is to use a technique known as remote detection of NMR, in which polarization, evolution, and detection occur in physically separated regions [31–34] (see Chapter 13 and Section 14.4). This technique has the advantage that these processes can be independently optimized, as discussed in more depth in the next section.

Piercing solenoid: One can create two regions of magnetic field by employing a "piercing" solenoid [35], so that the resonance frequencies of the magnetometer and the nuclear spins can be tuned independently. This was the technique employed in Refs. [29, 36], to detect proton NMR signals in water at frequencies of about 20 Hz and 62 kHz, respectively. More details are discussed in Section 14.5.

Flux transformer: Similar to the case of SQUIDs, one can employ a flux transformer to couple flux from the nuclear magnetization to the alkali-vapor spins. This is simply two coils with appropriate inductances hooked in parallel, one acting as a pickup from the nuclear spins and the other acting as an input to the atomic magnetometer.

Zero- and near-zero-field NMR: In zero or very-near-zero magnetic field, the Larmor precession frequencies of alkali spins are sufficiently small that the magnetic field generated by the nuclei falls within the magnetometer bandwidth. Chemical shifts are nearly nonexistent, but scalar couplings remain and can generate observable magnetization. Operating at zero magnetic field is appealing because of the exceptional temporal and spatial magnetic field homogeneity, which yields narrow lines, facilitating accurate determination of coupling constants. This is the approach used in Refs. [37–39], and discussed in more depth in Section 14.8.

14.4 Remote detection

In conventional NMR, the processes of polarization, excitation/encoding, and detection all occur in the same physical location, using the same coil to apply RF pulses and to detect the nuclear spins. In many cases, the requirements for the excitation and the detection differ, leading to nonoptimal performance in one of these stages. The idea of remote detection [31] solves this problem by spatially separating the processes of polarization, excitation, and detection, enabling independent optimization of each stage. A schematic of this concept is presented in Fig. 14.2. In this case, the sample flows first into a prepolarizing magnet,

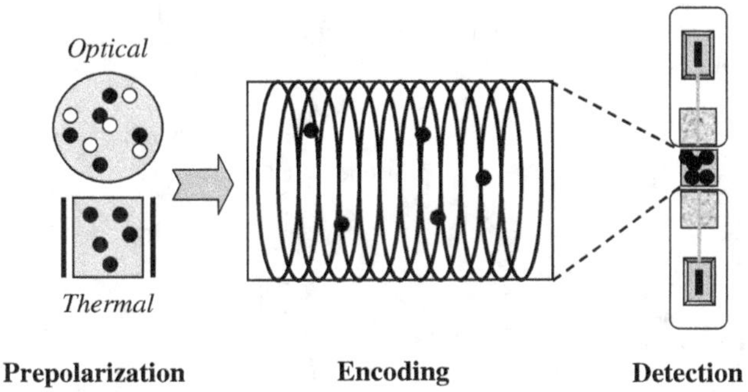

Figure 14.2 Remote-detection NMR and MRI, showing spatially separated polarization, encoding, and detection stages. In this example, detection is performed using two alkali vapor cells acting as a gradiometer. (From Ref. [40].)

with fields of 1–2 T easily obtainable using a permanent magnet without concern about the field homogeneity; the sample then flows into the encoding region in the Earth's field, where appropriate pulses and imaging gradients are applied to encode spectroscopic or spatial information in the longitudinal magnetization of the spins; finally, the sample flows into the detection region, where the longitudinal magnetization is measured (in this case, by a magnetometer, potentially at or near zero field). Information is encoded point-by-point, for instance, by considering individual points of a free-induction decay or individual points in the k-space of an image, and the encoded information is read out by taking the Fourier transform of the time-varying sample magnetization. An in-depth review of remote-detection NMR is given in Ref. [41].

Remote detection solves some of the practical problems of directly detected NMR performed with atomic magnetometers. Encoding can be conducted in the Earth's field while the magnetometer, in a shielded environment, is unaffected by the pulses and gradients applied to the sample in the encoding region. Imaging can be performed inside objects where direct placement of a sensor is impractical or impossible. On the other hand, the necessity of conducting measurements point-by-point can significantly increase the duration of an experiment. In addition to the applications described here, remote detection has been used with inductive coils for imaging and flow characterization in microfluidic chips [42, 43].

A magnetometer designed for remote-detection NMR is described in detail in Ref. [40]. It is based on nonlinear magneto-optical rotation [44, 45] (see Chapter 6) and consists of two rubidium cells each with volume 1 cm^3 acting as a gradiometer, with the difference signal giving a sensitivity of about 80 $fT/\sqrt{Hz}$ inside magnetic shielding. Water flows first into a 0.7 T prepolarization magnet and then into an encoding region with volume about 0.5 cm^3, before flowing into a detection region with volume 0.5 cm^3 positioned between the rubidium cells (Fig. 14.3). The small field produced by the water magnetization has

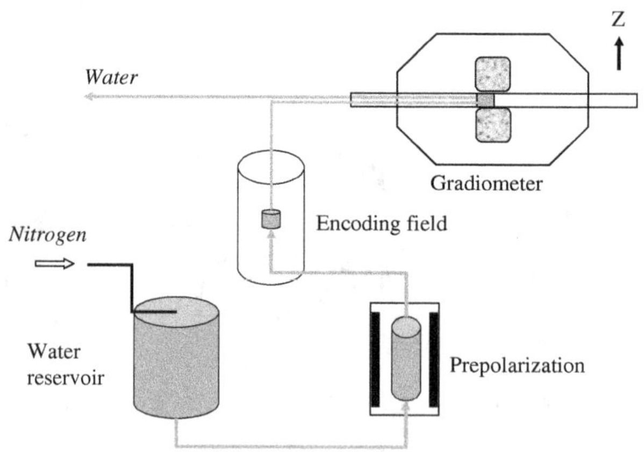

Figure 14.3 Flow of water in a remote-detection NMR experiment, with water magnetization detected by a rubidium magnetometer acting as a gradiometer. (From Ref. [40].)

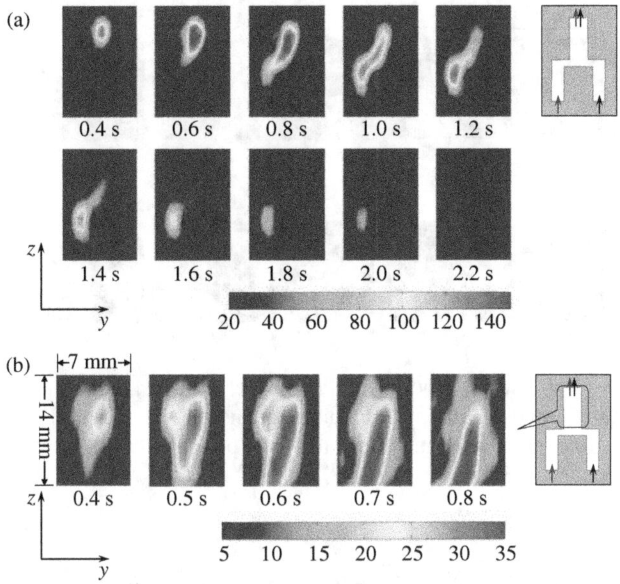

Figure 14.4 (a) Remotely detected images showing mixing of polarized and unpolarized water entering from the lower left and right channels, respectively. (b) Zoomed-in image of the mixing channel with resolutions of 0.7 mm and 3 mm along the y and z axes, respectively. (Adapted from Ref. [33].)

opposite projection onto the ~mG bias field of the magnetometer at the locations of the two cells, so that subtraction of the magnetometer signals to remove common-mode noise has the effect of adding the NMR signals. Not shown is a long "piercing" solenoid that extends through the magnetic shields, providing a leading field of approximately 0.5 G to ensure adiabatic flow of the water spins without significantly affecting the field experienced by the rubidium atoms.

Imaging was initially demonstrated using an encoding field of 31 G [32], but imaging in the Earth's field is possible due to its relatively high homogeneity. One concern when imaging at low fields is that concomitant gradients can result in severe image distortion if the strength of the imaging gradients is comparable to the magnitude of the ambient field [46,47], although this can be compensated for with appropriate pulse sequences [48]. Nevertheless, at 0.5 G, remotely detected images have been acquired using the magnetometer without special compensation for concomitant gradients, with resolution as fine as 0.7 mm, as shown in Fig. 14.4 [33]. Here, mixing between two input channels is observed, with prepolarized spins flowing into one channel and giving a clear magnetization signal, and unpolarized spins flowing into the second channel and giving no observable signal. Time-resolved flow images are possible because of different times required for flowing water to reach the detector from different parts of the encoding region.

Remote detection using Earth's field with magnetometers also allows for characterization and imaging inside metallic objects, similar to the demonstrations with SQUIDs. Water (or

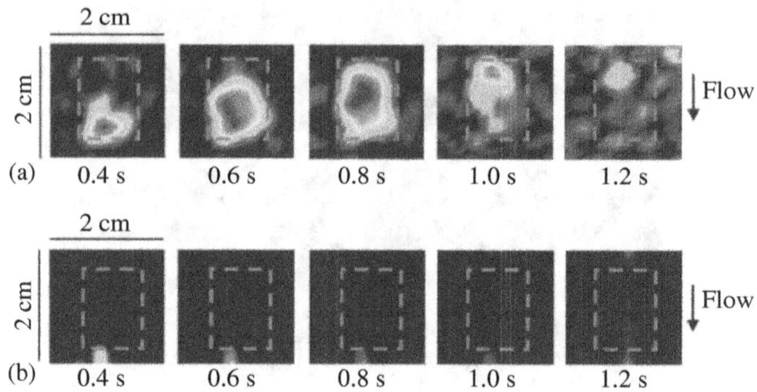

Figure 14.5 Images of flow inside a porous stainless steel cylinder, taken with (a) remote detection with encoding at 0.5 G and detection by a rubidium magnetometer and (b) direct detection by a 7 T NMR spectrometer. Imaging inside metallic materials is possible using Earth's field but not at high magnetic fields. (From Ref. [34].)

another fluid) flowing inside the metal can be encoded using audio-frequency pulses without significant attenuation, and without concern for the susceptibility artifacts that can plague high-field experiments, and then flowed into the detection region elsewhere. Figure 14.5(a) shows images of water flow inside a porous stainless steel structure, recorded using remote detection with a magnetometer (the encoding region is in a field ~0.5 G). Images of the same phantom [Fig. 14.5(b)], recorded with a conventional high-field spectrometer, show no features [34]. One-dimensional imaging inside metal pipes has also been demonstrated [49] without even using RF magnetic field pulses.

14.5 Solenoid matching of Zeeman resonance frequencies

As mentioned above, the problem of overcoming the different resonance frequencies of the alkali magnetometer's spins and those of the nuclear spins in the sample can be achieved by winding a solenoid around the sample [29,36]. The solenoid nominally produces a magnetic field only inside itself, enabling the two different spin species to experience different magnetic fields, thus allowing the two different resonance frequencies to be matched. The same technique was employed to measure hyperpolarized xenon magnetization in Ref. [35], but in that work no resonance phenomena were observed, and the solenoid served primarily to maintain a quantization axis for the nuclear spins.

This technique works well when the solenoid field is relatively small. For example, in Ref. [36], the authors employed a spin-exchange-relaxation-free magnetometer (see Chapter 5) to detect proton NMR with Larmor precession frequency of about 20 Hz, obtaining $T_2^* = 1.7$ s. Modifying the configuration slightly so that the magnetometer operated in RF mode and increasing the solenoid current so that the proton resonance frequency was ≈ 60 kHz [29]

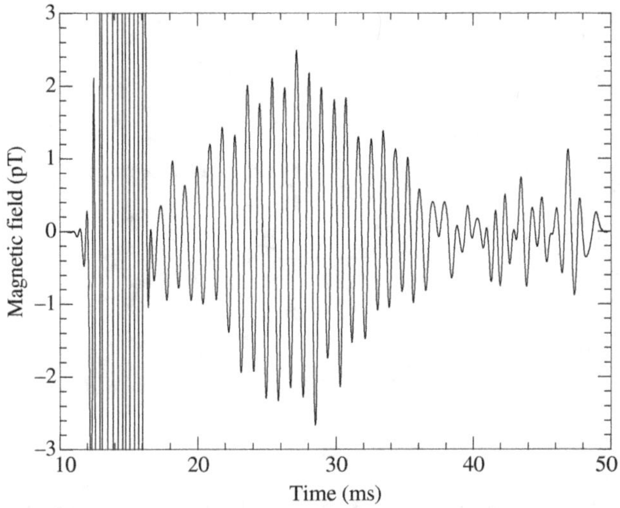

Figure 14.6 Spin-echo detected at 60 kHz proton precession frequency, detected using an RF atomic magnetometer and a solenoid to match proton and alkali resonance frequencies. Data presented here have been mixed down to about 1 kHz. Nuclear spins are tipped into the transverse plane at $t = 0$ with a $\pi/2$ pulse, a refocusing π pulse is applied at about 12 ms (the ringing from 12 to 16 ms is due to saturation of the magnetometer by the π pulse), and the echo peaks at about 12 ms. (From Ref. [29].)

resulted in much shorter lifetimes (Fig. 14.6), on the order of 9 ms, corresponding to NMR lines with half-width at half-max (HWHM) of about $1/(2\pi T_2^*) = 17$ Hz. The authors attributed the additional broadening to inhomogeneities associated with variations in the thickness of the wire.

14.6 Flux transformer

Recent work [50] has demonstrated magnetic resonance imaging with atomic magnetometers using room-temperature copper pickup and input coils to couple the flux due to the nuclear spins with the atomic magnetometer. The experimental setup is shown in the top panel of Fig. 14.7. The pickup and input coils are physically isolated, connected only through copper wires. This enables the user to independently control the magnetic field in either region, and provides some degree of enhancement of the magnetic field felt by the magnetometer due to the nuclear spins. The enhancement depends on the geometry and inductance of the two coils. A magnetic resonance image of a water phantom obtained using this technique is shown on the bottom right of Fig. 14.7. For comparison, an image of the same phantom obtained using SQUIDs is shown in the bottom left. The proton precession frequency was 3.2 kHz, and both images were acquired in 12 minutes.

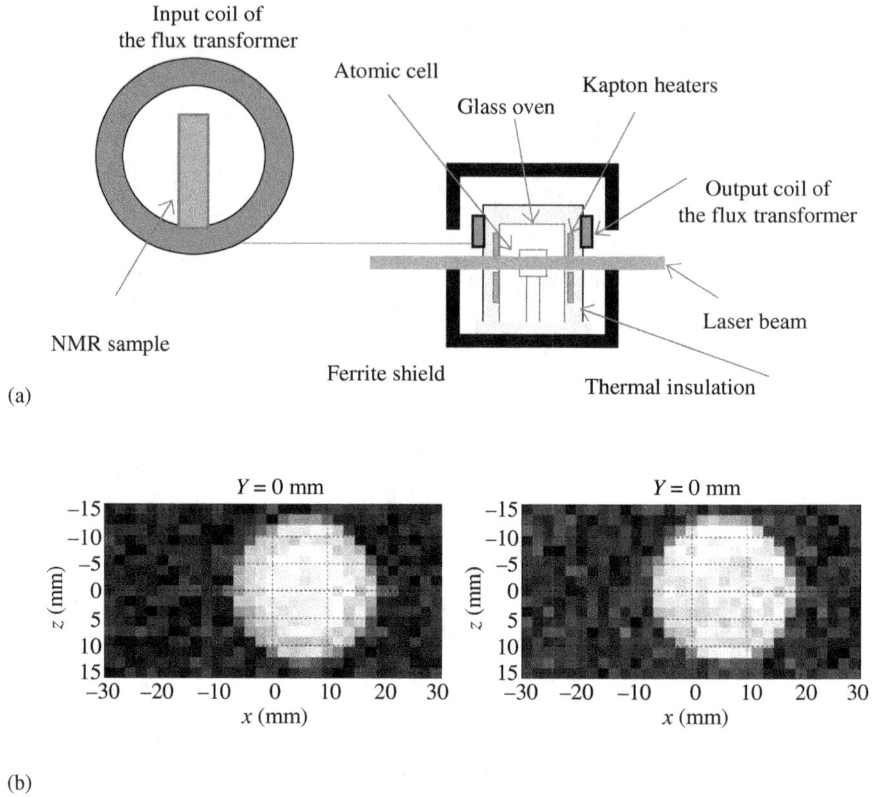

Figure 14.7 (a) Experimental setup for imaging with an atomic magnetometer and a flux transformer. (b) Images of a water phantom obtained with an atomic magnetometer and flux transformer (right panel) and SQUID magnetometer (left panel). (From Ref. [50].)

14.7 Nuclear quadrupole resonance

As mentioned in Section 14.2, nuclear quadrupole resonance (NQR) arises due to interaction of a nucleus with spin $I \geq 1$ with the local electric field gradient. The primary application of NQR is in detection of explosives, since many explosive compounds contain ^{14}N with nuclear spin $I = 1$ [30]. NQR frequencies for ^{14}N span a wide range, from kHz to several MHz. For such low frequencies, atomic magnetometers offer potentially large improvements in sensitivity compared to traditional pickup-coil detection. Reference [51] reports detection of NQR in ammonium nitrate (NH_4NO_3) using spin-locking pulses and an RF magnetometer tuned to the NQR resonance frequency at 423 kHz. There is roughly 10-fold improvement in sensitivity compared to that achieved with a tuned pickup coil.

14.8 Zero-field nuclear magnetic resonance

Zero-field NMR has a long history, starting with Pines' group in the 1980s [52, 53]. In this early work samples were shuttled back and forth between high field (for polarization and readout with inductive pickup coils) and near-zero field (where spin evolution occurred). The anecdote Alex Pines relates is that an unnamed graduate student quenched a magnet, after which it was not possible to achieve an acceptable shim. As a reward, Pines declared that this student's thesis would be on zero-field NMR. This technique has not found widespread use as it requires indirect detection; each point in a free-induction decay required shuttling the sample in and out of the magnet. Nevertheless, it was quickly realized that an important advantage of operating in zero field is that some forms of inhomogeneous broadening can be eliminated. For example, broadening due to random orientation of crystallites in powder samples, or due to chemical shift anisotropy is eliminated in zero field.

There has recently been a resurgence of interest in zero-field NMR, driven by the development of high-sensitivity spin-exchange-relaxation-free atomic magnetometers. We now turn to a brief discussion of recent work in the area of liquid-state zero-field NMR [37–39], where the only contribution to the NMR Hamiltonian is the J-coupling interaction, Eq. (14.4).

14.8.1 Thermally polarized zero-field NMR J spectroscopy

In Ref. [37] direct detection of hetero- and homonuclear scalar coupling in zero magnetic field using an optical-atomic magnetometer was demonstrated. It was shown that characteristic functional groups have distinct spectra, with straightforward interpretation for molecular structure identification, allowing extension to larger molecules and to higher-dimensional Fourier NMR spectroscopy. In that work, a magnetically shielded, zero-field environment provided high absolute field homogeneity and temporal stability, allowing the authors to obtain 0.1 Hz linewidths without using spin echoes, and to determine scalar coupling parameters with a statistical uncertainty of 4 mHz. This can be compared to the case of high field, where one has to work very hard to achieve part-per-billion homogeneity, which, in a 400 MHz spectrometer, would correspond to 0.5 Hz linewidths.

How does NMR at zero magnetic field work? Assuming a liquid state, the Hamiltonian for a network of spins coupled through scalar interactions is H_J (Eq. 14.4). We define the z axis to be the direction in which the magnetometer is sensitive to magnetic fields. In this case, the observable is the z component of the magnetization of the sample: $M_z(t) = n\hbar \mathrm{Tr} \sum_j \gamma_j I_{j,z} \rho(t)$ where n is the number density of molecules, γ_j is the gyromagnetic ratio of the jth spin, and $\rho(t)$ is the density matrix. In general, the evolution of an arbitrary system of spins can be determined by diagonalizing the Hamiltonian to find the eigenstates $|\phi_a\rangle$ and eigenvalues E_a, and expressing the initial density matrix as a sum of the operators $|\phi_a\rangle\langle\phi_b|$, each of which evolves as $e^{i\omega_{ab}t}$, where $\omega_{ab} = (E_a - E_b)/\hbar$.

Because $I_{j,z}$ are vector operators with magnetic quantum number zero, observable coherences are those between states that differ by zero or one quantum of total angular momentum

F, $\Delta F = 0, \pm 1$ with $\Delta M_F = 0$. This selection rule can be used for prediction of the positions of peaks and for interpretation of spectra. For instance, consider the case of $^{13}CH_N$, where the J coupling, J_{CH}, between the ^{13}C and all N protons pairs is identical. Since the protons are all equivalent, the homonuclear J couplings can be ignored [1]. Denoting the total proton spin by K and the ^{13}C spin by S, Eq. (14.4) can be rewritten as $H_J = \hbar J_{HC} \mathbf{K} \cdot \mathbf{S}$, which has eigenstates determined by the quantum numbers F, K, S, and F_z, with eigenvalues

$$E_{F,K} = \frac{\hbar J_{HC}}{2}[F(F+1) - K(K+1) - S(S+1)]. \tag{14.5}$$

In the case where all the protons are equivalent, there is an additional constraint on the allowed transitions, $\Delta K = 0$, since one cannot produce coherences between states with different K via magnetic-field pulses. The selection rules above yield the observable quantum-beat frequencies $\omega_K = (E_{K+1/2,K} - E_{K-1/2,K})/\hbar = J_{HC}(K+1/2)$. For the methyl group, $^{13}CH_3$, there are two lines, one at J_{HC} and another at $2J_{HC}$, corresponding to coupling of the ^{13}C nucleus with the proton doublet and quadruplet states. For the methylene group, $^{13}CH_2$, there is a single line at $3J_{HC}/2$ due to coupling with the proton triplet state. In more complicated molecules, long range homo- and heteronuclear couplings can result in a splitting of the lines—however, the positions of the multiplets can be determined by the above argument. Exact spectra can be calculated via numerical diagonalization of the Hamiltonian. In the case where one coupling is much stronger than all others, perturbation theory can also be employed to obtain approximate solutions.

The zero-field NMR spectrometer used in Ref. [37] based on a microfabricated atomic magnetometer is shown in Fig. 14.8. A microfabricated ^{87}Rb alkali-vapor cell with $\sim$1300 torr of N_2 was placed inside a set of magnetic shields and optically pumped via a single circularly polarized laser, tuned to the center of the pressure-broadened D1 transition.

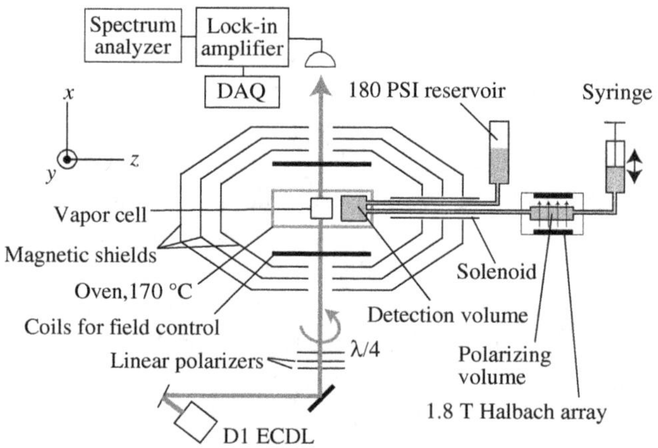

Figure 14.8 Experimental setup for detection of zero-field NMR using a single beam SERF magnetometer. ECDL = external cavity diode laser. (From Ref. [37].)

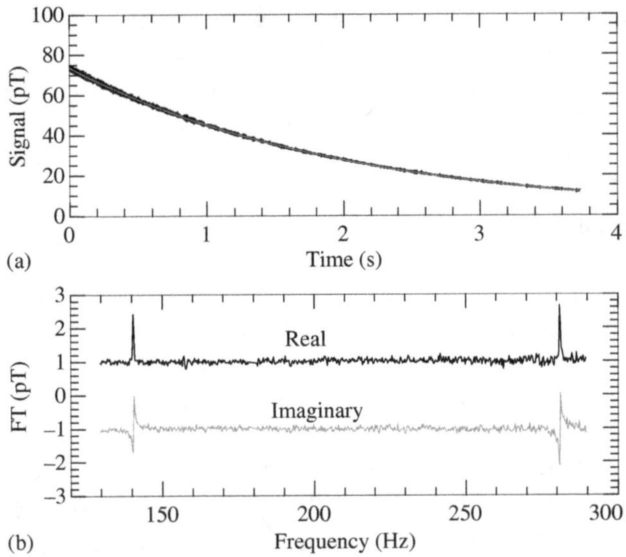

Figure 14.9 Raw magnetometer signal (a) and Fourier transform (b) for ^{13}C labeled methanol, $^{13}CH_3OH$. The raw signal in (a) has a large slowly decaying part corresponding to T_1 relaxation, and a smaller high-frequency component. The Fourier transform reveals two lines at J and $2J$, as expected for a $^{13}CH_3$ group. (From Ref. [37].)

The magnetometer operated at near zero magnetic field, in the spin-exchange-relaxation-free regime [54], using the technique demonstrated by V. Shah and co-workers [55] (Chapter 5). At zero magnetic field, the absorption of light as a function of B_z has a roughly absorptive Lorentzian profile. In order to obtain a dispersive signal, B_z was modulated with an amplitude of roughly 150 mG at about 1.8 kHz. The first harmonic of the absorption is then proportional to the static component of B_z. A set of coils was used to control and zero the magnetic fields inside the shields.

In the measurements described below, a syringe pump was used to cycle fluid back and forth between a polarization region inside a 1.8 T Halbach [56] array, and an 80 mL detection volume adjacent to the atomic magnetometer. An additional set of coils was used to apply pulses to manipulate the nuclear spins. Transport of the sample was adiabatic since no signal was observed without application of an excitation pulse.

The time domain signal (a) and zero-field NMR spectrum (b) for ^{13}C methanol, $^{13}CH_3OH$, are shown in Fig. 14.9. The time domain signal features a large slowly decaying component, corresponding to T_1 relaxation, and a higher-frequency signal due to J coupling. In the presence of OH exchange, OH coupling to the $^{13}CH_3$ part of the molecule rapidly averages to zero, effectively yielding an isolated $^{13}CH_3$ molecule. Thus, one expects two lines at J and $2J$, as discussed above, exactly as observed in Fig. 14.9(b). Either line in the spectrum is approximately 0.1 Hz (half-width at half-maximum).

Attaching additional spins to the molecule produces more complex spectra; however, the gross features of the spectra can be predicted from the discussion above. Zero-field spectra of ethanol, with a ^{13}C on either the CH_3 group (ethanol-2) or the CH_2 group (ethanol-1), are shown in Fig. 14.10. In either case, the protons on the unlabeled group couple weakly via two or three bonds to either the ^{13}C or the protons on the labeled group. Therefore, one expects that the additional spins will produce splittings of the peaks due to the labeled group. Hence, for ethanol-2, one expects two multiplets centered about $^1J_{\text{HC}}$ and $2 \times {}^1J_{\text{HC}}$, each of which is split by the long-range couplings $^2J_{\text{HC}}$ and $^3J_{\text{HH}}$, and an additional signal at low frequencies, also due to the long-range couplings. The spectrum shown in Fig. 14.10(a) displays exactly these features, and full numerical simulation, given by the smooth, lower trace, is in good agreement with experimental data. In the case of ethanol-1, the observed spectrum in Fig. 14.10(b) displays the expected multiplet centered about $3J/2$, split by the long range couplings, and an additional signal at low frequencies. The spectrum is in good agreement with the full numerical simulation, given by the smooth trace. These results indicate that a perturbative treatment of the effects of long-range couplings is promising for understanding and analyzing zero-field spectra arising from complex molecules, a subject of ongoing work.

14.8.2 Parahydrogen-enhanced zero-field NMR

In the previous section, we discussed zero-field NMR using thermally polarized samples. To an experienced NMR spectroscopist, an obvious limitation is that the signal-to-noise ratio is somewhat worse than what is usually observed in high-field NMR in the case of neat liquids. One way around this limitation is to use any number of hyperpolarization tricks, such as dynamic nuclear polarization [7], spin-exchange optical pumping [8], or parahydrogen-induced polarization (PHIP) [9–12]. In Ref. [38], zero-field NMR using parahydrogen-induced polarization was reported, resulting in dramatic enhancements in the signal-to-noise ratio.

Excellent reviews of the properties of parahydrogen and its use in nuclear magnetic resonance can be found in Refs. [10, 11], so we present only a very brief discussion here. Parahydrogen is the antisymmetric nuclear spin isomer of molecular hydrogen with zero nuclear spin angular momentum (orthohydrogen refers to the symmetric nuclear spin triplet). In the Born–Oppenheimer rigid rotor approximation, the total wave function of molecular hydrogen is given by the product of electronic, vibronic, rotational, and nuclear spin wave functions, $\Psi = \psi_e \psi_v \psi_r \psi_n$. In the ground state, the electrons have zero orbital angular momentum and are in the antisymmetric spin-singlet wave function, thereby satisfying the Pauli exclusion principle for electrons. Nuclear vibrational levels are symmetric under interchange of nuclei [57]. Hence the product of the rotational and nuclear spin wave functions must be antisymmetric. The symmetry of the rotational wave functions is given by $(-1)^j$, where j is the rotational angular momentum quantum number, and hence rotational levels with even j must have a nuclear spin wave function that is odd under interchange of nuclei, i.e., the para state. The energy associated with rotational angular

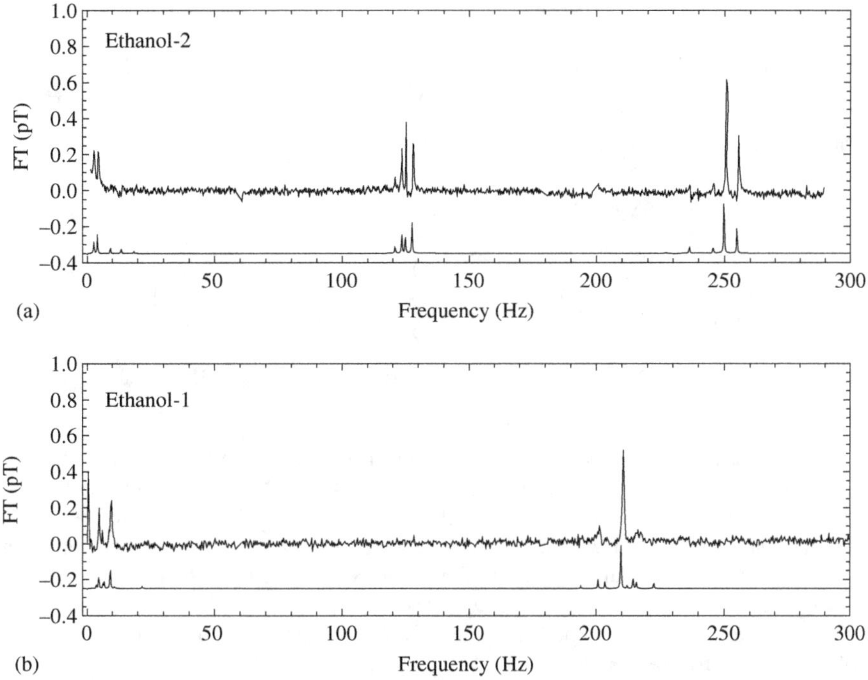

Figure 14.10 Top panel: spectrum obtained with singly labeled $^{13}CH_3{-}CH_2{-}OH$ (ethanol-2). Bottom panel: spectrum obtained with singly labeled $CH_3{-}^{13}CH_2{-}OH$ (ethanol-1). The trace in the bottom of each panel shows the result of numerical simulations, in good agreement with the data. (From Ref. [37].)

momentum is given by $E_r = j(j+1)\hbar^2/(2\mu r_{12}^2)$, where $\mu = m_p/2$ is the reduced mass, m_p is the proton mass, and r_{12} is the internuclear separation. In units of temperature $\hbar^2/(2\mu r_{12}^2)/k = 87.3$ K, and so the separation between the lowest two rotational energy levels corresponds to a temperature of 174.6 K. In thermal equilibrium at 30 K, nearly 99% of H_2 molecules are in the lowest rotational state with $j = 0$, and hence the nuclear spin para state. Transitions between the ortho and para states are magnetically forbidden, so that once polarized, the lifetime of parahydrogen in an aluminum tank can be on the order of weeks. In the presence of a paramagnetic catalyst, however, the magnetic equivalence of the two protons can be broken, enabling rapid thermal equilibration. Cooling H_2 in the presence of such a catalyst thus provides a mechanism to rapidly produce large quantities of parahydrogen.

The Pauli-enhanced order of parahydrogen is very appealing when compared to the thermal polarization of protons in a 10 T magnet, which is on the order of 10^{-5}. However, parahydrogen by itself is useless in the context of NMR, as the singlet state has no magnetic moment. Bowers and Weitekamp [9] realized some time ago that the singlet order of parahydrogen can be harnessed by catalytic addition of parahydrogen to a substrate molecule. If

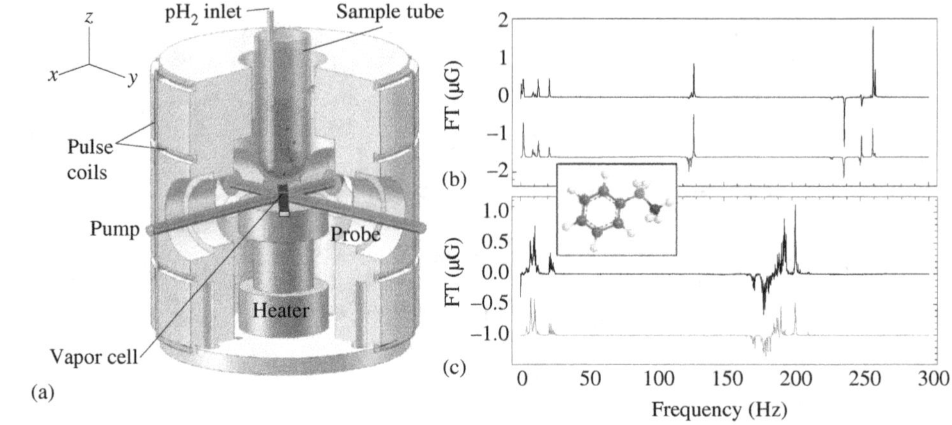

Figure 14.11 (a) Experimental setup used for performing zero-field NMR experiments with parahydrogen-induced polarization. (b) Experimental zero-field NMR spectrum arising from ethylbenzene labeled with ^{13}C in the methyl (CH_3) group (upper trace). The ethylbenzene was polarized via addition of parahydrogen to labeled styrene. The lower trace shows the result of a numerical simulation, in good agreement with the data. (c) Zero-field NMR spectrum arising from ethylbenzene, labeled with ^{13}C in the methylene (CH_2) group (upper trace). Again, the lower trace shows the result of a numerical simulation, also in good agreement with the data. Spectra shown in (b) and (c) are the result of just a single transient. (From Ref. [38].)

the reaction occurs such that both hydrogen atoms attach to the same substrate molecule, the parahydrogen-derived spins remain in the singlet state. In the presence of chemical shift differences or asymmetries in scalar couplings to a heteronucleus, symmetry is broken, yielding an observable signal with very large enhancement compared to that of thermal polarization. Since Bowers and Weitekamp's initial demonstration, parahydrogen-induced polarization has been investigated in a variety of molecules and magnetic fields. In most cases, observation of the resulting polarization and free-induction decay are carried out in a high-field magnet, although the details of the resulting spectra depend on whether the hydrogenation is carried out in the Earth's field or high field, the so-called ALTADENA (adiabatic longitudinal transport after dissociation engenders net alignment) or PASADENA (parahydrogen and synthesis allows dramatically enhanced nuclear alignment) experiments, respectively.

In Ref. [38], direct observation of PHIP in a zero-field environment using an atomic magnetometer was reported. The resulting signal intensities were enhanced dramatically with respect to thermal polarization in a 1 T magnet, enabling observation of high-resolution, zero-field spectra in samples with ^{13}C in natural abundance. These measurements were performed using the apparatus shown in Fig. 14.11. Parahydrogen was bubbled through a test tube containing a substrate molecule and catalyst. The measurements presented here used styrene as the substrate, which forms ethylbenzene upon addition of two hydrogen atoms. The test tube was located adjacent to a microfabricated ^{87}Rb vapor cell (also containing about 1300 torr of N_2). A circularly polarized laser beam, tuned to the center of the D1 resonance, optically pumped the alkali vapor. A linearly polarized probe beam, tuned off resonance (by about twice the pressure broadened linewidth) was used to probe the alkali spins, resulting in magnetic sensitivity of about 15 $\text{fT}/\sqrt{\text{Hz}}$. A set of coils was used to null the residual fields inside the magnetic shields, and an additional set of coils was used to apply DC pulses to the spins in order to generate NMR coherences.

Figure 14.11(b) and (c) shows the zero-field PHIP spectra arising from ethylbenzene isotopomers with ^{13}C in either the methyl (CH_3) or methylene (CH_2) groups, the β and α labels, respectively. For the case of the β label, shown in Fig. 14.11(a), the spectrum is qualitatively similar to that observed with thermally polarized ethanol with the CH_3 group labeled (Fig. 14.10, top panel): One observes multiplets at $^1J_{HC}$ and $2\times(^1J_{HC})$. The splitting in these multiplets is determined by long-range couplings, which also give rise to the signal in the low-frequency part of the spectrum. Aside from the dramatic improvement in signal-to-noise ratio compared to the thermally polarized case of ethanol-2 shown in Fig. 14.10(a), the zero-field PHIP (ZF-PHIP) spectrum of ethylbenzene also differs in the relative phase of the peaks, owing to the different initial conditions. The ZF-PHIP spectrum of ethylbenzene-α is qualitatively similar to that of ethanol-1, with a multiplet centered about $3/2\times(^1J_{HC})$ and an additional signal at low frequency due to the long-range couplings. The structure of the multiplets is considerably more complicated than that of ethylbenzene-β, due to large additional couplings to spins on the benzene ring. For both isotopomers, numerical simulations, presented in the lower traces in each panel, are in good agreement with the experimental data.

14.8.3 Zeeman effects on J-coupled multiplets

The question arises whether the magnetic fields really are zeroed? For most of what we have discussed so far, the data are completely consistent with all fields being zeroed. However, it is natural to ask what is the effect of fields that are close to, but not quite, zero? This question is addressed in Ref. [39]. For magnetic fields that are sufficiently small that the Zeeman-induced energy splittings are much smaller than the J couplings, the effect of the magnetic field can be treated as a small perturbation. This results in splitting patterns that are easy to understand and that can be calculated via first-order perturbation theory. The case of a CH_3 group is shown in Fig. 14.12. For these data, a magnetic field is applied in the z direction, and the x component of the magnetization is measured using the atomic magnetometer. The bottom trace shows the spectrum for zero magnetic field, consisting of a zero-frequency peak, and peaks at J and $2J$. Applying a small magnetic field produces splittings of these lines as follows: The peak at J, arising due to transitions between states with total angular momentum $F=1$ and $F=0$ and total proton spin $K=1/2$, is split into a doublet, corresponding to transitions with $\Delta M_F=\pm1$. The peak at $2J$, arising due to transitions between states with total angular momentum $F=2$ and $F=1$ and total proton spin $K=3/2$, is split into six lines, corresponding to transitions with $\Delta M_F=\pm1$ for each of the three states in the $F=1$ manifold. Finally, the zero-frequency peak splits into three

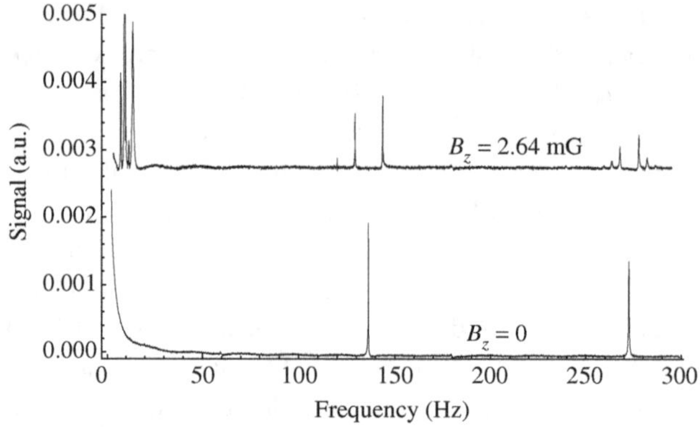

Figure 14.12 Zero- and near-zero-field NMR spectrum of 2-acetonitrile, $^{13}CH_3CN$. The splitting patterns in the near-zero-field case are discussed in the text. (Figure from Ref. [39].)

lines, corresponding to transitions with $\Delta F = 0$ and $\Delta M_F = \pm 1$ for the three different manifolds ($F = 2$, $K = 3/2$, $F = 1$, $K = 3/2$, and $F = 1$, $K = 1/2$). The positions of all the lines in the spectrum can be accurately determined by first-order perturbation theory.

These results indicate that near-zero-field NMR (NZF NMR) can be used to gain additional information about the degeneracy of the zero-field levels, and also about the gyromagnetic ratios of the spins, and is potentially a useful diagnostic for understanding pure zero-field spectra. There are two additional points to emphasize: (1) This type of spectroscopy shares a great deal in common with atomic spectroscopy, where internal energy levels typically dominate the Hamiltonian and Zeeman splittings are a small perturbation. (2) A potential concern regarding NMR spectroscopy in zero field is that one loses chemical shift information. The presence of a J coupling to a heteronucleus can produce resolved line splittings, so that chemical shifts can be observed, even when they are smaller than the linewidth. This has already been demonstrated in magnetic fields of about 40 G [58], and ongoing work based on atomic magnetometers is attempting to detect such effects in mG magnetic fields.

14.9 Conclusions

Low- and zero-field NMR have recently attracted considerable attention with the advent of (1) high-sensitivity atomic magnetometers and (2) methods for pre- or hyperpolarization of samples. This chapter has briefly covered some of the more recent contributions in which atomic magnetometers have been applied to detection of low- and zero-field NMR. These techniques are a promising tool for chemical analysis and imaging. NMR in general is a very mature field, but the vast majority of research in the field has been performed in very high magnetic fields. Hence, there may be a wealth of phenomenology to uncover in the world of low- and zero-field NMR.

Acknowledgments

The work of Igor Savukov was sponsored by NIH Grant 5 R01 EB009355.

References

[1] C. P. Slichter, *Principles of Magnetic Resonance*, 3rd ed. (Springer-Verlag, New York, Berlin, Heidelberg, 1990).
[2] R. R. Ernst, G. Bodenhausen, and A. Wokaun, *Principles of Nuclear Magnetic Resonance in One and Two Dimensions* (Oxford University Press, New York, 1987).
[3] J. W. Akitt and B. E. Mann, *NMR and Chemistry: An Introduction to Modern NMR Spectroscopy,* 4th ed. (Stanley Thomas, Cheltenham, UK, 2000).
[4] M. H. Levitt, *Spin Dynamics* (Wiley, Chichester, UK, 2001).
[5] E. Fukushima and S. B. Roeder, *Experimental Pulse NMR: A Nuts and Bolts Approach* (Addison-Wesley, London, 1981).
[6] W. C. Griffith *et al.*, Phys. Rev. Lett. **102**, 101601 (2009).
[7] A. Abragam and M. Goldman, Rep. Prog. Phys. **41**, 395 (1978).
[8] T. G. Walker and W. Happer, Rev. Mod. Phys. **69**, 629 (1997).
[9] C. R. Bowers and D. P. Weitekamp, Phys. Rev. Lett. **57**, 2645 (1986).
[10] J. Natterer and J. Bargon, Prog. in Nucl. Magn. Reson. Spectrosc. **31**, 293 (1997).
[11] C. R. Bowers, in *Encyclopedia of Nuclear Magnetic Resonance*, edited by D. M. Grant and R. K. Hascis (Wiley, Chichester, 2002), vol. 9, p. 750.
[12] D. Canet *et al.*, Concepts Magn. Reson. Part A **28A**, 321 (2006).
[13] S. Appelt, H. Kühn, F. W. Häsing, and B. Blümich, Nature Physics **2**, 105 (2006).
[14] J. N. Robinson, A. Coy, R. Dykstra, C. D. Eccles, M. W. Hunter, and P. T. Callaghan, J. Magn. Reson. **185**, 343 (2006).
[15] M. E. Halsea, A. Coy, R. Dykstra, C. Eccles, M. Hunter, R. Ward, and P. T. Callaghan, J. Magn. Reson. **182**, 75 (2006).
[16] J. Clarke and A. I. Braginski, ed., *The SQUID Handbook*, Vol. 1 (Wiley-VCH, Weinheim, 2004).
[17] Ya. S. Greenberg, Rev. Mod. Phys. **70**, 175 (1998).
[18] H. C. Seton, J. M. S. Hutchison, and D. M. Bussell, Meas. Sci. Technol. **8**, 198 (1997).
[19] S. Saxena, A. Wong-Foy, A. J. Moule, J. A. Seeley, R. McDermott, J. Clarke, and A. Pines, J. Am. Chem. Soc. **123**, 8133 (2001).
[20] R. McDermott, A. H. Trabesinger, M. Mück, E. L. Hahn, A. Pines, and J. Clarke, Science **295**, 2247 (2002).
[21] R. McDermott, S. K. Lee, B. ten Haken, A. H. Trabesinger, A. Pines, and J. Clarke, Proc. Nat. Acad. Sci. USA **101**, 7857 (2004).
[22] J. Clarke, M. Hatridge, and M. Mössle, Annu. Rev. Biomed. Eng. **9**, 389 (2007).
[23] A. N. Matlachov, P. L. Volegov, M. A. Espy, J. S. George, and R. H. Kraus, Jr., J. Magn. Reson. **170**, 1 (2004).
[24] M. Mössle, S.-I. Han, W. R. Myers, S.-K. Lee, N. Kelso, M. Hatridge, A. Pines, and J. Clarke, J. Magn. Reson. **179**, 146 (2005).
[25] M. Pannetier, C. Fermon, G. Le Goff, J. Simola, E. Kerr, Science **304**, 1648 (2004).
[26] H. Dyvorne, C. Fermon, M. Pannetier-Lecoeur, H. Polovy, and A.-L. Walliang, IEEE Trans. Appl. Supercon. **19**, 819 (2009).
[27] M. Pannetier-Lecoeur, C. Fermon, H. Dyvorne, J. F. Jacquinot, H. Polovy, and A. L. Walliang, J. Magn. Magn. Mater. **322**, 1647 (2010).

[28] M. Pannetier-Lecoeur, C. Fermon, N. Biziere, J. Scola, and A. L. Walliang, IEEE Trans. Appl. Supercon. **17**, 598 (2007).
[29] I. M. Savukov, S. J. Seltzer, and M. V. Romalis, J. Magn. Reson. **185**, 214 (2006).
[30] A. N. Garroway *et al.*, IEEE Trans. Geosci. Remote Sensing **39**, 1108 (2001).
[31] A. Moulé, M. Spence, S.-I. Han, J. Seeley, K. Pierce, S. Saxena, and A. Pines, Proc. Nat. Acad. Sci. USA **100**, 9122 (2003).
[32] S. Xu, V. V. Yashchuk, M. H. Donaldson, S. M. Rochester, D. Budker, and A. Pines, Proc. Nat. Acad. Sci. USA **103**, 12668 (2006).
[33] S. Xu, C. W. Crawford, S. Rochester, V. Yashchuk, D. Budker, and A. Pines, Phys. Rev. A **78**, 013404 (2008).
[34] S. Xu, E. Harel, D. J. Michalak, C. W. Crawford, D. Budker, and A. Pines, J. Magn. Reson. Imaging **28**, 1299 (2008).
[35] V. V. Yashchuk *et al.*, Phys. Rev. Lett. **93**, 160801 (2004).
[36] I. M. Savukov and M. V. Romalis, Phys. Rev. Lett. **94**, 123001 (2005).
[37] M.P. Ledbetter *et al.*, J. Magn. Reson. **199**, 25 (2009).
[38] T. Theis *et al.*, Nature Physics **7**, 571 (2011).
[39] M. P. Ledbetter *et al.*, Phys. Rev. Lett. **107**, 107601 (2011).
[40] S. Xu, S. M. Rochester, V. V. Yashchuk, M. Donaldson, and D. Budker, Rev. Sci. Instrum. **77**, 083106 (2006).
[41] E. Harel, L. Schröder, and S. Xu, Annu. Rev. Anal. Chem. **1**, 133 (2008).
[42] E. Harel and A. Pines, J. Magn. Reson. **193**, 199 (2008).
[43] V. S. Bajaj, J. Paulsen, E. Harel, and A. Pines, Science **330**, 1078 (2010).
[44] D. Budker, D. F. Kimball, S. M. Rochester, V. V. Yashchuk, and M. Zolotorev, Phys. Rev. A **62**, 043403 (2000).
[45] D. Budker, D. F. Kimball, V. V. Yashchuk, and M. Zolotorev, Phys. Rev. A **65**, 055403 (2002).
[46] D. G. Norris and J. M. S. Hutchinson, Magn. Reson. Imaging **8**, 33 (1990).
[47] D. A. Yablonskiy, A. L. Sukstanskii, and J. J. H. Ackerman, J. Magn. Reson. **174**, 279 (2005).
[48] N. Kelso, S.-K. Lee, L.-S. Bouchard, V. Demas, M. Mück, A. Pines, and J. Clarke, J. Magn. Reson. **200**, 285 (2009).
[49] C. W. Crawford, S. Xu, E. J. Siegel, D. Budker, and A. Pines, Appl. Phys. Lett. **93**, 092507 (2008).
[50] I. M. Savukov *et al.*, J. Magn. Reson. **199**, 188 (2009).
[51] S. K. Lee, K. L. Sauer, S. J. Seltzer, O. Alem, and M. V. Romalis, Appl. Phys. Lett. **89**, 214106 (2006).
[52] D. B. Zax, A. Bielcki, K. W. Zilm, A. Pines, and D. P. Weitekamp, J. Chem. Phys. **83**, 4877 (1985).
[53] C. J. Lee, D. Suter, and A. Pines, J. Magn. Reson. **75**, 110 (1987).
[54] I. K. Kominis, T. W. Kornack, J. C. Allred, and M. V. Romalis, Nature **422**, 596–599 (2003).
[55] V. Shah, S. Knappe, P. D. D. Schwindt, and J. Kitching, Nature Photonics **1**, 649 (2007).
[56] K. Halbach, Nucl. Instrum. Methods **169**, 1 (1980).
[57] J. M. Brown, *Molecular Spectroscopy* (Oxford University Press, Oxford, 1998).
[58] S. Appelt *et al.*, Chem. Phys. Lett. **485**, 217 (2010).

15

Space magnetometry

B. Patton, A. W. Brown, R. E. Slocum, and E. J. Smith

15.1 Introduction

Magnetometry has been an invaluable tool in all stages of space exploration, from the first ionospheric sounding rockets to the most modern interplanetary probes. Our solar system is fundamentally a magnetically active environment – indeed, one might define the extent of the solar system by its heliopause, as it is the magnetic influence of the Sun which separates us from the interstellar medium. The interactions between the solar wind and the bodies of the solar system are varied and complex, and they have strong implications for the past and future of these bodies. Most importantly, a planet's magnetic field is one of the few characteristics which can be measured from space to yield information about the nature and dynamics of its interior. Recognizing these scientific imperatives, mission designers have included precise magnetometers on nearly all the spacecraft used to explore our solar system; this in turn has driven advances in magnetometer technology over the past fifty years.

15.1.1 Achievements of space magnetometry

Discoveries made by space magnetometers have been among the most profound achievements of space exploration. Rocket-borne magnetometers gave the first definitive evidence of electrical currents in the Earth's ionosphere and their effect on diurnal variations of the geomagnetic field [1]. These data not only shed light on the interaction between the solar wind and the Earth; they also complemented radiation studies which mapped out the Van Allen belts and thus paved the way for manned space flight. Later spacecraft magnetometers advanced dynamo theory by confirming the lack of a planet-scale dipolar field on Venus [2,3] and discovering, to much surprise, a still-active dynamo within Mercury [4]. The axisymmetric dipole of Saturn still challenges our understanding of the dynamo process, as do the

Optical Magnetometry, ed. Budker, D. and Jackson Kimball, D. F. Published by Cambridge University Press.

strongly tilted and non-dipolar fields of Uranus and Neptune [5]. Mars has no internally generated field at present, but its strong and spatially varying remanent field is most likely the result of a dynamo combined with plate tectonics early in its history [6, 7].[1] Perhaps the most profound discovery made by space magnetometers is the strong evidence for liquid water oceans under the icy surfaces of Jupiter's moons Europa and Callisto [10, 11]. These oceans arguably present the most hospitable environment for life elsewhere in the Solar System, and their detection serves as a strong motivation for future magnetic explorations.

15.1.2 Challenges unique to space magnetometers

In addition to the standard requirements of precision and accuracy, any magnetometer incorporated in a space mission must contend with difficulties arising from its exceptional environment:

- A spacecraft-borne magnetometer is subject to stringent mass, volume, and power constraints.
- The strength of the interplanetary magnetic field is typically less than 10 nT [12], four to five orders of magnitude weaker than planetary fields. In a single mission, a space magnetometer may be required to make high-precision measurements over several decades of field strength.
- Many satellites and space probes are spin-stabilized, in which case the magnetometer must cope with a continuously changing orientation.
- The temperature ranges and radiation levels faced by a space magnetometer are more extreme than typically encountered in terrestrial applications.
- A magnetometer sent on a space mission cannot be repaired or re-calibrated. Long-term calibration errors and DC offsets must be inferred from in-flight measurements and corrected remotely.
- The satellite or spacecraft carrying the magnetometer often itself generates substantial and time-varying fields.

Many of these issues can be mitigated by careful sensor design. To increase dynamic range, many spacecraft magnetometers operate at the *null point*. In this configuration the magnetic field at the sensor is canceled by current-carrying coils which are controlled by the magnetometer's output [13]. The measurement of the external magnetic field then becomes a problem of quantifying the current in the coils, a conversion potentially fraught with its own set of errors. Spacecraft spin has motivated the development of scalar magnetometers with minimal heading error and high-bandwidth vector sensors able to measure rapidly changing field components. The temperature range experienced by spacecraft instruments has been a driving force in the search for saturable-core sensors with low temperature coefficients [14] and remains a concern for alkali-based atomic magnetometers. Long-term calibration and DC offset issues can be resolved with the use of atomic sensors or by reorienting the axes of

[1] Upon losing this dynamo, Mars lost most of its atmosphere to solar-wind sputtering and transformed from a warm and water-covered world to the desert planet we see today [8, 9].

a three-component vector sensor to cross-check zero levels. Ultimately it is stray spacecraft fields which effectively limit magnetometric sensitivity in most space missions, despite efforts to subtract these fields from the raw magnetometer data [15]. This issue requires great attention to detail during the mission planning phase and, if costs permit, extensive testing prior to launch [16].

No single magnetometer technology is uniquely suited to addressing all of the above challenges. It is therefore typical to include multiple magnetic sensors in each space mission.

15.1.3 Magnetic sensors used in space missions

The simplest magnetometer which has been deployed in space is the search-coil magnetometer, consisting of a coil or solenoid wrapped around a magnetically permeable core. These magnetometers are innately sensitive to AC fields; DC fields can be measured by mounting the sensor with its axis perpendicular to the spin axis of the satellite or rocket [17]. Due to their low sensitivity, search-coil magnetometers have generally not been used for precision magnetometry. Fluxgate magnetometers were the first magnetic sensors to be used in sub-orbital rocket flights [18, 19] and within a few years became the preferred technology for satellite missions [14, 20]. The vector nature, low cost of manufacture, and low power consumption of fluxgate magnetometers has made them the default choice in many space applications. The principal drawback of fluxgates is not their sensitivity, but rather systematic drifts of their zero levels [13, 20]. These drifts, while small, can be comparable to the fields being measured.

Quantum magnetometers – i.e., proton-precession, alkali-vapor, and helium magnetometers – are termed absolute sensors because they detect quantum transitions whose frequencies are related to the applied magnetic field directly by fundamental constants. As such, they are often used to calibrate zero offsets in relative sensors such as fluxgates. Proton-precession magnetometers measure the nuclear magnetic resonance (NMR) frequency of ^{1}H nuclei in a sample such as water or kerosene. This concept arose shortly after the first NMR experiments [21], but it was the invention of the Overhauser magnetometer which dramatically boosted the sensitivity of these devices [22,23]. Despite this advance, Overhauser magnetometers remain limited to Earth's-field applications because of the low gyromagnetic ratio of the proton. Alkali-vapor and helium magnetometers were developed shortly before the first space flights and have played a crucial role in space magnetometry research. Their principles of operation have been discussed extensively in Chapters 1, 4, and 10. Here we cover these sensors in the context of space exploration, with particular focus on the design of spacecraft sensors and the missions to which these magnetometers have contributed.

15.2 Alkali-vapor magnetometers in space applications

15.2.1 Initial development of Earth's-field alkali magnetometers

Mere months after the first demonstration of atomic magnetometry [24, 25], laboratory alkali-vapor magnetometers were being used to measure Earth's field with unprecedented precision [26–28]. Encouraged by these successes, researchers at Varian Associates and

NASA began collaborating on the first alkali-vapor magnetometers intended for operation in space [29].

By this point not only were the advantages of alkali-vapor magnetometers abundantly clear, but so were some of their intrinsic challenges. These magnetometers were fundamentally scalar sensors, and they suffered from "dead zones" – orientations in an external field at which a magnetometer becomes magnetically insensitive. A related issue was the one of heading error, a false dependence of the field measurement on the relative orientation of the sensor to the field (see Chapters 6, 10, and 17). This error can be minimized in a laboratory-based magnetometer, but it is nearly impossible to control in spacecraft operation. Particular attention was devoted to mitigation of heading error and dead zones in the early years of alkali-vapor magnetometers, especially during their adaptation for space flight.

15.2.2 Sensor design

Although all-optical magnetometry had been demonstrated by Bell and Bloom in 1961 [30], magnetometers constructed through the 1970s generally relied upon radiofrequency coils to excite a magnetic resonance within the alkali vapor. Moreover, early alkali-vapor magnetometers relied on alkali discharge lamps to produce pump and probe light.[2] Given these design constraints, three alkali-vapor magnetometer designs were commonplace in the early 1960s: the M_z magnetometer, the M_x magnetometer (often self-oscillating), and the dual-cell self-oscillating M_x magnetometer [31] (see Chapter 4). Of these designs, the dual-cell magnetometer quickly became the instrument of choice for space missions due to its resilience to heading error and dead zones [29].

Figure 15.1 shows a schematic of the dual-cell magnetometer. Each arm of the device essentially acts like a single M_x self-oscillating sensor: modulation of the pump light intensity is detected by a photocell whose signal is amplified and used to drive the RF coil at the Larmor frequency ω_L. In each arm the pump light is circularly polarized, but because the light propagates in opposite directions in the two arms, one cell is pumped with σ_+ light and the other with σ_-. The novel quirk of this particular design, first conceived by J. T. Arnold at Varian, is that each photocell drives the RF coil of the other arm. This eliminates the need of an added $\mp 90°$ electronic phase shift which would be required in a single-cell self-oscillating magnetometer pumped with $\sigma_\pm$ light. More crucially, each cell is optically pumped to maximize the magnetic-resonance amplitude at one extreme of the nonlinear Zeeman spectrum, but RF-driven at the frequency corresponding to the other extreme of the spectrum. Thus the maximum gain of the self-oscillating feedback loop lies in the center of the overall spectrum, reducing heading error by an order of magnitude or more [29].

[2] Due to the rigorous lifetime requirements of flight hardware, discharge lamps are still preferred over lasers in space-borne atomic clocks. This situation will likely change in the coming decade due to advances in laser technology.

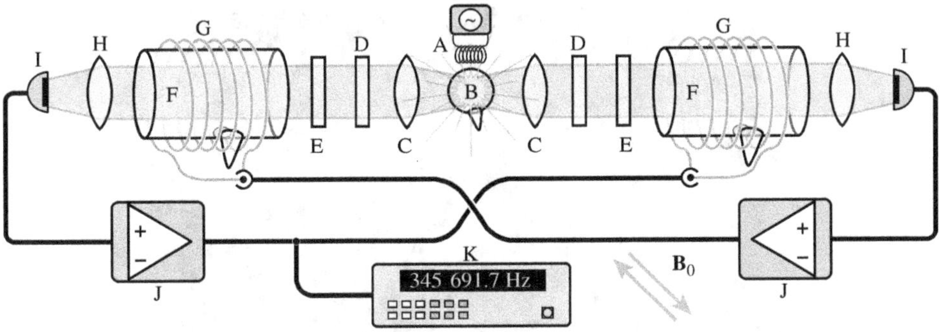

Figure 15.1 Schematic of a dual-cell self-oscillating magnetometer. An RF oscillator (A) drives a discharge in an alkali lamp (B). The light is collimated (C), sent through an interference filter (D) to remove the D1 line, and circularly polarized (E). The light then travels through the alkali-vapor cell (F) surrounded by an RF coil (G). The transmitted light, now modulated at ω_L, is focused (H) onto a photocell (I) whose signal is amplified (J) and used to drive the RF coil around the other cell. A frequency counter (K) reads the self-oscillation frequency. The optimal sensitivity occurs when the external field $\mathbf{B}_0$ is oriented at 45° to the sensor axis (as shown).

The sensor shown in Fig. 15.1 experiences dead zones when the field is aligned to within a few degrees of the its axis or near the equatorial plane. To avoid signal dropout, mission designers in the early 1960s began incorporating two such sensors into a single instrument, with the sensor axes aligned at a mutual angle of 45–55°. The two sensor outputs are mixed before processing, and the resulting dead zone is the very small angular intersection of the individual dead zones of the two dual-cell sensors [32]. A photograph of such a four-cell magnetometer is shown in Fig. 15.2.

15.2.3 NASA missions employing alkali-vapor magnetometers

The advent of alkali-vapor atomic magnetometers coincided fortuitously with the development of spacecraft technology, with rubidium-vapor magnetometers being deployed in rocket-borne measurements as early as 1960 [33]. In the United States, rubidium magnetometers were used in satellite missions regularly in the 1960s before being supplanted by cesium-based magnetometers in the mid-1970s. Cesium magnetometers were the primary focus of Soviet magnetometric research during this era. What follows is an overview of American space missions which employed alkali-vapor magnetometers.[3]

The Explorer Program

The first alkali-vapor magnetometer to be launched into space was aboard the Explorer X satellite (also known as Explorer 10 or P-14) in March of 1961. The sensor was a

[3] Less information is available on the topic of alkali-vapor magnetometers aboard Soviet spacecraft; such missions are discussed in the online supplement (www.cambridge.org/9781107010352).

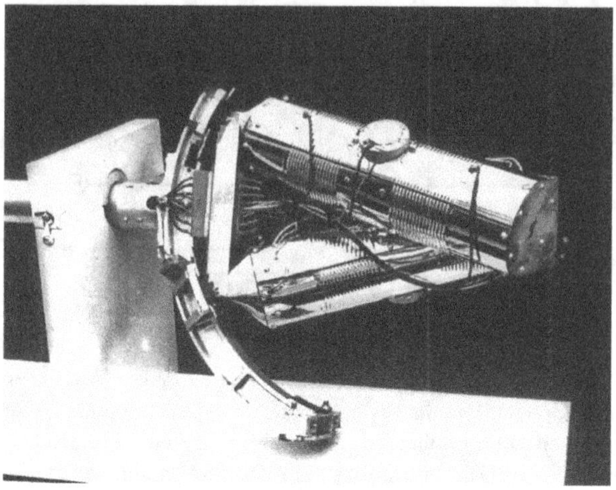

Figure 15.2 Photograph of the four-cell self-oscillating rubidium magnetometer used in the OGO missions. Each of the two cylinders contains a two-cell magnetometer configured as in Fig. 15.1. The two sensors are aligned at an angle of 55° to minimize the intersection of their respective dead zones. Reprinted with permission from Ref. [13].

dual-cell ^{87}Rb self-oscillating magnetometer, as shown in Fig. 15.1 and first described by Ruddock [29]. The alkali magnetometer was complemented by two vector fluxgate magnetometers located at opposite ends of the spacecraft [34]. The rubidium magnetometer housing included a bias coil which could generate a calibrated field of 10 nT at an angle of 54° 45′ with respect to the spacecraft spin axis. Due to this geometry the sensor could be biased along three orthogonal axes during a single revolution and thus detect vector field components; the bias field also allowed the magnetometer to measure fields below its limit of self-oscillation (~1 nT) [35]. The self-oscillation frequency of the magnetometer modulated the phase of a 108 MHz transmitter; the observed sideband separation was then used to determine the magnetic field strength.[4] Unfortunately the magnetometer began to fail about two hours into the planned 60-hour mission due to a steady increase in temperature in the satellite housing [36]; nevertheless it succeeded in calibrating the vector fluxgate sensors in low field before operation ceased altogether. Explorer X detected the Earth's magnetopause and the correlation between solar plasma influx and geomagnetic field disturbances [35].

Explorer 18 (also known as the Interplanetary Monitoring Platform, IMP 1, or IMP-A) was launched in November 1963 to explore the transition between the Earth's magnetosphere and the interplanetary magnetic field. It was equipped with a single-cell ^{87}Rb self-oscillating magnetometer similar to that designed for the failed Ranger 1 and 2 missions [29], with the added feature of a bias coil configuration similar to that used in Explorer X. Explorer 18 was the first spacecraft to make accurate measurements of the interplanetary

[4] This was a fairly common method of data transmission in early satellite missions before on-board digitization and recording, and it continued to be used extensively in sounding-rocket missions.

magnetic field and the Earth's bow shock [12]. Longer-term analysis of the Explorer 18 data also confirmed the solar origin of the interplanetary magnetic field [37].

Explorer 21 (IMP 2/B) and Explorer 28 (IMP 3/C) were launched in October 1964 and May 1965, respectively, and included magnetometers similar to that of Explorer 18. These missions helped to solidify understanding of the boundaries of the Earth's magnetosphere, its interactions with the solar wind, and the magnetohydrodynamics of the interplanetary magnetic field [38–40]. The ^{87}Rb scalar magnetometer used in IMP-2 and -3 weighed 1.4 kg and consumed a mere 3.5 W of power [17].

In December of 1964, satellite 1964-83C (equivalently, Transit 5E5) was launched. Its somewhat unconventional design included a strong permanent magnet which magnetically stabilized the satellite axis along the local magnetic field [41]. A single-cell rubidium magnetometer was included at the end of a 4.88 m boom. Although this magnetometer malfunctioned at fields above $\sim$31 000 nT, the mission provided low-latitude magnetic-field data with fair (20 nT) precision. These data helped to confirm earlier data on the South Atlantic Anomaly taken by the Soviet Kosmos-26 and Kosmos-49 missions and also helped to refine the spherical harmonic components and secular variation terms in contemporary geomagnetic field models [41].

The Orbiting Geophysical Observatories

In the mid-1960s, advances in launch vehicles allowed heavier payloads and thus more complex spacecraft to be launched into Earth orbit. The Orbiting Geophysical Observatory (OGO) series of satellites, a direct outcome of this evolution, were the first spacecraft devoted to extensive geophysical monitoring [42]. The OGO missions were divided into two classes: the *eccentric* OGO satellites OGO-1, -3, and -5 (also known as EOGO-1, -3, and -5) and the *polar* OGO satellites OGO-2, -4, and -6 (POGO-1, -2, and -3). The former were launched in 1964, 1966, and 1968 into high-eccentricity orbits capable of exploring the environment within and beyond the Earth's magnetosphere. The latter (launched in 1965, 1967, and 1969) had low-altitude, low-eccentricity polar orbits optimized for near-Earth measurements, viz. the first World Magnetic Survey.[5]

Each OGO mission included two dual-cell self-oscillating rubidium magnetometers configured as in Figs. 15.1 and 15.2. The POGO missions explored higher-field regions near the Earth's poles and thus chose ^{85}Rb over ^{87}Rb due to the lower gyromagnetic ratio of the former [44].[6] The EOGO missions needed to measure fields ranging from 40 μT down to only a few nT; these sensors therefore opted for the higher gyromagnetic ratio of ^{87}Rb [45]. A particular concern in the EOGO magnetometer was proper self-oscillation behavior at

[5] Up until the OGO missions, geophysicists relied upon data from ground-based surveys to calculate the multipole expansion of Earth's internal field, a technique fraught with poor coverage and decades-long acquisition. The POGO orbiters in particular were designed to make a global map of the Earth's magnetic field within a short time interval, allowing a far more accurate determination of the field and, eventually, its secular variation [43].

[6] Phase shifts in any self-oscillating feedback loop will cause spurious frequency shifts. The high-frequency signal of a self-oscillating ^{87}Rb magnetometer along the POGO orbit would have been too difficult to amplify without such errors.

extremely low magnetic fields where the self-oscillation frequency would approach the magnetic-resonance linewidth. The EOGO rubidium magnetometers were further equipped with triaxial bias coils to allow vector field measurements.[7] The EOGO magnetometer weighed approximately 6 kg and required 6–7 W of power; the POGO magnetometers weighed 7.7 kg and required 7.5 W [17].

Although the ^{87}Rb magnetometer on OGO-1 remained inoperative due to a boom deployment failure, the fluxgate magnetometer returned extensive data over a 20-month period, helping to map out Earth's bow shock [46]. OGO-3 and OGO-5 operated successfully and returned a wealth of data on the structure of the Earth's magnetosphere and ring-current systems [47, 48]. The ^{85}Rb magnetometers on all three POGO missions operated successfully; the resulting data elucidated the equatorial electrojet [49], provided measurements of secular variation in the geomagnetic field [50], and most importantly, provided the basis for the first International Geomagnetic Reference Field (IGRF) [51, 52] and the first global magnetic-anomaly map [53].

Magsat and beyond

Following the success of the World Magnetic Survey, scientists at NASA and the US Geological Survey began planning a dedicated satellite mission which would make higher-resolution scalar field measurements, map the vector components of the Earth's field,[8] and correct the IGRF for secular variation. The resulting satellite, Magsat, was launched in late 1979. Its four-cell self-oscillating cesium magnetometer was a direct evolution of the rubidium magnetometers carried aboard POGO; it provided absolute field calibration for the three-axis fluxgate magnetometer, whose output drifted by ~20 nT over the mission lifetime [57]. Cesium was chosen for the Magsat magnetometer because of its narrower nonlinear Zeeman structure (and thus lower heading error) than either isotope of rubidium [32]. This magnetometer had similar construction to that shown in Fig. 15.2, with added feedback loops for more precise temperature control and regulation of the output intensity of each cesium discharge lamp. A new feature of the magnetometer digital data-processing unit was an error bit which would report a valid field measurement only if (i) the tracking filter was phase-locked throughout the measurement, (ii) the signal amplitude was large enough, and (iii) the cesium lamp brightness was within a specified range [32]. This error bit proved to be crucial in the later data analysis.

In order for Magsat to achieve a vector field measurement accurate to a few nanotesla, the absolute orientation of the fluxgate at the end of its 6 m boom needed to be determined to within 7 arcsec RMS [58]. This goal was achieved with an "attitude transfer system" which employed passive reflectors on the magnetometer to compare its orientation

[7] Published data focus exclusively on scalar measurements, so it is not clear what became of this capability.

[8] In early literature one often encounters the claim that scalar field maps are sufficient to reconstruct uniquely the magnetic scalar potential of the Earth's field. In 1970 this was demonstrated by Backus to be false [54]; subsequent analysis revealed the need for full vector measurements [55, 56].

to the star cameras aboard the main spacecraft [59].[9] Magsat also required unprecedented orbital tracking in order to ensure that uncertainty in the satellite's position did not induce substantial errors in the resulting IGRF.

The Magsat cesium magnetometer met its objective of calibrating the fluxgate data, despite lamp intensity fluctuations which occasionally caused the tracking filter to lose lock [32]. The error bit in the data stream was critical for diagnosing this issue and allowing only the most reliable data to be selected. Magsat data were used to generate the most precise geomagnetic reference field created up to that date, with spherical harmonic expansion to degree and order 23. The power spectrum of these coefficients drops off dramatically with order up to order 14, at which point it levels off. This implies that core sources dominate at lower order and lithospheric sources at higher [61]. Refinements in data processing continued for more than a decade after the Magsat mission, resulting in successively more precise magnetic anomaly maps [62–64]. Magsat also allowed much more precise measurements of secular variation [65] and ionospheric fields [66, 67]. Despite its seven-month life span, Magsat contributed a wealth of magnetic-field data whose analysis continued for nearly two decades. It was not until the Ørsted, CHAMP, and SAC-C missions in the early 2000s that this feat was reproduced [68].

Magsat was equipped with the last alkali-vapor atomic magnetometer to be launched into space. Nevertheless, alkali-vapor magnetometer technology has progressed dramatically in recent years, with the advent of space-certified alkali lasers [69], new schemes for laser-pumped high-bandwidth magnetometers [70–72], sub-femtotesla precision [73], and drastically reduced heading error [74,75]. These advances may once again place alkali-vapor magnetometers at the farthest reaches of human exploration.

15.3 Helium magnetometers in space applications

15.3.1 Introduction

Helium magnetometers have played an important role in space magnetometry for the last fifty years. The first vector helium magnetometer (VHM) for space applications was described by Slocum and Reilly in 1963 [76]. The success of this instrument led to a family of lamp-pumped ^{4}He magnetometers developed by Dr. E. J. Smith's team at NASA's Jet Propulsion Lab in collaboration initially with Texas Instruments and later with Polatomic, Inc. The early VHM instruments were designed for vector component measurements [77, 78], but the design was later modified to include scalar field measurement capability. Principles of operation for both scalar and vector helium magnetometers are described in detail in Chapter 10. The helium magnetometers shown in Table 15.1 have undoubtedly flown more scientific survey miles than any other optical magnetometer.

[9] Such precision remains a challenge in spacecraft magnetometers; one of the principal achievements of the 1999 Ørsted mission was the successful construction of a nonmagnetic star tracker which could be mounted directly onto the vector magnetometer [60].

Table 15.1. *Summary of helium space magnetometers and their missions.*

Spacecraft	Launch year	Type	Primary destination(s)
Mariner 4	1964	Vector	Mars
Mariner 5	1967	Vector	Venus
Pioneer 10	1972	Vector	Jupiter
Pioneer 11	1973	Vector	Jupiter, Saturn
ISEE-3	1978	Vector	Earth–Sun libration point L_1
ISEE-3 (ICE)	1978	Vector	Comet Giacobini–Zinner
Ulysses	1990	Vector	Sun (polar orbit)
Cassini	1997	Vector/Scalar	Saturn
SAC-C	2000	Scalar	Earth orbit

Mariner 4

The Mariner 4 Mars payload included a low-field VHM shown in Fig. 15.3 and was launched by NASA in 1964. The VHM was jointly developed by JPL and Texas Instruments. The VHM consisted of a single ^{4}He RF lamp, a single ^{4}He cell, collimating and circular polarizing optics and a photodetector. The cell was located at the center of a triaxial Helmholtz coil set, and vector field information was extracted using the bias field nulling (BFN) method described in Chapter 10. The low-field vector helium magnetometer had a sensitivity of 200 pT, bandwidth of 1 Hz, dynamic range of ±150 nT, and a zero-field offset of ±1 nT [76]. The instrument used 3 W power and had a mass of 0.45 kg. The low dynamic range and the stability of the VHM parametric resonance obviated the need for in-flight calibration. The Mariner 4 VHM successfully measured the interplanetary magnetic fields between Earth and Mars and the planetary magnetic fields near Mars. The notable result of the magnetic measurements was the absence of a detectable planetary magnetic field for Mars [79].

Mariner 5

The Mariner 5 Venus mission, launched by NASA in 1967, included a VHM that was a spare flight unit originally built as a backup for the Mariner 4 mission (see Fig. 15.3). The VHM successfully measured the magnetic field between Earth and Venus and the planetary field of Venus [80].

Pioneer 10

The Pioneer 10 Jupiter mission payload included a full-range VHM shown in Fig. 15.4. It was launched on a NASA spacecraft in 1972 and became the first space probe to encounter Jupiter [81]. The vector helium magnetometer has the same sensor configuration as the Mariner series VHM, but modified to cover the full high-field range encountered on the mission. The Pioneer 10 VHM had seven dynamic range settings from ±4 nT up to $\pm140\,000$ nT.

Figure 15.3 Mariner 4 vector helium magnetometer. The sensor (middle) is shown next to a six-inch (15 cm) ruler. The electronics units are shown on either side.

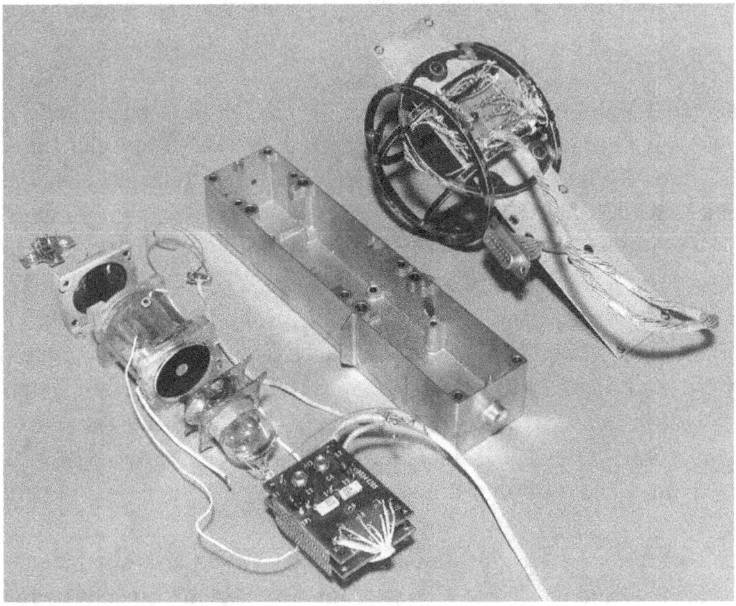

Figure 15.4 The Pioneer 10 and Pioneer 11 vector helium magnetometers were identical instruments. The sensor components are shown here in an exploded view. The lamp, cell, and photodetector are on the left. The triaxial bias-field nulling vector coils are on the right.

The accuracy of the instrument was 25 pT, and the noise limit was 10 pT/$\sqrt{\mathrm{Hz}}$ for the bandwidth of 0–10 Hz when operated in the lowest dynamic range setting. The Pioneer 10 VHM sensor had dimensions 9.9 cm $\times$ 10.6 cm $\times$ 22.4 cm with a mass of 0.57 kg. The electronics had dimensions 9.6 cm × 14.6 cm × 19.7 cm with a mass of 2.15 kg. The VHM power requirements were 0.65 W for the sensor and 3.45 W for the electronics for a total of 4.45 W after RF cable losses were included. The Pioneer 10 magnetometer successfully measured the interplanetary field between Earth and Jupiter, the Asteroid Belt, the Jovian field seen in a fly-by trajectory, and the fields in the outer regions of the Solar System.

Pioneer 11

The Pioneer 11 Jupiter/Saturn mission payload was launched on a NASA spacecraft in 1973 to become the first space probe to encounter Saturn [82]. The spacecraft itself was a duplicate of Pioneer 10 [81]. Like the Pioneer 10 VHM, the key sensor elements included the lamp with a parabolic reflector, an arsenic trisulfide focusing lens, an iodine film polarizer, low-pressure helium cell, and a lead sulfide photodetector. These components were located in a box-like housing made from fiberglass flashed with gold for thermal stability. The sensor was heated by a radioactive plutonium dioxide source to counter the very low temperatures anticipated for the mission. The Pioneer 11 VHM measured the interplanetary field beyond the orbit of Mars in the Asteroid Belt, the magnetic field of Jupiter, the magnetic field of Saturn, and continued to observe the magnetic fields in the outer Solar System.

ISEE-3 (ICE)

The International Sun/Earth Explorer 3 (ISEE-3) mission payload included a full-range VHM and was launched on a spacecraft sponsored by NASA and the European Space Agency (ESA) in 1978. ICEE-3 became the first space probe to be placed at the Earth–Sun libration point (the L_1 Lagrangian point) where it was continuously upstream from Earth with respect to the solar wind. One of its primary purposes was to examine the magnetic interaction between the solar wind and Earth's outer magnetosphere and, later, the distant geomagnetic tail near the L_2 Lagrangian point. The VHM on board the ISEE-3 satellite was a modified Pioneer VHM flight spare. The eight selectable dynamic range settings covered the range from ± 4 nT up to $\pm 140\,000$ nT. The sensitivity of the magnetometer in the lowest dynamic range setting was 10 $\mathrm{pT}/\sqrt{\mathrm{Hz}}$ with a bandwidth of 0–3 Hz. The sensor had a mass of 0.72 kg and a volume of 2351 cm^3, and used 0.77 W of power. The electronics and data-processing units had a combined mass of 2.68 kg and volume of 3644 cm^3, and used 3.65 W [83, 84].

The ISEE-3 satellite was assigned a new mission in 1982 and was renamed the International Cometary Explorer (ICE). This new mission had the objective of studying the interaction between the solar wind and the atmosphere of Comet Giacobini–Zinner. After successfully flying through the comet's plasma tail, the spacecraft later went on to aid in the study of Comet Halley. The ICE spacecraft was then placed in a heliocentric orbit, where it contributed to the study of coronal mass ejections [85, 86].

Ulysses

The Ulysses spacecraft payload included a low-field VHM and was launched on a NASA/ESA spacecraft in 1990 to the Sun. The Ulysses VHM was similar to the Pioneer 10/11 VHM but incorporated some improved design features. The sensor housing was changed to a cylindrical carbon-fiber–epoxy material that avoided thermo-electric currents and provided RF shielding of the lamp and cell driver signals. A large-area silicon detector replaced the lead sulfide detector, and a wire-grid polarizer was used in place of the iodine film polarizer. The vector helium magnetometer on the Ulysses satellite was one of several instruments designed to examine the heliosphere. Data obtained from this mission indicated unexpected asymmetries in the heliosphere and helped refine the understanding of the interaction between solar magnetic fields and the Solar System [87, 88].

Cassini

The Cassini spacecraft payload included a Dual Technique Helium Magnetometer (MAG) and was launched on a NASA spacecraft in 1997. It became the first spacecraft to enter orbit around Saturn. This mission was supported by NASA, ESA, and the Italian Space Agency (ASI). The MAG was a modified version of a Ulysses flight spare VHM sensor with the important added feature of scalar magnetic field measurements [89]. Scalar or vector measurements were selectable by command. The Cassini spacecraft provided information about magnetic fields generated in the core and interior of Saturn, along with measurements of magnetic fields due to the moon Titan and other natural satellites. The Cassini instrument scalar mode has a dynamic range of 256–16 384 nT, a sensitivity of 40 $pT/\sqrt{Hz}$, and an accuracy of <1 nT. The vector mode has two range settings (±32 nT and ±256 nT) with a sensitivity of 5 $pT/\sqrt{Hz}$. The sensor has a mass of 0.93 kg and uses 2.6 W of power. The electronics have a mass of 2.25 kg and use 4.2 W of power [90, 91]. During a fly-by of Earth in 1999, the Cassini MAG obtained measurements of Earth's magnetosphere which indicated a non-negligible offset from the field values predicted by the IGRF model. These data were subsequently used to improve the model parameters of the IGRF [91, 92].

SAC-C

The SAC-C Earth satellite payload included a scalar helium magnetometer (SHM) shown in Fig. 15.5. It was launched in 2000 as a joint endeavor between Argentina's National Space Activities Commission (CONAE) and NASA. The SHM is part of the Magnetic Mapping Payload (MMP) onboard this satellite and has a dynamic range of 20 000–70 000 nT, an accuracy of <1 nT, and a bandwidth of 0–1 Hz. The SAC-C magnetometer uses a single helium lamp to optically pump two ^{4}He cells in an orthogonal alignment. Since the spacecraft orientation keeps one face pointing Earthward, the SHM is able to perform scalar measurements throughout the highly inclined polar orbit using just two axes. The objectives of the magnetic measurements are to examine secular variations in Earth's magnetic field, study the physical properties of the core and mantle, and investigate the interactions between Earth's magnetic field and the solar wind [93–95]. In addition, the scalar field measurements are used to calibrate the vector fluxgate magnetometer (CSC) that is also part of the MMP on this satellite [96].

Figure 15.5 SAC-C scalar magnetometer sensor with two orthogonal helium cells pumped by a single helium lamp.

15.3.2 Future helium space magnetometers

For the twenty-first century, laser-pumped ^{4}He magnetometers will play an increasingly important role in planetary science and Earth science missions. Two instruments currently under development that illustrate the technical advances possible with laser-pump sources are the self-calibrating vector helium-4 magnetometer (SVHM) at Polatomic, Inc. and the absolute scalar magnetometer (ASM) at CEA/LETI (France) which will be flown on the upcoming SWARM mission.

SWARM absolute scalar magnetometer (ASM)

The SWARM mission is sponsored by the European Space Agency and will place three satellites in three separate polar orbits around Earth. The objective of the SWARM mission is to provide the most advanced survey of the geomagnetic field and its temporal evolution in order to improve understanding of Earth's interior and climate [97,98]. Launch is scheduled for 2013. The ASM is a laser-pumped single-axis ^{4}He magnetometer and is included in each SWARM satellite payload for calibration of the three fluxgate vector magnetometers. The ASM instrument was developed by CEA/LETI in collaboration with the French space agency CNES.

The ASM laser pump source is a diode-pumped Yb-doped fiber Bragg grating (FBG) laser. The laser pumps a single ^{4}He cell with linearly polarized radiation which improves accuracy by eliminating light shifts in the resonance frequency. A piezo-actuated rotation apparatus orients the linear polarizer and magnetic excitation coils with respect to the magnetic field direction in order to maintain a strong resonance signal. The scalar measurements have a sensitivity of 1 $pT/\sqrt{Hz}$ averaged over the bandwidth of DC to 100 Hz with an accuracy of $<$ 45 pT [99]. The instrument is also equipped with an optional vector measurement capability. It has inferior sensitivity (1 $nT/\sqrt{Hz}$) and bandwidth (0.4 Hz) to the primary vector fluxgate instruments, but vector measurements are obtained simultaneously with the scalar measurements [100].

Self-calibrating vector helium magnetometer (SVHM)

The SVHM is a laser-pumped ^{4}He magnetometer developed by Polatomic under a NASA-sponsored space magnetometer research and development program. The SVHM design is for a single dual-function instrument that includes a triaxial vector sensor and a scalar magnetometer, and the instrument is electronically multiplexed to sample the three vector component values and the scalar field value. The SVHM provides accurate scalar field measurements in order to calibrate the vector component measurements.

The full-range vector mode design has evolved from the Mariner 4 low-field vector helium magnetometer design and employs the bias field nulling technique (see Chapter 10 for a detailed description) to measure the three vector components. An InGaAs semiconductor laser replaces an RF helium lamp as the circularly polarized optical pumping source. The scalar mode of the SVHM employs three orthogonal miniaturized ^{4}He cells to achieve full sensitivity for all orientations of the magnetic field without using any moving parts. The SVHM prototype presently exhibits a scalar sensitivity of $<$5 $pT/\sqrt{Hz}$ at 1 Hz and a dynamic range of 1000–100 000 nT. The vector sensitivity is $<$50 $pT/\sqrt{Hz}$ at 1 Hz with a dynamic range of $\pm$75 000 nT. Ongoing improvements to the vector measurement technique are expected to improve the vector sensitivity to $<$10 $pT/\sqrt{Hz}$. The bandwidth is presently 0–8 Hz in the multiplexed mode, but either measurement mode can be selected individually with an expanded bandwidth up to 215 Hz.

Acknowledgments

B. Patton would like to thank Professor N. F. Ness, Dr. W. H. Farthing, and Professor A. N. Kozlov for their helpful input.

References

[1] S. F. Singer, E. Maple, and W. A. Bowen, J. Geophys. Res. **56**, 265 (1951).
[2] E. J. Smith, L. Davis, Jr., P. J. Coleman, Jr., *et al.*, J. Geophys. Res. **70**, 1571 (1965).
[3] S. S. Dolginov, E. G. Yeroshenko, and L. N. Zhuzgov, Kosm. Issled. **6**, 562 (1968).
[4] N. F. Ness, K. W. Behannon, R. P. Lepping, *et al.*, J. Geophys. Res. – Space Phys. **80**, 2708 (1975).
[5] J. E. P. Connerney, J. Geophys. Res. **98**, 18659 (1993).
[6] S. S. Dolginov, E. G. Yeroshenko, and L. N. Zhuzgov, J. Geophys. Res. **78**, 4779 (1973).
[7] J. E. P. Connerney, M. H. Acuña, P. J. Wasilewski, *et al.*, Science **284**, 794 (1999).

[8] D. M. Kass and Y. L. Yung, Science **268**, 697 (1995).
[9] E. Chassefiere, F. Leblanc, and B. Langlais, Planet. Space Sci. **55**, 343 (2007).
[10] M. G. Kivelson, K. K. Khurana, C. T. Russell, *et al.*, Science **289**, 1340 (2000).
[11] C. Zimmer, K. K. Khurana, and M. G. Kivelson, Icarus **147**, 329 (2000).
[12] N. F. Ness, J. B. Seek, and C. S. Scearce, J. Geophys. Res. **69**, 3531 (1964).
[13] M. H. Acuña, Rev. Sci. Instrum. **73**, 3717 (2002).
[14] D. I. Gordon, R. H. Lundsten, R. A. Chiarodo, *et al.*, IEEE Trans. Magn. **MAG4**, 397 (1968).
[15] N. F. Ness, K. W. Behannon, R. P. Lepping, *et al.*, J. Geophys. Res. **76**, 3564 (1971).
[16] M. H. Acuña, J. L. Scheifele, P. Stella, *et al.*, in *Photovoltaic Specialists Conference, Conference Record of the Twenty Fifth IEEE*, pp. 325–328 (1996).
[17] N. F. Ness, Space Sci. Rev. **11**, 459 (1970).
[18] S. F. Singer, E. Maple, and W. A. Bowen, Phys. Rev. **77**, 398 (1950).
[19] E. Maple, W. E. Bowen, and S. F. Singer, J. Geophys. Res. **55**, 115 (1950).
[20] P. Ripka, Sensors and Actuators A: Physics **33**, 129 (1992).
[21] G. S. Waters and G. Phillips, Geophys. Prospect. **4**, 1 (1956).
[22] A. W. Overhauser, Phys. Rev. **92**, 411 (1953).
[23] A. Salvi, Rev. Phys. Appl. **5**, 131 (1970).
[24] H. G. Dehmelt, Phys. Rev. **105**, 1924 (1957).
[25] W. E. Bell and A. L. Bloom, Phys. Rev. **107**, 1559 (1957).
[26] T. L. Skillman and P. L. Bender, J. Geophys. Res. **63**, 513 (1958).
[27] J. T. Arnold, W. E. Bell, A. L. Bloom, *et al.*, J. Geophys. Res. **65**, 2472 (1960).
[28] S. H. Ward and K. A. Ruddock, J. Geophys. Res. **67**, 1889 (1962).
[29] K. A. Ruddock, *Space Research II (Proceedings of Second International Space Science Symposium, Florence 1961)*, 692 (1961).
[30] W. E. Bell and A. L. Bloom, Phys. Rev. Lett. **6**, 280 (1961).
[31] A. L. Bloom, Appl. Op. **1**, 61 (1962).
[32] W. H. Farthing, Johns Hopkins APL Technical Digest **1**(3), 205 (1980).
[33] NASA Technical Report 19710070062, NASA-TM-X-67377 (1964).
[34] E. D. Angulo and R. K. Browning, NASA Technical Report 19620000922, NASA-TN-D-1175 (1962).
[35] J. P. Heppner, N. F. Ness, C. S. Scearce, and T. L. Skillman, J. Geophys. Res. **68**, 1 (1963).
[36] A. R. Timmins and K. L. Rosette, NASA Technical Report 19630008779, NASA-TN-D-1748 (1963).
[37] N. F. Ness and J. M. Wilcox, Phys. Rev. Lett. **13**, 461 (1964).
[38] D. H. Fairfield and N. F. Ness, J. Geophys. Res. **72**, 2379 (1967).
[39] K. H. Schatten, J. M. Wilcox, and N. F. Ness, Solar Phys. **6**, 442 (1969).
[40] J. M. Wilcox, Space Sci. Rev. **8**, 258 (1968).
[41] A. J. Zmuda, W. E. Radford, F. T. Heuring, *et al.*, J. Geophys. Res. **73**, 2495 (1968).
[42] G. H. Ludwig and W. E. Scull, NASA Technical Report 1965070482, NASA-TM-X-56155 (1963).
[43] J. P. Heppner, in *Publications of Goddard Space Flight Center 1963, Volume I: Space Sciences*, pp. 887–915 (1963); also NASA Technical Report 19660022716, NASA-TM-X-57737 (1963).
[44] W. H. Farthing and W. C. Folz, Rev. Sci. Instrum. **38**, 1023 (1967).
[45] B. G. Ledley, Rev. Phys. Appl. **5**, 164 (1970).
[46] J. P. Heppner, M. Sugiura, T. L. Skillman, *et al.*, J. Geophys. Res. **72**, 5417 (1967).
[47] M. Sugiura, B. G. Ledley, T. L. Skillman, *et al.*, J. Geophys. Res. **76**, 7552 (1971).

[48] M. Sugiura and D. J. Poros, Planet. Space Sci. **21**, 1763 (1973).
[49] J. C. Cain and R. E. Sweeney, J. Atmos. Terr. Phys. **35**, 1231 (1973).
[50] J. C. Cain, Rev. Geophys. Space Phys. **9**, 259 (1971).
[51] J. C. Cain, S. J. Hendricks, R. A. Langel, *et al.*, J. Geomagn. Geoelectr. **19**, 335 (1967).
[52] American Geophysical Union, J. Geophys. Res. **74**, 4407 (1969).
[53] R. D. Regan, J. C. Cain, and W. M. Davis, J. Geophys. Res. **80**, 794 (1975).
[54] G. E. Backus, J. Geophys. Res. **75**, 6339 (1970).
[55] D. P. Stern and J. H. Bredekamp, J. Geophys. Res. **80**, 1776 (1975).
[56] P. Alberto, O. Oliveira, and M. A. Pais, Geophys. J. Int. **159**, 548 (2004).
[57] R. Langel, G. Ousley, J. Berbert, *et al.*, Geophys. Res. Lett. **9**, 243 (1982).
[58] F. F. Mobley, L. D. Eckard, G. H. Fountain, *et al.*, IEEE Trans. Magn. **16**, 758 (1980).
[59] F. W. Schenkel and A. Finkel, Johns Hopkins Applied Physics Laboratory Report CP 055 (1977).
[60] C. C. Liebe and J. L. Joergensen, Proc. SPIE **2810**, 239 (1996).
[61] R. A. Langel and R. H. Estes, Geophys. Res. Lett. **9**, 250 (1982).
[62] R. A. Langel, C. C. Schnetzler, J. D. Phillips, *et al.*, Geophys. Res. Lett. **9**, 273 (1982).
[63] J. Arkani-Hamed and D. W. Strangway, J. Geophys. Res. **91**, 8193 (1986).
[64] D. Ravat, R. A. Langel, M. Purucker, *et al.*, J. Geophys. Res. – Solid Earth **100**, 20111 (1995).
[65] R. A. Langel and R. H. Estes, J. Geophys. Res. **90**, 2495 (1985).
[66] R. A. Langel, M. Purucker, and M. Rajaram, J. Atmos. Terr. Phys. **55**, 1233 (1993).
[67] R. A. Langel, T. J. Sabaka, R. T. Baldwin, *et al.*, Phys. Earth Planet. Inter. **98**, 235 (1996).
[68] N. Olsen, G. Hulot, and T. Sabaka, Space Sci. Rev. **155**, 65 (2010).
[69] E. Luvsandamdin, G. Mura, A. Wicht, *et al.*, in 2011 *CLEO EUROPE/EQEC* (2011).
[70] P. D. D. Schwindt, L. Hollberg, and J. Kitching, Rev. Sci. Instrum. **76**, 126103 (2005).
[71] J. M. Higbie, E. Corsini, and D. Budker, Rev. Sci. Instrum. **77**, 113106 (2006).
[72] J. Belfi, G. Bevilacqua, V. Biancalana, *et al.*, J. Opt. Soc. Am. B **26**, 910 (2009).
[73] H. B. Dang, A. C. Maloof, and M. V. Romalis, Appl. Phys. Lett. **97**, 151110 (2010).
[74] E. B. Alexandrov, Phys. Scr. **T105**, 27 (2003).
[75] A. Ben-Kish and M. V. Romalis, Phys. Rev. Lett. **105**, 193601 (2010).
[76] R. E. Slocum and F. N. Reilly, IEEE Trans. Nucl. Sci. **10**, 165 (1963).
[77] R. E. Slocum, Rev. Phys. Appl. **5**, 109 (1970).
[78] R. E. Slocum and B. I. Marton, IEEE Trans. Magn. **MAG9**, 221 (1973).
[79] E. J. Smith, L. Davis, P. J. Coleman, *et al.*, Science **149**, 1241 (1965).
[80] B. V. Connor, IEEE Trans. Magn. **MAG4**, 391 (1968).
[81] E. J. Smith, B. V. Connor, and G. T. Foster, IEEE Trans. Magn. **11**, 962 (1975).
[82] T. G. Northrop, A. G. Opp, and J. H. Wolfe, J. Geophys. Res. **85**, 5651 (1980).
[83] A. M. A. Frandsen, B. V. Connor, J. Vanamersfoort, *et al.*, IEEE Trans. Geosci. Remote Sens. **16**, 195 (1978).
[84] B. T. Tsurutani and T. T. Vonrosenvinge, Geophys. Res. Lett. **11**, 1027 (1984).
[85] E. J. Smith, B. T. Tsurutani, J. A. Slavin, *et al.*, Science **232**, 382 (1986).
[86] M. Neugebauer, Phys. Today **38**, 38 (1985).
[87] E. J. Smith, Adv. Space Res. **13**, 5 (1993).
[88] J. Glanz, Science **270**, 912 (1995).
[89] S. Kellock, A. Balogh, D. Southwood, *et al.*, Proc. SPIE **2803**, 141–152 (1996).
[90] M. W. Dunlop, M. K. Dougherty, S. Kellock, *et al.*, Planet. Space Sci. **47**, 1389 (1999).

[91] E. J. Smith, M. K. Dougherty, C. T. Russell, *et al.*, J. Geophys. Res. – Space Phys. **106**, 30129 (2001).
[92] R. E. Slocum and E. J. Smith, Contrib. Geophys. Geodesy **31**, 99 (2001).
[93] R. Colomb, C. Alonso, and I. Nollmann, Acta Astronaut. **52**, 995 (2003).
[94] C. A. Raymond, I. K. Kulikov, J. B. Jewell, *et al.*, American Geophysical Union, Spring Meeting 2004, Abstract #GP13A-05 (2004).
[95] M. Purucker and N. Olsen, *OIST-4 Proceedings*, 319 (2003).
[96] J. M. G. Merayo, P. Brauer, F. Primdahl, *et al.*, Meas. Sci. Technol. **11**, 120 (2000).
[97] R. Haagmans, *ESA SP-1279(6)*, edited by B. Battrick (Noordwijk: ESA Publications Division, 2004).
[98] R. Haagmans. *Annex to ESA SP-1279(6)*, edited by B. Battrick (Noordwijk: ESA Publications Division, 2004).
[99] T. Jager, J.-M. Léger, F. Bertrand, *et al.*, IEEE Sensors 2010, 2392–2395 (2010).
[100] O. Gravrand, A. Khokhlov, J. L. Le Mouel, *et al.*, Earth Planets Space **53**, 949 (2001).

16

Detection of biomagnetic fields

A. Ben-Amar Baranga, T. G. Walker, and R. T. Wakai

16.1 Sources of biomagnetism

Magnetic fields arise from many parts of the body and are produced by two distinct types of sources–ionic currents and magnetic tissues. Although most of the body is weakly diamagnetic, the magnetic tissues of greatest interest are paramagnetic or ferromagnetic. The major source of paramagnetism in the body is the liver, which contains iron compounds [1]. The strongest sources of ferromagnetism in the body are ingested or inhaled ferromagnetic substances, which can be detected even in trace amounts [2].

Like other electrical currents, ionic currents generated by nerve and muscle tissue produce magnetic fields and potential differences, which can be detected at the body surface or even outside the body. The ionic current has two components – a "primary" current and a "volume" current [3] (Fig. 16.1). Primary currents result directly from physiological activity and are confined largely within the electrically excited cells. Volume currents are passive return currents that extend far into the surrounding medium in response to the electromotive forces that drive the primary currents. Bioelectric signals are conducted to the body surface by volume currents and are the main source of electroencephalographic (EEG) and electrocardiographic (ECG) signals. Bioelectric signals are strongly influenced by the highly inhomogeneous electrical conductivity of the body. In contrast, biomagnetic signals arise predominantly from primary currents and are affected to a much lesser extent. This accounts for a key advantage of biomagnetic versus bioelectric fields – their simpler signal-transmission properties – which enables a more accurate determination of source location. Magnetic signals have been recorded from the spinal cord, peripheral nerves, skeletal and smooth muscle; however, the great majority of investigation has focused on the magnetic fields of the brain and heart, i.e., the magnetoencephalogram (MEG [4]) and magnetocardiogram (MCG [5]), respectively. However, bioelectric and biomagnetic

Optical Magnetometry, ed. Budker, D. and Jackson Kimball, D. F. Published by Cambridge University Press.

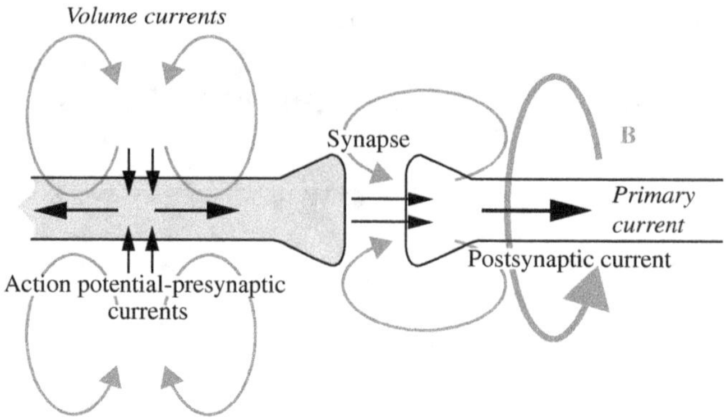

Figure 16.1 Physiological activity of nerve and muscle cells generates intracellular current, known as primary current, and passive return current that extends far into the surrounding medium, known as volume current. Magnetic signals arise primarily from the primary current, while electric signals are conducted to the surface by the volume current.

signals are strongly correlated, and EEG (ECG) and MEG (MCG) are often considered complementary technologies.

16.2 Development of biomagnetic field detection

The study of biomagnetism presents enormous technical challenges because the signals are so small (Fig. 16.2). The MCG, a relatively large biomagnetism signal, is approximately 10^{-10} T at the skin. Evoked MEG signals are of order 10^{-13} T, while spontaneous brain rhythms are of order 10^{-12} T. These signals are dwarfed by the clutter of magnetic interference in a typical hospital or laboratory environment from such sources as power lines, elevators, electric motors, and other electrical equipment. For this reason, the signals are recorded within magnetically shielded chambers, consisting of multiple layers of aluminum and high-magnetic-permeability alloy. (See Chapter 12 for details on magnetic shielding.)

The study of biomagnetism began in 1963 with the first measurement of magnetic fields produced by the human heart by Baule and MaFee [6] followed by Cohen's 1967 measurement of the heart field [7] and the first measurement of the brain's magnetic field in 1968 by Cohen [8]. Using a copper induction coil as detector, this first measurement of the brain's magnetic field was enabled by the efficient multilayer magnetically shielded chamber at MIT [9]. It was in 1970 that Cohen introduced Zimmerman's recently developed superconducting magnetometers (SQUIDs [10]) into his heavily shielded room to measure the heart's magnetic field with a significantly better signal-to-noise ratio and opened the way wide to the emerging new field of biomagnetism [11]. Years later, their paper was recognized as "the Magna Carta of biomagnetism" [12]. It took Cohen less than two years to measure also the brain's magnetic field with his SQUID setup [13], opening the way to magnetoencephalography (MEG). Improvements in SQUID sensitivity, magnetic

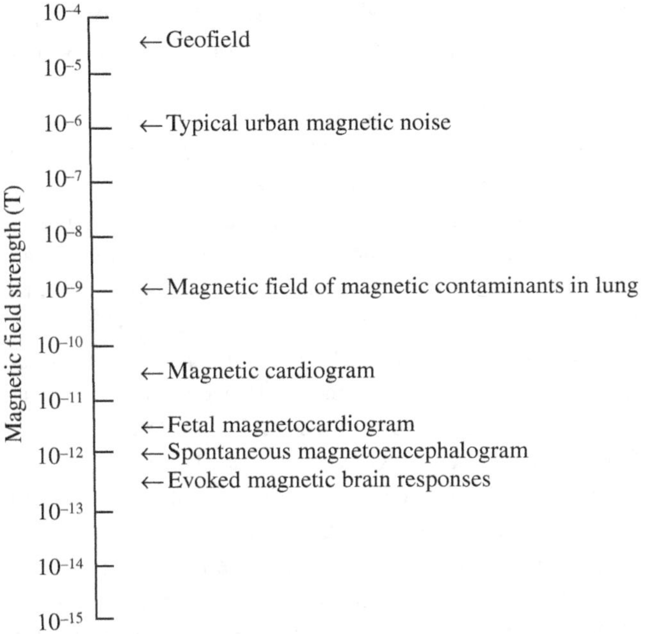

Figure 16.2 Amplitudes of various biomagnetic signals and environmental interference.

shielding and data processing allowed J. Wikswo *et al.* to perform in 1980 the first measurement of an isolated frog nerve [14]. From its first appearance, the SQUID dominated biomagnetism measurements due to its high sensitivity at low frequency (<100 Hz). Today there are commercially available, or as prototypes, SQUID-based systems for magnetoencephalography (MEG), magnetocardiography (MCG), fetal-magnetocardiography (fMCG), magnetogastrography (MGG), magnetomyography (MMG), magnetorelaxometry (MRX), and magnetopneumography (MPG).

The high price of the systems based on cryogenic cooling to liquid-helium temperature (4.2 K) required by the highly sensitive "low-temperature superconducting" SQUIDs has encouraged the development of alternative technologies. Optical magnetometers (including atomic magnetometers and lamp- and laser-pumped magnetometers) have been developed for tens of years and have recently reached the high sensitivity sufficient to monitor biomagnetic fields such as the MCG. In an alkali-atom-based optical magnetometer an alkali vapor (usually K, Cs, or Rb) is generated in a glass cell either at room temperature or at elevated temperatures (yielding higher densities) through heating. In a typical optical atomic magnetometer, the electron spins (electron magnetic dipoles) of the alkali atoms are oriented along an incoming circularly polarized laser beam tuned to a specific atomic transition by optical pumping. If a magnetic field perpendicular to the spins is applied, the spins rotate at the Larmor precession frequency, which is proportional to the magnetic field strength. There are various methods to measure the Larmor frequency, resulting in high-sensitivity magnetometers [15]. (For details on optical magnetometry techniques see Chapters 1 and 4–9.)

The first measurements of biological magnetic fields using an atomic magnetometer were reported by Livanov *et al.* in 1978 [16]. They used a scalar Cs magnetometer with a sensitivity of a few pT to map out heart magnetic fields and observe other biological magnetic signals. The interest in biological applications of atomic magnetometers was revived in 2003 when Bison *et al.* presented a two-dimensional mapping of a human-heart magnetic field performed at room temperature with a laser-pumped magnetometer [17]. The quality of their maps is similar to that obtained by SQUID MCG systems. Lately, a multichannel MCG system based on optical Cs vapor magnetometers has been developed by Bison *et al.* [18]. S. Knappe *et al.* carried out simultaneous MCG and MRX measurements with a chip-scale atomic magnetometer and a SQUID-based magnetometer, demonstrating similar levels of performance for the two types of sensors [19].

Optical magnetometer sensitivity can be limited by spin-polarization relaxation due to collisions of the polarized alkali atoms with buffer-gas atoms, with the walls of the cell, and with one another. The latter, known as spin-exchange-relaxation, is the main sensitivity-limiting factor in traditional atomic magnetometers. As conceptualized by Happer and Tang in their 1973 paper [20], this phenomenon is almost completely eliminated in a spin-exchange relaxation-free (SERF) magnetometer (Chapter 5) by using a high-density alkali vapor in a low magnetic field along with a high-pressure buffer gas which prevents diffusion to the walls [21,22]. SERF magnetometers have an estimated fundamental sensitivity limit below ≈ 0.01 fT/$\sqrt{\text{Hz}}$ for a measurement volume of only 0.3 cm^3, and a demonstrated sensitivity of 160 aT/$\sqrt{\text{Hz}}$ in a gradiometer arrangement with a measurement volume of 0.45 cm^3 at low frequencies (< 100 Hz) [23], challenging the SQUID's undisputed dominance. Xia *et al.* performed the first measurement of the brain's magnetic field using a SERF magnetometer opening the way to a new generation of MEG systems based on non cryogenic magnetometers [24]. A compact fiber-coupled MEG detector was recently demonstrated by Johnson *et al.* [25]. Recent SERF demonstrations of adult [26] and fetal [27] MCG were carried out by the Wisconsin group.

In a SERF magnetometer a magnetic field perpendicular to the pump beam rotates the spins by a small angle proportional to the field magnitude. This angle is measured by the polarization rotation of a laser probe beam perpendicular to the pump beam [21], or parallel to it [19,28]. The fundamental sensitivity of an atomic magnetometer is (see also Chapter 1, Eq. (1.1) and Chapters 2 and 3)

$$\delta B_{\text{SNL}} \approx \frac{1}{\gamma} \frac{1}{\sqrt{nVT_2\tau}}, \tag{16.1}$$

where n is the number density of atoms, γ is the gyromagnetic ratio, T_2 is the transverse spin relaxation time, V is the measurement volume and τ is the measurement time. Due to the high-pressure buffer gas in a SERF magnetometer, the diffusion of the alkali atoms is drastically reduced and the spin rotation is a local effect of the magnetic field. Thus, interrogating any small volume in the cell, termed a "voxel," measures the local magnetic field in that volume [29]. Therefore, the SERF atomic magnetometer is capable of multichannel operation, gradiometer operation, and multiple-site three-dimensional measurements of the magnetic field inside a single vapor cell [30]. Although the sensitivity of an atomic magnetometer relates to the number of alkali atoms in the measuring volume, nV, its high sensitivity allows the measuring volume to be reduced to a few mm^3 while maintaining an acceptable signal-to-noise ratio. Thus, some sensitivity can be sacrificed in order to enable a local measurement of the magnetic field.

A significant step forward towards a multichannel MEG system based on atomic magnetometers has been made by Schwindt, Johnson, and Weisend who developed a single optical-axis, two-color probe/pump SERF magnetometer with Rb vapor [25,31]. They used a unique optical-fiber coupling scheme with a pump laser at the Rb D1 line (795 nm) and a probe laser at the Rb D2 (780 nm). The design proved the possibility of replacing MEG SQUID detectors with SERF atomic magnetometers, by reducing the overall footprint of the magnetometer on the head. This technology lets single atomic magnetometers be densely arrayed around the head as required for a whole-head MEG system. With two such atomic magnetometers placed on the two sides of the head, they measured evoked responses from both nerve and auditory stimulation and compared the signals to those obtained with an Elekta-Neuromag SQUID-based MEG system. Although the signals are not identical (the atomic magnetometer measures the magnetic field tangential to the scalp and the SQUID detects fields perpendicular to it), the first 100 ms are comparable, proving the atomic magnetometers' capability to perform MEG measurements. The evident conclusion, however, is the necessity to further increase the atomic-magnetometer bandwidth.

Besides noncryogenic operation, atomic magnetometers offer several potential benefits over SQUID magnetometers: lower operating cost, no Johnson currents in dewar and radiation shields close to the measuring site, well-developed multichannel photodetector technology, independent and simultaneous measurement of the three magnetic field components, three-dimensional simultaneous measurement of field distribution inside a single large vapor cell, high spatial resolution, and no danger of flux trapping. There are drawbacks too, such as a much smaller bandwidth (though the bandwidth is generally sufficient for biomagnetism applications).

MEG or MCG system design requires more than ultra-sensitive magnetometers in order to measure biomagnetic fields. Effective magnetic shielding including passive and active shielding is required (Chapter 12). With sufficient signal-to-noise ratio (SNR) and sophisticated control loops to actively cancel distant magnetic-noise sources, it may prove possible to make sensitive biomagnetic measurements with minimal shielding. This is a challenge for the field of biomagnetism, and emphasizes the importance of developing sensors with large dynamic range and high bandwidth.

The simplicity of the atomic-magnetometer systems along with their high sensitivity, low-cost operation, and long-time stability have allowed researchers to start investigating biomagnetism in small animals [32] and even in plants [33]. S. Taue *et al.* used a SERF magnetometer to investigate magnetocardiograms in rats and compared them to measurements performed with a 306-channel MEG SQUID system [32]. Corsini *et al.* searched for plant biomagnetism [33]. The plant biomagnetism is unique in its relatively slow characteristic time spanning from minutes to days. The biomagnetism generated by ionic flows and time-varying ionic distributions is expected to be significantly smaller in plants than in the faster processes in animals or humans, imposing a major challenge to biomagnetic measurements. Plants' magnetic field detection therefore requires high-sensitivity magnetometers and long measurement times, making the noncryogenic atomic magnetometers a preferred choice. The choice for this pioneering measurement was the huge titan arum (*Amorphophallus titanum*) plant at the University of California at Berkeley's Botanical Garden. They used two Geometrics G858 scalar atomic magnetometers, establishing an upper limit to the surface plant magnetic field of 0.6 μG. Higher-sensitivity magnetometers in a shielded environment might be therefore necessary in order to measure the tiny magnetic fields of plants.

16.3 Medical applications

Although the heart signal is the easiest to detect and interpret, MCG has had more difficulty transitioning to clinical application due to the large number of competing technologies [34]; however, fetal MCG has become one of the most interesting and compelling applications of biomagnetism. Routine investigation of fetal electrophysiology has not been possible because signal quality of the fetal electrocardiogram is so poor [35] (Fig. 16.3). It is believed that this is due to the formation of the *vernix caseosa*, a waxy substance that forms on the fetal skin and impedes the transmission of the electrical signal to the maternal surface. Because fetal magnetic signals are less affected by the vernix, they provide recordings of much higher quality. The fetal MCG offers important advantages over other methods for assessment of fetal heart rate and rhythm. This application underscores another key advantage of biomagnetic measurement: its safety and noninvasiveness.

Detection of magnetic tissues also has practical utility. Using magnetic susceptometry, paramagnetic-iron compounds in the liver can be assayed noninvasively to diagnose such serious disorders as thalassemia and hemochromatosis, which result in iron overload. Ingested ferromagnetic tracers allow assessment of gastrointestinal function, including orocaecal transit time – the time for food to pass through the gastrointestinal system – and gastric motility – the mixing motion of the stomach [36, 37].

Understanding the workings of the human brain is a foremost challenge for modern science [38]. For this reason, and despite significant achievements in other areas of biomagnetism, the use of MEG as a means of imaging brain function has been the main driving force behind the development of biomagnetism as a new medical technology. Magnetic-source locations derived from MEG need to be registered with anatomical images in order to associate the magnetic source with a specific part of the brain. The fusion of MEG source localization and anatomical imaging is referred to as magnetic source imaging (MSI). In comparison to the other main functional neuroimaging techniques, namely,

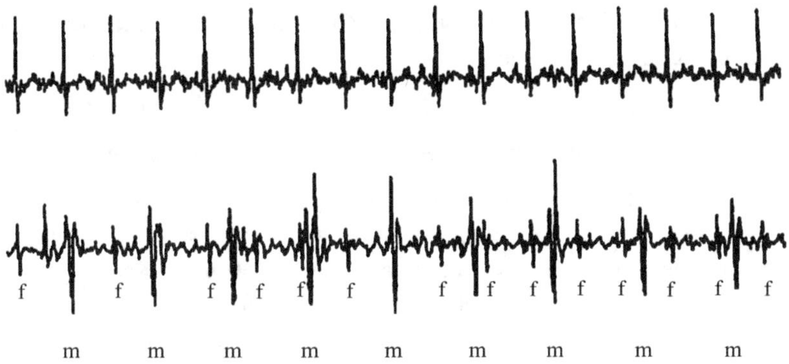

Figure 16.3 Simultaneous fetal magnetocardiogram (fMCG) (top) and fetal electrocardiogram (fECG) (bottom) recording at 38 weeks' gestation. Notice that the fECG contains both fetal (f) and maternal signals. (From Ref. [35].)

functional magnetic resonance imaging (fMRI), single photon emission computed tomography (SPECT), and positron emission tomography (PET), MSI has a much higher temporal resolution; however, a fundamental drawback of MSI is the ill-posed nature of the magnetic inverse problem – i.e., the estimation of the neural current distribution, based on measurement of the scalp magnetic field. This limits the reliability of the source-location estimates.

Most MEG studies employ a stimulus to evoke a brain response. A wide range of brain function can be investigated with the proper choice of stimulus. Typically, the evoked response is smaller than the background spontaneous brain activity, so averaging of many trials is required to increase the signal-to-noise ratio. For a simple stimulus, such as a brief tone burst, the evoked response consists of several components, or deflections, that arise from the auditory cortex. For a more complex stimulus, such as a spoken word, the evoked response will contain components that arise from language-specific areas in addition to auditory areas. By localizing the source of each component, one can determine when and where activity takes place in the brain.

The main clinical application of MEG/MSI is presurgical evaluation of patients with intractable epilepsy, i.e., epilepsy that is not well controlled with drug therapy. MSI is especially useful in cases when no anatomical lesion is evident. The goal in these cases is to identify the seizure focus so that the surgeon can remove it, potentially curing the patient. This is accomplished by localizing abnormal spontaneous MEG activity, such as spikes and slow waves that arise from the vicinity of the seizure focus. The high spatial and temporal resolution of MSI makes it ideal for this application. Another important clinical application of MSI is presurgical functional mapping of such vital brain functions as tactile sensation, motor control, and language. During brain surgery, it is critical for the surgeon to avoid damage to the brain regions that control these functions. In patients with brain tumors, which can markedly distort brain anatomy, or lesions such as arteriovenous malformations, which can cause reorganization of brain function, anatomical landmarks are unreliable and functional mapping becomes essential.

16.4 Magnetocardiography with atomic magnetometers

The magnetic field produced by a human heart consists typically of three dominant temporal features, illustrated in Fig. 16.4. In adults, the P- and T-waves are typically of order 5–10 pT in magnitude. These embrace the much larger (~50 pT) QRS complex. Thus, to measure such signals in real time with high fidelity in the absence of backgrounds requires magnetometers with better than 100 $\mathrm{fT}/\sqrt{\mathrm{Hz}}$ sensitivity over a 100 Hz bandwidth. This sensitivity level is attainable with atomic magnetometers of various designs, and several groups have reported both single and multichannel MCG detection.

Following early work by Livanov *et al.* [16], MCG with an atomic magnetometer was developed in Fribourg in 2003 [17]. The magnetometer was a room-temperature optically pumped Cs M_x resonance magnetometer (Chapter 4), operated as a phase-locked oscillator. The feedback flattened the amplitude/phase response of the magnetometer and allowed the magnetometer to respond to signals outside its nominal ~10 Hz bandwidth. Using a

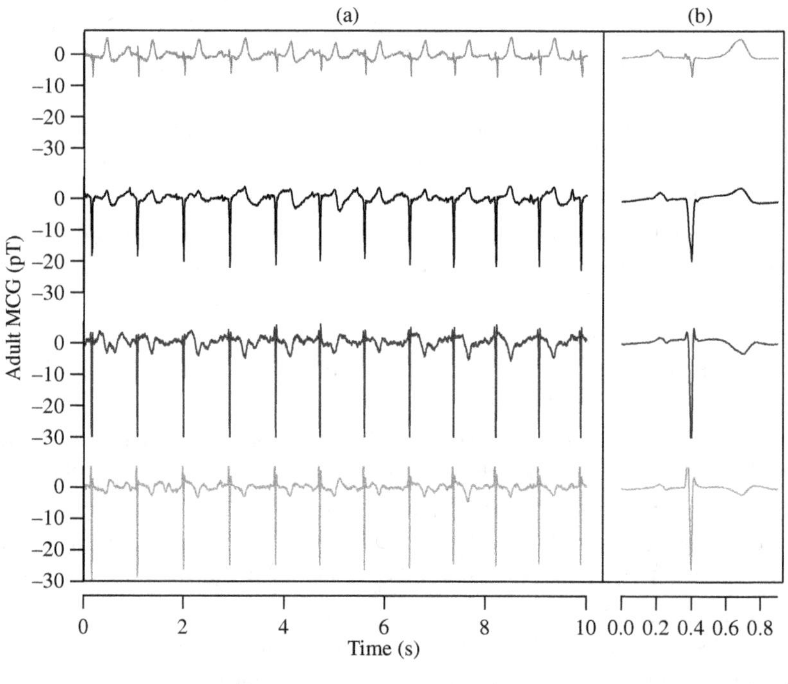

Figure 16.4 (a) MCG traces taken from a 4-channel SERF magnetometer in a magnetically shielded room, with narrow-band backgrounds subtracted. (b) 10-heart-beat averages. Note that since the signals come from different parts of the heart, the phase of the P- and T-complexes changes between different channels. (From Ref. [26].)

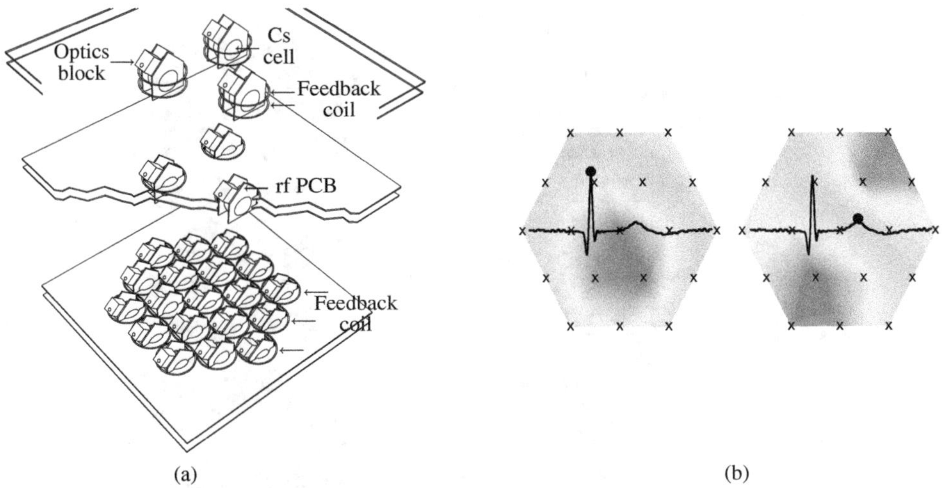

Figure 16.5 (a) Fribourg 19-channel magnetometer array, with additional 3-sensor planes for active field and gradient noise suppression. (b) Spatial maps of the R- and T-wave signals. (From Ref. [18].)

second magnetometer to form a gradiometer, they were able to make a spatial map of the MCG signals by slowly scanning the gradiometer across the subject's chest. The magnetometer sensitivity of 100 $\mathrm{fT}/\sqrt{\mathrm{Hz}}$ was close to the shot-noise level for this type of atomic magnetometer. Although a SERF magnetometer has greater sensitivity, the room temperature M_x magnetometer is simpler and still attains sufficient sensitivity for adult MCG. A detailed analysis of the magnetometer design is presented in Ref. [39].

Taking advantage of this simplicity, the Fribourg group recently demonstrated a 19-channel array of second-order gradiometers operated in a weakly shielded environment [18]. The geometry of the device is shown in Fig. 16.5. Each magnetometer in the array consisted of a vapor cell, a fiber-optic receiver, collimation optics, a photodiode, and coils for generating the RF drive and feedback compensation parallel to the bias-field direction. Magnetic-field and field-gradient noise were suppressed by using six additional magnetometers in two planes parallel to the array to actively stabilize the bias magnetic field and null its gradient along the bias direction. Finally, the individual feedback coil currents were adjusted to keep each channel resonant at the RF frequency. Tight stabilization of the external fields and their gradients allowed the device to operate satisfactorily in an eddy-current shield. Measured spatial maps of the MCG signal are shown in Fig. 16.5. In related work, the group of L. Moi detected MCG in an unshielded environment using a coherent population trapping (CPT) magnetometer [40]. Recently, the Krakow group has also detected MCG in a partially shielded environment using simultaneous ECG detection (S. Pustelny, private communication, 2012).

Multichannel detection is an essential component of high-quality MCG recordings, especially when spatial information is critical. To this end, miniaturization of the individual magnetometer channels is an important goal. One approach is to use a single cell with

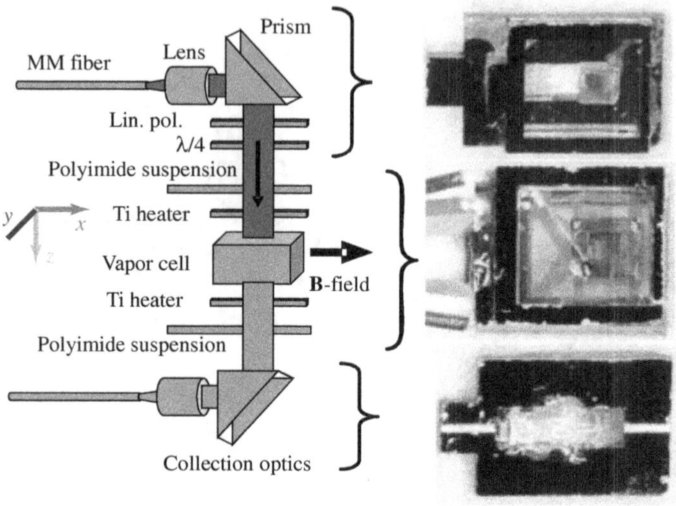

Figure 16.6 Microfabricated 0.75 cm^3 magnetometer sensor for MCG. (From Ref. [41].)

multiple probe beams. This approach lends itself to high-quality gradiometry with small (mm) baselines and will be discussed further in the MEG section below. For optimum signal size and background suppression, however, the gradiometer baselines should be cm scale and the entire array should be 10–20 cm in size. At these sizes, it becomes advantageous, especially when system portability is important, to have multiple pump and/or probe beams as opposed to a single large-diameter laser beam. To this end, it is interesting to explore sub-cm^3 volume magnetometers, as the one recently demonstrated by the NIST group [41] (see Chapter 7). They used a microfabricated silicon/glass cell with fiber-coupled optics contained in a total channel volume of 0.75 cm^3, shown in Fig. 16.6. The magnetometer, run in a SERF resonance mode, achieved a few 100s fT/$\sqrt{\text{Hz}}$ sensitivity, producing good SNR MCG signals in a high-performance shielded room at the Physikalisch-Technische Bundesanstalt (PTB), Braunschweig, Germany. The MCG signals recorded with the SERF magnetometer were similar to those recorded by a SQUID magnetometer under nearly identical conditions. Another approach was demonstrated by the Wisconsin group [26]: a 4-channel SERF magnetometer array with a 5 cm baseline and a sensitivity of 10 fT/$\sqrt{\text{Hz}}$ limited by the residual noise of the magnetically shielded room. The magnetometers were run open loop and therefore the amplitude and phase responses were separately measured and corrected for. Adult MCG signals obtained with this setup are shown in Fig. 16.4.

Fetal MCG (fMCG) is of considerable interest for *in utero* diagnosis and treatment of cardiac arrhythmia. For such an application, the temporal behavior of the heart signal is of primary interest, and spatial information is not necessary in principle. However, because the fMCG field is much smaller than the maternal MCG field, it is still necessary to have sufficient spatial information to suppress the interfering maternal fields. Recently, the Wisconsin group demonstrated real-time fMCG detection using a 4-channel SERF array [27].

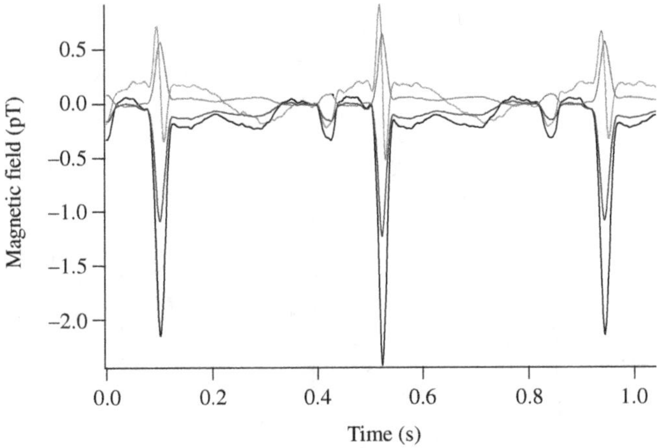

Figure 16.7 Fetal MCG signal obtained from an atomic magnetometer array after application of a spatial filter to remove the maternal MCG interference. (From Ref. [27].)

A spatial filter was used to eliminate the maternal MCG signal, and the isolated fMCG signal (Fig. 16.7) was shown to have nearly the same timing features as measured with a commercial SQUID system.

16.5 Magnetoencephalography with an atomic magnetometer

The first noncryogenic MEG system based on a SERF magnetometer was developed at Princeton University [24] and is depicted in Fig. 16.8. We include here the details of this experimental setup as illustrative of typical practical issues encountered in applying atomic magnetometers to biomagnetic measurements. The core of this MEG is a SERF magnetometer with a cubic Pyrex vapor cell of 7.5 cm on the side filled with less than 1 atm of He buffer gas, 50 torr of nitrogen and a drop of potassium metal. The cell is warmed up to a stabilized temperature of 180°C by hot air flow to obtain a K vapor density of 3×10^{13} atoms/cm^3. The total pressure in the cell at working temperature is less than 1 atm so as to not endanger the subject. The vapor cell is located in a compact "oven." Three oven-walls are transparent, optical-flat vacuum cells for thermal insulation. Their windows are antireflective-coated at 770 nm. The other three walls are insulated by 7-mm-thick microporous ceramic insulation. The surface of the oven facing the subject is maintained at room temperature by a thin medical cooling pad. The oven and the in-shield optics are assembled on a common base, mountable in ten different positions to cover the subject's head for various brain-function investigations.

The magnetometer is placed in a three-layer mumetal cylindrical shield of 1 m inner diameter and 2.6 m length. The cylindrical enclosure has a 0.6 m diameter hole on one end for a sliding bed with a human subject to enter the system. The hole remains open during the magnetometer operation with negligible impact on the measurement performed close to the

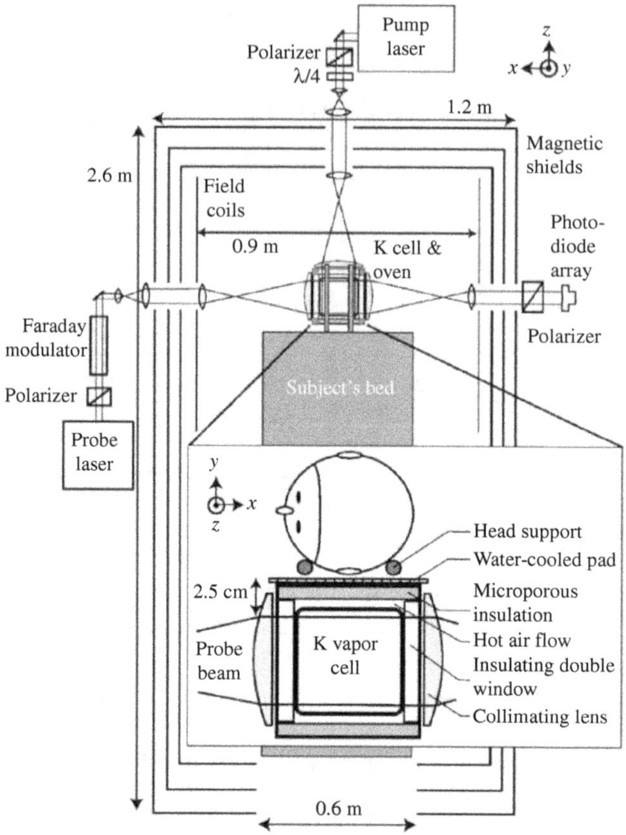

Figure 16.8 SERF-MEG at Princeton University. The closest distance between the probe beam and the room temperature surface is 2.5 cm. (From Ref. [24].)

cylinder's opposite end, due to the attenuation of field penetration through an opening (see Chapter 12). The overall size of this magnetic shield is much smaller than the usual shielded rooms used for SQUID-MEG as there is no need to accommodate a large cryogenic dewar. Still, relatively comfortable room is left for the subject. The smaller shield improves the shielding factor by a scaling factor of $1/L$ while reducing shielding material weight and cost by a scaling factor of L^2, with L being the characteristic size of the shield. Besides the three mumetal layers of passive shielding, there are 18 computer-controlled coils for active shielding. The 18 coils are mounted on an insulating cylindrical shell and compensate for residual constant and linear-gradient fields. The laser beams entering the shield cover the entire vapor.

This SERF-MEG system uses a 0.5 W circularly polarized pump laser beam resonantly tuned to the potassium D1 line at 770 nm, and a 0.5 W linearly polarized probe beam detuned from the resonance by approximately 1 nm. The probe-beam linear polarization is modulated

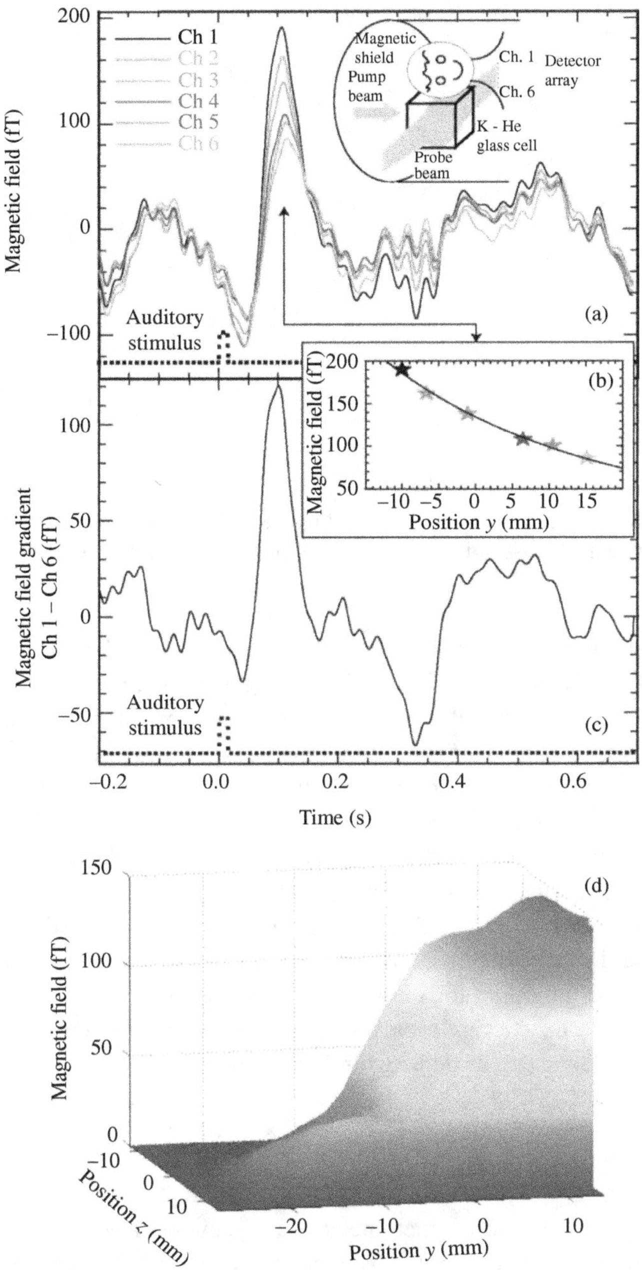

Figure 16.9 Auditory evoked magnetic fields obtained with an atomic magnetometer after averaging 600 stimuli presentations. Magnetic fields recorded with a six-channel linear photodetector [panel (a)] and magnetic field gradient determined from the difference between the two outermost channels [panel (c)] as a function of time. The magnetic field background fluctuations [seen, for example, before the auditory stimulus in panel (a)] are significantly reduced in the gradient measurement. The spatial dependence of the magnetic field at 100 ms after the stimulus is shown in panel (b) with a dipolar field fit. 2D spatial distribution of magnetic fields recorded with a 16×16 photodiode array at 100 ms after the stimulus is shown in panel (d). (From Ref. [24].)

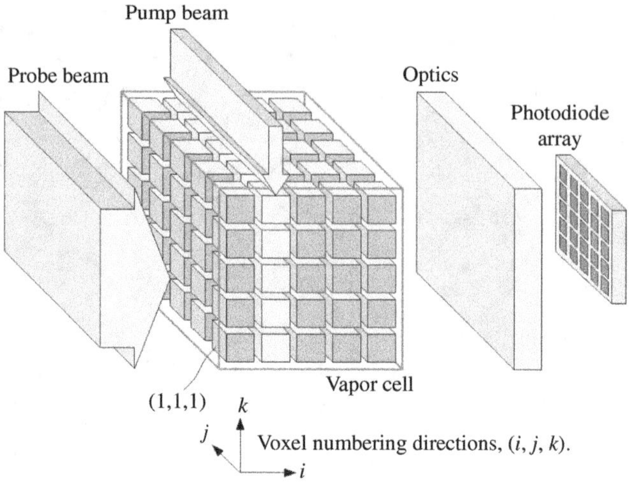

Figure 16.10 Three-dimensional operation of a single SERF magnetometer cell. The pump beam is sequentially illuminating consecutive layers in the vapor cell. (From Ref. [30].)

with a Faraday rotator at 3 kHz. The optical rotation of the probe-beam polarization due to the investigated magnetic field is visualized with a crossed polarization analyzer and a photodiode linear array of six elements or a two-dimensional array of 16×16 elements. The signals are demodulated by field programmable gate arrays on data acquisition boards. A two-dimensional mapping of the magnetic field in the cell is obtained after calibration. Figure 16.9 depicts evoked brain signals following auditory stimuli applied to the subject's ear opposite to the magnetometer cell by a pneumatic earphone. The N100m peak observed 100 ms after the stimuli is clearly seen [38]. The six channels monitored by the linear photodiode array are separated by 5 mm inside the magnetometer cell. Although this was not done in these first experiments, pumping the cell in successive layers perpendicular to the probe, instead of pumping the entire cell simultaneously, allows the measurement of polarization rotation due to a single "voxel," producing a three-dimensional mapping of the magnetic field inside the cell as depicted in Fig. 16.10 [30].

16.6 Summary

In the last few years, atomic magnetometers have demonstrated sufficient sensitivity to join SQUID detectors in being limited by residual magnetic-field noise in shielded rooms. Challenges facing atomic-magnetometer technology include improving the bandwidth and dynamic range and extensions to large numbers of channels for imaging applications. These goals appear to be technically feasible and may result in a substantially reduced detector cost/channel for clinical applications. They may also allow for biomagnetic measurements in minimally shielded environments.

References

[1] P. Nielsen, R. Engelhardt, *et al.*, Transfus. Sci. **23**, 257 (2000).
[2] V. Le Gros, D. Lemaigre, *et al.*, Eur. Respir. J. **2**, 149 (1989).
[3] S. J. Williamson and L. Kaufman, J. Magn. Magn. Mater. **22**, 129 (1981).
[4] M. S. Hämäläinen, Brain Topography **5**, 95 (1992).
[5] R. Fenici, D. Brisinda, *et al.*, Expert Rev. Mol. Diagn. **5**, 291 (2005).
[6] G. M. Baule and R. McFee, Am. Heart J. **66**, 95 (1967).
[7] D. Cohen, Science **156**, 652 (1967).
[8] D. Cohen, Science **161**, 784 (1968).
[9] D. Cohen, Rev. Phys. Appl. **5**, 53 (1970).
[10] J. E. Zimmerman, P. Theine and J. T. Harding, J. Appl. Phys. **41**, 1572 (1970).
[11] D. Cohen, E. A. Edelsack and J. E. Zimmerman, Appl. Phys. Lett. **16**, 278 (1970).
[12] R. Crease, Science **245**, 1041 (1989).
[13] D. Cohen, Science **175**, 664 (1972).
[14] J. P. Wikswo Jr., J. P. Barach and J. A. Freeman, Science **208**, 53 (1980).
[15] D. Budker and M. Romalis, Nature Physics **3**, 227 (2007).
[16] M. N. Livanov, A. N. Kozlov, A. V. Koronevskii, V. P. Markin, S. E. Sinelnikova, and Yu. A. Kholodov, Dokl. Akad. Nauk USSR **238**, 253 (1978).
[17] G. Bison, R. Wynands and A. Weis, Appl. Phys. B **76**, 325 (2003).
[18] G. Bison, N. Castagna, A. Hofer, P. Knowles, J.-L. Schenker, M. Kasprzak, H. Sudan, and A. Weis, Appl. Phys. Lett. **95**, 173701 (2009).
[19] S. Knappe, T. H. Sander, O. Kosch, F. Wiekhorst, J. Kitching, and L. Trahms, Appl. Phys. Lett. **97**, 133703 (2010).
[20] W. Happer and H. Tang, Phys. Rev. Lett. **31**, 273 (1973).
[21] J. C. Allred, R. N. Lyman, T.W. Kornack and M.V. Romalis, Phys. Rev. Lett. **89**, 130801 (2002).
[22] I. K. Kominis, T. W. Kornack, J. C. Allred and M. V. Romalis, Nature **422**, 596 (2003).
[23] H. B. Dang, A. C. Maloof, and M. V. Romalis, Appl. Phys. Lett. **97**, 151110 (2010).
[24] H. Xia, A. Ben-Amar Baranga, D. E. Hoffman, and M. V. Romalis, Appl. Phys. Lett. **89**, 211104 (2006).
[25] C. N. Johnson, P. Schwindt, and M. Weisend, Appl. Phys. Lett. **97**, 243703 (2010).
[26] R. Wyllie, M. Kauer, G. Smetana, R. Wakai, and T. Walker, arXiv:1106:4479 (2011).
[27] R. Wyllie, M. Kauer, R. T. Wakai, and T. G. Walker, arXiv:1202.5019 (2012).
[28] V. Shah and M. Romalis, Phys. Rev. A **80**, 013416 (2009).
[29] A. Ben-Amar Baranga, S. Appelt, C. J. Erickson, A. R. Young and W. Happer, Phys. Rev. A **58**, 2282 (1998).
[30] A. Gusarov, D. Levron, E. Paperno, R. Shuker and A. Ben-Amar Baranga, IEEE Trans. Magn. **45**, 4478 (2009).
[31] P. D. D. Schwindt and C. N. Johnson, Sandia Report **SAND201O-8443** (2010).
[32] S. Taue, S. Ichihara, Y. Sugihara, K. Ishikawa, H. Sugioka, N. Mizutani, L. Shujie, Y. Hirai, O. Tabata, and T. Kobayashi, in *Proceedings of Biomag 2008* **9–11** (Sapporo, Japan, 2008).
[33] E. Corsini, V. Acosta, N. Baddour, J. Higbie, B. Lester, P. Licht, B. Patton, M. Prouty, and D. Budker, J. Appl. Phys. **109**, 074701 (2011).
[34] H. Koch, J. Electrocard. **37** (supplement), 117 (2004).
[35] R. T. Wakai, M. Wang, *et al.*, Am. J. Obstet. Gynecol. **170**, 770 (1994).
[36] R. B. Oliveira, O. Baffa, *et al.*, Eur. J. Gastroenterol. Hepatol. **8**, 491 (1996).
[37] A. Irimia, W. O. Richards, *et al.*, Phys. Med. Biol. **51**, 1347 (2006).

[38] M. Hämäläinen, R. Hari, R. Ilmoniemi, J. Knuutila, and O. Lounasmaa, Rev. Mod. Phys. **65**, 413 (1993).
[39] G. Bison, R. Wynands, and A. Weis, J. Opt. Soc. Am. B **22**, 77 (2005).
[40] J. Belfi, G. Bevilacqua, V. Biancalana, S. Cartaleva, and L. Moi, J. Opt. Soc. Am. B **24**, 2357 (2007).
[41] R. R. Mhaskar, S. Knappe, and J. Kitching, *Proceedings of the IEEE International Frequency Control Symposium, Long Beach, CA*, 376 (2010).

17
Geophysical applications

M. D. Prouty, R. Johnson, I. Hrvoic, and A. K. Vershovskiy

17.1 Airborne magnetometers and gradiometers

Along with electromagnetic (EM), gravity, and radiation detection methods, magnetometry is a basic method for geophysical exploration for minerals, including diamonds and oil. Fixed-wing and helicopter-borne magnetometers and gradiometers are generally used for assessment explorations, with ground and marine methods providing for follow-up mapping of interesting areas.

Magnetometers have been towed by or mounted on airborne platforms for resource exploration since the 1940s [1]. Mapping of the Earth's magnetic field can illuminate structural geology relating to rock contacts, intrusive bodies, basins, and bedrock. Susceptibility contrasts associated with differing amounts of magnetite in the subsurface can identify areas that are good candidates for base and precious metal mineral deposits or diamond pipes. Existing magnetic anomalies associated with known mineralization are often extrapolated to extend drilling patterns and mining activities into new areas.

After World War II fluxgate sensors, originally employed for submarine detection, replaced dipping needle and induction coil magnetic field sensing systems as the airborne magnetometer of choice. While the fluxgate and induction magnetometers could measure the components of the Earth's field rapidly (100 Hz or faster), their sensitivity to orientation made them a poor choice for installation on moving platforms. Experiments by Packard and Varian in 1953 on nuclear magnetic resonance resulted in the invention of the orientation-independent total-field proton precession magnetometer and total-field magnetometers replaced vector magnetometer systems in mobile platforms.

Basic research on the cesium atomic clock during the 1960s resulted in the development of high-performance magnetometers with much higher sample rates and sensitivities than those using proton precession methods. These optically pumped magnetometers replaced

Optical Magnetometry, ed. Budker, D. and Jackson Kimball, D. F. Published by Cambridge University Press.

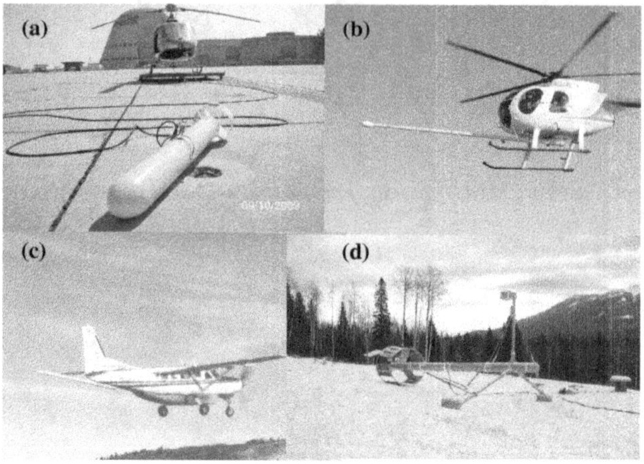

Figure 17.1 Various airborne magnetometers. (a) Helicopter-towed bird sensor; (b) sensor mounted on a nose stinger on a helicopter; (c) tail stinger-mounted sensor; (d) gradiometer bird for helicopter surveys.

proton precession systems in the 1970s and 1980s. Since that time most airborne surveys have been conducted with high-speed, high-sensitivity cesium vapor or potassium magnetometers.

The primary purpose of an airborne magnetic survey is to delineate magnetic features of a survey area in an economical manner. Installation of one to four sensors allows for up to three gradient measurements (along- and across-track and vertical). Airborne magnetometers can be installed in wing-tip or tail-stinger housings, externally mounted in "booms," or towed in streamlined "birds." Up to 100 readings per second are utilized by fixed-wing installations, providing enough sampling to allow for anti-alias pre-filtering of the data. Figure 17.1 shows several examples of airborne systems.

Measurements along-track cover the terrain adequately with one reading every 10 to 30 m at a flying altitude of 30 to 50 m. However, across-track measurements, determined by line spacing, are generally under-sampled and subject to the aliasing of geologic information. Aliasing can be reduced by the installation of one magnetometer on each wing tip to obtain smaller line spacing or to measure an across-track gradient.

Fixed-wing aircraft allow for high survey speeds and, therefore, faster coverage of desired areas. Sensors are typically mounted in nose or tail stingers and wing-tip housings to minimize interference with flight dynamics or pilot practices in the way a towed bird would. However, sensors mounted close to the airframe are greatly influenced by the magnetic signature of the aircraft itself. The proximity of the conductive aircraft structure, engines, and magnetic components for control of the aircraft, along with electrical currents from the avionics, require data compensation either in real time or in post processing. Through compensation, the signals created by the aircraft itself may be reduced by a factor or 20 or more.

Compensation systems must also subtract the complicated induced and permanent magnetic fields of the aircraft and eddy currents associated with motion or acceleration of the conducting aircraft skin in the Earth's field. Final data noise levels of 100 pT/$\sqrt{\text{Hz}}$ RMS are common, and may approach 1 pT using high-sensitivity vapor magnetometers, magnetically clean aircraft, and modern compensation solutions.

Signal from the airframe may be much reduced, and compensation may be entirely avoided, by mounting the magnetometers in towed birds. Towed birds are difficult to utilize with a fixed-wing aircraft, but are often used from helicopters. The highest quality geophysical data can often be obtained from sensors mounted in helicopter-towed birds. Direct installations on helicopter skids or stingers are rare in mineral exploration but are used for ordnance detection.

A typical helicopter installation utilizes a bird towed 30 m or more below and behind the helicopter. Due to reduced interference from the aircraft, system sensitivities are superior, reaching 10 pT or lower total-field noise level and few pT/m gradient sensitivities at 10 readings per second. While there is no need for active compensation in helicopter installations, higher operational costs of rotary wing aircraft make helicopter magnetometer surveys most suitable for mountainous or rough terrain or detailed surveys.

17.2 Ground magnetometers/gradiometers

Portable magnetometers are used for ground follow-up of airborne surveys for minerals, diamonds, and hydrocarbon structures, as well as to aid interpretation of seismic and electromagnetic surveys (Fig. 17.2). They are also used for detection and characterization of unexploded ordnance (UXO) and of utilities (underground storage tanks, pipelines, telecommunication cables) and for rapid nonintrusive survey of archaeological sites. Alkali magnetometers are also suitable for forensic science investigations, since most excavations, such as grave sites, create a magnetic anomaly.

Single-axis fluxgate magnetometers were used in ground magnetic surveys until the invention of the proton precession magnetometer in the late 1960s. In the early 1970s proton precession magnetometers became small and lightweight enough to be carried as portable systems. The total-field measurement of those sensors eliminated the orientation dependence of fluxgate magnetometers, which measure a single component of the vector field.

Proton precession magnetometers have a sample interval of approximately 0.5 to 3 s and require substantial power to polarize the proton-rich hydrocarbon fluid prior to relaxation. Incorporating a fluid with a small number of unpaired electrons and utilizing RF pumping resulted in the dynamic-polarization (Overhauser-effect) magnetometer, which requires much less power and provides faster sampling of as little as 0.2 s. Still, performance is limited due to the proton's low-frequency precession signal and the requirement of a deflection pulse, which briefly interrupts the field-reading process.

Self-oscillating optically pumped cesium vapor magnetometers became available in the 1980s and 1990s and provided a much higher Larmor frequency and continuous

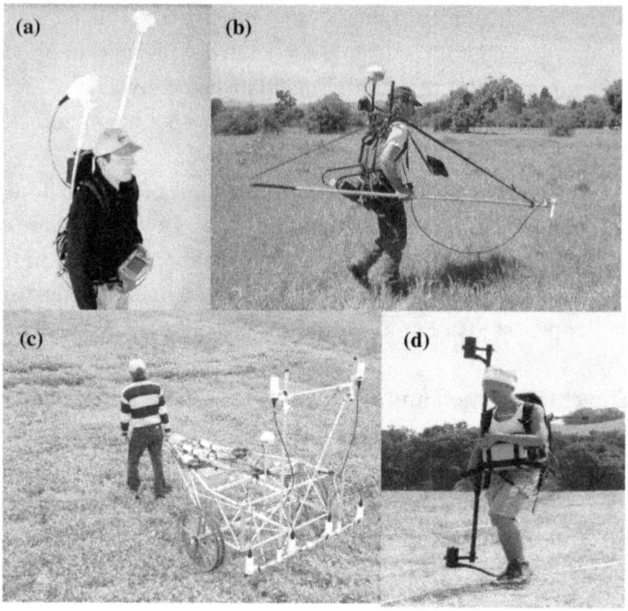

Figure 17.2 Land magnetometers in various configurations. (a) System for mineral exploration with a high-mounted sensor and GPS antenna from Geometrics; (b) system for shallow object detection, with the sensor close to the ground; (c) 6-sensor system for gathering high-density data; (d) portable gradiometer from GEM Systems.

readings. Cesium or potassium magnetometers used in portable instruments typically sample at 10 Hz. Optically pumped magnetometers also have improved tolerance of undesirable environmental conditions such as gradients and high-frequency or large-amplitude interferences.

Today, ground magnetic surveys are usually performed by making a differential measurement between a roving magnetometer and a stationary base station. This technique eliminates the temporal variations of the magnetic field, which are about 10–30 nT in amplitude over the course of a day. For good performance, base-station readings must be synchronous with the roving magnetometer readings. A variety of sensor configurations are used, including a single sensor, vertical or horizontal across-track gradiometers, or multisensor arrays on carts, snowmobiles, or other vehicles.

Measurement of position is generally done by GPS (Global Positioning System). The previous method of surveying and clearing lines and marking measurement points with flags is now largely abandoned.

Ground surveys face a number of difficulties, some of which do not exist in airborne installations. Magnetic contamination from the electronics and operator may result in significant heading errors; power lines and other man-made magnetic fields create interfering signals; and the presence of high gradients places greater requirements on the positioning of the sensors.

The advantages of optically pumped magnetometers greatly enhance the utility of magnetic-gradient surveys, which are more susceptible to all the problems mentioned. Gradiometric measurements are helpful in more precisely determining target depth and shape. Gradient maps define the position and shape of anomalies more sharply than magnetic maps. In addition, shallow targets are emphasized in gradient maps. The gradient must be calculated from strictly synchronous readings of the sensors so that measured gradients are not influenced by temporal variations.

Archaeological surveys are usually performed with magnetic gradiometers. One sensor is placed close to the ground and the other about 2 m above. This configuration attenuates the influence of deeper structures, allowing better resolution of shallower objects. Archaeological surveys require high sensitivity as the amplitudes of expected anomalies are low [2].

17.3 Marine magnetometers/gradiometers

Oceanographers quickly adapted early airborne magnetometer survey tools for marine use. Using such instruments, Scripps Institution of Oceanography discovered magnetic stripes in the seafloor in the 1950s. This led to the understanding of seafloor spreading and the reversal of the Earth's magnetic field approximately every 100 000 years [3].

Applications broadened in the 1970s and 1980s to include surveys for marine archaeology and wreck location. Most major treasure shipwrecks were found using marine magnetometers. During the 1980s and 1990s, as oil drilling reached farther offshore and into deeper environments, route surveys for pipelines, drill rig emplacements, and fiber-optic cables became routine. Atomic vapor magnetometers were brought to the marine market in the late 1980s and early 1990s to increase the detection efficiency and range. They are now supplemented by Overhauser magnetometers with fully omnidirectional sensors (no dead zones), very low power consumption, and tridirectional gradiometers. Figure 17.3 shows several commercial systems.

Offshore ordnance cleanup has become a major use of high-speed marine magnetometers. It is estimated that as much as 10% of artillery-fired ordnance does not detonate, meaning that there are thousands of targets in near offshore locations as diverse as Hawaii, Hong Kong, the Philippines, Iraq, Kuwait, France, the UK, Southern California, and Japan. Marine magnetometers and gradiometers are used to delineate and map marine ordnance, which may range from a 20-mm anti-aircraft round to a 1000-lb bomb.

Transverse gradiometers provide additional clarity for marine UXO mapping. Analytic signal maps have been used to make the target locations easier to visualize and newer inversion techniques resolve more subtle anomalies making ordnance and other target remediation more efficient and successful.

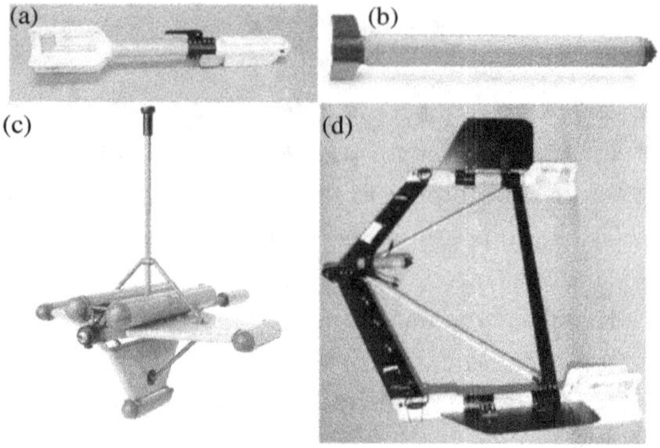

Figure 17.3 Commercial marine magnetometers. (a) Single sensor towfish from Geometrics; (b) towfish from Marine Magnetics; (c) gradiometer system from Marine Magnetics; (d) transverse gradiometer system from Geometrics.

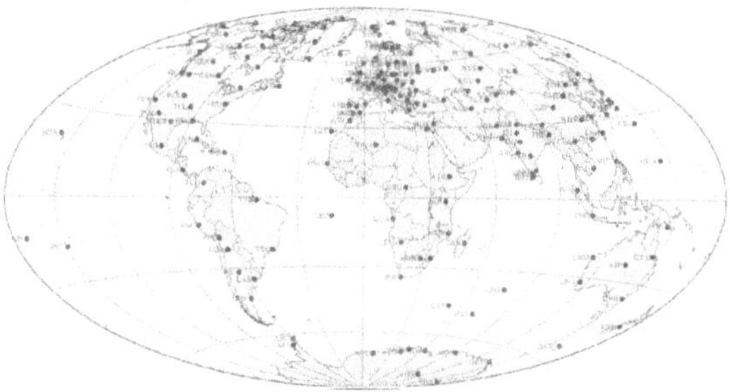

Figure 17.4 Magnetic observatories of the world.

17.4 Vector magnetometry with optically pumped magnetometers

Short- and long-term temporal variations of the vector Earth's magnetic field are measured in magnetic observatories around the world. There are approximately 200 such stations scattered on all continents, as well as many islands. There is a concentration of magnetic observatories in Western Europe and a relative scarcity on the Southern Hemisphere (Fig. 17.4).

Monitoring the Earth's magnetic field (EMF) in a magnetic observatory requires measuring the individual components of the magnetic induction vector **B**, as well as the magnetic induction modulus $B = |\mathbf{B}|$. Component measurements must be made in real-world coordinates. The EMF field vector may be described in several ways. The horizontal projection of **B** is typically denoted as **H**, with modulus $H = |\mathbf{H}|$ and the vertical component as Z. The inclination, I, is defined as the angle between **B** and **H**. Declination, D, is defined as the angle between **H** and geographic north. The induction modulus is often denoted as F. Two common methods of specifying the magnetic field vector are (H,Z,D) and (F,I,D).

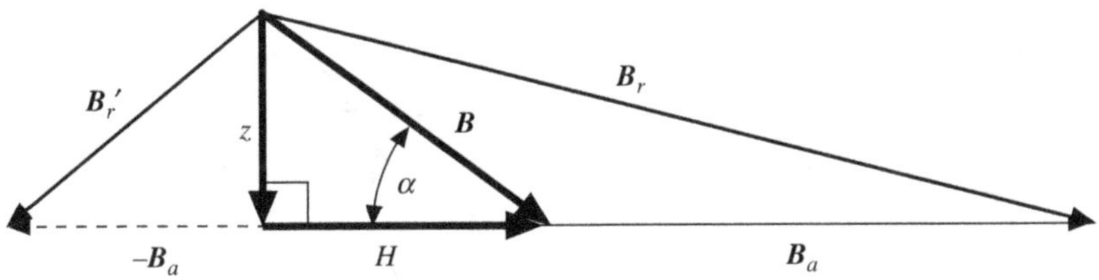

Figure 17.5 Scheme of measuring a field vector with a scalar sensor.

Initially, mechanical magnetometers were used for measuring EMF components, but they have been replaced by fluxgate magnetometers. These devices are characterized by a good (on the level of several pT/$\sqrt{\text{Hz}}$) sensitivity, but also by large (on the level of several nT) drifts, therefore requiring periodic correction. Using a fluxgate for an absolute measurement of EMF vector components requires making a set of measurements using a nonmagnetic theodolite in at least four different directions, allowing the correction of errors such as the lack of alignment between the optical axis of the theodolite and the magnetic axis of the fluxgate as well as the fluxgate magnetization. Recently this procedure was automated [4,5], but taking a single absolute measurement with a fluxgate still takes several minutes.

These factors have caused researchers to look for other methods of measuring EMF components that could combine absolute accuracy with a short measurement time. It was found that these demands could be satisfied by the use of optically pumped quantum magnetometers (OPQMs, see Chapter 4). All OPQMs are scalar instruments, whose readings are nearly independent of the direction of the magnetic field. Interestingly, this property has applicability to the measurement of magnetic field vector components.

Most of the well-known vector schemes use a scalar sensor integrated into a system of magnetic coils, creating a succession of artificial fields nearly perpendicular to the field being measured. This way the field can be calculated after several (three in an idealized case, but typically five) subsequent scalar measurements. These methods are reviewed in Ref. [6].

Shown in Fig. 17.5 is a plane given by $\mathbf{B}$ and an additional field $\mathbf{B}_a$. The resulting vector $\mathbf{B}_r$ is measured by a scalar magnetometer:

$$B_r^2 = B^2 + B_a^2 + 2BB_a \cos\alpha = B^2 + B_a^2 + 2B_a H \qquad (17.1)$$

Methods allowing calculation of Z and H from Eq. (17.1) without knowledge of the exact value of B_a include: (1) finding a B_a which provides a minimal value of B_r, then $B_r(\text{min}) = Z$; (2) using two additional fields, B_a and B'_a; (3) using one additional field and its opposite, B_a and $-B_a$. The last method allows the calculation of H after two successive measurements. B must be measured separately with the additional fields switched off.

All schemes based on these or similar methods require that the scalar sensor must be (1) able to measure the scalar value of the magnetic field independently of its direction at least in some angular zone, and (2) fast, allowing the whole cycle of vector measurements to be

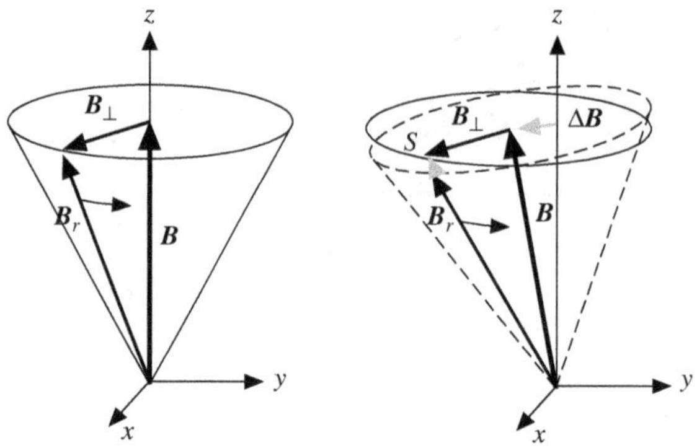

Figure 17.6 A vector diagram of vector magnetometer-variometer principles: **B** is the Earth's magnetic field (EMF) vector, $\Delta\mathbf{B}$ is the EMF vector variation, $\mathbf{B}_\perp$ is the transverse additional field, and S is the error signal.

done while the EMF vector does not change appreciably. A proton magnetometer, which is often used in schemes of this kind, satisfies the first condition, but fails to satisfy the second. Overhauser magnetometers are somewhat faster, with a rate of 5 readings/s, allowing for the cycle to be completed in 1 s.

Because of their fast response, OPQMs (see Chapter 4), with their typical bandwidth of several kHz, are much better suited for use in these schemes. The first OPQM-based component device of this kind [7] was based on an M_z sensor. Its working cycle consisted of five successive measurements – four intended for determining two transverse field components, and one for measuring $|\mathbf{B}|$.

Another example of a successful implementation of an OPQM-based vector magnetometer is reported in Ref. [8]. It involves a ^{4}He vector magnetometer with laser pumping (see Chapter 10), in which the helium sensor is placed in a system of magnetic coils, generating three orthogonal alternating fields varying with three different frequencies (of the order of 10 Hz). The three field components are deduced from responses at their respective frequencies. The magnetometer has a sensitivity of 1 $\mathrm{nT}/\sqrt{\mathrm{Hz}}$ and an operational speed of 1 Hz. Noting that the helium scalar sensor has a sensitivity of 1 $\mathrm{pT}/\sqrt{\mathrm{Hz}}$, the vector measurement is characterized by a sensitivity reduced by 1000 times relative to the scalar measurement.

Two schemes for a vector magnetometer-variometer (VMV) using an alkali OPQM are described in Refs. [9, 10]. This apparatus uses a fast M_x sensor, and fast continuous rotation of the transverse additional field component $\mathbf{B}_a$. The coil systems of both devices produce a transverse field rotating at a frequency f (several hundred Hz) in the plane perpendicular to the main device axis. As a result, an alternating magnetic induction $\mathbf{B}_r$ in the sensor is created, whose vector rotates around the EMF vector, as depicted in Fig. 17.6.

While the EMF vector is parallel to the coil-system axis, the length B_r remains constant during the rotation period (Fig. 17.6, diagram on the left). Upon a change of transverse EMF components, the rotation axis of $\mathbf{B}_r$ deviates from the coil axis (Fig. 17.6, diagram on the right). This causes periodic modulation of $|\mathbf{B}_r|$ at frequency f with phase depending on the direction and amplitude of $\Delta\mathbf{B}$. The signal at f is detected with two quadrature detectors, and the information is used to induce fields, fully compensating for $\Delta\mathbf{B}$. The magnitude of currents in the coils generating these fields is regarded as a measure of the transverse components of the magnetic field B.

The signal amplitude can be estimated as $S = k\Delta B$, where $k = B_\perp/\sqrt{B^2 + B_\perp^2}$. This means that the vector sensitivity of the device grows with the amplitude of the rotating field; unfortunately, this amplitude is limited by the stability and sensitivity of measuring the z-component of the field $\mathbf{B}$ on one hand, and by the scalar-sensor temporal response (to the modulated $|\mathbf{B}_r|$) on the other.

Fast M_x-sensors allow a relatively large amplitude of $B_\perp$, ensuring k on the level of 1/10. The variant of the VMV described in Ref. [9] uses additional 90–95% compensation of the EMF, which permits an increase in the sensitivity of measuring the transverse field components by almost an order of magnitude ($k \approx 1/1.6$). This instrument is characterized by a sensitivity of about 0.015 nT RMS at a 0.1 s sample rate, and reproducibility of the z-channel at a level of ± 0.15 nT.

The instruments described above are classified as variometers because their readings are not absolute: the field they measure is influenced by the magnetic coils. The method for the absolute measurement of magnetic field components, widely used in magnetic observatories, consists of zeroing two field components and measuring the third with the help of a scalar sensor. It does not require high nullification accuracy because the contribution of small residual transverse components to the field modulus is suppressed by a few orders of magnitude.

A scheme employing a fast Cs M_x-OPQM together with relatively fast (5 Hz) switching of additional fields was reported in Ref. [11]. The measurement cycle consists of five stages, ensuring an accuracy of 0.1 nT and a cycle time of 1 s. For further accuracy a slow correction of the readings of the Cs sensor may be done with a Cs–He magnetometer (see Chapter 4).

A way to measure the three components of the EMF vector with high absolute accuracy using OPQM integrated into a symmetric coil system was proposed in Ref. [10]. In this method, the compensating fields are harmonically modulated so that the resultant magnetic field vector B_r in the sensor rotates, with constant magnitude, around the initial field direction, and passes in each rotation cycle through the three positions shown in Fig. 17.7. At each position, two components of the magnetic field B are compensated with high accuracy, while the third one is fully uncompensated and amenable to measurement. It has been shown theoretically that this method provides an absolute accuracy of 0.1 nT with 0.1 s time response. The main drawback of this method is related to the strict requirements which it sets on the orthogonality of the coil system, as well as on the stability of its geometric parameters.

In the 1970s, projects involving the use of the angular dependence of the M_x signal for direct vector measurements started to appear. Indeed, the amplitude of a magnetic resonance shows a dependence on the direction of the pump and probe beams (see Chapter 4). However, this dependence is in itself too weak to be used for practical purposes.

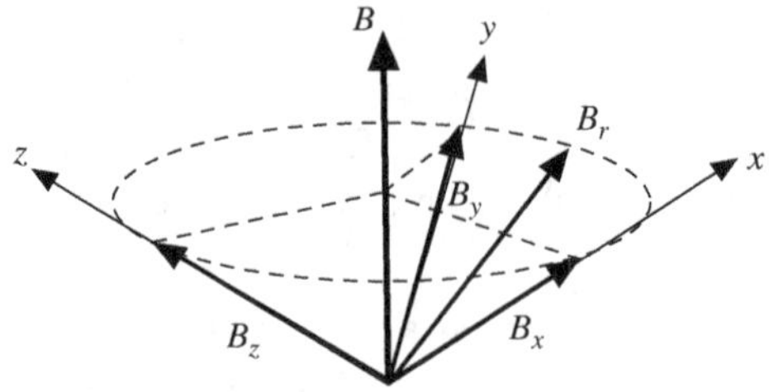

Figure 17.7 An absolute measurement of EMF vector components: B_x, B_y, B_z are the projections of B. (x,y,z) is the local coordinate frame, bound to the coil-system axes.

A more promising possibility is suggested by the dependence of the phase and amplitude of the modulation of the probe-beam on its direction with respect to the magnetic field. A variant of such a measuring scheme is suggested in Ref. [12]. It uses two beams, L_1 parallel to **B**, and L_2 normal to L_1. The first beam provides optical pumping of alkali atoms, and the second serves as the probe. As a result of resonant RF-field action on the system, L_2 undergoes modulation known as the M_x signal due to the interaction with the transverse component of the atomic moment; this signal is used in the magnetometer's self-oscillating loop.

L_1 is not subject to modulation, provided it is exactly parallel to the magnetic field. Deviation of the field from the direction of the beam by an angle θ produces a modulation signal with an amplitude proportional to $\sin(\theta)$ and a phase dependent on the direction of the field's deviation (this signal is also the M_x signal, probed with the transverse component of L_1). This signal can be expanded by respective phase detectors into two orthogonal components corresponding to two components of **B**. The detectors' outputs are fed into additional magnetic coils, nullifying transverse field variations. Thus, this scheme measures the modulus of the field **B** and its two transverse components in its own system of coordinates. According to Ref. [12], the device is characterized by transverse sensitivity of 0.1 nT and drifts of 2–3 nT per day.

The disadvantages of this scheme are common to nearly all devices employing systems of additional artificial magnetic fields; they are (1) the presence of additional magnetic fields influencing the process of absolute measurements, and (2) absence of easy means to define the coordinate system of the device relative to the geophysical coordinate system.

A significant modification of the Fairweather–Usher scheme was proposed in Ref. [13]. This scheme does not stipulate using any additional magnetic fields and, therefore, is capable of an absolute measurement of the magnetic-field vector components. Instead of the forced return of the measured magnetic-field vector **B** toward the z-direction, it is suggested varying the direction of the laser beam L_1 so that it would follow the vector of the measured magnetic field. This would make it possible to relate the measurement to the Earth's coordinate system by merely measuring the L_1 beam direction, which can be done with extremely high precision.

Preliminary calculations show that this scheme can ensure the measurement of the modulus, $|\mathbf{B}|$, with an absolute accuracy of 0.02 nT, and of two angles of deviation of **B** with an absolute accuracy and sensitivity of no worse than 0.1 nT for a measurement time of 1 s. The accuracy of this scheme does not depend on the precision of its positioning in space. Its additional advantage is the absence of generated magnetic fields, which makes possible its use in magnetometric observatories concurrently with other devices.

Most of the instruments described in this section exist only as laboratory prototypes, whereas almost all vector measurements worldwide are being done with fluxgates or proton-based magnetometers. Nevertheless, OPQMs have proven their advantages in this realm, so one can expect that commercial devices of this sort will eventually emerge.

17.5 Earthquake studies

A dire need for the ability to warn of impending earthquakes has led to studies with a variety of methods of measurement including magnetics. Observations with proton magnetometers, search coils, Overhauser, and optical magnetometers have led to many reports (some more reliable than others) of earthquake precursors. The results have been only somewhat promising at best. The inability to repeat the measurements, rare occurrences of high-magnitude earthquakes, the very localized extent of expected magnetic anomalies, and the spread of the study areas around the world are all reasons for the moderate success of this endeavor to date.

Historically, a significant precursor was observed in Japan at the Kawazu observatory in 1978 with a moderate earthquake having an epicenter just beneath the magnetic observatory. Sheldon Breiner (the founder of Geometrics) studied magnetic phenomena around the San Andreas Fault with rubidium optical magnetometers in 1967 [14]. He showed several anomalies: precursors preceded the San Andreas Fault creep by tens of hours and preceded local earthquakes by several days. However, he denies consistent correlation of magnetic signals and earthquakes. The Loma Prieta earthquake near San Francisco in 1989 produced a larger precursor [15], which was, however, disputed by a number of scientists [16], sometimes with unconvincing arguments [17].

Magnetic precursors, if they exist, will have some limitations. First of all, as we expect a dipolar type of anomaly, the precursors will be localized (a dipolar field falls off with the third power of distance). Secondly, detectable signals must be of a low frequency as the hypocenters are deep in the Earth and, with even a low conductivity of the ground above, the skin effect will attenuate higher frequencies and prevent their detectability. Millihertz frequencies seem to be the upper limit. Indeed, the Loma Prieta precursor (Fig. 17.8) was of mHz frequency. Measurements at higher frequencies would be unlikely to indicate such a precursor.

Short-base or gradiometer methods of measurement can eliminate distant sources of anomalies (including diurnal changes, ionosphere interference, etc.). Since the gradient signal of a dipole falls off with the fourth power of the distance, anomalies from all but the closest sources are attenuated. Gradiometer measurements require extreme sensitivities.

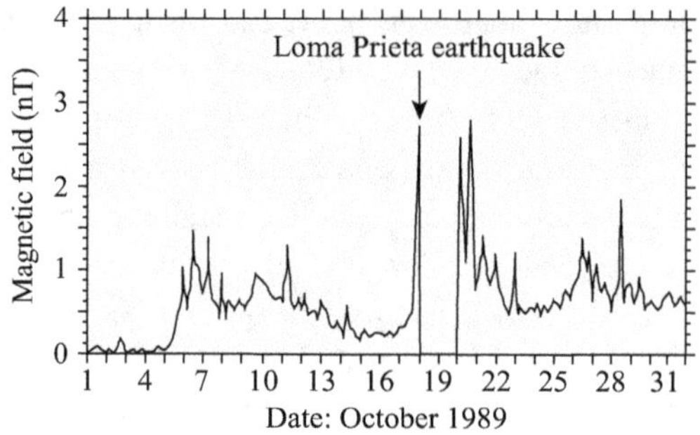

Figure 17.8 Loma Prieta 1989 precursor [15].

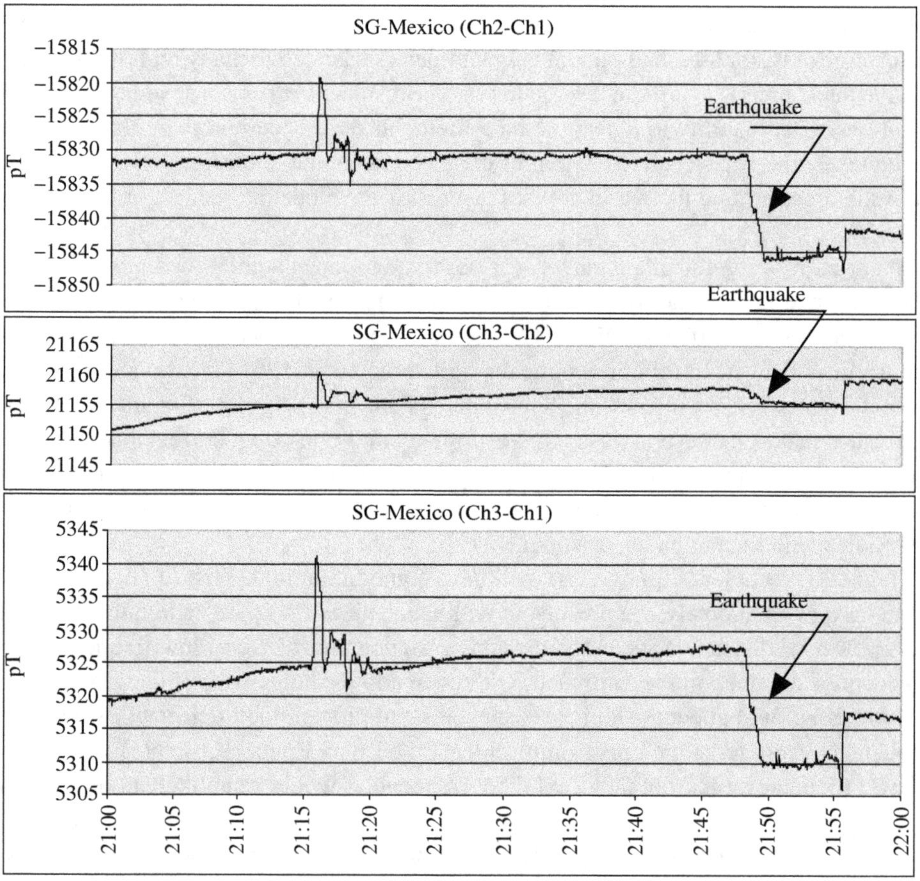

Figure 17.9 Oaxaca 2009 precursors measured with a potassium supergradiometer.

Table 17.1. *Examples of possible magnetic precursors to earthquakes. Distance refers to the distance of sensor from epicenter*

Location (Date)	Mag.	Distance	Freq. (Hz)	Sig. (pT)	Ref.
San Juan Bautista, CA (2000)	5.1	9.4 km	0.0–0.01	20	[18]
Spitak, Armenia (1992)	6.9	129 km	0.1–1.0	50–200	[19]
Guam (1993)	8.0	65 km	0.02–0.05	100	[20, 21]
Oaxaca, Mexico (2009)	4.7	21.8 km		130	[22]

Mag. = magnitude, Freq. = frequency, Sig. = signal amplitude.

Large potassium sensors achieving about 30 fT RMS noise at one reading per second seem to be the best for short-base (a few hundred meters) systems. Preliminary deployment of those sensors in the active region of Southern Mexico (Oaxaca Province) has so far produced one precursor to a magnitude M4.8 earthquake 6 km away from the epicenter and 21 km deep 33 minutes before the earthquake (Fig. 17.9).

With all the magnetotelluric, atmospheric, diurnal, and other interferences it appears that the gradiometer systems must have relatively dense networks to cover an area under examination. Some case histories of reported precursors are summarized in Table 17.1.

17.6 Applications of magnetometers to detecting unexploded ordnance (UXO)

17.6.1 Introduction to the problem

Unexploded ordnance (UXO) are munitions items left over from gunnery practice, ill-conceived disposal, or war-time conflicts that may contain explosive materials posing a hazard if disturbed. Items may range in size from 37-mm shells to 1000-lb bombs.

Detecting UXO is distinctly easier than detecting land mines. UXO were never intended to be difficult to detect, nor to be triggered by weak vibration or low-pressure contact. UXO usually contain significant amounts of conductive, ferrous material, in contrast to the medium in which they are found. Therefore, geophysical instruments may be successfully applied to their detection. Magnetometers, in particular, are among the key sensors used to detect, identify, and locate UXO.

In the United States prior to 1990, UXO remained largely ignored, as most were buried on military sites with little or no public access. In the early 1990s, as the breakup of the former Soviet Union left the United States with a surplus of military bases, the US government began closing many former military bases and other sites. As these lands were turned toward uses with greater public access, the significant environmental damage left behind by the military soon became evident. UXO contamination was one of the significant environmental hazards associated with the former usage of the sites, and cleanup efforts were required to remediate the dangers posed by these munitions.

Figure 17.10 "Mag and flag" method of locating UXO.

UXO are commonly found near former gunnery targets and practice ranges. In the United States alone, over 11 million acres were determined to contain, or potentially contain, UXO [23]. In addition, surplus ordnance was sometimes dumped into disposal sites on land or underwater near military harbors. Record keeping of the locations of such activity was not a high priority. This further complicates the problem by greatly increasing the amount of land that must be investigated to determine whether explosive hazards may exist in those locations.

Initial cleanup efforts used simple fluxgate gradiometer instrumentation in a method known as "mag and flag." Fluxgate gradiometers are good at detecting UXO, and operators carrying them would traverse ranges containing UXO and place flags where magnetic anomalies were detected by the instrument. Other crews would then dig up the anomalies at each flag. Such an operation is shown in Fig. 17.10.

While this method is simple, there are several drawbacks to using such a method of predicting UXO locations. First of all, a great number of anomalies are often detected, most of which are not hazardous UXO items. Yet, to best ensure the safety of the ordnance-disposal personnel, each anomaly must be carefully exposed and extracted, under the assumption that it might be a dangerous explosive. This greatly increases the cost of investigation. In order to make site remediation more cost effective, new methods for detecting UXO needed to be developed.

17.6.2 Using magnetometers for UXO detection

Magnetometers work by detecting the influence a buried ferrous object has on the magnitude and direction of the magnetic field near the object. The first task in performing a magnetometer survey is to gather magnetic field data across the site. This is done by moving a magnetometer sensor across the area of interest, recording both the magnetic field strength and the positions at which the readings were recorded.

This may be done using a single sensor, or, for greater productivity, an array of several sensors. A few platforms for recording magnetometer data over areas of several or tens of acres are shown in Figs. 17.11 and 17.12.

Figure 17.11 Land-based magnetometer array developed by the US Naval Research Laboratory (NRL), known as the Multi-Sensor Towed Array Detection System (MTADS).

Figure 17.12 Helicopter-deployed magnetometer arrays. Left: a single sensor array from Sky Research. Right: a gradiometer array from Battelle.

Total-field magnetometers, as opposed to vector devices, are particularly suited for deployment on mobile survey platforms. For these platforms, maintaining a constant orientation is impossible, and the uncertainty introduced in the measurement from vector sensors is difficult to remove. Anomalies are typically a few nT in magnitude, while the background magnetic field is typically 50 000 nT. Certain locations, such as Hawaii, contain highly magnetic soils. Some areas in Hawaii also have a lot of UXO, making magnetic detection of UXO particularly difficult there.

Obviously, the nature of the geography of the site will greatly influence the methods used to gather data. In general, geophysical methods for addressing the UXO problem are most suitable when the sites are relatively flat and contain less vegetation. However, such sites are typically the most desired from a development standpoint, so these sites are in general the highest priority for cleanup.

Once the data are gathered over a site, they are filtered, plotted, and analyzed. Methods for doing so are discussed in the following section.

17.6.3 Mathematics of UXO detection

Magnetometers are capable of detecting anomalies caused by the effect of ferrous material on the background Earth's magnetic field. The Earth's field at the surface varies from about

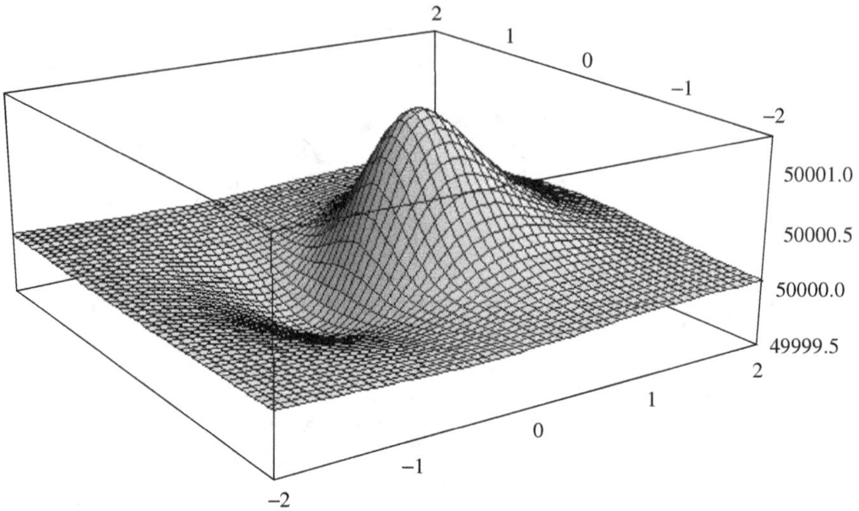

Figure 17.13 Anomaly in the total magnetic field above a magnetic dipole source.

20 000 nT to 80 000 nT. In the small regions under consideration for UXO detection, the field may be treated as a constant. However, its variation with time must typically be taken into account as a survey utilizes data taken at different positions and at different times. Typically, a base-station magnetometer is established, and its readings are subtracted from those of the survey magnetometer(s).

The magnetic field in a source-free region is a potential field, and thus may be expressed as the gradient of a scalar function [24]. Anomalies created by UXO may be decomposed into multipole moments. For the geometries typically used for UXO detection, the dipole moment dominates the response. The magnetic field from a dipole falls off as the cube of the distance between the observation point and the object.

Ideally, it would be possible to completely characterize the object responsible for the magnetic field observations at a set of points above the object. However, sources of potential fields are not uniquely determined by such observations. Hence, it is not possible to uniquely determine the source producing observed distortions in the magnetic field. Any anomaly may, for example, be generated by a layer of dipole sources creating an identical magnetic field.

However, since UXO objects are small and localized, and since the fields they produce may be well approximated by a dipole source (Fig. 17.13), we may determine many useful parameters about the object producing the field. The method typically used is to use a least-squares optimization method to fit a dipole field to the observed field [25].

By adjusting the position and moment of the source dipole to minimize the difference between the field due to the dipole and the field readings taken with the magnetometer, the

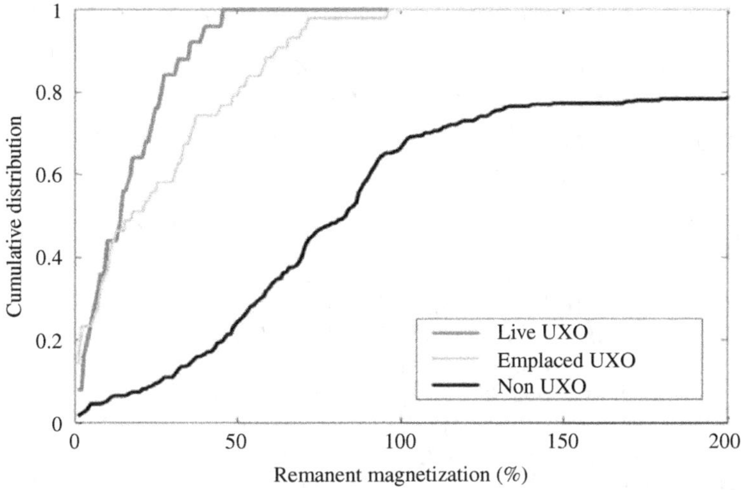

Figure 17.14 Remanent and induced components of the magnetic field.

position and magnetic moment of the target may be estimated. For single targets whose fields do not overlap, this method produces excellent results [26].

In addition, some of the parameters extracted from the dipole model may be used to discriminate unexploded ordnance from other scrap metal and fragments. The magnetic moment estimated in the dipole-fitting method is made up of two components. An object will in general have both an induced component and a remanent magnetic dipole moment (Fig. 17.14). The induced component is a result of the magnetic susceptibility of the object, while the remanent magnetization is a function of the history of the object. It has been found that intact UXO items generally have little remanent magnetic moment.

Therefore, the possible magnitude and direction of magnetic moments from particular ordnance items may be compared to the unknown targets [26]. This method works best when only a few items are expected at a particular site. In such a case, the magnetic moments of those particular items may be plotted as a function of their orientation. This gives a map of the possible directions and magnitudes of magnetic moments produced by the ordnance items. If a target yields an estimated magnetization far from that which would be expected from one of the ordnance items, that item may not require excavation.

References

[1] W. M. Telford, *Applied Geophysics* (Cambridge University Press, Cambridge, 1976).
[2] T. Smekalova, O. Voss, S. Smekalov, V. Myts, and S. Koltukhov, Geoarchaeology **20**, 461 (2005).
[3] M. N. Nabighian, V. J. S. Grauch, R. O. Hansen, T. R. LaFehr, Y. Li, J. W. Peirce, J. D. Phillips, and M. E. Ruder, Geophysics **70**, 33ND (2005).
[4] J. L. Rasson, *Proceedings of the XIII IAGA Workshop on Geomagnetic Observatory Instruments* **2009-1226** (U.S. Geological Survey Open-File Report), 220 (2009).

[5] H. U. Auster, M. Mandea, A. Hemshorn, E. Pulz, and M. Korte, Earth Planets Space **59**, 1007 (2007).
[6] R. J. Lamden, J. Phys. E, **2**, 125 (1969).
[7] J. L. Rasson, Geophys. Trans. **36**, 187 (1991).
[8] O. Gravrand, A. Khokhlov, J. L. Le Mouël, and J. M. Léger, Earth Planets Space **53**, 949 (2001).
[9] E. B. Alexandrov, M. V. Balabas, V. N. Kulyasov, A. E. Ivanov, A. S. Pazgalev, J. L. Rasson, A. K. Vershovski, and N. N. Yakobson, Meas. Sci. Technol. **15**, 918 (2004).
[10] A. K. Vershovskii, M. V. Balabas, A. E. Ivanov, V. N. Kulyasov, A. S. Pazgalev, and E. B. Aleksandrov, Tech. Phys. **51**, 112 (2006).
[11] E. Pulz, *Proceedings of the XIII IAGA Workshop on Geomagnetic Observatory Instruments* **2009-1226** (U.S. Geological Survey Open-File Report), 216 (2009).
[12] A. J. Fairweather and M. J. Usher, J. Phys. E **5**, 986 (1972).
[13] A. K. Vershovskii, Technical Phys. Lett. **37**, 140 (2011).
[14] S. Breiner, *The Piezomagnetic Effect in Seismically Active Areas*, Ph.D. thesis, Stanford University (1967).
[15] A. C. Fraser-Smith, A. Bernardi, P. R. McGill, M. E. Ladd, R. A. Helliwell, and O. G. Villard, Jr., Geophys. Res. Lett. **17**, 1465 (1990).
[16] W. H. Campbell, J. Geophys. Res. **114**, A05307 (2009).
[17] A. C. Fraser-Smith, P. R. McGill, and A. Bernardi, J. Geophys. Res. **116**, A08228 (2011).
[18] D. Karakelian, in *Proc 3rd Conference on Tectonics Problems of the San Andreas Fault System* **XXI** (2000).
[19] O. A. Molchanov, Yu. A. Kopytenko, P. M. Voronov, E. A. Kopytenko, T. G. Matiashvili, A. C. Fraser-Smith, and A. Bernardi, Geophy. Res. Lett. **19**, 1495 (1992).
[20] M. Hayakawa, R. Kawate, O. A. Molchanov, and K. Yumoto, Geophys. Res. Lett. **23**, 241 (1996).
[21] M. Hayakawa, T. Ito, and N. Smirnova, Geophys. Res. Lett. **26**, 2797 (1999).
[22] I. Hrvoic, in *Proceedings of the AGU Annual Meeting* (2010).
[23] Report of the Defense Science Board Task Force on Unexploded Ordnance (2004).
[24] R. J. Blakely, *Potential Theory in Gravity and Magnetic Applications* (Cambridge University Press, New York, 1995).
[25] M. Tchernychev and D. D. Snyder, J. Appl. Geophys. **61**, 184 (2007).
[26] S. D. Billings, IEEE Trans. Geosci. Remote Sens. **42**, 1241 (2004).

Part III

Broader impact

18

Tests of fundamental physics with optical magnetometers

D. F. Jackson Kimball, S. K. Lamoreaux, and T. E. Chupp

18.1 Overview and introduction

At present, we know of four fundamental forces, three of which (electromagnetism, the strong force, and the weak force) are well described by what has come to be known as the Standard Model, a theory developed in the 1960s by Glashow, Weinberg, Salam, and others [1–3]. The fourth, gravity, is well understood at macroscopic scales in terms of Einstein's theory of general relativity [4, 5]. In spite of the spectacular agreement between these theoretical descriptions and numerous experimental measurements, it has been exceedingly challenging to develop a consistent theory of gravitation at the quantum scale, primarily because of the extreme difference between the mass and distance scales at which experimental tests of the two theories are performed. Furthermore, there are a variety of observations that have defied satisfactory explanation within this framework, prominent among them the matter–antimatter asymmetry of the universe [6], evidence for dark matter [7], and the accelerating expansion of the universe, attributed to a mysterious "dark energy" permeating spacetime [8]. It is always of great interest to carry out experiments testing the agreement between theory and experiment beyond the frontier of present precision, and the abundant mysteries confronting our modern understanding of fundamental particles and interactions make the present era an especially auspicious time for the discovery of new physics.

The techniques of optical magnetometry are ideally suited for experimental tests of fundamental physical laws involving atomic spins. For example, a variety of optical magnetometry techniques are being used to search for heretofore undiscovered spin-dependent forces that would indicate the existence of new fundamental interactions. The basic concept of such experiments is to search for spin-dependent energy shifts caused by an anomalous interaction as opposed to a magnetic field. To succeed, it is critical that such experiments are carefully designed to rule out or control the myriad known causes of spin-dependent

Optical Magnetometry, ed. Budker, D. and Jackson Kimball, D. F. Published by Cambridge University Press.

Table 18.1. *Experimental and atomic parameters and their discrete symmetry properties under parity (P) and time-reversal (T) transformations, where + indicates even and – indicates odd. Note that the imaginary number i, which appears as a relative complex phase in transition rates, is odd with respect to T. (Adapted from lecture notes of P. G. H. Sandars.)*

Symbol	Property	Transformation P	T
S (**I**), **L** (**J**, **F**)	Intrinsic spin, angular momentum	+	–
k	Wave vector (photon propagation)	–	–
$\hat{\sigma}$	Circular polarization	+	–
$\hat{\epsilon}$	Linear polarization	–	+
E	Electric field	–	+
B	Magnetic field	+	–
g	Earth's gravitational field	–	+
i	Imaginary number ↔ complex phase	+	–
χ	Chirality of handed molecule	–	+

energy shifts such as the Zeeman and Stark effects, gyroscopic effects, and spin-exchange collisions, to name but a few.

Another important class of fundamental experiments are those testing discrete symmetries (symmetries related to noncontinuous changes), the most important examples being parity (P, spatial inversion or reflection of coordinate axes through the origin), time-reversal (T), and charge conjugation (C, the transformation from particle to antiparticle). For example, if nature is invariant under the parity transformation P, then for any allowed physical process the mirror image of that process is equally allowed. Experimental discoveries of discrete symmetry violations gave key insights into the nature of fundamental interactions: in 1957, an experiment by Wu and collaborators [9] established that the charged weak interaction violated parity symmetry; in 1964, Christenson, Cronin, Fitch, and Turlay discovered violation of the combined *CP* symmetry [10]; and measurements of atomic parity violation helped establish the existence of parity-violating neutral weak currents [11, 12].

Experimental observables in atomic tests of discrete symmetries are either energy shifts (monitored, for example, through a quantum-beat frequency – see Chapter 1, Section 1.1) or, as in the case of atomic parity-violation experiments, atomic transition rates (see, for example, Ref. [13]). The observable is related to the Hamiltonian governing the atomic system, which depends on experimental variables such as the magnetic field **B**, electric field **E**, and light polarization $\boldsymbol{\epsilon}$. Each experimental and atomic parameter can be classified according to its transformation properties under spatial inversion (P-odd quantities change sign and P-even quantities are invariant) or time reversal (T-odd quantities change sign and T-even quantities are invariant), see Table 18.1. Consider, for example, orbital angular momentum **L**. Since $\mathbf{L} = \mathbf{r} \times \mathbf{p}$, where **r** is the displacement from the origin and **p** is the momentum, the

Table 18.2. *Rotational invariants describing measured signals corresponding to various experimental geometries and their discrete symmetry properties. (Adapted from lecture notes of P. G. H. Sandars.)*

P- and T-invariant	*P- not T-violating*
$\hat{\sigma}\cdot\mathbf{B}$, $\mathbf{J}\cdot\mathbf{B}$, $\mathbf{F}\cdot\mathbf{B}$, $(\hat{\epsilon}\cdot\mathbf{E})^2$	$\hat{\sigma}\cdot\mathbf{k}$, χ, $(\hat{\epsilon}\cdot\mathbf{B})[(\mathbf{E}\times\hat{\epsilon})\cdot\mathbf{B}]$ $\hat{\sigma}\cdot(\mathbf{E}\times\mathbf{B})$, $\mathbf{F}\cdot\mathbf{k}$, $\mathbf{k}\cdot\mathbf{B}$
T- not P-violating	*T- and P-violating*
$\mathbf{k}\cdot\mathbf{E}$, $(\mathbf{J}_1\times\mathbf{J}_2)\cdot\mathbf{B}$, $(\hat{\sigma}_1\times\hat{\sigma}_2)\cdot\mathbf{B}$	$\mathbf{S}\cdot\mathbf{E}$, $\mathbf{I}\cdot\mathbf{E}$, $\hat{\sigma}\cdot\mathbf{E}$, $\mathbf{E}\cdot\mathbf{B}$

symmetry properties of **L** can be found from the symmetry properties of **r** and **p**. Position and momentum change sign upon spatial inversion, and so **r** and **p** are classified as polar or ordinary vectors (P-odd), but evidently **L** is invariant under spatial inversion and is consequently classified as an axial vector or pseudovector (P-even). Similarly, **r** is T-even and **p** is T-odd, making **L** a T-odd quantity. The discrete-symmetry properties of the Hamiltonian and, consequently, the experimental observable, can be succinctly summarized in terms of a *rotational invariant*, a scalar (P-even) or pseudoscalar (P-odd) quantity proportional to the Hamiltonian built from relevant independent experimental variables (see Table 18.2).

Based on rather general arguments in the framework of quantum field theory, assuming Lorentz invariance and locality, the combined *CPT* symmetry is expected to be valid. Thus, for example, a search for a T-violating interaction also indirectly tests CP-violation, assuming *CPT* invariance. However, as the history of science has repeatedly demonstrated, no matter how general the arguments or elegant the principles, all theories should be subjected to rigorous experimental tests – and indeed there has recently been a concerted effort by many groups to test the assumption of local Lorentz invariance.

In this chapter we review a variety of experiments using optical magnetometry techniques to search for new physics, focusing on three of the most active areas of research: searches for P- and T-violating permanent electric dipole moments (EDMs) of electrons and nuclei, tests of *CPT* or local Lorentz invariance, and searches for anomalous spin-dependent interactions.

18.2 Searches for permanent electric dipole moments

The permanent electric dipole moment (EDM) of a particle or composite system is a displacement between negative and positive charges for a neutral object such as an atom or the neutron, or a displacement between the center of mass and the center of charge for a non-neutral object such as the electron (see Ref. [14] for a comprehensive review). In all cases, this displacement must lie along the total angular momentum **F** of the system (a fact that follows from the Wigner–Eckart theorem, see Ref. [15]), and in fact must be proportional to **F** otherwise additional quantum numbers would be required to describe the object.[1] Thus the EDM $\mathbf{d} = d\mathbf{F}$, where d is analogous to the magnetic moment, $\boldsymbol{\mu} = \mu\mathbf{F}$. Under P, **d** changes sign, but **F** does not, and under T, **F** changes sign, but **d** does not. Thus the existence of an EDM violates P and T symmetries. We can think of the EDM as arising from the electrical polarization of the system along **F** induced by elementary particle interactions that violate P and T, and, assuming CPT invariance, also violate CP.

Searches for EDMs of atoms and molecules probe CP-violating interactions in the background of the much stronger CP-conserving interactions that bind the atom and its nucleus. CP violation arises naturally in the Standard Model of strong, weak, and electromagnetic interactions through two mechanisms: (1) the phase of a complex amplitude in the weak interactions, known as the CKM phase, which is constrained by measurements with neutral K and B mesons [16], and (2) the strong interaction parameter related to the vacuum expectation value of the gluon field, θ_{QCD}, which is constrained by EDM measurements [17]. The Standard Model does not, however, appear to be a complete theory. For example, the predominance of matter over antimatter in the universe is not compatible with Standard Model mechanisms of baryogenesis [18]. CP-violating physics beyond the Standard Model's interactions could produce the observed baryon asymmetry as well as detectable EDMs. Thus the contemporary view of EDM measurements is that they probe new physics beyond the Standard Model of elementary particle interactions. The CP violation that arises naturally with any new physics scenario may give rise to EDMs of fundamental particles, i.e., leptons and quarks, as well as to composite systems (neutrons, protons, nuclei, atoms, and molecules) due to the constituent EDMs and CP-violating interactions among the constituents.

For more than five decades, experimenters have been searching for EDMs, first in the neutron-beam experiments of Ramsey and Purcell [19] and subsequently in experiments on atomic and molecular systems as well as elementary particles. Over this time the experiments have become increasingly more sensitive for many reasons including, for example, improved understanding and control of systematic errors and their uncertainties, new technologies and techniques such as lasers and ultra-cold neutrons, and new theoretical insights that have uncovered systems or observables that are potentially much more sensitive to CP violation. The chase has been a long one, but the enthusiasm has been sustained and has even grown, and the motivations for such studies provided by new developments in elementary-particle physics have become stronger.

EDMs have been measured in a number of systems including the neutron (the original subject), atoms (both paramagnetic and diamagnetic), molecules, and elementary particles, and the EDM of each system can be viewed as potentially arising from one or more fundamental sources. For the neutron and diamagnetic atoms or molecules, θ_{QCD} and quark EDMs are most important with an additional contribution in atoms and molecules from tensor-type CP-violating electron-nuclear forces. For paramagnetic systems, the intrinsic

[1] Such additional quantum numbers are experimentally ruled out by wave function symmetry, hence statistical, requirements (i.e., only two electrons are allowed in an atom's $1s$ level – an additional degree of freedom would permit four (or more!) electrons to occupy the $1s$ level).

Table 18.3. *Summary of most precise direct EDM results for a number of systems. Here d_e refers to the intrinsic electron EDM, C_S and C_T are the coupling strength of the scalar and tensor electron–nucleus interaction and $g_{\pi NN}$ is the CP-violating contribution of the pion–nucleon nuclear binding interaction. Statistical and systematic errors, when reported separately by the authors, are combined in quadrature.*

System	Primary sensitivity	d ($e \cdot$cm)	Ref.
Cs	d_e, C_S	$(-1.8 \pm 6.9) \times 10^{-24}$	[20]
Tl	d_e, C_S	$(-4.0 \pm 4.3) \times 10^{-25}$	[21]
YbF	d_e, C_S	$(-2.4 \pm 5.8) \times 10^{-23}$	[22]
muon	d_μ	$(\ 0.0 \pm 0.9) \times 10^{-19}$	[23]
neutron	θ_{QCD}, quark-EDMs	$(\ 0.2 \pm 1.7) \times 10^{-26}$	[24]
TlF	θ_{QCD}, $g_{\pi NN}$, d_p, C_T	$(-1.7 \pm 2.9) \times 10^{-23}$	[25]
^{129}Xe	θ_{QCD}, $g_{\pi NN}$, C_T	$(\ 0.7 \pm 3.3) \times 10^{-27}$	[26]
^{199}Hg	θ_{QCD}, $g_{\pi NN}$, C_T	$(\ 0.5 \pm 1.5) \times 10^{-29}$	[27]

EDM of the electron d_e contributes along with an electron–nuclear interaction of scalar nature. Table 18.3 provides a summary of the most precise results from a number of important systems. A history of EDM experiments is presented in the online supporting material (www.cambridge.org/9781107010352). The situation can be summarized as follows:

- Results of direct measurements published over the past 50 years for the neutron, atoms, and molecules are all consistent with no EDM for any of these systems.
- There are several possible sources of EDMs in each composite system. These include θ_{QCD}, intrinsic EDMs of the quarks or electrons, and P-odd and T-odd interactions among the constituents. No single measurement has isolated or set a limit on just one phenomenological or fundamental source of an EDM.
- Results for the neutron, TlF, ^{199}Hg and the paramagnetic systems Cs, Tl, and YbF suggest that either most sources of *CP* violation considered so far are small or that fortuitous cancelations have conspired to make all these EDMs smaller than current experimental sensitivities.

For atomic and molecular EDM measurements, significant future advances in sensitivity require new systems and new techniques. New systems that may be of interest for EDM experiments gain sensitivity through enhancements that arise, in essence, because they are significantly more polarizable by *CP*-violating forces, which induce larger T-odd observables. New techniques include cold atoms in fountains or lattices, cold molecular beams, and condensed-matter systems (for example, polarized liquids and solid-state systems).

18.2.1 Basic experimental setup for an EDM experiment

To detect an EDM **d** of a particle (such as the electron, the neutron, or a particular nucleus), the particle of interest is placed in an electric field **E** and the energy shift $-\mathbf{d}\cdot\mathbf{E}$ is measured (or, equivalently, spin precession due to the $\mathbf{d}\times\mathbf{E}$ torque is measured). For charged particles such as the electron, measurements are typically conducted with neutral composite systems (atoms or molecules). One might naively expect the charged constituents of the atom or molecule to rearrange themselves in such a way as to screen the effects of the applied external electric field, so that EDMs would be undetectable in such systems. However, as originally pointed out by Schiff [28] and Sandars [29], the screening can be avoided due to relativistic effects in heavy (high-Z) atoms and molecules with heavy atoms (see also the detailed explanation of this phenomenon by Commins *et al.* [30]). In fact the EDMs of high-Z atoms can greatly exceed the EDMs of the constituent particles [29]: for example, in paramagnetic atoms an electron EDM d_e induces an atomic EDM $d_a \propto \alpha^2 Z^3 d_e$, where α is the fine-structure constant. In heavy polar molecules, strong mixing of opposite parity states with small energy separation generates large internal electric fields along the internuclear axis with which electron and nuclear EDMs can interact [31], leading to energy shifts orders of magnitude larger than in paramagnetic atoms [32,33]. A similar situation can be found in some lanthanide (rare-earth) and actinide atoms [34].

Generally, in EDM experiments both electric (**E**) and magnetic (**B**) fields are applied to an ensemble of the particles of interest, so that the interaction Hamiltonian H can be written

$$H = \boldsymbol{\mu}\cdot\mathbf{B} + \mathbf{d}\cdot\mathbf{E} = \mu\mathbf{F}\cdot\mathbf{B} + d\mathbf{F}\cdot\mathbf{E}\,. \tag{18.1}$$

The first term is T-even and P-even while the second term is T-odd and P-odd (see Table 18.2). Experiments measure the energy separation ΔE or frequency Ω of transitions among Zeeman sublevels labeled by their magnetic quantum number M_F:

$$\Delta E = \hbar\Omega = (\mu B \pm dE)\frac{\Delta M_F}{F}\,, \tag{18.2}$$

where the $+$ and $-$ signs refers to **E** parallel and antiparallel to **B**, respectively. It can be seen from Eq. (18.2) that the experimental observable in EDM experiments is essentially identical to that in optical magnetometry, and hence the significant overlap in tools and techniques.

In Eqs. (18.1) and (18.2) we have neglected higher-order electromagnetic moments which can also be of interest. It should be noted that **B** has contributions from a number of sources, including the desired applied field ($\mathbf{B}_0$) which is often used to establish a quantization axis; spurious laboratory noise fields which include at least the noise generated by magnetic shielding (see Chapter 12) and noise in $\mathbf{B}_0$; the relativistic $\mathbf{E}\times\mathbf{v}$ magnetic field due to particle motion; spurious and systematic fields due to displacement and leakage currents that occur with the application of high voltage to establish an electric field; and the sample magnetization itself. Furthermore, there are magnetic interactions at, for example, the walls of the cell used to contain the sample. Basically, one is faced with all of the problems associated with high-precision magnetometry with the nontrivial additional complication of the need to apply a reversible electric field.

The art of EDM measurements is in the separation of spurious electric-field effects from a true EDM effect. The spurious electric-field effects can often be made small. An important feature of experiments where an electric field is applied is that the EDM effect can be modulated, and thereby discriminated from background drift and understood offsets.[2]

18.2.2 Sensitivity to EDMs

We see from Eq. (18.2) that the sensitivity of an EDM experiment is determined by the precision with which the precession frequency Ω can be measured, weighted by the electric field. F is usually a number of order unity, and most typically $1/2$.[3] Spin-$1/2$ systems exhibit no dilution of the EDM energy shift due to distribution among the Zeeman sublevels.

Another potential advantage of spin-1/2 entities is that they have no electric-quadrupole moment, which in many cell experiments enables longer spin coherence times (see Chapter 14). A dominant spin-relaxation mechanism for vapors of spin $>1/2$ atoms arises from electric-field gradients at the cell walls; this effect is pronounced in, for example, ^{199}Hg [ground state (nuclear) spin $I = 1/2$] where relaxation times in excess of 1000 s have been observed, compared to ^{201}Hg ($I = 3/2$) which in similar situations has significantly shorter relaxation times.[4]

Because EDM measurements can in most cases be reduced to a measurement of a frequency shift,[5] the usual fundamental atomic shot-noise limit, analogous to that discussed in Chapter 1 [Eq. (1.1)] for magnetometric measurements, applies in the determination of the experimental sensitivity:

$$\delta(d)_{\mathrm{SNL}} \approx \frac{\hbar}{E}\sqrt{\frac{\Gamma_{\mathrm{rel}}}{N\tau}}, \tag{18.3}$$

where E is the magnitude of the applied electric field (maximum conceivable laboratory fields are $\sim 10^4$ to $\sim 10^5$ V/cm), N is the number of particles measured (taking into account efficiency and other loss mechanisms), Γ_{rel} is the Zeeman coherence relaxation rate, and τ is the measurement time ($\tau \gg \Gamma_{\mathrm{rel}}{}^{-1}$). The sensitivity $\delta(d)_{\mathrm{SNL}}$ can also be written in terms

[2] This illustrates an advantage of EDM experiments over T-violation studies involving β decay or neutron transmission [35] where the sought-after T-violation signal cannot be turned on and off and appears alongside other allowed processes.

[3] Indeed, all nuclear EDM experiments of significant sensitivity have used spin-1/2 entities.

[4] Note, however, that there are cases of atoms with spin $>1/2$, such as ^{21}Ne, where spin relaxation times can be on the order of hours.

[5] A notable exception, discussed in Section 18.2.3 and in detail in Ref. [36], is the case of solid-state EDM experiments, which do not follow the sensitivity scaling given by Eq. (18.3).

of the counting rate dN/dt or the total number of particles $N_{\rm tot}$ counted:

$$\delta(d)_{\rm SNL} \approx \frac{\hbar\Gamma_{\rm rel}}{E}\sqrt{\frac{1}{(dN/dt)\tau}} = \frac{\hbar\Gamma_{\rm rel}}{E}\sqrt{\frac{1}{N_{\rm tot}}}\,. \tag{18.4}$$

Of course, the actual sensitivity of the experiment must take into account imperfect polarization, photon-detection efficiency, and other technical issues.

Different composite systems (atoms, molecules, nucleons), depending on their particular internal structure, are sensitive to different fundamental sources of *CP* violation, as noted in the discussion surrounding Table 18.3. As an example, the sensitivity of an atomic system to the electron EDM d_e is often parameterized in terms of an enhancement factor, where $d = \eta d_e$: for paramagnetic atoms, $\eta \gtrsim 100$, while for a diamagnetic atom such as ^{199}Hg $\eta \approx 1/100$. However, in the case of heavy polar molecules, the language of "enhancement factors" is inadequate for describing the relationship between external and effective internal fields. In this case, the effective internal electric field is best described in the form $E_{\rm eff} = Q\mathcal{P}$, where Q is a factor accounting for relativistic effects and details of the molecular structure, and $\mathcal{P}$ is the degree of polarization of the molecule by the external field [37]. For atomic systems in achievable laboratory fields, $\mathcal{P} \ll 1$, and so $E_{\rm eff}$ is proportional to the externally applied field E. In contrast, modest laboratory fields (tens of V/cm) can fully polarize polar molecules such that $\mathcal{P} \sim 1$ and $E_{\rm eff} \sim 10^9$ V/cm. Furthermore, molecular experiments can be designed to be intrinsically insensitive to external magnetic fields by choice of the internal molecular states, and promising approaches to comagnetometry with internal molecular states have been developed [38]. These factors have led to concerted efforts by a number of groups to measure EDMs in heavy polar molecules such as YbF [22], TlF [25], PbO [38, 39], ThO [40], WC [41], PbF [42], and also in trapped molecular ions [43].

18.2.3 Electric fields and coherence times for various systems

Once the particle, atom, or molecule for an EDM experiment is chosen, the sensitivity to fundamental *CP*-violating interactions (e.g., η) is determined and, to a large degree, N is also determined. Of course, one of sound mind would not consider an EDM experiment unless N, E, η, and $\Gamma_{\rm rel}^{-1}$ were large enough to extend our knowledge of the fundamental sources of T violation required for an EDM. It also bears remembering that sometimes a system looks good on paper but unforeseen technical problems limit the ultimate sensitivity.

A useful exercise is to compare N, $\Gamma_{\rm rel}$, and E for typical beam and cell experiments, and for specificity let us consider the case of paramagnetic atoms used to search for the electron EDM. For a beam experiment, N is almost always smaller (except in the case of neutrons). In addition, for a beam, the voltage that can be applied across a gap in vacuum, depending on the electrode material, is about 300 kV, and is roughly independent of the separation. The relatively large electric field of $\sim$100 kV/cm achieved in the Berkeley Tl beam EDM experiment, together with an $\eta \approx -550$, enabled a remarkable sensitivity to d_e [21] that has only recently been challenged by a molecular (YbF) beam experiment [22]. For a thermal beam experiment without cooling, $\Gamma_{\text{rel}}{}^{-1}$ can be of order 10^{-2} s, depending on the beam velocity and length of the interaction region. For a paramagnetic atom in a cell, $\Gamma_{\text{rel}}{}^{-1} \gtrsim 1$ s is possible, however, the electric field is limited to a few kV/cm for a variety of reasons. Furthermore, an element such as Tl would need to be heated to a high temperature to have an appreciable vapor pressure in a cell, and such heating is incompatible with the need for high-voltage insulation. The upshot is that beam experiments have dominated the search for the electron EDM to date.

As a comparative case study for a cell experiment employing paramagnetic atoms, the Amherst Cs electron EDM experiment ($E = 4$ kV/cm, $\Gamma_{\text{rel}}{}^{-1} = 20$ ms) obtained a limit of $|d_e| < 1 \times 10^{-25}$ e $\cdot$ cm [20]. The Berkeley Tl beam experiment ($E = 100$ kV/cm, $\Gamma_{\text{rel}}{}^{-1} =$ 2.5 ms) obtained a limit of $|d_e| < 2 \times 10^{-27}$ e $\cdot$ cm [21]. Noting that the sensitivity in Eqs. (18.3) and (18.4) scales with $\sqrt{\Gamma_{\text{rel}}}$ and inversely with E, and further noting that η is five times larger for Tl compared to Cs, we see that the shot-noise-limited sensitivity of the Berkeley experiment is about 50 times better than that of the Amherst experiment for comparable N_{tot}, corresponding to the final relative experimental sensitivities of the two experiments.[6]

In order to apply large electric fields without electric breakdown, an experiment must either employ high vacuum (as in the case of the Berkeley Tl beam experiment) or use a buffer gas (the strategy of the Amherst Cs experiment). The presence of a buffer gas in a paramagnetic-atom magnetometer cell can also help reduce spin relaxation by "hiding" relaxing regions of the cell surface. However, there is a limit to the voltage that can be applied to a cell containing buffer gas: typically, for nitrogen, the breakdown voltage is about 100 V/torr of buffer gas, for pressures of $\sim$1 torr. Noble gases have considerably lower breakdown voltages. On the other hand, for paramagnetic atoms, collisions with the buffer gas lead to spin disorientation, which limited the coherence time in the Amherst experiment to $\sim$20 ms [20].

For a molecular experiment, the presence of a buffer gas would lead to an almost immediate decoherence of the delicate state due to collisions, and as such beam experiments typically employ active pumping to maintain high vacuum. There are instances where a high-vacuum system that is not actively pumped can be stable under the application of high voltage. Old-fashioned electron tubes and cathode-ray tubes are prime examples, but all of these have a passive or active gettering system that physically and/or chemically absorbs the gases that are inevitably emitted by all materials, particularly when heated. The presence of such gettering materials in a magnetometry cell would almost certainly ruin its favorable spin-relaxation properties, and furthermore would probably absorb the atoms of interest. There are two exceptions: (1) the ultra-cold-neutron (UCN) EDM experiments that have been performed to date employ a storage cell that is actively pumped [24], as does (2) a molecular PbO experiment currently being performed at Yale [38, 39]. In these situations,

[6] This agreement, although instructive, is somewhat coincidental as neither experiment reached the shot-noise-limit of sensitivity, and the Amherst Cs EDM experiment had four times larger N_{tot} than the Berkeley Tl EDM experiment.

the cells are "leaky" in that the cells comprise plate electrodes pressed onto an insulating cylinder without any gas seal, and this assembly is placed in a vacuum chamber. For the PbO experiment, molecules that collide with the walls are completely depolarized in any case, and there is a chuck of PbO in the cell which replenishes that which leaks out. For the ultra-cold-neutron experiment, the cell is constructed in the same way, but the leaks are not a serious limitation compared to UCN loss on the walls. The cell is filled and emptied of UCN through an aperture which is closed by a sliding cover during the free-precession phase of the measurement.

The upshot is that for optical magnetometry employing sealed cells, buffer gases that provide high-voltage insulation are generally necessary. For paramagnetic atoms, gases that are good high-voltage insulators (nitrogen) do not have small spin-depolarization cross-sections; however, one gains in sensitivity linearly in E, and loses only as $1/\sqrt{\Gamma_{\rm rel}}$. On the other hand, diamagnetic atoms such as ^{199}Hg and ^{129}Xe do not have their nuclear spins affected in any appreciable way by the presence of a nitrogen buffer gas (oxygen, not surprisingly, depolarizes ^{199}Hg rapidly). The presence of a buffer gas also turns out to be important in limiting several systematic effects associated with the $\mathbf{E}\times\mathbf{v}$ motional magnetic field.

While the above arguments and recent results point to some general advantages of beam experiments compared to cell experiments for electron EDM searches with paramagnetic atoms (best limits to date on the electron EDM come from Tl and YbF beam experiments [21, 22]), the best limit on an atomic EDM in general ($d < 3.1\times10^{-29}$ e$\cdot$cm at a 95% confidence level) has been achieved in a cell experiment using the diamagnetic atom ^{199}Hg [27]. The experimental sensitivity benefitted from a high atomic density ($\sim 4\times10^{13}$ atoms/cm^3) leading to a large N and much longer coherence times than can be achieved in beam experiments ($T_2 \approx 100$–200 s). It can be seen from Eqs. (18.3) and (18.4) that the disadvantage of smaller E (by a factor of 10) achievable in cell experiments relative to beam experiments can be overcome if the factor $N/\Gamma_{\rm rel}$ can be made large enough. Although this is more readily achievable for diamagnetic atoms, recent advances such as spin-exchange-relaxation-free magnetometry (Chapter 5) and new antirelaxation coatings for alkali vapor cells (Chapter 11), as well as the development of novel approaches such as those employed in the neutron and PbO experiments, offer hope that cell-based EDM experiments with paramagnetic atoms may yet again become competitive with beam experiments.

It is possible to use optical trapping to hold an ensemble of atoms to be used for an EDM measurement [44,45]. In this case, the experimental volume is necessarily pumped to ultra-high vacuum, so it is possible to apply fields of many tens of kV/cm. However, electric field gradients will lead to a displacement of the center of mass of the ensemble, and the magnetic and electric gradients therefore need to be carefully controlled. More insidious are motional-field effects related to geometric phases that become especially important when particles can move for long times between direction-randomizing collisions, as is the case for optically trapped ensembles [46].

The possibility of performing an EDM experiment in a liquid has been discussed from time to time. The most prominent example is an ultra-cold-neutron EDM experiment performed directly in a superfluid helium bath, where an increase in electric field by a factor of 5 compared to existing experiments appears feasible [14]. Another case of interest is liquid xenon [47], where the high number density of liquid-state ^{129}Xe ($\sim 10^{22}$ atoms/cm^3), relatively small values of Γ_{rel}, and high electric-field breakdown strength ($\approx$400 kV/cm) offer the possibility of extremely favorable shot-noise-limited sensitivity [Eq. (18.3)]. Furthermore, liquid ^{129}Xe can be hyperpolarized via spin-exchange optical pumping with Rb [48]. Although complications due to nonlinear dipolar interactions create significant challenges for such an experiment [49], recent developments in low-field nuclear magnetic resonance (NMR) may offer new paths forward [50].

One might expect the largest fields of all in a solid system. A solid-state electron EDM experiment, very different in nature from the aforementioned EDM experiments, was proposed by F. L. Shapiro in 1968 [51]. The idea is that if an electric field is applied to a paramagnetic insulating solid sample, the electron spins would tend to become polarized due to their interaction with the field. This polarization increases at lower temperatures, as determined by the Boltzmann factor. Because the electron also carries a magnetic moment, this polarization leads to a net sample magnetization. This magnetization is then measured by an external magnetometer, and the most recent experiments [52] use a pickup coil that "feeds a SQUID"; however, the magnetization could be measured with any magnetometer (although coupling any given system to a cryogenic environment is an engineering challenge). The required magnetometric sensitivity is at the fT/$\sqrt{\mathrm{Hz}}$ level to be competitive with molecular electron EDM experiments using beams. Similar techniques can be applied to search for nuclear EDMs [36]. There have also been proposals to embed alkali atoms in a solid noble-gas matrix, where Cs polarization lifetimes (T_1) of order 1 s have been observed [53].

18.2.4 Magnetometry and comagnetometry in EDM experiments

From the foregoing discussion, it is apparent that there are multiple ways that optical magnetometry can be applied to EDM experiments.

First, the magnetometer itself is modified to allow the application of the strongest electric field possible. A shift in Larmor frequency correlated with, for example, the quantity $\mathbf{E} \cdot \mathbf{B}_0$ would be evidence for an EDM, barring systematic effects. An EDM interaction causes the magnetometer to respond to an electric field in nearly the same manner as a magnetic field, but with near-zero effect for even the largest conceivable fields.

Second, magnetometry can provide measurement of the EDM signal. In an early ^{129}Xe EDM experiment [54], optically pumped Rb was used to polarize a Xe sample through spin exchange, while the subsequent Xe free precession was monitored by its back-effect on the Rb, which then served as a sensitive spin-exchange enhanced magnetometer for the Xe. In the case of the Shapiro solid-state EDM proposal [51], the magnetometer provides a direct measure of the EDM-induced magnetization.

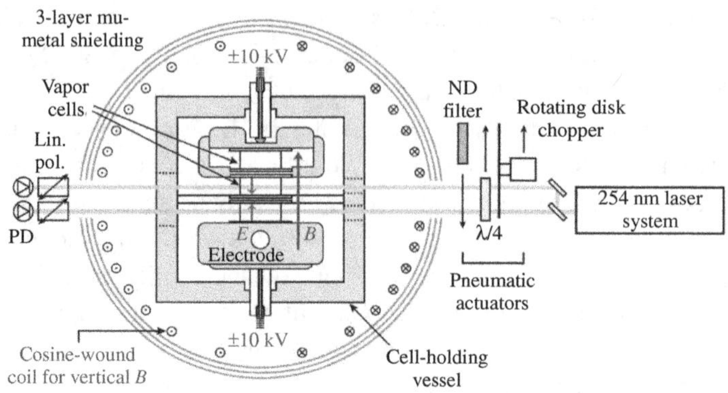

Figure 18.1 Simplified diagram of the experimental setup for the ^{199}Hg EDM experiment, from Ref. [27]. The middle two cells have oppositely directed electric fields, resulting in EDM frequency shifts of opposite sign; the top and bottom cells enclosed by the high voltage (HV) electrodes and thus placed at $\mathbf{E} = 0$, are free of EDM effects and serve to cancel B-field gradient noise and provide checks for spurious HV-correlated B-field shifts. (The outer cell light beams are not shown but travel along the magnetic-shield axis, perpendicular to the middle-cell beams.) Synchronous optical pumping with circularly polarized light spin polarizes the Hg atoms, and spin precession is monitored by measuring optical rotation of linearly polarized probe light.

Third, magnetometry can be used in conjunction with an experiment to test for the presence of systematic effects. In the case of the recent ^{199}Hg experiment [27], extra magnetometer cells without applied electric fields were used in conjunction with similar cells that did have an electric field (Fig. 18.1), and together the system was relatively immune from magnetic effects associated with the application of high voltage.

The term "comagnetometer" first appears in print in 1989 [55], about the time when it was realized that magnetometers separated from the main EDM experimental region often do not provide an adequate measure of magnetic fields correlated with the application of an electric field (Fig. 18.2). A comagnetometer is a spin-polarized gas, distinct from the species used for the EDM measurements but occupying the same volume. The use of a comagnetometer was first advocated by Ramsey in regard to the UCN neutron EDM experiment, and has now been recognized by nearly all workers in the field as a highly desirable feature. Ramsey points out the features required for a comagnetometer species: it must not have an EDM of its own and it must occupy the same volume as the species of interest and provide an identical time average of the field. (Another possibility, as in the PbO experiment [38, 39] and ThO [40] molecular EDM experiments, is to find two internal states with opposite EDM-induced shifts.)

An example of an EDM experiment utilizing a comagnetometer occupying the same volume as the particles of interest is the EDM experiment reported in Ref. [26], where overlapping ensembles of ^{129}Xe and ^{3}He operated as nearly independent spin-exchange pumped masers. Because the sensitivity of an atom to fundamental sources of CP-violation typically scales as Z^2 or Z^3, where Z is the atomic number, the EDM of ^{129}Xe induced by

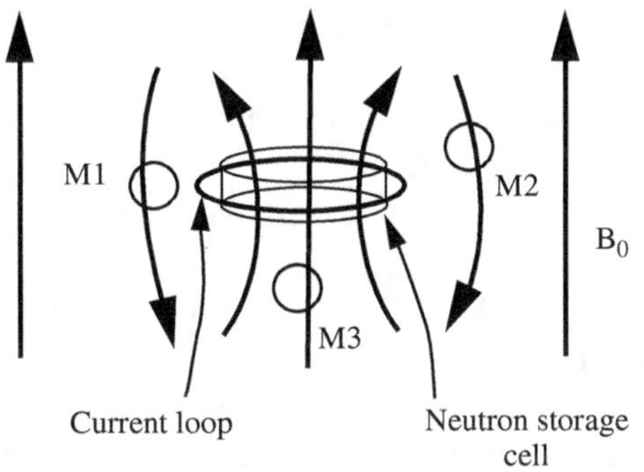

Figure 18.2 The external magnetometer problem. Leakage currents associated with the application of a high voltage to the measurement cell can flow in a loop (or some fraction thereof) around the cell, creating a magnetic field that is correlated with the direction of the electric field. Depending on the location of the magnetometer, the field from the loop can add or subtract to the applied static field B_0.

new CP-violating interactions is expected to be much greater than that of ^{3}He. This allows the ^{3}He maser to operate as a comagnetometer for an EDM search using the ^{129}Xe maser.

A detailed discussion of the large variety of subtle systematic effects that plague EDM experiments, as well as the limitations of comagnetometry, is given in the online supporting material (www.cambridge.org/9781107010352). In the context of our discussion of comagnetometry, here we consider a particular important example: leakage currents. Whenever high voltages are applied to a system, small leakage currents invariably flow through insulators, and these currents generate magnetic fields which are correlated with the electric-field direction and generate spin-precession frequency shifts indistinguishable from those caused by an EDM. The leakage-current magnetic field is a function of the electric field and adds a term to the Hamiltonian Eq. (18.1),

$$H = -\left[\mu\left(\mathbf{B} + \hat{\mathbf{z}}[\beta\hat{\mathbf{z}}\cdot\mathbf{E}]\right) + d\mathbf{E}\right]\frac{\mathbf{F}}{F}\,, \tag{18.5}$$

where β represents the average projection of the magnetic field generated by the current density $\mathbf{j} = \sigma\mathbf{E}$ along the static magnetic field direction ($\hat{\mathbf{z}}$), with σ representing the electrical conductivity.[7] A nonzero β implies some helicity of $\mathbf{j}$ along $\hat{\mathbf{z}}$.

For beam experiments where the insulators and conductors leading to the high-voltage plates can be spatially separated from the sensitive measurement area, leakage current fields are not so troublesome as in the case of cell experiments where the cell walls generally serve a secondary purpose as the high-voltage electrode spacer.

A worst-case scenario is when all the leakage current flows in a closed loop around the cell; given that such a current flow is unlikely, to estimate a possible systematic effect,

[7] The apparent T-odd character of this new term is the result of the irreversible "macroscopic" process(es) related to the Second Law of Thermodynamics which lead to $\mathbf{j}$.

one-quarter of the field at the center of a loop is sometimes taken. Note that if such a helical current exists due to some imperfection in the cell walls, reversing the cell orientation does not distinguish this effect because the leakage-current helicity and hence magnetic-field direction is a fixed property of the cell.

Interesting case studies are two ultra-cold-neutron experiments, one at the Institut Laue-Langevin (ILL) in Grenoble, France [56], and one at the Petersburg Nuclear Physics Institute (PNPI) in Gatchina, Russia [57]. Both reported results in the mid-1980s that were unknowingly affected by systematics associated with the application of high voltage. The EDM results announced by these two groups were nonzero, agreed in sign and magnitude, and were at the 90% statistical confidence level. This announcement created a worldwide sensation. However, problems with the data were found and reported in later publications [58, 59].

The key problem was uncovered when the correlations between the individual magnetometer measurements of systematic magnetic fields (associated with application of the high voltage) and the neutron frequency were shown to be statistically significant, invalidating any possibility of reliably detecting a nonzero neutron EDM at the level of the intrinsic sensitivity of the experiment.[8]

The point of this cautionary tale is to emphasize the importance of monitoring local magnetic fields with the highest possible sensitivity in EDM experiments, for as we know the past is usually prologue.

18.3 Anomalous spin-dependent forces

18.3.1 Background

Of the four known fundamental forces, only two have a macroscopic range (extending beyond the nuclear or atomic scale): electromagnetism (mediated by the spin-1 photon) and gravity (a tensor field presumably mediated by a spin-2 graviton, see for example Ref. [61]). It remains possible that more sensitive measurements could reveal the existence of additional macroscopic-range forces. Even if new experiments produce null results, more stringent experimental constraints are useful for guiding theoretical extensions of the Standard Model and general relativity. There is a long history of experiments searching for additional spin-independent macroscopic forces, reviewed for example in Refs. [62–64]. Experimental searches for anomalous spin-dependent forces are less numerous; the first test was carried out by Ramsey in 1979 [65]. Over the past few decades, searches for anomalous spin-dependent forces have drawn interest primarily as a means to search for the possible existence of axions [66], hypothetical particles [67,68] arising out of a theoretical framework aimed at resolving the strong-*CP* problem [69, 70].

In general, given basic assumptions within the context of quantum field theory (e.g., rotational invariance, energy-momentum conservation, locality), interactions mediated by

[8] The correlation technique is described in detail in Ref. [60] (Section 7.3.4) and in Ref. [14] (Section 4.5.2).

new bosons can generate a variety of spin-dependent macroscopic potentials. Recently, Dobrescu and Mocioiu [71] enumerated possible macroscopic potentials between fermions from exchange of new bosons in the nonrelativistic limit (small fermion velocity and momentum transfer), finding 16 independent, long-range couplings. Since the nonexotic matter typically employed in laboratory experiments consists of electrons (e) and nucleons [either protons (p) or neutrons (n)], there are 6 fermion pairs (ee, ep, en, pp, nn, np) that can couple with 16 different potentials (leading to $6 \times 16 = 96$ different parameters to constrain!).[9] In general, because of their different quark content, the couplings of protons and neutrons can be expected to differ. Each of the 16 potentials has a coupling strength f_i^{XY}, possibly zero, between different fermions (where XY denotes the possible fermion pairs: $X, Y = \mathrm{e,n,p}$ and $i = 1, 2, \ldots, 16$ labels the corresponding potential), to be determined by experiments. The two potentials most commonly searched for are the dipole–dipole potential,

$$\begin{aligned}\mathcal{V}_3(r) = f_3^{XY} \frac{\hbar^2}{m_X m_Y c^2} \frac{1}{r^3} \Big[& \boldsymbol{\sigma}_X \cdot \boldsymbol{\sigma}_Y \left(1 - r\frac{d}{dr}\right) \\ & - 3(\boldsymbol{\sigma}_X \cdot \hat{\mathbf{r}})(\boldsymbol{\sigma}_Y \cdot \hat{\mathbf{r}}) \left(1 - r\frac{d}{dr} + \frac{1}{3} r^2 \frac{d^2}{dr^2}\right)\Big] y(r) \,, \end{aligned} \tag{18.6}$$

and the monopole–dipole potential,

$$\mathcal{V}_{9,10}(r) = f_{9,10}^{XY} \frac{\hbar}{m_X c} \frac{1}{r^2} \boldsymbol{\sigma}_X \cdot \left[\hat{\mathbf{r}} \left(1 - \frac{d}{dr}\right) y(r) \right], \tag{18.7}$$

where $m_{X,Y}$ are the fermion masses, $\boldsymbol{\sigma}_{X,Y}$ are the fermion intrinsic spins in units of $\hbar$, $y(r)$ is a dimensionless function of the fermion separation r, and we have adopted the enumerating convention of Dobrescu and Mocioiu [71] for the potentials and coupling strengths. Note the discrete symmetry properties of the potentials (18.6) and (18.7): $\mathcal{V}_3(r)$ is P- and T-even, whereas $\mathcal{V}_{9,10}$ is P- and T-odd, having a form similar to the EDM potential (Section 18.2).

Assuming one-boson exchange within a Lorentz-invariant quantum field theory, $y(r)$ takes on a Yukawa-like form:[10]

$$y(r) = \frac{1}{4\pi} e^{-r/\lambda} \,, \tag{18.8}$$

where $\lambda = \hbar / m_0 c$ is the Compton wavelength of the new boson of mass m_0, which sets the scale of the new interaction. Thus an experimental setup characterized by a distance scale L is sensitive to new bosons of mass $m_0 \lesssim \hbar / cL$. If particular spin and parity properties of the new boson are specified, correlations between the coupling strengths are found. For

[9] Although it should be noted that a nonnegligible part (a fraction of a percent) of the mass content of ordinary matter arises from the nuclear binding energy, and thus, for example, searches for new interactions may also be sensitive to electron-gluon couplings, etc.

[10] If there is multiboson exchange or Lorentz invariance is violated, other forms of $y(r)$ can arise, but the spin dependence of the potential functions is preserved.

example, if the new boson is a pseudoscalar particle such as the axion, $f_3^{XY} = g_p^X g_p^Y/4$, where $g_p^{X,Y}$ parameterizes the vertex-level coupling to the fermions. If the new interaction possesses both scalar and pseudoscalar couplings, $f_{9,10}^{XY} = g_p^X g_s^Y/2$, where the subscripts s and p denote the scalar and pseudoscalar couplings, respectively. Thus for spin-0 boson exchange assuming Lorentz invariance, the potentials in Eqs. (18.6) and (18.7) can be written in the forms commonly encountered in the literature:

$$\mathcal{V}_3(r) = \frac{g_p^X g_p^Y \hbar^2}{16\pi m_X m_Y c^2}\left[\boldsymbol{\sigma}_X \cdot \boldsymbol{\sigma}_Y \left(\frac{1}{\lambda r^2} + \frac{1}{r^3}\right) - (\boldsymbol{\sigma}_X \cdot \hat{\mathbf{r}})(\boldsymbol{\sigma}_Y \cdot \hat{\mathbf{r}})\left(\frac{1}{\lambda^2 r} + \frac{3}{\lambda r^2} + \frac{3}{r^3}\right)\right] e^{-r/\lambda}, \tag{18.9}$$

and

$$\mathcal{V}_{9,10}(r) = \frac{g_p^X g_s^Y \hbar}{8\pi m_X c} \boldsymbol{\sigma}_X \cdot \hat{\mathbf{r}} \left(\frac{1}{r\lambda} + \frac{1}{r^2}\right) e^{-r/\lambda}. \tag{18.10}$$

Monopole–dipole and dipole–dipole potentials, and indeed the vast majority of the possible potentials enumerated in Ref. [71], can also be generated by exchange of spin-1 bosons, although the description of the coupling strengths is somewhat more complicated.

Another theoretical framework for considering anomalous spin-dependent interactions comes from extensions of general relativity. General relativity describes the interactions between mass/energy and a dynamical tensor field. Covariant derivatives in general relativity are taken with respect to what is known as the Christoffel connection, for which a property called *torsion* is defined to be zero. This choice, motivated to simplify the theory, is somewhat arbitrary, and in general there is no fundamental underlying reason why the connection should be torsion-free. In fact the connection can easily account for both curvature and torsion of spacetime. The introduction of a torsion tensor into general relativity generates new scalar and vector components of gravity [72–75], which manifest themselves as new spin–mass and spin–spin couplings analogous to the potentials discussed above. Depending on the specific model, these spin-dependent couplings can be long or short range [76–79]. General relativity with torsion has the additional attractive feature of naturally incorporating particles with intrinsic spin, a nonclassical physical property not included in general relativity.

Beyond these two broad frameworks, exotic spin-dependent interactions are predicted by a variety of novel theories, for example, para-photons [80], unparticles [81], and theories with spontaneous Lorentz violation [82]. Of particular interest is the observation that a massless or nearly massless spin-0 component of gravity manifests as dark energy over cosmological distances [83–86]. The pseudoscalar component of such a field leads to a P- and T-violating interaction which has the nonrelativistic form [86]

$$H_g = k\frac{\hbar}{c}\boldsymbol{\sigma} \cdot \mathbf{g}, \tag{18.11}$$

where k is a dimensionless parameter setting the scale of the new interaction and $\boldsymbol{g}$ is the Earth's gravitational field. For a pseudoscalar coupling equal to the usual tensor component of gravity the parameter $k \approx 1$ [86]. At present, the best experimental limits [87, 88] rule out $k \gtrsim 10$, and new experiments are underway targeting the parameter space $k \sim 1$ [89]. The existence of such an interaction would impart a "gravitational dipole moment" (GDM) $k\hbar\boldsymbol{\sigma}/c$ to elementary particles. One could envision this as a separation between the center of inertial mass and the center of gravitational mass by a distance $k\hbar/mc$, where m is the fermion mass, consequently violating the equivalence principle. If k is of order unity then the separation between the centers of inertial mass and gravitational mass is on the order of the Compton wavelength.

18.3.2 Experiments

Recent experiments have significantly improved constraints on anomalous dipole–dipole interactions between neutrons [90,91] and monopole–dipole interactions between electrons and nucleons [88]. Experiments carried out in the early 1990s have provided the best constraints on monopole–dipole interactions between nuclei [87] and anomalous dipole–dipole interactions between electrons [92, 93] and electrons and nuclei [94]. Parameter exclusion plots showing direct experimental constraints on the monopole–dipole interaction [Eq. (18.10)] and dipole–dipole interaction [Eq. (18.9)] are presented in Figs. 18.3 and 18.4, respectively. (There are also indirect constraints from gravitational inverse-square-law tests discussed in Refs. [95, 96].) At the laboratory scale, the best limits on dipole–dipole interactions [91] and monopole–dipole interactions [87] of neutrons are obtained using optical magnetometry techniques.[11]

In the experiment of Venema *et al.* [87], the spin precession frequencies of two isotopes of Hg were measured simultaneously as the orientation of a magnetic field $\mathbf{B}$ was changed relative to the Earth's gravitational field $\boldsymbol{g}$. Since the ground electronic state of Hg is $^1S_0$, the ground-state polarization is entirely due to the nuclear spin I, with ^{199}Hg having $I = 1/2$ and ^{201}Hg having $I = 3/2$. Suppose there exists a heretofore undiscovered long-range, monopole–dipole interaction: for example, a spin–gravity interaction of the type described by Eq. (18.11). In the presence of only $\mathbf{B}$ and $\boldsymbol{g}$, the spin precession frequencies for the two Hg isotopes are

$$\Omega_{199} = \gamma_{199}B + \chi_{199}g\cos\phi\ , \tag{18.12}$$

$$\Omega_{201} = \gamma_{201}B + \chi_{201}g\cos\phi\ , \tag{18.13}$$

[11] The best limits on long-range monopole–dipole interactions of electrons come from the spin-polarized torsion pendulum experiments carried out at the University of Washington [88]. The best limits on short-range dipole–dipole interactions of nucleons come from spectroscopy of molecular hydrogen [65] and deuterated molecular hydrogen [97]. Also of note is a recent analysis of proton–neutron dipole–dipole interactions based on spin-exchange collisions [98].

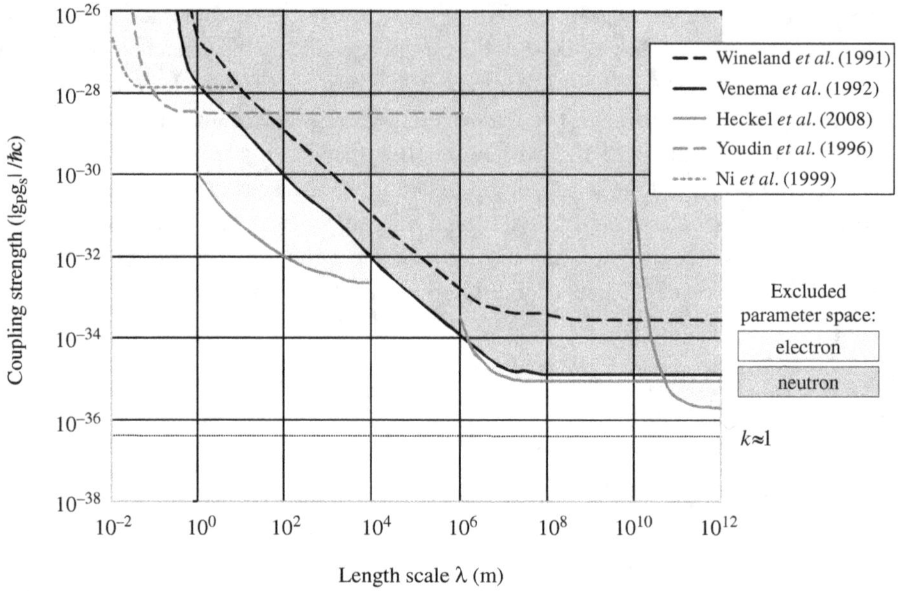

Figure 18.3 Existing experimental constraints (at the 1-σ level) on anomalous monopole–dipole (scalar–pseudoscalar) couplings $|g_p g_s|/\hbar c$ as a function of the range λ of the interaction. Constraints for the electron are from Youdin *et al.* (1996) [99], Ni *et al.* (1999) [100], and Heckel *et al.* (2008) [88]; constraints for the neutron are from Wineland *et al.* (1991) [94] and Venema *et al.* (1992) [87]. Dashed horizontal line near bottom of plot indicates coupling strength for pseudoscalar component of gravity with $k \approx 1$, see Eq. (18.11). Figure adapted from Ref. [88].

where γ_i is the gyromagnetic ratio and χ_i is the so-called gyrogravitational ratio parameterizing the new interaction (where the subscripts i denote the respective isotopes), and ϕ is the angle between $\mathbf{B}$ and $\boldsymbol{g}$. As long as $\chi_{199}/\chi_{201} \neq \gamma_{199}/\gamma_{201}$ (as generally expected), the ratio $\mathcal{R} = \Omega_{199}/\Omega_{201}$ acquires a dependence on B and ϕ if the χ_i's are nonzero, enabling a search for the spin–gravity (or spin–mass) coupling.

The experimental setup, essentially a dual-isotope optical magnetometer, is shown in Fig. 18.5. The Hg atoms were contained in a cylindrical vapor cell situated at the center of a three-layer cylindrical mumetal shield with internal coils to apply controlled magnetic fields to the atoms. The axes of the concentric cylinders of the shield system (defined to be z) and quadrupole axis of the vapor cell ($\hat{\boldsymbol{Q}}_c$), as well as the magnetic field during spin precession, were oriented along the Earth's rotation axis ($\hat{\boldsymbol{\Omega}}_E$). The experiment consisted of pump and probe stages. During the pump stage, the atoms were optically pumped in the presence of a small magnetic field along x ($B_x \lesssim 10$ mG) by circularly polarized light propagating along $\hat{\mathbf{x}}$. In the probe stage, lasting 200 s, the magnetic field was re-directed along $\pm\hat{\mathbf{z}}$, the light intensity was reduced, and a photoelastic modulator (PEM) rapidly alternated the light polarization between right- and left-circular. The detected signal was demodulated at the PEM frequency (42 kHz) and the free-precession-decay signal analyzed to extract the precession frequencies (Fig. 18.6).

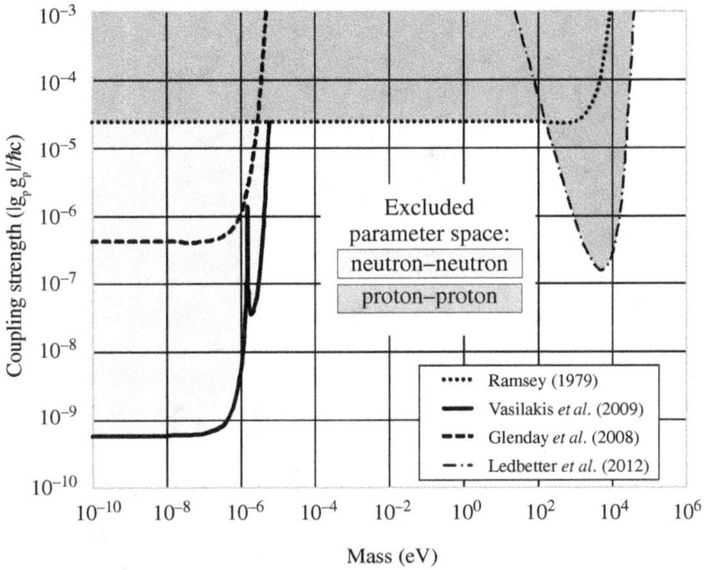

Figure 18.4 Existing direct experimental constraints (at the 1-σ level) on anomalous dipole–dipole (pseudoscalar) couplings of the neutron and proton spins $|g_P g_P|/\hbar c$ as a function of the pseudoscalar boson mass. Proton constraints are from Ramsey (1979) [65] and Ledbetter, Romalis, and Jackson Kimball (2012) [97] (which also constrains neutron–proton couplings at a similar level). Neutron constraints are from Glenday *et al.* (2008) [90] and Vasilakis *et al.* (2009) [91].

Two important systematic errors required special consideration. The first arose due to a collisional interaction of the ^{201}Hg atoms with the walls of the cylindrical vapor cell [101, 102], causing a ~50 mHz quadrupolar shift. The shift led to resolved splitting of the Zeeman frequencies for ^{201}Hg (Fig. 18.6). The second arose because the experimental apparatus was attached to the Earth, while the Hg spins were effectively decoupled from the Earth's rotation during the probe stage. Consequently, the Hg isotopes exhibited apparent precession at the rotation rate of the Earth, $\Omega_E = 2\pi \times 11.6$ μHz. This effect, known as the gyro-compass effect [88], can be understood as the result of viewing an inertial system, the Hg spins, from a noninertial frame, the surface of the rotating Earth. Both systematic effects were constrained at or below the statistical sensitivity of the experiment by orienting the apparatus so that uncertainty in the effects were quadratic in the misalignment angles. Performing over 6000 individual measurements over the course of several months led to an upper limit on the anomalous spin-precession frequency of < 0.5 μHz at the 95% confidence limit, leading to the constraints on scalar–pseudoscalar couplings shown in Fig. 18.3.

The best limit on anomalous dipole–dipole interactions between neutrons was achieved in the recent experiment of Vasilakis *et al.* [91] using the setup shown in Fig. 18.7. The measurement technique is based on the principles of spin-exchange-relaxation-free (SERF)

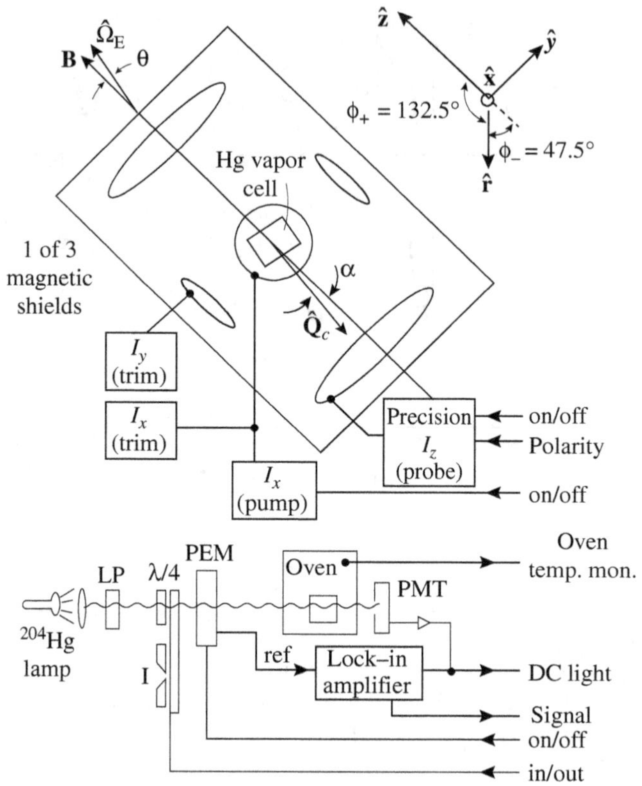

Figure 18.5 Experimental setup from Ref. [87]. LP = linear polarizer, $\lambda/4$ = quarter-wave plate, I = iris, PMT = photomultiplier tube, PEM = photoelastic modulator. Arrows on right-hand side indicate computer control and data acquisition. The angles $\phi_\pm$ indicate the projection of $\hat{\mathbf{r}}$ (parallel to $\boldsymbol{g}$) along $\hat{\mathbf{z}}$ (parallel to $\mathbf{B}$) for the two precession field orientations.

magnetometry (discussed in Chapter 5). The atomic sample consists of overlapping ensembles of potassium (K) and ^{3}He at relatively high vapor densities[12] (He density ≈ 12 amagat and K density $\approx 10^{14}$ atoms/cm^3). The *K* sample is polarized through optical pumping and the ^{3}He sample is polarized through spin-exchange collisions with K (^{3}He polarization $\approx$ 3%). The vapor cell is located within a five-layer mumetal shield fitted with internal coils used to cancel residual magnetic fields and create a small field **B** parallel to the propagation direction of the pump beam.

Under these experimental conditions, for which the Larmor frequencies are comparable to or smaller than the frequency of spin-exchange collisions, the spin-exchange interaction between K atoms and the polarized ^{3}He vapor strongly couples the two spin ensembles [103, 104]. In a spherical cell this coupling can be represented as an effective magnetic field

[12] The cell also contained 46 torr of nitrogen to quench excited state K atoms in order to improve optical pumping efficiency.

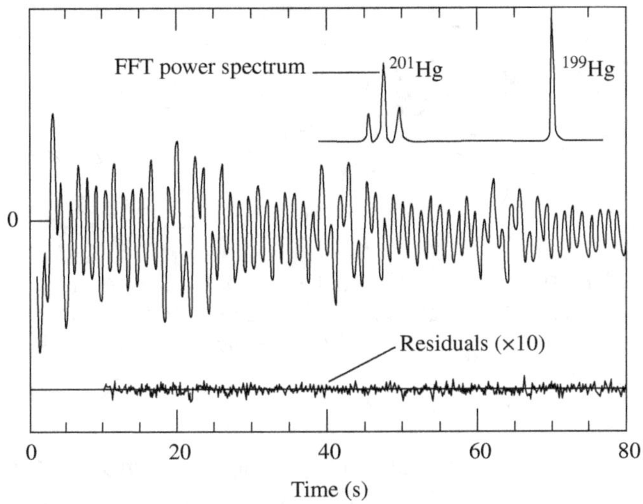

Figure 18.6 Typical free-precession signal from Ref. [87]. Residuals (multiplied by 10) from a non-linear least-squares fit of the data to a sum of exponentially decaying cosines are shown at the bottom of the plot. A portion of the fast Fourier transform of the data is shown in the upper right-hand corner.

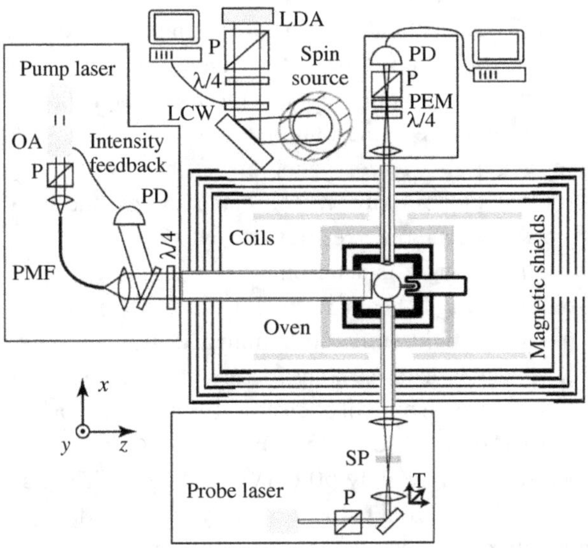

Figure 18.7 Experimental setup from Ref. [91]. PD = photodiode, SP = stress plate to control polarization of the probe beam, T = translation stage to shift the probe beam, P = polarizer, PMF = polarization maintaining fiber, OA = optical amplifier, LCW = liquid crystal wave plate, PEM = photoelastic modulator, $\lambda/4$ = quarter-wave plate, LDA = laser diode array.

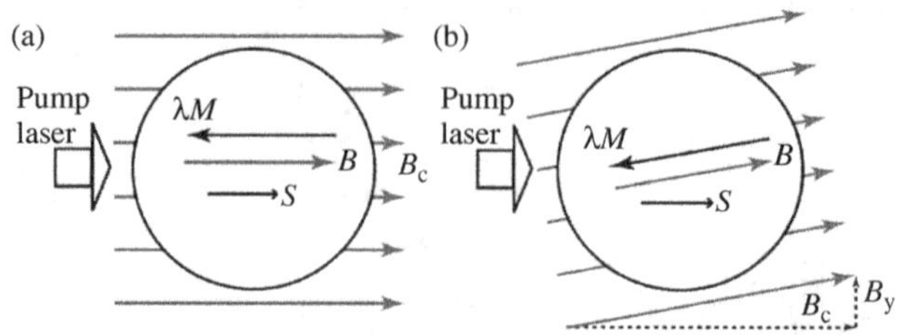

Figure 18.8 (Figure from Ref. [104].) Basic operation of the self-compensating K-^{3}He comagnetometer. (a) K and ^{3}He atoms are polarized along the compensation field $B_c \approx \lambda M$. (b) ^{3}He magnetization adiabatically follows the ambient magnetic field B canceling the effect of transverse fields on the K spins S.

$\mathbf{B}_{\text{eff}}$ experienced by one spin species due to the average magnetization $\mathbf{M}$ of the other, due to enhancement of the K valence electron density at the He nucleus,

$$\mathbf{B}_{\text{eff}} = \lambda \mathbf{M}, \tag{18.14}$$

where $\lambda \approx (8\pi/3) \times 8.5$ is a parameter determined by the particular properties of K–^{3}He spin-exchange [105]. The applied field $\mathbf{B}$ is tuned so that it approximately cancels $\mathbf{B}_{\text{eff}}$ experienced by the K atoms. The K atoms are then in a zero-field environment, and because the ^{3}He magnetization $\mathbf{M}$ adiabatically follows $\mathbf{B}$, components of $\mathbf{B}$ transverse to $\hat{\mathbf{z}}$ are automatically compensated by $\mathbf{B}_{\text{eff}}$ to first order (Fig. 18.8). Such cancelation only occurs for interactions that couple to spins in proportion to their magnetic moments, leaving the K–^{3}He comagnetometer sensitive to inertial rotation [104] and anomalous spin couplings. Thus the self-compensating K–^{3}He comagnetometer enables one to use high-sensitivity SERF magnetometry techniques with K for detection of anomalous spin-dependent interactions causing precession about axes transverse to $\hat{\mathbf{z}}$.

The K spin-polarization along $\hat{\mathbf{x}}$ is determined by measuring optical rotation of an off-resonant, linearly polarized light beam from a distributed feedback (DFB) laser. After residual magnetic fields and light shifts are eliminated using zeroing routines described in Ref. [104], the K spin-polarization along $\hat{\mathbf{x}}$ can only arise due to nonmagnetic, spin-dependent interactions – offering a highly sensitive probe of such anomalous interactions. The spin source consisted of a dense ($\sim 3 \times 10^{20}$ cm^{-1}), highly polarized ($\sim$15% polarization) ^{3}He gas located approximately 50 cm from the cell. The nuclear spin direction of the ^{3}He sample was reversed at a 0.18 Hz rate by adiabatic fast passage (AFP). After approximately one month of data acquisition, no anomalous effect was detected at a level corresponding to a magnetic field value less than an attotesla (10^{-14} G).

The experiment of Vasilakis *et al.* [91] is notable in several regards. The apparatus achieved a record energy resolution of $\sim 10^{-25}$ eV, demonstrating the tremendous potential of these optical-magnetometry techniques for precision measurements. Furthermore, the coupling strength of a new massless spin-1 boson to neutrons, such as would exist in some models of torsion gravity [77], was constrained below 0.2% of the strength of the gravitational interaction, providing the first meaningful experimental limits on torsion gravity.

18.4 *CPT* and local Lorentz invariance tests

A system is said to be Lorentz invariant if the laws of physics governing that system are unaffected by Lorentz transformations, which include boosts (transforming to a frame moving relative to the system) and rotations. Any quantum field theory that observes locality and Lorentz invariance, such as the Standard Model and quantum electrodynamics, is also *CPT* invariant (see Section 18.1) – this is known as the *CPT* theorem [106]. Furthermore, in the case where *CPT* violation arises from spontaneous symmetry breaking and locality is preserved, the converse has been proven, namely that *CPT* invariance implies Lorentz invariance [107]. Therefore, under rather general assumptions, tests of local Lorentz invariance (LLI) test *CPT* invariance and vice versa.[13] Thus LLI tests provide a method to constrain *CPT* violation without the use of antiparticles.

In the early 1960s, Hughes *et al.* [112] and Drever [113] carried out the first searches for anomalous spin couplings to an anisotropy of space[14] using what has come to be known as a clock-comparison test, where the frequencies of different atomic "clock" transitions are compared as the clocks rotate with the Earth. The Hughes–Drever tests were inspired by the suggestion of Cocconi and Salpeter [115] that it might be possible, based on Mach's principle, for inertial mass to acquire a tensor character due to anisotropic distribution of matter in the universe. This would cause a particle's inertial mass to depend on the orientation of its orbit with respect to the matter anisotropy, which in turn would generate energy shifts in atoms and nuclei. Based on the anisotropy of matter in the Milky Way galaxy, Cocconi and Salpeter estimated that fractional inertial mass anisotropies of $\Delta m/m \sim 10^{-10}$ or larger were possible. Hughes *et al.* and Drever searched for sidereal variation of nuclear magnetic resonance (NMR) lines in ^{7}Li. Their null results produced the constraint $\Delta m/m \lesssim 5 \times 10^{-23}$ [113], which ruled out the suggestion of Ref. [115] by many orders of magnitude.

Since the original Hughes–Drever experiments, a series of ever more precise clock-comparison tests of LLI have been performed [88, 116–124] (see Table 18.4). Comparison of the LLI tests listed in Table 18.4 to experiments searching for EDMs (Section 18.2, Table 18.3) and to those searching for anomalous spin-dependent interactions (Section 18.3, Figs. 18.3 and 18.4) shows that there is considerable overlap between the techniques used in all three classes of experiments. For example, the Hg LLI experiment of Lamoreaux *et al.* [117] is based on the same apparatus used in Hg EDM measurements [27, 125, 126] and the spin–mass coupling search described in Ref. [87]. Many of the clock-comparison tests

[13] Important exceptions are theories with purely kinematic tensor Lorentz violation which is *CPT*-even and does not require explicit particle spin coupling at the fundamental level [108], nonlocal quantum field theories [109], and theories with non-finite-dimensional representations of the Lorentz group [110]. Curved spacetime may also introduce complexities in the proof of the *CPT* theorem [111].

[14] Also of note is a closely related but less sensitive experiment by Sherwin *et al.* using gamma-ray scattering [114].

Table 18.4. *Summary of limits on sidereal variation of clock transition frequencies. These clock-comparison (and closely related) experiments test local Lorentz invariance and CPT invariance. Techniques: ESR = electron spin resonance, NMR = nuclear magnetic resonance, opt. mag. = optical magnetometer, SEOP = spin-exchange optical pumping, SERF comag. = spin-exchange-relaxation-free comagnetometer.*

Year	Species	Limit (Hz)	Technique	Ref.
1960	^{35}Cl	$\lesssim 3 \times 10^{3}$	ESR	[112]
1960	^{7}Li	$\lesssim 8$	NMR	[112]
1961	^{7}Li	$\lesssim 2 \times 10^{-2}$	NMR	[113]
1985	^{9}Be^{+}, ^{1}H	$\lesssim 10^{-4}$	Ion clock vs. maser	[116]
1986	^{199}Hg, ^{201}Hg	$\lesssim 10^{-6}$	Opt. mag.	[117]
1987	e^{-}	$\lesssim 2 \times 10^{-3}$	Torsion pendulum	[118]
1989	^{21}Ne, ^{3}He	$\lesssim 10^{-6}$	NMR with SEOP	[119]
1995	^{199}Hg, ^{133}Cs	$\lesssim 10^{-7}$ (nucl) $\lesssim 2 \times 10^{-4}$ (e^{-})	Opt. mag.	[120]
2000	^{129}Xe, ^{3}He	$\lesssim 2 \times 10^{-8}$	Maser	[121]
2008	e^{-}	$\lesssim 4 \times 10^{-8}$	Torsion pendulum	[88]
2010	^{39}K, ^{3}He	$\lesssim 5 \times 10^{-10}$	SERF comag.	[122]
2010	^{129}Xe, ^{3}He	$\lesssim 10^{-8}$	Maser	[123]
2011	^{21}Ne, ^{87}Rb	$\lesssim 5 \times 10^{-10}$	SERF comag.	[124]

listed in Table 18.4 employ optical magnetometry techniques such as SERF magnetometry (Chapter 5).

The variety of different experimental tests of LLI, many of which do not involve clock comparison techniques (not listed in Table 18.4, such as tests of the isotropy of the speed of light [127, 128] akin to the famous Michelson–Morley experiment [129]), raises a natural question: are various LLI tests constraining the same type of new physics, or are the results of various tests largely orthogonal to one another? Recently a consistent theoretical framework for Lorentz and *CPT* violation, known as the Standard Model Extension (SME) [130], has been developed to answer precisely this question. The SME provides a phenomenological model based on quantum field theory that parameterizes Lorentz and *CPT* violation in the Standard Model and general relativity, enabling quantitative comparative analysis of different experiments [131]. It has been shown that the SME must emerge from any underlying theory that generates the Standard Model and contains spontaneous Lorentz violation [132, 133]. A recent tabulation of constraints on different SME parameters from a wide variety of experiments is given in Ref. [134].

Lorentz violation in the SME can be understood as arising from the interaction of particles with expectation values of vacuum fields possessing tensor characteristics. This is analogous to the case of a charged particle passing through a crystal, where electromagnetic interactions

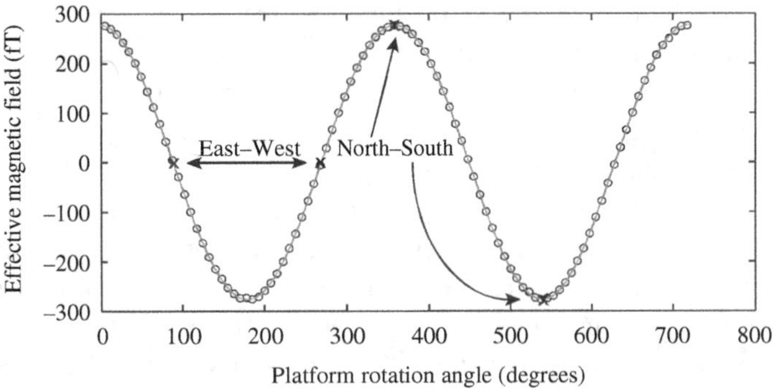

Figure 18.9 (From Ref. [122].) Change of K–^{3}He self-compensating SERF comagnetometer signal for 180° platform rotation as a function of the initial platform angle.

between the particle and the crystal break rotational and boost symmetries. Anisotropic interaction of spins with Lorentz-violating vacuum fields is predicted by a number of novel theories which can be related to the SME parametrization [135–138].

The most stringent clock-comparison tests of LLI [122, 124] have recently been carried out using the self-compensating SERF comagnetometry scheme [103, 104], which was also used for constraining anomalous dipole–dipole interactions [91] as discussed in detail in Section 18.3.

The experiment described in Ref. [122] employs overlapping ensembles of K and ^{3}He coupled via spin-exchange collisions. The atoms are in the low-magnetic-field SERF regime where broadening of the Zeeman resonances due to spin-exchange collisions is eliminated (Chapter 5). The magnetic field along $\hat{\mathbf{z}}$ is tuned to the compensation point where the K–^{3}He comagnetometer is insensitive to magnetic fields but highly sensitive to anomalous interactions that do not scale with the magnetic moments (see Fig. 18.8 and surrounding discussion). The $\hat{\mathbf{x}}$ polarization of the K atoms, probed via optical rotation with a linearly polarized laser beam propagating along x, is given to leading order by

$$P_x^e = P_z^e \frac{\gamma_e}{\Gamma_{\text{rel}}} \left(\beta_y^N - \beta_y^e + \frac{\Omega_y}{\gamma_N} \right), \tag{18.15}$$

where β_y^N and β_y^e describe the phenomenological SME background fields along $\hat{\mathbf{y}}$ coupling to the ^{3}He nucleus (N) and valence electron (e) of K, respectively, P_z^e is the K electron-spin polarization along z, Γ_{rel} is the relaxation rate for K polarization, γ_e and γ_N are the gyromagnetic ratios for electrons and ^{3}He nuclei, respectively, and Ω_y is the rotation rate of the apparatus. A number of steps are taken to eliminate various sources of noise and systematic error. For example, the signal described by Eq. (18.15) depends explicitly on the rotation rate of the apparatus, so the gyro-compass effect (discussed in Section 18.3 and Refs. [87, 88]) due to the Earth's rotation generates a nonzero P_x^e. To deal with this effect, the experiment is mounted on a rotary platform. Figure 18.9 shows the change in the comagnetometer signal for a 180° rotation as a function of the initial platform angle, demonstrating the significant effect of Earth's rotation on P_x^e. In order to test LLI, the orientation of the apparatus is alternated between north–south or east–west (marked by crosses on the plot of Fig. 18.9) every 22 s over the course of many days. Nonzero β_y^N or β_y^e would lead to sidereal oscillation of the amplitude of the difference between the north–south or east–west signals. The results of the measurements, carried out over 143 days, are consistent with no LLI violation. Combined with the constraints on electron couplings to SME background fields from Ref. [88], the results of this experiment probe neutron couplings to SME background fields at energy scales $\sim 10^{-25}$ eV [122].

The closely related experiment of Ref. [124], employing a ^{21}Ne–Rb–K SERF comagnetometry scheme, has an order-of-magnitude better shot-noise-limited sensitivity to LLI violating effects than the K–^{3}He comagnetometer. Additionally, because the nuclear spin $I = 3/2$ for ^{21}Ne, the ^{21}Ne–Rb–K comagnetometer is sensitive to tensor anisotropies as well as the vector anisotropies probed by the K–^{3}He comagnetometer ($I = 1/2$ for ^{3}He).

18.5 Conclusion

Experiments using optical-magnetometry techniques to search for parity- and time-reversal-violating electric dipole moments, anomalous spin-dependent interactions, and violations of Lorentz invariance have achieved remarkable precision: in some cases constraining anomalous energy shifts to $\lesssim 10^{-25}$ eV and achieving sensitivity to frequency shifts at the $\sim 10^{-10}$ Hz level. These experiments are complementary to the research efforts in high-energy particle physics, astrophysics, and cosmology that continue to offer glimpses of new physics beyond the Standard Model. Given the diverse variety of novel theories probed by these techniques, optical magnetometers will undoubtedly continue to be among the most important tools for testing fundamental physical laws.

Acknowledgments

The authors are grateful to M. V. Romalis for important early contributions to the chapter. D. F. Jackson Kimball gratefully acknowledges the support of the National Science Foundation.

References

[1] S. Glashow, Nucl. Phys. **22**, 579 (1961).
[2] S. Weinberg, Phys. Rev. Lett. **19**, 1264 (1967).
[3] A. Salam, *Elementary Particle Theory* (Almqvist and Wicksell, Stockholm, 1968).
[4] A. Einstein, Ann. Phys. **49**, 769 (1916).
[5] A. Einstein, *The Principle of Relativity* (Dover, New York, 1923).
[6] M. Dine and A. Kusenko, Rev. Mod. Phys. **76**, 1 (2004).
[7] G. Bertonea, D. Hooper, and J. Silk, Phys. Rep. **405**, 279 (2005).
[8] S. Matarrese, V. Gorini, and U. Moschella, *Dark Matter and Dark Energy* (Springer, New York, 2011).

[9] C. S. Wu, E. Ambler, R. W. Hayward, D. D. Hoppes, and R. P. Hudson, Phys. Rev. **105**, 1413 (1957).
[10] J. H. Christenson, J. W. Cronin, V. L. Fitch and R. Turlay, Phys. Rev. Lett. **13**, 138 (1964).
[11] L. M. Barkov and M. S. Zolotorev, Pis'ma Zh. Eksp. Teor. Fiz. **27**, 357 (1978) [JETP Lett. **27**, 379 (1978)].
[12] R. Conti, P. Bucksbaum, S. Chu, E. Commins, and L. Hunter, Phys. Rev. Lett. **42**, 343 (1979).
[13] M.-A. Bouchiat and C. Bouchiat, Rep. Prog. Phys. **60**, 1351 (1997).
[14] I. B. Khriplovich and S. K. Lamoreaux, *CP Violation Without Strangeness* (Springer-Verlag, Berlin, 1997).
[15] D. Budker, D. F. Kimball, and D. DeMille, *Atomic Physics: An Exploration Through Problems and Solutions*, 2nd ed. (Oxford University Press, Oxford, 2008).
[16] J. Charles, A. Höcker, H. Lacker, S. Laplace, F. R. Le Diberder, J. Malclés, J. Ocariz, M. Pivk and L. Roos, Eur. Phys. J. C **41**, 1 (2005).
[17] M. Pospelov and A. Ritz, Ann. Phys. **318**, 119 (2005).
[18] M. E. Shaposhnikov, Nucl. Phys. B **287**, 757 (1987).
[19] E. M. Purcell and N. F. Ramsey, Phys. Rev. **78**, 807 (1950).
[20] S. A. Murthy, D. Krause, Jr., Z. L. Li, and L. R. Hunter, Phys. Rev. Lett. **63**, 965 (1989).
[21] B. C. Regan, E. D. Commins, C. J. Schmidt, and D. DeMille, Phys. Rev. Lett. **88**, 071805 (2002).
[22] J. J. Hudson, D. M. Kara, I. J. Smallman, B.E. Sauer, M. R. Tarbutt, E. A. Hinds, Nature **473**, 493 (2011).
[23] G. W. Bennett *et al.* [Muon (g-2) Collaboration], Phys. Rev. D **80**, 052008 (2009).
[24] C. A. Baker *et al.*, Phys. Rev. Lett. **97**, 131801 (2006).
[25] D. Cho, K. Sangster, and E. A. Hinds, Phys. Rev. A **44**, 2783 (1991).
[26] M. A. Rosenberry and T. E. Chupp, Phys. Rev. Lett. **86**, 22 (2001).
[27] W. C. Griffith, M. D. Swallows, T. H. Loftus, M. V. Romalis, B. R. Heckel, and E. N. Fortson, Phys. Rev. Lett. **102**, 101601 (2009).
[28] L. I. Schiff, Phys. Rev. **132**, 2194 (1963).
[29] P. G. H. Sandars, Phys. Lett. **14**, 194 (1965); **22**, 290 (1966).
[30] E. D. Commins, J. D. Jackson, and D. P. DeMille, Am. J. Phys. **75**, 532 (2007).
[31] P. G. H. Sandars, Phys. Rev. Lett. **19**, 1396 (1967).
[32] O. P. Sushkov and V. V. Flambaum, Zh. Eksp. Teor. Fiz. **75**, 1208 (1978) [Sov. Phys. JETP **48**, 608 (1978)].
[33] M. G. Kozlov and L. N. Labzowsky, J. Phys. B **28**, 1933 (1995).
[34] V. A. Dzuba, V. V. Flambaum, and I. B. Khriplovich, Z. Phys. D **1**, 243 (1986).
[35] S. K. Lamoreaux and R. Golub, Phys. Rev. D **50**, 5632 (1994).
[36] D. Budker, S. K. Lamoreaux, A. O. Sushkov, and O. P. Sushkov, Phys. Rev. A **73**, 022107 (2006).
[37] E. A. Hinds, Phys. Scr. **T70**, 34 (1997).
[38] D. DeMille, F. Bay, S. Bickman, D. Kawall, D. Krause, Jr., S. E. Maxwell, and L. R. Hunter, Phys. Rev. A **61**, 052507 (2000).
[39] S. Bickman, P. Hamilton, Y. Jiang, and D. DeMille, Phys. Rev. A **80**, 023418 (2009).
[40] A. C. Vutha, W. C. Campbell, Y. V. Gurevich, N. R. Hutzler, M. Parsons, D. Patterson, E. Petrik, B. Spaun, J. M. Doyle, G. Gabrielse and D. DeMille, J. Phys. B **43**, 074007 (2010); **44**, 079803 (2011).

[41] J. Lee, E. R. Meyer, R. Paudel, J. L. Bohn, and A. E. Leanhardt, J. Mod. Opt. **56**, 2005 (2009).

[42] N. E. Shafer-Ray, Phys. Rev. A **73**, 034102 (2006).

[43] A. E. Leanhardt, J. L. Bohn, H. Loh, P. Maletinsky, E. R. Meyera, L. C. Sinclair, R. P. Stutz, and E. A. Cornell, J. Mol. Spectrsc. **270**, 1 (2011).

[44] C. Chin, V. Leiber, V. Vuletic, A. J. Kerman, and S. Chu, Phys. Rev. A **63**, 033401 (2001).

[45] F. Fang and D. S. Weiss, Opt. Lett. **34**, 169 (2009).

[46] A. L. Barabanov, R. Golub, and S. K. Lamoreaux, Phys. Rev. A **74**, 052115 (2006).

[47] M. V. Romalis and M. P. Ledbetter, Phys. Rev. Lett. **87**, 067601 (2001).

[48] K. L. Sauer, R. J. Fitzgerald, and W. Happer, Chem. Phys. Lett. **277**, 153 (1997).

[49] M. P. Ledbetter, I. M. Savukov, and M. V. Romalis, Phys. Rev. Lett. **94**, 060801 (2005).

[50] M. Ledbetter, S. Pustelny, D. Budker, M. Romalis, J. Blanchard, and A. Pines, arXiv:1201.4438 (2012).

[51] F. L. Shapiro, Sov. Phys. Usp. **11**, 345 (1968).

[52] B. J. Heidenreich, O. T. Elliott, N. D. Charney, K. A. Virgien, A. W. Bridges, M. A. McKeon, S. K. Peck, D. Krause, Jr., J. E. Gordon, L. R. Hunter, and S. K. Lamoreaux, Phys. Rev. Lett. **95**, 253004 (2005).

[53] M. Arndt, S. I. Kanorsky, A. Weis, and T. W. Hänsch, Phys. Rev. Lett. **74**, 1359 (1995).

[54] T. G. Vold, F. J. Raab, B. Heckel, and E. N. Fortson, Phys. Rev. Lett. **52**, 2229 (1984).

[55] S. K. Lamoreaux, Nucl. Instrum. Methods Phys. Res. A **284**, 43 (1989).

[56] J. M. Pendelbury, *Proc. Ninth Symposium on Grand Unification* (France, Aix-Les Bains, 1988).

[57] I. S. Altarev *et al.*, Sov. Phys. JETP Lett. **44**, 460 (1986).

[58] K. F. Smith *et al.*, Phys. Lett. B **234**, 33 (1990).

[59] I. S. Altarev *et al.*, Phys. At. Nucl. **59**, 1152 (1996).

[60] R. Golub, D. J. Richardson, and S. K. Lamoreaux, *Ultracold Neutrons* (Adam Hilger, Bristol, 1991).

[61] J. F. Donoghue, Phys. Rev. D **50**, 3874 (1994).

[62] E. G. Adelberger, B. R. Heckel and A. E. Nelson, Annu. Rev. Nucl. Part. Sci. **53**, 77 (2003).

[63] J. C. Long and J. C. Price, C. R. Phys. **4**, 337 (2003).

[64] E. G. Adelberger, J. H. Gundlach, B. R. Heckel, S. Hoedl, S. Schlamming, Prog. Part. Nucl. Phys. **62**, 102 (2009).

[65] N. F. Ramsey, Physica **96A**, 285 (1979).

[66] J. E. Moody and F. Wilczek, Phys. Rev. D **30**, 130 (1984).

[67] S. Weinberg, Phys. Rev. Lett. **40**, 223 (1978).

[68] F. Wilczek, Phys. Rev. Lett. **40**, 279 (1978).

[69] R. D. Peccei and H. R. Quinn, Phys. Rev. Lett. **38**, 1440 (1977).

[70] R. D. Peccei and H. R. Quinn, Phys. Rev. D **16**, 1791 (1977).

[71] B. A. Dobrescu and I. Mocioiu, J. High Energy Phys. **11**, 5 (2006).

[72] F. W. Hehl, P. von der Heyde, G. D. Kerlick, and J. M. Nester, Rev. Mod. Phys. **48**, 393 (1976).

[73] I. L. Shapiro, Phys. Rep. **357**, 113 (2002).

[74] R. T. Hammond, Rep. Prog. Phys. **65**, 599 (2002).

[75] V. A. Kostelecky, N. Russell, and J. D. Tasson, Phys. Rev. Lett. **100**, 111102 (2008).

[76] D. E. Neville, Phys. Rev. D **21**, 2075 (1980).
[77] D. E. Neville, Phys. Rev. D **25**(2), 573 (1982).
[78] S. M. Carroll and G. B. Field, Phys. Rev. D **50**(6), 3867 (1994).
[79] R. T. Hammond, Phys. Rev. D **52**, 6918 (1995).
[80] B. A. Dobrescu, Phys. Rev. Lett. **94**, 151802 (2005).
[81] H. Georgi, Phys. Rev. Lett. **98**, 221601 (2007).
[82] N. Arkani-Hamed, H.-C. Cheng, M. Luty and J. Thaler, J. High Energy Phys. **7**, 29 (2005).
[83] B. Ratra and P. J. E. Peebles, Phys. Rev. D **37**, 3406 (1988).
[84] C. Wetterich, Nucl. Phys. B **302**, 668 (1988).
[85] R. R. Caldwell, R. Dave, and P. J. Steinhardt, Phys. Rev. Lett. **80**, 1582 (1998).
[86] V. Flambaum, S. Lambert, and M. Pospelov, Phys. Rev. D **80**, 105021 (2009).
[87] B. J. Venema, P. K. Majumder, S. K. Lamoreaux, B. R. Heckel, and E. N. Fortson, Phys. Rev. Lett. **68**, 135 (1992).
[88] B. R. Heckel, E. G. Adelberger, C. E. Cramer, T. S. Cook, S. Schlamminger, and U. Schmidt, Phys. Rev. D **78**, 092006 (2008).
[89] D. F. Jackson Kimball, L. R. Jacome, S. Guttikonda, E. J. Bahr, and L. F. Chan, J. Appl. Phys. **106**, 063113 (2009).
[90] A. G. Glenday, C. E. Cramer, D. F. Phillips, and R. L. Walsworth, Phys. Rev. Lett. **101**, 261801 (2008).
[91] G. Vasilakis, J. M. Brown, T. W. Kornack, and M. V. Romalis, Phys. Rev. Lett. **103**, 261801 (2009).
[92] T. C. P. Chui and W.T. Ni, Phys. Rev. Lett. **71**, 3247 (1993).
[93] V. F. Bobrakov *et al.*, JETP Lett. **53**, 294 (1991).
[94] D. J. Wineland, J. J. Bollinger, D. J. Heinzen, W. M. Itano, and M. G. Raizen, Phys. Rev. Lett. **67**, 1735 (1991).
[95] E. Fischbach and D. E. Krause, Phys. Rev. Lett. **82**, 4753 (1999).
[96] E. G. Adelberger, B. R. Heckel, S. Hoedl, C. D. Hoyle, D. J. Kapner, and A. Upadhye, Phys. Rev. Lett. **98**, 131104 (2007).
[97] M. P. Ledbetter, M. V. Romalis, and D. F. Jackson Kimball, arXiv:1203.6894 (2012).
[98] D. F. Jackson Kimball, A. Boyd, and D. Budker, Phys. Rev. A. **82**, 062714 (2010).
[99] A. N. Youdin, D. Krause, Jr., K. Jagannathan, L. R. Hunter, and S. K. Lamoreaux, Phys. Rev. Lett. **77**, 2170 (1996).
[100] W. T. Ni, S. S. Pan, H. C. Yeh, L. S. Hou, and J. Wan, Phys. Rev. Lett. **82**, 2439 (1999).
[101] Z. Wu, W. Happer, M. Kitano, and J. Daniels, Phys. Rev. A **42**, 2774 (1990).
[102] P. K. Majumder, B. J. Venema, S. K. Lamoreaux, B. R. Heckel, and E. N. Fortson, Phys. Rev. Lett. **65**, 2931 (1990).
[103] T. W. Kornack and M. V. Romalis, Phys. Rev. Lett. **89**, 253002 (2002).
[104] T. W. Kornack, R. K. Ghosh, and M. V. Romalis, Phys. Rev. Lett. **95**, 230801 (2005).
[105] T. G. Walker, Phys. Rev. A **40**, 4959 (1989).
[106] R. Jost, Helv. Phys. Acta **30**, 409 (1957).
[107] O. W. Greenberg, Phys. Rev. Lett. **89**, 231602 (2002).
[108] S. Coleman and S. L. Glashow, Phys. Rev. D **59**, 116008 (1999).
[109] M. Chaichian, A. D. Dolgov, V. A. Novikov, A. Tureanu, Phys. Lett. B **699**, 177 (2011).
[110] A. I. Oksak and I. T. Todorov, Comments Math. Phys. **11**, 125 (1968); D. T. Stoyonov and I. T. Todorov, J. Math. Phys. **9**, 2146 (1968).
[111] R. M. Wald, Phys. Rev. D **21**, 2742 (1980).

[112] V. W. Hughes, H. G. Robinson, and V. Beltran-Lopez, Phys. Rev. Lett. **4**, 342 (1960).
[113] R. W. P. Drever, Phils. Magn. **6**, 683 (1961).
[114] C. W. Sherwin, H. Fraunfelder, E. L. Garwin, E. Luscher, S. Margulies, and R. N. Peacock, Phys. Rev. Lett. **4**, 399 (1960).
[115] G. Cocconi and E. E. Salpeter, Nuovo Cimento **10**, 646 (1958).
[116] J. D. Prestage, J. J. Bollinger, W. M. Itano, and D. J. Wineland, Phys. Rev. Lett. **54**, 2387 (1985).
[117] S. K. Lamoreaux, J. P. Jacobs, B. R. Heckel, F. J. Raab, and E. N. Fortson, Phys. Rev. Lett. **57**, 3125 (1986).
[118] P. R. Phillips, Phys. Rev. Lett. **59**, 1784 (1987).
[119] T. E. Chupp, R. J. Hoare, R. A. Loveman, E. R. Oteiza, J. M. Richardson, and M. E. Wagshul, Phys. Rev. Lett. **63**, 1541 (1989).
[120] C. J. Berglund, L. R. Hunter, D. Krause, Jr., E. O. Prigge, and M. S. Ronfeldt, Phys. Rev. Lett. **75**, 1879 (1995).
[121] D. Bear, R. E. Stoner, R. L. Walsworth, V. A. Kostelecký, and C. D. Lane, Phys. Rev. Lett. **85**, 5038 (2000); Phys. Rev. Lett. **89**, 209902(E) (2002).
[122] J. M. Brown, S. J. Smullin, T. W. Kornack, and M. V. Romalis, Phys. Rev. Lett. **105**, 151604 (2010).
[123] C. Gemmel *et al.*, Phys. Rev. D **82**, 111901(R) (2010).
[124] M. Smiciklas, J. M. Brown, L. W. Cheuk, S. J. Smullin, and M. V. Romalis, Phys. Rev. Lett. **107**, 171604 (2011).
[125] S. K. Lamoreaux, J. P. Jacobs, B. R. Heckel, F. J. Raab, and N. Fortson, Phys. Rev. Lett. **59**, 2275 (1987).
[126] J. P. Jacobs, W. M. Klipstein, S. K. Lamoreaux, B. R. Heckel, and E. N. Fortson, Phys. Rev. A **52**, 3521–3540 (1995).
[127] Ch. Eisele, A.Yu. Nevsky, and S. Schiller, Phys. Rev. Lett. **103**, 090401 (2009).
[128] S. Herrmann, A. Senger, K. Möhle, M. Nagel, E. V. Kovalchuk, and A. Peters, Phys. Rev. D **80**, 105011 (2009).
[129] A. A. Michelson and E. W. Morley, Am. J. Science **34**, 333 (1887).
[130] D. Colladay and V. A. Kostelecký, Phys. Rev. D **58**, 116002 (1998).
[131] V. A. Kostelecký and C. D. Lane, Phys. Rev. D **60**, 116010 (1999).
[132] V. A. Kostelecký and S. Samuel, Phys. Rev. Lett. **63**, 224 (1989); Phys. Rev. Lett. **66**, 1811 (1991); Phys. Rev. D **39**, 683 (1989); Phys. Rev. D **40**, 1886 (1989).
[133] V. A. Kostelecký and R. Potting, Nucl. Phys. **B359**, 545 (1991); Phys. Lett. B **381**, 89 (1996).
[134] V. A. Kostelecký and N. Russell, Rev. Mod.Phys. **83**, 11 (2011); arXiv:0801.0287 (2012).
[135] P. Horăva, Phys. Rev. D **79**, 084008 (2009).
[136] R. C. Myers and M. Pospelov, Phys. Rev. Lett. **90**, 211601 (2003).
[137] I. Mocioiu, M. Pospelov, and R. Roiban, Phys. Rev. D **65**, 107702 (2002).
[138] P. A. Bolokhov, S. Groot Nibbelink, and M. Pospelov, Phys. Rev. D **72**, 015013 (2005).

19

Nuclear magnetic resonance gyroscopes

E. A. Donley and J. Kitching

19.1 Introduction

Gyroscopes measure rotation of a platform with respect to an inertial system [1]. The classic example of a mechanical gyroscope is a spinning massive object in a gravitational field, with one point of the object on its rotation axis held fixed. In this system, rotation ("precession") about the gravitational field axis will be observed at a rate proportional to the field magnitude and inversely proportional to the angular momentum of the object. Changes in the precession frequency indicate rotation with respect to inertial space.

As described in previous chapters, atomic spins in a magnetic field precess in a manner analogous to spinning mechanical systems in a gravitational field, with the precession rate proportional to the field magnitude and inversely proportional to the atom's angular momentum. Rotation of a measurement platform within the inertial frame can be detected by monitoring the apparent change in the Larmor frequency of the atoms.

In an inertial frame, the equation of motion for the spin polarization $\mathbf{P}$ of an ensemble of spins is $\partial\mathbf{P}/\partial t = \gamma\mathbf{B}\times\mathbf{P}$. When the laboratory frame containing the instruments used to measure the Larmor precession is rotating with respect to inertial space, the Larmor frequency is shifted by Ω, the rotation rate of the apparatus about the direction of the applied field. In this case, the equation of motion for $\mathbf{P}$ becomes

$$\frac{\partial\mathbf{P}}{\partial t} = (\Omega + \gamma\mathbf{B})\times\mathbf{P}. \tag{19.1}$$

The concept of rotation sensing by a measurement of the Larmor frequency is illustrated in Fig. 19.1.

The problem therefore becomes one of measuring the rotation-induced precession-frequency shift with high precision, while simultaneously controlling all other effects that lead to precession-frequency shifts. Electrons (and atoms with a single valence electron

Optical Magnetometry, ed. Budker, D. and Jackson Kimball, D. F. Published by Cambridge University Press.

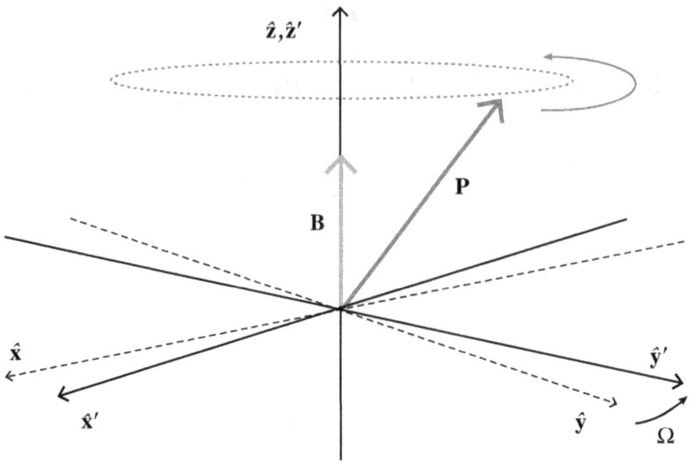

Figure 19.1 Determining rotation rate based on a measurement of the Larmor frequency. Here the laboratory frame is shown as the $(\hat{\mathbf{x}}', \hat{\mathbf{y}}', \hat{\mathbf{z}}')$ coordinate system and the inertial coordinate system is shown as $(\hat{\mathbf{x}}, \hat{\mathbf{y}}, \hat{\mathbf{z}})$. The magnetic field **B** is applied along the $\hat{\mathbf{z}}$-axis in both coordinate systems. **P** precesses in the $(\hat{\mathbf{x}}, \hat{\mathbf{y}}, \hat{\mathbf{z}})$ coordinate system with the "absolute" frequency γB. The lab frame, which contains the lasers and coils that measure the precession frequency, is rotating about the magnetic field with respect to the inertial frame at a frequency Ω. In transforming the precession from the inertial to the laboratory frame, the observed frequency is shifted by Ω.

whose magnetic moment is dominated by the electron magnetic moment) precess in a magnetic field at rate of approximately 1 MHz/G. Atoms with closed electron shells have a magnetic moment dominated by the nucleus, and typically precess 1000 times more slowly. A nuclear magnetic resonance gyroscope (NMRG) is one based on the precession of nuclear magnetic moments.

Table 19.1 summarizes the nuclear spin angular momentum values, gyromagnetic ratios, and nuclear quadrupole moments for several isotopes that have been used in NMR gyroscopes. Also included is an effective field for a rotation rate of 1°/h as determined by γ.[1] Larger values of γ put stricter requirements on the field stability in order to reach the same level of rotation-rate uncertainty. Table 19.1 also gives the gyromagnetic ratios for bare electrons as well as for ^{87}Rb in weak fields. For a free electron, γ is equal to the magnetic moment divided by the spin angular momentum. For an electron bound to a nucleus, γ is reduced by a factor of $1/q$, where $q = 2I + 1$ is the nuclear slowing-down factor, and I is the nuclear spin.

The most established gyroscope application is inertial navigation, in which the position of a moving platform is determined from quasi-continuous monitoring of the platform acceleration and rotation. To get a sense of the requirements for gyroscope performance, one can consider an example requiring high-performance inertial navigation. For a platform

[1] 1°/h is the minimum resolution needed for short-flight navigation applications, see Table 19.2.

Table 19.1. *Nuclear spin, gyromagnetic ratio, and the effective field equivalent to a 1°/h rotation frequency for atoms commonly used in NMR gyroscopes [2]. The final column contains the nuclear quadrupole moment, Q_n. For comparison, data for electron paramagnetic resonance (EPR) species are also shown.*

NMR species	I (h)	γ (MHz/T)	B_{eff} for 1°/h (fT)	Q_n (fm^2)
^{3}He	1/2	−32.44	23.8	
^{21}Ne	3/2	−3.363	229	10.16
^{83}Kr	9/2	−1.644	469	25.9
^{129}Xe	1/2	−11.86	65.1	
^{131}Xe	3/2	3.516	219	−11.4
^{199}Hg	1/2	7.712	100	
^{201}Hg	3/2	−2.847	271	38.6
EPR species		γ (MHz/T)	B_{eff} for 1°/h (fT)	
e^-		28030	0.028	
^{87}Rb	3/2	6998	0.110	

Table 19.2. *Required gyroscope uncertainty for navigation applications [3].*

Gyroscope uncertainty °/h	Systems applications
> 10	Automotive sensing
1	Attitude heading, short-flight devices
0.1	Some aircraft
0.01	Commercial airliners
< 0.001	Ships, submarines, spacecraft

moving at 800 km/h, inertial positioning of the platform to within 1 km over 2000 km of movement requires a rotation uncertainty of order $2\pi \times 10^{-8}$ rad/s, or 0.01°/h. This is equivalent to a magnetic field uncertainty of about 10 fT for nuclear species and 10 aT for electrons. A gyroscope based on alkali electronic spin precession alone would put very strict requirements on the magnetic-field stability, which is why NMR is the method of choice for spin-based gyroscopes. Nuclear spins also tend to have long coherence times since they interact only weakly with their environment. Table 19.2 summarizes the required gyroscope uncertainty for several navigation applications.

The operation of most NMR gyroscopes can be thought of as proceeding in three distinct steps: first the nuclear spins are polarized; then they are allowed to precess/rotate about the magnetic field direction during which time the rotation information is encoded in the spin orientation; finally the spin orientation is read out. Often these three steps happen simultaneously and continuously.

The nuclei of some atoms (such as ^{199}Hg or ^{201}Hg) can be polarized through direct optical pumping with resonant circularly polarized light. The orientation of the nuclei after precession can be detected in a similar manner. The earliest implementations of NMR gyroscopes were based on Hg atoms. More frequently, the nuclei of other atoms (noble gases, in particular) can be polarized through spin-exchange collisions with another optically polarized species, such as alkali atoms, contained in the same cell. In this case, the alkali atoms can also be used as an in situ magnetometer to read out the polarization orientation of the noble gas through its effective magnetic field.

A more detailed description of the NMR gyroscope based on spin-exchange optically pumped noble-gas nuclei is as follows. A vapor cell contains one or more noble-gas isotopes with nonzero nuclear spin (such as ^{129}Xe and ^{131}Xe), and an alkali metal (such as ^{87}Rb). Circularly polarized light resonant with an optical transition polarizes the alkali species, and the alkali electronic spin polarization is transferred to the noble-gas nuclei through collisions via spin-exchange optical pumping [4]. The alkali and noble-gas spins are continuously driven at their respective resonant frequencies and they precess about an applied magnetic field. In addition to the applied field, the alkali spins sense a field arising from the noble-gas polarization such that the total field as sensed by the alkali atoms rotates on a cone centered on the static field at the noble-gas precession frequency. All of the precession frequencies can be measured by analyzing the signal of a transverse probe beam that senses the direction of the alkali spin polarization.

Over short measurement times, noise processes inherent to the NMR gyroscope system generate an angle random walk (ARW), or white frequency noise, in the rotation rate. The rotation rate uncertainty of an NMR gyroscope due to this white frequency noise is determined by the transverse nuclear spin relaxation time, T_2, the measurement time, τ, and the measurement signal-to-noise ratio averaged over a time T_2 (S/N) [5,6]:

$$\delta\Omega \sim \frac{1}{T_2(\mathrm{S/N})\sqrt{\tau/T_2}}. \tag{19.2}$$

This can be understood as follows. The linewidth of the transition is given by the inverse of the relaxation time. This line can be resolved over a measurement time T_2 to a statistical fractional uncertainty given by the measurement signal-to-noise ratio. Repeated measurements over a time τ improve the resolution further by an amount $\sqrt{\tau/T_2}$. The ARW noise specification for an NMR gyroscope is determined from the rotation rate statistical uncertainty as ARW = $\delta\Omega\sqrt{\tau}$, and is usually specified in the units of degrees per root hour.

In the case of a spin-exchange optically pumped NMR gyroscope, it is useful to view the signal-to-noise ratio as being equal to the magnetic field produced by the noble-gas nuclei as seen by the alkali atoms divided by the sensitivity of the alkali magnetometer. Thus, a high noble-gas density, large polarization fraction, and high-precision measurement of the induced magnetic field lead to improved rotation sensing.

The bias instability expresses the limit of the gyroscope uncertainty on long time scales as determined by drift, $1/f$ noise and other non-Gaussian or nonstationary noise processes. For NMR gyroscopes, the most critical factor influencing the bias instability is the extent to which the magnetic field can be controlled, since changes in this parameter are indistinguishable from changes in rotation rate for a single atomic species. An elegant and powerful technique for reducing the effects of magnetic-field instability is to introduce a second noble-gas species, with a gyromagnetic constant different from the first, into the cell. In this configuration, changes in magnetic field cause different frequency shifts for the two species, while rotations cause identical shifts. As a result, the effects of changing magnetic fields can be separated from the true rotation signal and the bias instability from field changes is reduced. This method is discussed in more detail in Section 19.4. A second method for reducing the effects of magnetic fields, the alkali–noble-gas comagnetometer is discussed in Section 19.5.

A second important factor influencing bias instability is the effect of the nuclear electric quadrupole moment. Atomic nuclei are not perfectly round, and the nonspherical electric charge distribution interacts with electric field gradients (for example near the wall of the container) to produce a shift in the precession frequency. The effects of these shifts are discussed in more detail in Section 19.2.2.

The first proposals for NMR gyroscopes appeared in the 1950s [7, 8]. Optically pumped NMR gyroscopes were introduced in the 1960s with the advent of gyroscopes based on mercury nuclei [9] and continued into the 1980s with the development of NMR gyroscopes based on the noble gases [10]. Work on NMR gyroscopes mostly stopped in the 1980s, but NMR gyroscopes have seen a revival recently, in part because of advances in microfabrication and laser technology, and also because of the development of new NMRG methods based on comagnetometry [11].

19.2 NMR frequency shifts and relaxation

As with most resonant instruments, relaxation of the NMR spin resonance plays a critical role in determining the gyroscope short-term instability or ARW. The closely related issue of shifts in the NMR resonant frequency affects the bias instability and offset. In this section, we discuss the physical phenomena that underlie relaxation and frequency shifts in NMR gyroscopes.

Note that in the following, we generally speak of relaxation in terms of the longitudinal spin relaxation time T_1. The extent to which T_1 is dominated by spin-exchange collisions determines the signal-to-noise ratio, since it determines the achievable noble-gas polarization through spin-exchange optical pumping. T_1 also provides an upper bound on the transverse relaxation time T_2, which determines the ARW, with T_2 being reduced from T_1 by effects like random field fluctuations.

19.2.1 Spin exchange

Spin exchange [2] is an interaction between two atoms in which angular momentum is conservatively transferred from one atom to the other through a collision. Spin exchange can occur between atoms of the same species or of different species and typically results in a change in the equilibrium polarization of the ensemble, as well as both relaxation and a shift of one or both resonance frequencies.

When one atomic species, polarized through optical pumping, interacts with a second atomic species, the second species can become polarized. This is known as spin-exchange optical pumping (SEOP), and is one of the many interesting and useful effects that arise from spin-dependent interactions in atomic samples. A review of spin-dependent interactions can be found in [2].

SEOP in alkali–noble-gas mixtures occurs through binary collisions, as well as through the formation of van der Waals molecules through three-body collisions [12, 13]. The third body can either be a buffer-gas molecule, like N_2, or another noble-gas atom if the noble-gas density is sufficiently high [14].

For binary collisions, the longitudinal spin-relaxation rate can be written simply as the product of the alkali density n_{al}, a spin-exchange cross-section, σ, and the mean relative thermal velocity $\bar{v} = \sqrt{8k_B T/\pi \mathrm{m}}$, where m is the reduced mass in the alkali–noble-gas collision [15]. The longitudinal spin-relaxation rate from the formation of van der Waals molecules is inversely proportional to the noble-gas density and is characterized by the rate γ_V. Combining the effects of binary collisions and van der Waals molecule formation, the spin relaxation rate is [12, 14]

$$\frac{1}{T_1} = n_{\mathrm{al}} \left(\frac{\gamma_V}{n_{\mathrm{ng}}} + \sigma v \right) + \gamma' = R_{\mathrm{se}} + \gamma'. \tag{19.3}$$

Here, n_{ng} is the noble-gas number density and γ' is the relaxation rate from mechanisms other than spin exchange, which includes effects from wall collisions and magnetic field gradients.

Binary collisions dominate the longitudinal spin-exchange rate for the lighter noble gases, but the formation of van der Waals molecules can play a dominant role for heavy noble gases such as Xe, and can significantly shorten the period required to spin-polarize the noble-gas nuclei. At high noble-gas and/or buffer-gas densities, the lifetime of the weakly bound van der Waals molecules can be reduced to the point that the molecules do not live long enough for spin exchange to occur efficiently [12], and the spin-exchange rate due to binary collisions can exceed the spin-exchange rate due to van der Waals molecules [14].

The equilibrium noble-gas polarization can be expressed in terms of the alkali–noble-gas spin exchange rate R_{se}, the equilibrium spin polarization of the alkali atoms, P^{e}, and γ', as

$$P^{\mathrm{ng}} = P^{\mathrm{e}} \frac{R_{\mathrm{se}}}{R_{\mathrm{se}} + \gamma'}. \tag{19.4}$$

Spin-exchange also causes frequency shifts in each interacting species, which are manifest as an apparent enhancement of the magnetic field arising from the magnetization of the spin-exchange partner. The enhancement is caused by the hyperfine contact interaction [16, 17], which arises from an attraction between the alkali electron and the noble-gas nucleus. This effect enhances the ability for in situ alkali magnetometry to sense the noble-gas magnetization, since the field generated by the noble-gas nuclei can be magnified by orders of magnitude over what would be expected classically [18].

Expressed as an effective magnetic field arising from the polarization **P** of the other species, we can write an effective magnetic field

$$\mathbf{B}_{\text{eff}} = \frac{8\pi\kappa_0}{3}\mathbf{M} = \frac{8\pi\kappa_0}{3}\mu n\mathbf{P}. \tag{19.5}$$

Here $\mathbf{M} = \mu n\mathbf{P}$ is the magnetization of the other species, μ is the magnetic moment, n is the density, and κ_0 is the spin-exchange enhancement factor. Spin-exchange shifts were studied in detail in [19], and estimates of κ_0 were made in [15]. The enhancement κ_0 tends to be larger for the heavier noble gases. Depending on the alkali species, κ_0 ranges from $6 < \kappa_0 < 10$ for ^{3}He to $300 < \kappa_0 < 900$ for ^{129}Xe, with the higher values for a particular noble-gas nucleus corresponding to the heavier alkali atoms.

While the enhanced magnetic field improves NMRG sensitivity, spin-exchange also causes magnetization-dependent shifts of the NMR resonances that depend on cell temperature and laser power. A method of removing NMRG bias instability introduced by drifts in the alkali polarization was proposed in [20], and would require the use of three noble-gas isotopes instead of two. The shift is proportional to the alkali polarization, and having measurements of three Larmor frequencies allows for cancelation of the shift.

19.2.2 Quadrupole surface frequency shifts

Nuclei that have electric quadrupole moments are affected by electric-field gradients, which can cause relaxation and shifts of the NMR frequency. The magnitude of such shifts is proportional to the dot product between the electric field gradient and the nuclear quadrupole moment, Q_n. All nuclei that have a nuclear spin of $I > 1/2$ have nuclear quadrupole moments; these nuclei include ^{201}Hg and all $I \neq 0$ noble gases except for ^{3}He and ^{129}Xe (see Table 19.1). For all of the demonstrated dual-isotope NMR gyroscopes referenced below (Section 19.4), at least one of the NMR isotopes has a nuclear quadrupole moment.

In the limit where the nuclear quadropole resonance (NQR), shifts are small compared to the Zeeman shifts, the NQR interaction causes each NMR line to split into an array of $2I$ lines (Fig. 19.2). A nucleus with $I = 3/2$, for example, has four Zeeman energy levels. Three $\Delta m = 1$ transitions are allowed. When the electric-field gradient is zero, all the transitions between the Zeeman levels have the same frequency. However, when an electric field gradient is present, nonidentical shifts of the Zeeman energy levels result in a line triplet, with the spacing between the lines determined by the mean quadrupole interaction energy. The NQR shift also depends on the relative orientation of the principal axis of the

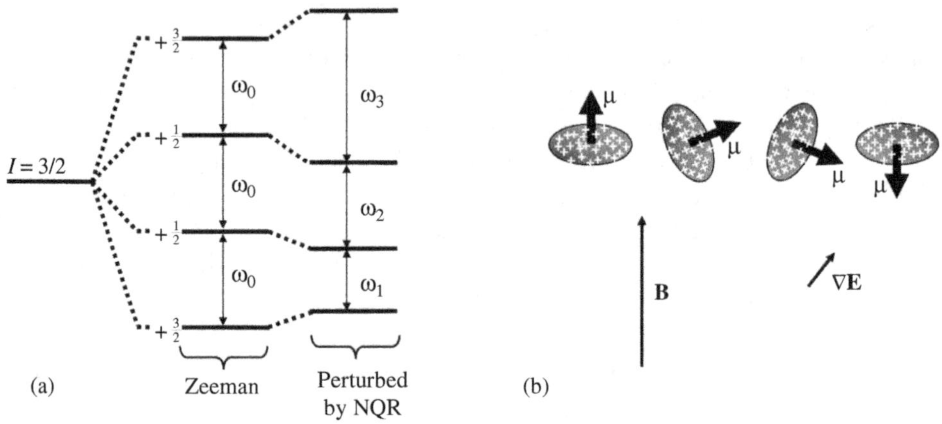

Figure 19.2 (a) Shifts of the Zeeman lines for an $I = 3/2$ nucleus perturbed by a weak electric-field gradient. (b) Pictorial view of the states of the nucleus in the fields. Nuclear quadrupole moments arise from asphericity of the nuclear charge distribution.

electric field gradient (a tensor) and the magnetic field axis, and it vanishes at the "magic angle" of 54.7°. Without careful experimental design, NQR interactions shift the lines and complicate the NMR spectrum.

For gases in vapor cells, electric-field gradients can arise through distortions in the electron cloud that occur during wall collisions [21], with the size of the NQR shift being proportional to the time-averaged electric-field gradient. The averaged electric-field gradients vanish for spherical and cubic cells; however, significant electric-field gradients can arise in cells with asymmetries in their shape or material construction. Smaller cells, all else being equal, result in more wall collisions and a correspondingly larger NQR shift.

Early studies on NQR shifts were performed for ^{83}Kr [22], ^{201}Hg [23, 24], and ^{131}Xe [25]. In these studies, the sample cells were nearly spherically symmetric. The shifts caused line broadening and low-frequency beating of the NMRG signals but were not large enough to be well resolved.

An enhancement of $\sim$100$\times$ in the size of the shifts was demonstrated by use of highly asymmetric cells [26], which allowed for clear resolution of the NQR lines in ^{131}Xe. A detailed microscopic theory of the wall interactions was also developed [27] and compared to measurements of the NQR splittings versus cell asymmetry [28]. In the asymmetric cells, collisions with the walls along the axis of the narrower cell dimension are more frequent than collisions with the walls along the other directions, which results in a large average electric-field gradient. The large NQR interactions allowed for detailed quantitative studies of the electric-field gradients and the activation energies for surface adsorption. Similar studies with asymmetric cell shapes were later performed on ^{131}Xe [29] and ^{83}Kr [30]. NQR shifts for ^{201}Hg have also been studied in detail in the context of precision-measurement tests of spatial isotropy [31, 32].

NQR resonances of ^{131}Xe in the limit of zero applied magnetic field and deviations from Berry's adiabatic geometric phase under rotation have also been studied [33]. The studies were performed in a highly asymmetric cylindrical cell of 2 cm diameter and 2 mm height. In this regime, the NQR shifts were larger than the Zeeman frequency, and mixing of the nuclear spin energy levels through rotation made all six transitions between nuclear energy levels allowed ($\Delta m = 1$, 2, and 3).

More recently, an NQR study on ^{131}Xe nuclei in a microfabricated cell of 1 mm^3 volume was performed [34]. Although the cells had cubic symmetry, the NQR shifts were larger than the Zeeman frequency for magnetic fields of less than 50 nT. The strong electric-field gradient ($\sim$30 GV/cm^2) was generated by anisotropy in the cell-wall materials as opposed to the cell shape, since two of the cell walls were formed of borosilicate glass and four of the cell walls were formed of silicon.

19.2.3 General wall relaxation

Even for nuclei that do not have a nuclear quadrupole moment (^{3}He and ^{129}Xe), relaxation from wall collisions can limit the longitudinal spin lifetime, particularly for small cells. The mechanisms for this relaxation are poorly understood at present [2], but wall relaxation is stronger when atoms spend a longer time adhered to the cell wall during wall collisions. The wall relaxation rate is inversely proportional to the cell diameter [2], and decreases exponentially with increasing temperature, since the wall-adhesion time also decreases.

The use of buffer gases can slow the rate of diffusion across the cell and reduce the relaxation rate from wall collisions. Silicone- and hydride-based wall coatings can also be used to lower the adhesion time and the wall-relaxation rate [35, 36]. Borosilicate is commonly used for glass-blown cells because of the relative ease with which it can be blown. More exotic cell-wall materials like aluminosilicate can also be used to minimize wall relaxation, particularly for ^{3}He, for which other relaxation rates are very low [37]. The He diffusivity in aluminosilicate is orders of magnitude lower than in borosilicate such that the ^{3}He wall adhesion time on aluminosilicate is much shorter.

19.2.4 Magnetic-field gradients

The spatial diffusion of atoms in the presence of inhomogeneous magnetic fields is another important source of relaxation [17, 38, 39]. At high pressures, when the spin precession period is short compared to the cell diffusion period, the longitudinal relaxation rate arising from gradients can be expressed as [17]

$$\frac{1}{T_1} = D\frac{|\nabla B|^2}{B_z^2}. \tag{19.6}$$

The resulting relaxation can be reduced by applying a large uniform bias field, B_z. D is the diffusion constant, which is inversely proportional to the pressure.

At low pressures, when the diffusion period is much shorter than the spin precession period, the longitudinal relaxation rate is proportional to the pressure and motional narrowing is observed [39].

Field gradients can arise from sources external to the cell as well as from the alkali and noble-gas atoms in the cell. For cells that are perfect spheres, the self-field as sensed by the nuclear spins is zero, and the field arising from the uniform magnetization of the alkali atoms is constant. For cells that are not spherical, the nuclei can sense gradients from both their own polarization and the alkali polarization [40]. Field gradients from cell asphericity can be the primary limit on the longitudinal relaxation rate of ^{3}He, for which relaxation rates from other mechanisms can be small. For small cells, effective field gradients from gradients in the alkali polarization can be significant because of depolarizing wall collisions.

19.2.5 Noble-gas self-relaxation

Self-relaxation of the nuclear polarization also occurs through self-collisions, and the mechanism for the relaxation depends on the isotope under study. Noble-gas electron clouds are distorted during noble-gas self-collisions, and these distortions produce electric-field gradients. For $I > 1/2$ isotopes, the interaction between the induced electric-field gradients and the nuclear quadrupole moments generate torques on the spins that cause longitudinal relaxation. The relaxation rate is proportional to the noble-gas pressure, and can be the dominant relaxation mechanism for $I > 1/2$ isotopes – particularly at high noble-gas pressure, and in large cells where the wall effects are relatively weaker. Quadrupolar relaxation from self-collisions can be significant for any NMR gyroscope that uses an isotope that has a nuclear quadrupole moment.

Quadrupolar relaxation causes the maximum achieved noble-gas polarization in optical-pumping experiments to depend on the noble-gas partial pressure [41, 42]. In large cells, the quadrupolar relaxation rate for ^{131}Xe is proportional to the noble-gas density, with the proportionality constant 3.96×10^{-2} amagat^{-1} s^{-1} [42],[2] which limits the spin lifetime to 25 s for a 1-amagat ^{131}Xe sample. For ^{21}Ne, the relaxation rate is much lower, and the lifetime as limited by quadrupolar relaxation is 210 minutes for a 1-amagat sample [43].

For ^{3}He, which does not have a nuclear quadrupole moment ($I = 1/2$), self-relaxation from dipole–dipole interactions can dominate the longitudinal spin relaxation in carefully prepared spherical sample cells. When dipolar spin relaxation is the limiting factor for the spin lifetime, the lifetime can be as long as 75 h for a 10 amagat ^{3}He sample [44].

For ^{129}Xe self-collisions, the dipole–dipole interaction is negligible, and bulk relaxation times are limited by the collisional spin–rotation interaction to 55 h for a 1-amagat sample [4].

[2] The amagat is defined as the ideal-gas density at standard temperature and pressure.

19.3 Alkali shifts and relaxation mechanisms

Many varieties of NMR gyroscopes depend on alkali atoms for optical pumping and/or detection of the magnetic field produced by the noble gas; therefore the S/N ratio depends strongly on the relaxation mechanisms and frequency shifts that affect the alkali atoms. Since these effects are reviewed in detail in Chapters 1–6 and 10–12, they are mentioned only briefly here.

At low alkali densities, the alkali relaxation is usually dominated by a combination of wall collisions and collisions with any buffer gas (including the NMR gas) in the cell. Since the wall-induced relaxation rate is inversely proportional to the buffer gas pressure, while the buffer-gas-induced relaxation rate is proportional to the pressure, an optimum pressure occurs for any given cell size [45] (see also Chapter 7).

Naturally, better signal-to-noise ratios for the measurement of magnetic fields are obtained with higher densities of alkali atoms because of statistical averaging of a larger number of atoms. However, at higher alkali densities, spin-exchange collisions between alkali atoms usually limit the relaxation time. In certain regimes of operation (notably for the comagnetometer approach described below), alkali–alkali spin-exchange-relaxation can be strongly suppressed. This spin-exchange-relaxation-free (SERF) operation [46–50] allows much higher magnetic field sensitivities to be achieved for alkali magnetometers (see Chapter 5), and correspondingly better sensitivity to the magnetization of the noble gas and hence rotation rate.

Collisional spin destruction from the spin–rotation interaction transfers polarization from the electron spin to the atoms' translational degrees of freedom, causing alkali spin relaxation and putting a fundamental limit on the efficiency of spin-exchange optical pumping [4]. This process therefore limits the maximum nuclear polarization that can be achieved and hence the rotation signal magnitude.

Finally, the light shift induced by the optical pumping light results in an effective magnetic field [51,52]. This effective field varies throughout the cell if the laser beam is not perfectly collimated and if the optical density of the alkali atoms is high. This can lead to broadening and relaxation similar to magnetic-field gradients.

19.4 Two-spin NMR gyroscope

As pointed out in Section 19.1, the magnetic-field stability requirements are greatly relaxed when two NMR isotopes are used in an NMRG. NMR gyroscopes developed in the 1960s and 1970s based on two isotopes demonstrated bias instabilities lower than 0.1°/h. The history of NMR gyroscopes is somewhat masked by commercial and military secrecy, but several reviews of this early work have been published [53–57].

The first two-spin NMRG was based on mercury isotopes [9,58,59]. A simplified diagram of an NMR gyroscope based on mercury is presented in Fig. 19.3. The vapor cell contains ^{199}Hg, ^{201}Hg, and a buffer gas. The Hg nuclear spins are optically pumped and probed with light at 253.7 nm emitted by two ^{204}Hg lamps. Coherent spin precession at the atomic

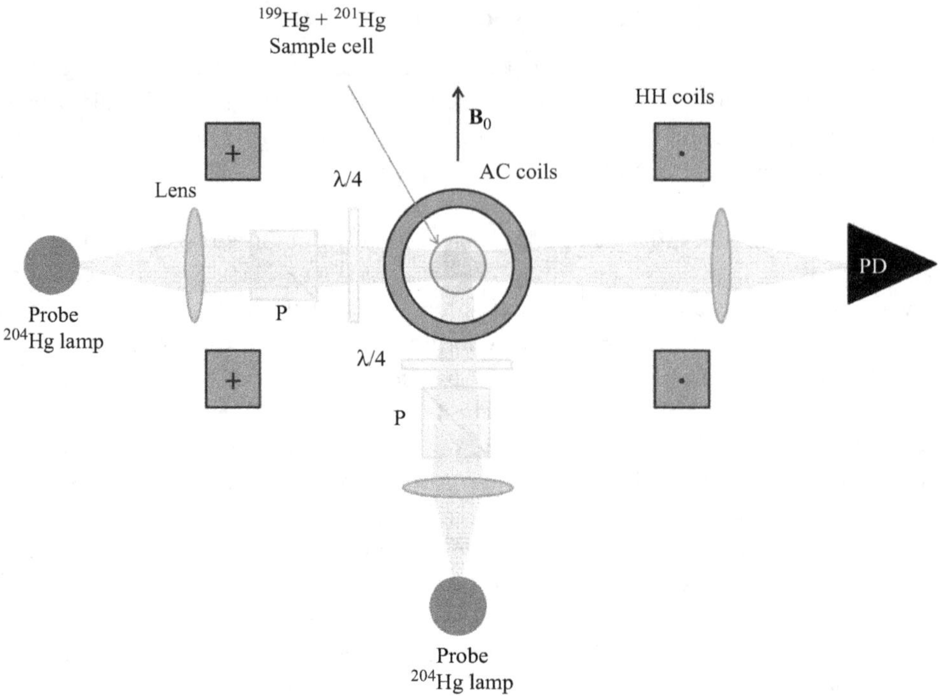

Figure 19.3 A mercury NMRG. A vapor cell contains ^{199}Hg, ^{201}Hg, and some buffer gas. The Hg nuclear spins are optically pumped and probed with light from ^{204}Hg lamps at 253.7 nm that is circularly polarized by polarizers (P) and quarter-wave plates ($\lambda/4$). The 253.7 nm line from the ^{204}Hg lamps overlaps optical pumping transitions in both ^{199}Hg and ^{201}Hg, making ^{204}Hg lamps a convenient light source. A DC bias field $\mathbf{B}_0$ is generated by a pair of Helmholtz coils (HH). Coherent spin precession is generated with applied AC magnetic fields perpendicular to $\mathbf{B}_0$. The Hg spin-precession frequencies are detected as modulations of the probe-beam transmission through the cell as measured by the photodiode (PD).

Larmor frequency is driven by applied AC magnetic fields perpendicular to a static field, $\mathbf{B}_0$. The precession frequency is detected optically as a modulation of the probe-beam transmission [60].

When the NMRG is rotating about the axis of the static field B_0, the measured Larmor frequencies are shifted by the rotation rate Ω:

$$\omega_{199} = \gamma_{199} B_0 + \Omega, \tag{19.7}$$

$$\omega_{201} = |\gamma_{201}| B_0 - \Omega. \tag{19.8}$$

Note that the signs of γ are opposite for ^{199}Hg and ^{201}Hg (they precess in opposite directions; see Table 19.1), so the rotational shifts for the two isotopes have the same magnitude but opposite sign. However, the two Hg isotopes have different magnetic-field shifts, owing to their different gyromagnetic ratios. Solving Eqs. (19.7) and (19.8) for Ω gives

$$\Omega = \frac{\gamma_{199}\omega_{201} - |\gamma_{201}|\,\omega_{199}}{\gamma_{199} + |\gamma_{201}|}, \tag{19.9}$$

thus the rotation rate can be determined from measurements of ω_{199} and ω_{201}. Note also that the magnetic field can be servoed by stabilizing the sum of the NMR frequencies, since the sum $\omega_{199} + \omega_{201} = (\gamma_{199} + |\gamma_{201}|)\,B_0$ is independent of the rotation rate. When the total magnetic field is servoed by the sum of the NMR frequencies, the rotation rate and angle can be found from the frequency and phase shifts between the signal from one of the NMR isotopes and a reference oscillator [61], which can simplify signal processing. As of the early 1980s, this dual-species approach with Hg had achieved $0.053°/\sqrt{\text{h}}$ for the ARW [53] and a bias instability of 0.02°/h [57].

Spin-exchange optical pumping (SEOP) [4] in mixtures of alkali and noble-gas atoms can also be used to realize NMR gyroscopes based on spin-polarized noble-gas nuclei [10], making it possible to optically pump the alkali atoms with near-infrared lasers or lamps. Beginning in the 1970s, NMR gyroscopes based on noble-gas isotopes were introduced [10, 61, 62]. The alkali atoms that pump the noble-gas nuclei were simultaneously mechanized as an in-cell magnetometer to sense noble-gas nuclear precession. NMR gyroscopes based on ^{129}Xe and ^{131}Xe and spin-exchange optically pumped by Rb achieved a bias instability of 0.01°/h and an ARW of $0.002°/\sqrt{\text{h}}$ [63]. These systems had a total volume of several hundred cubic centimeters per axis of rotation.

19.5 Comagnetometer

Recently, a novel version of an NMR gyroscope based on a potassium–^{3}He comagnetometer was demonstrated that achieved performance comparable to dual species NMR gyroscopes, but with a single nuclear isotope [11]. In this configuration, the K spins act as the rotation sensors, with their response to rotation enhanced by the presence of the ^{3}He and their response to magnetic-field drifts suppressed by the presence of the ^{3}He. The comagnetometer was developed to perform sensitive tests of physics beyond the standard model [64] (see Chapter 18, Sections 18.3 and 18.4).

The comagnetometer consisted of a vapor cell of 2.5 cm diameter containing ~7 amagat of ^{3}He, K metal, and about 20 torr of nitrogen gas to optimize optical pumping by quenching the K excited state. A circularly polarized pump laser beam polarized the K vapor and hence the ^{3}He nuclei through collisions. A diagram of the cell and relevant magnetic fields is shown in Fig. 19.4.

Through the spin-exchange interaction [Eq. (19.5)], the ^{3}He nuclei generate an effective field of order 100 nT as sensed by the K spins. The K effective field as sensed by the ^{3}He is of order 0.1 nT. Therefore the two spin ensembles effectively sit in different magnetic fields. In spite of their vastly different gyromagnetic ratios, the spin-exchange interaction makes it possible for the K electronic spins and the ^{3}He nuclear spins to have nearly the same Larmor frequency, such that their spin dynamics are strongly coupled [65]. The required applied

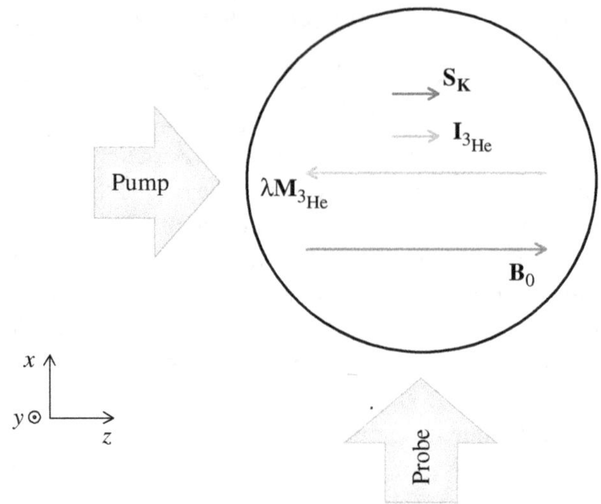

Figure 19.4 Field configuration for the comagnetometer gyroscope. Here $\mathbf{S_K}$ denotes the K electron spin, $\mathbf{I}_{^3\mathrm{He}}$ denotes the ^{3}He nuclear spin, and $\lambda\mathbf{M}_{^3\mathrm{He}}$ is the effective magnetic field felt by the K atoms due to the ^{3}He polarization, where $\lambda = 8\pi\kappa_0/3$. The linearly polarized probe beam detects any K spin projection along the x-axis via the polarization rotation technique. The instrument is sensitive to rotations along the y-axis.

field to reach this regime of strong coupling is nearly equal in magnitude but opposite in direction to the ^{3}He effective field.

In this field configuration, the total field as sensed by the K spins nearly vanishes, and the K–K spin-exchange rate exceeds the K Larmor frequency such that the comagnetometer operates in the spin-exchange-relaxation-free regime [48, 49], thus enhancing the magnetometer sensitivity. The K effective field as sensed by the ^{3}He is large enough to reduce relaxation from magnetic-field gradients [Eq. (19.6)].

The gyroscope signal arises from the x-component of the K polarization that results from a rotation about the y-axis. Because of the strong coupling between the ^{3}He and K spins, the x-component of the K polarization is enlarged by a factor of γ_K/γ_{3He}, which is nearly a factor of 1000.

The comagnetometer configuration reduces the sensitivity of the gyroscope to magnetic field drifts, transients, and gradients since the effective field arising from the ^{3}He magnetization screens the K spins such that they sit in a near-zero magnetic field. For field changes that are slow compared to the ^{3}He Larmor frequency, the ^{3}He polarization adiabatically follows the total field and cancels slow changes in the applied field.

With the K–^{3}He comagnetometer, an ARW of $0.002°/\sqrt{\mathrm{h}}$ and a bias instability of 0.04°/h were achieved. A comagnetometer based on ^{21}Ne might exhibit improvement in instability by an order of magnitude, resulting from its smaller gyromagnetic ratio (1/10 that of ^{3}He).

19.6 Miniaturization

For many applications, the most important issue is not improving performance but, instead, maintaining performance while reducing size, weight, and power. Microfabrication techniques, described in detail in Chapter 7, can allow for substantial reductions in these instrument parameters. Instrument cost can also be reduced, especially if batch fabrication processes are used and volume manufacturing is possible. For some applications, NMR gyroscopes have the potential advantage over micromachined spinning or vibratory gyroscopes in that they contain no moving parts and they are relatively insensitive to vibrations.

The recent level of activity in NMR gyroscopes is indicated by the number of patents that have emerged [20,66–72] – most of which are related to miniaturization. Reference [66] is a scheme for using batch processing to develop a chip-scale atomic gyroscope. To enable low power consumption, the proposed device uses permanent magnets and low-power vertical-cavity surface emitting lasers. Two patents pertain to cell fabrication techniques [67, 68]. One patent describes ways to pattern electrical conductors to mostly cancel stray fields that arise from the cell heaters [69]. Two patents describe solenoid geometries that improve magnetic-field uniformity [70, 71]. One patent describes instrument geometries that make use of a diverging laser beam to simultaneously pump and probe the atoms and reduce common-mode noise through differential detection [72].

Wafer-level arrays of glass-blown spherical microcells have been developed [73, 74], as have methods to deposit mirrors onto angled cell walls etched in silicon [75, 76]. Several miniaturized NMR gyroscope components and integration methods were used for an NMR gyroscope prototype described in [77].

An advantage of miniaturization is that magnetic shielding is less expensive and works much better for smaller instruments [78]. A disadvantage of miniaturization is that cell-wall effects become much stronger, which results in larger NQR frequency shifts and shorter spin lifetimes.

In spite of the reduced spin lifetimes, the performance of microfabricated prototypes is approaching the requirements for the navigation of commercial aircraft [79]. So far, an ARW of $0.01°/\sqrt{\mathrm{h}}$ and a bias instability of 0.05°/h has been demonstrated in a 6-cm^3 physics package. Work is ongoing to reduce the power and size requirements while continuing to improve the performance.

19.7 Conclusion

NMR gyroscopes are a relatively old idea that is experiencing a revival – partly because of their potential for miniaturization and low-cost fabrication, and partly because of their potential for use in demanding navigation applications. We have presented a review of the basic physics involved with NMR gyroscopes, and have summarized the published performance results.

Acknowledgments

Enlightening discussions with M. Bulatowicz, M. Larsen, and M. V. Romalis are gratefully acknowledged. This article is a contribution of the US government, not subject to US copyright.

References

[1] H. Goldstein, *Classical Mechanics* (Addison-Wesley, New York, 1950).
[2] R. K. Harris *et al.*, Pure Appl. Chem. **73**, 1795 (2001).
[3] P. G. Groves, *Principles of GNSS, Inertial, and Multisensor Integrated Navigation Systems* (Artech House, Norwood, MA, 2008).
[4] T. G. Walker and W. Happer, Rev. Mod. Phys. **69**, 629 (1997).
[5] I. A. Greenwood and J. H. Simpson, *Proceedings of the IEEE 1977 National Aerospace and Electronics Conference (NAECON77)*, 1246 (1977).
[6] J. Kitching, S. Knappe, and E. A. Donley, IEEE Sensors **11**, 1749 (2011).
[7] B. D. Leete, United States Patent 2,720,625 (1955).
[8] A. Hansen, United States Patent 2,841,760 (1958).
[9] J. H. Simpson, J. T. Fraser, and I. A. Greenwood, IEEE Trans. Aerospace **1**, 1107 (1963).
[10] E. Kanegsberg, *Proceedings of the 2nd Digital Avionics Systems Conference*, 40 (1977).
[11] T. W. Kornack, R. K. Ghosh, and M. V. Romalis, Phys. Rev. Lett. **95**, 230801 (2005).
[12] W. Happer, E. Miron, S. Schaefer, D. Schreiber, W. A. van Wijngaarden, and X. Zeng, Phys. Rev. A **29**, 3092 (1984).
[13] R. J. Knize, Z. Wu, and W. Happer, Adv. Atomic Mol. Phys. **24**, 223 (1987).
[14] G. D. Cates, R. J. Fitzgerald, A. S. Barton, P. Bogorad, M. Gatzke, N. R. Newbury, and B. Saam, Phys. Rev. A **45**, 4631 (1992).
[15] T. G. Walker, Phys. Rev. A **40**, 4959 (1989).
[16] R. M. Herman, Phys. Rev. **137**, A1062 (1965).
[17] R. L. Gamblin and T. R. Carver, Phys. Rev. **138**, A946 (1965).
[18] B. C. Grover, Phys. Rev. Lett. **40**, 391 (1978).
[19] S. R. Schaefer *et al.*, Phys. Rev. A **39**, 5613 (1989).
[20] E. Kanegsberg, United States Patent 7,282,910 B1 (2007).
[21] C. Cohen-Tannoudji, J. Phys.-Paris **24**, 653 (1963).
[22] C. H. Volk, J. G. Mark, and B. C. Grover, Phys. Rev. A **20**, 2381 (1979).
[23] P. A. Heimann, Phys. Rev. A **23**, 1204 (1981).
[24] P. A. Heimann, I. A. Greenwood, and J. H. Simpson, Phys. Rev. A **23**, 1209 (1981).
[25] T. M. Kwon, J. G. Mark, and C. H. Volk, Phys. Rev. A **24**, 1894 (1981).
[26] Z. Wu, W. Happer, and J. M. Daniels, Phys. Rev. Lett. **59**, 1480 (1987).
[27] Z. Wu, S. Schaefer, G. D. Cates, and W. Happer, Phys. Rev. A **37**, 1161 (1988).
[28] Z. Wu, W. Happer, and J. M. Daniels, Phys. Rev. A **42**, 2774 (1990).
[29] R. Butscher, G. Wäckerle, and M. Mehring, J. Chem. Phys. **100**, 6923 (1994).
[30] R. Butscher, G. Wäckerle, and M. Mehring, Chem. Phys. Lett. **249**, 444 (1996).
[31] S. K. Lamoreaux, J. P. Jacobs, B. R. Heckel, F. J. Raab, and E. N. Fortson, Phys. Rev. Lett. **57**, 3125 (1986).
[32] S. K. Lamoreaux, J. P. Jacobs, B. R. Heckel, F. J. Raab, and E. N. Fortson, Phys. Rev. A **39**, 1082 (1989).

[33] S. Appelt, G. Wäckerle, and M. Mehring, Phys. Rev. Lett. **72**, 3921 (1994).
[34] E. A. Donley, J. L. Long, T. C. Liebisch, E. R. Hodby, T. A. Fisher, and J. Kitching, Phys. Rev. A **79**, 013420 (2009).
[35] X. Zeng, E. Miron, W. A. Van Wijngaarden, D. Schreiber and W. Happer, Phys. Lett. **96A**, 191 (1983).
[36] A. T. Nicol, Phys. Rev. B **29**, 2397 (1984).
[37] W. A. Fitzsimmons and G. K. Walters, Phys. Rev. Lett. **19**, 943 (1967).
[38] L. D. Schearer and G. K. Walters, Phys. Rev. **139**, A1398 (1965).
[39] G. D. Cates, S. R. Schaefer, and W. Happer, Phys. Rev. A **37**, 2877 (1988).
[40] M. V. Romalis and G. D. Cates, Phys. Rev. A **58**, 3004 (1998).
[41] K. F. Stupic, Z. I. Cleveland, G. E. Pavlovskaya, and T. Meersmann, J. Magn. Reson. **208**, 58 (2011).
[42] D. Brinkmann, E. Brun, and H. H. Staub, Helvetica Phys. Acta **35**, 431 (1962).
[43] R. K. Ghosh and M. V. Romalis, Phys. Rev. A **81**, 043415 (2010).
[44] N. R. Newbury, A. S. Barton, G. D. Cates, W. Happer, and H. Middleton, Phys. Rev. A **48**, 4411 (1993).
[45] W. Happer, Rev. Mod. Phys. **44**, 169 (1972).
[46] W. Happer and H. Tang, Phys. Rev. Lett. **31**, 273 (1973).
[47] W. Happer and A. C. Tam, Phys. Rev. A **16**, 1877 (1977).
[48] J. C. Allred, R. N. Lyman, T. W. Kornack, and M. V. Romalis, Phys. Rev. Lett. **89**, 130801 (2002).
[49] I. K. Kominis, T. W. Kornack, J. C. Allred, and M. V. Romalis, Nature **422**, 596 (2003).
[50] I. M. Savukov and M. V. Romalis, Phys. Rev. A **71**, 023405 (2005).
[51] W. Happer and B. S. Mathur, Phys. Rev. **163**, 12 (1967).
[52] C. Cohen-Tannoudji and J. Dupont-Roc, Phys. Rev. A **5**, 968 (1972).
[53] F. A. Karwacki, J. Inst. Navigation **27**, 72 (1980).
[54] K. F. Woodman, P. W. Franks, and M. D. Richards, J. Navigation **40**, 366 (1987).
[55] K. Liu, W. Zhang, W. Chen, K. Li, F. Dai, F. Cui, X. Wu, G. Ma, and Q. Xiao, J. Micromech. Microeng. **19**, 113001 (2009).
[56] E. A. Donley, *IEEE Sensors, Waikaloa, HI*, 17 (2010).
[57] M. S. Goldstein, I. A. Greenwood, M. M. Kuritsky, H. Lerman, J. E. McCarthy, T. Shanahan, M. Silver, and J. H. Simpson, Proc. IEEE **71**, 1156 (1983).
[58] J. H. Simpson, Astronaut. Aeronaut. **2**, 42 (1964).
[59] D. S. Bayley, I. A. Greenwood, and J. H. Simpson, United States Patent 3,778,700 (1973).
[60] H. G. Dehmelt, Phys. Rev. **105**, 1924 (1957).
[61] B. C. Grover, E. Kanegsberg, J. G. Mark, and R. L. Meyer, United States Patent 4,157,495 (1979).
[62] E. Kanegsberg, Proc. SPIE **157**, 73 (1978).
[63] L. K. Lam, E. Phillips, E. Kanegsberg, and G. W. Kamin, Proc. SPIE **412**, 272 (1983).
[64] J. M. Brown, S. J. Smullin, T. W. Kornack, and M. V. Romalis, Phys. Rev. Lett. **105**, 151604 (2010).
[65] T. W. Kornack and M. V. Romalis, Phys. Rev. Lett. **89**, 253002 (2002).
[66] H. C. Abbink, E. Kanegsberg, and R. A. Patterson, United States Patent 7,239,135 (2007).
[67] H. C. Abbink, E. Kanegsberg, K. D. Marino, and C. H. Volk, United States Patent 7,292,031 (2007).
[68] H. C. Abbink, W. P. Debley, C. E. Geosling, D. K. Sakaida, and R. E. Stewart, United States Patent 7,292,111 (2007).

[69] M. Bulatowicz, United States Patent 2009/0316753 A1 (2009).
[70] M. Bulatowicz, United States Patent 2010/0194506 A1 (2010).
[71] M. Bulatowicz, United States Patent 2011/0247414 A1 (2011).
[72] J. Kitching, E. A. Donley, E. Hodby, A. Shkel, and J. Eklund, United States Patent 7,872,473 (2011).
[73] E. J. Eklund, A. M. Shkel, S. Knappe, E. Donley, and J. Kitching, Sensor Actuat. A **16**, 232 (2007).
[74] E. J. Eklund, A. M. Shkel, S. Knappe, E. Donley, and J. Kitching, Sensor Actuat. A **143**, 175 (2008).
[75] M. A. Perez *et al.*, Sensor Actuat. A **154**, 295 (2009).
[76] M. A. Perez, J. Kitching, and A. M. Shkel, Sensor Actuat. A **155**, 23 (2009).
[77] E. J. Eklund, Ph.D. Thesis, University of California, Irvine (2008).
[78] E. A. Donley, E. Hodby, L. Hollberg, and J. Kitching, Rev. Sci. Instrum. **78**, 083102 (2007).
[79] M. Larsen, *42nd Annual Meeting of the APS Division of Atomic, Molecular, and Optical Physics* (Atlanta, GA, June 13-17, 2011).

20

Commercial magnetometers and their application

D. C. Hovde, M. D. Prouty, I. Hrvoic, and R. E. Slocum

20.1 Introduction

Applications such as geophysical exploration for minerals and oil, anti-submarine warfare, volcanology, earthquake studies, magnetic observatories, and the detection of buried objects at sites of archaeological significance require instruments that can measure small variations of the Earth's magnetic field in time and space. Other applications such as laboratory metrology, space exploration, and biomedicine may require measurements at lower or higher fields. The demand for precise measurements of magnetic fields is met in part by commercial atomic magnetometers supplied by a number of manufacturers over the last fifty years. Commercial optical magnetometers based on cesium, potassium, and helium are now firmly established. Commercial magnetometers designed for measuring anomalies in the Earth's field are typically operated as self-oscillating devices or VCO (voltage-controlled oscillator) lock-in devices with the oscillation or lock-in frequency proportional to the magnetic field. This chapter examines the specifications for atomic magnetometers, compares the most widely used approaches, describes some of the features demanded by different applications, and surveys the history of atomic magnetometers. (The online supplemental material contains a table listing many of the United States patents related to the field of atomic magnetometry.)

Earth's field at the surface ranges from 20 to 80 μT. Typically the user is seeking to generate a map of the local magnetic field upon which magnetic anomalies can be discerned. The map may be made by means of a walking survey, for instance in the case of archaeological sites. For mineral exploration much wider areas must be surveyed, so that typically airborne and ground surveys are required. The magnetic anomalies can be in the pT range or even smaller. For such applications, the most significant requirement for the magnetometer is its sensitivity. Because the fluctuations of Earth's field are in the range of 100 nT or even greater, applications may require differential methods such as gradiometers

Optical Magnetometry, ed. Budker, D. and Jackson Kimball, D. F. Published by Cambridge University Press.

(two or more magnetometers whose difference is measured during the survey) or a single moving magnetometer in tandem with a fixed "base station" that is used to correct the long-scale fluctuations seen by the mapping magnetometer. In either case, because the magnetometer moves as the map is generated, heading-error effects are also important. Other applications, such as magnetic observatories or long-term studies, require absolute accuracy and freedom from slow drifts. Often the orientation of the magnetometer is fixed for these applications. Reliability, cost, size, power, weight, and other features such as the ability to export data to standard mapping programs also impact the commercial success of atomic magnetometers.

20.2 Specifications

20.2.1 Noise

Noise, by definition, is any variation in the measurement not caused by actual variations in the external magnetic field. Sources of noise in the magnetometer include quantum shot noise, excess noise from the electronics, and eddy currents and other technical noise sources from the sensor. Noise is specified for a fixed sensor, so that the heading or motion errors are not included in the background system-noise specification.

Noise is normally specified in nanoteslas (10^{-9} T) per square root hertz ($\mathrm{nT}/\sqrt{\mathrm{Hz}}$) or in some cases picoteslas (10^{-12} T) per square root hertz ($\mathrm{pT}/\sqrt{\mathrm{Hz}}$). The frequency in hertz (Hz) refers to the measurement bandwidth rather than to the Larmor frequency. The bandwidth is not simply the inverse of the sampling rate or measurement cycle time, but includes electronic bandwidth, oversampling, and Nyquist effects. Although peak-to-peak noise measurements have been used, more typically the measurements are reported as root mean squared (RMS), as the RMS has a more rigorous statistical definition than peak-to-peak.

Magnetometer noise does not always scale as the square root of the sample rate or bandwidth. The actual scaling depends on the time scale of the measurement (compared to the time scale of the underlying atomic polarization being measured) and on the electronics used to measure the Larmor frequency. Magnetic field amplitude is measured by counting precession frequency cycles during the measurement time. Simultaneously, time is measured by counting a much higher frequency internal clock. The precession frequency can be inferred from the change in the phase $\Delta\phi$ over the counting interval Δt:

$$\Omega = \frac{\Delta\phi}{\Delta t} . \tag{20.1}$$

With the simplest counter circuitry, the start time is determined from a single zero crossing, and likewise for the stop time. The counter counts cycles that differ in phase by 2π. The precession frequency is given by

$$\Omega = \frac{2\pi n_{\mathrm{cycles}}}{\Delta t} . \tag{20.2}$$

The numerator is 2π times an integer with no uncertainty in normal operation (although missing counts may occur and can be corrected for). Over short time scales, uncertainty in the frequency therefore arises from estimating the start and stop times for the interval:

$$\left(\frac{\sigma_\Omega}{\Omega}\right)^2 = \left(\frac{\sigma_{\Delta t}}{\Delta t}\right)^2. \tag{20.3}$$

The uncertainty in time depends on the slope and noise of the Larmor waveform at the two crossings. These are independent of the measurement interval Δt; so the uncertainty in Ω, and hence magnetic field noise reported by the sensor, scales as $1/\Delta t$. Graphically, this is equivalent to determining the slope of the line $\phi(t)$ from the phase measurements at the beginning and end of the measurement time, without using any of the data in between.

However, it is possible to measure the frequency from all n zero crossings within the measurement time by finding the best-fit line to each of the n zero crossings within the measurement time. This could be implemented in hardware with a phase-locked loop limited in bandwidth. As a result of the additional phase data, the uncertainty in the magnetic field decreases by an extra factor of $1/\sqrt{\Delta t}$, for an overall factor of $1/\Delta t^{3/2}$. The overall result is that the noise of a commercial magnetometer at short measurement times scales as the measurement bandwidth to the 3/2 power. In this discussion the magnetic field is assumed to be constant over the course of the measurement, so that the fluctuations are due to noise alone. The effects of narrow-band filters such as phase-locked loops are discussed below in the section on bandwidth (Section 20.2.5).

While noise affects the counter directly, over longer time scales it also affects the operation of the self-oscillating or VCO lock-in loop. Noise near the precession frequency adds to the measured phase, and the feedback loop adjusts the frequency slightly to cancel the phase noise. The result is Brownian diffusion of the phase of the oscillator, which gives rise to noise proportional to $1/\sqrt{t}$ when the system is observed for times long compared to the atomic-coherence time [1].

When a manufacturer does not specify a measurement bandwidth for the noise, but instead provides an absolute noise in nT, the reader should assume the specification is valid only at the slowest cycle rate of the system, i.e., several seconds or more.

The noise of a magnetometer is typically determined from measurements taken over time scales appropriate for the frequency range of interest – usually on the order of a few minutes. In practice there are three methods used to measure the sensitivity of commercial magnetometers:

1. Comparison of readings of two co-located magnetometers in a magnetic field that is sufficiently quiet that the gradients are constant. Most manufacturers maintain a magnetically quiet test site in a rural area for purposes of testing magnetometers at the pT/$\sqrt{\text{Hz}}$ sensitivity and better. The time intervals of the readings of the two magnetometers are strictly synchronized. Typically a few hundred readings (when determining the short-term sensitivity) are recorded. The difference of readings is taken and the RMS

value of its scatter is calculated. Since this represents noise of two equal, independent, and uncorrelated magnetometers, the noise has to be divided by the square root of 2. This method assumes a stable gradient between the two sensor positions.

2. Measurement in a quiet magnetic field. If the magnetic field can be stabilized more accurately than the expected noise of the magnetometer under test, one can take a sufficient number of readings in this stable field and compute the RMS value of the scatter. Magnetic shields (Chapter 12) can be used to produce a magnetically quiet region, inside which a DC field can be generated by means of suitable coils. However, the noise of typical current sources makes this method difficult to apply to the best commercial magnetometers. A current source used to produce a field of 50 μT, typical of Earth's surface, must be stable to 1 part in 10^8 in order to keep current noise from dominating the measurement. Gradient measurements with two magnetometers inside the shields can ease the requirements on the current source.
3. Recording a time series and processing it to determine the power spectral distribution of the field can refine the measurements in a quiet field. Such a time series and related power spectral distribution are shown in Figs. 20.1 and 20.2, for signals from a single magnetometer recorded in a magnetic shield. Strong structure from power lines (at 60 Hz) is present, even in a well-shielded location. At the low-frequency end, below 1 Hz and lower frequencies, fluctuations in the applied field dominate the spectrum. White noise from the instrument is probably the limiting factor near 5 Hz.

For measurements of magnetic anomalies, long-term sensitivity or long-term stability of magnetometers is not important. For magnetic observatories, volcanology, earthquake

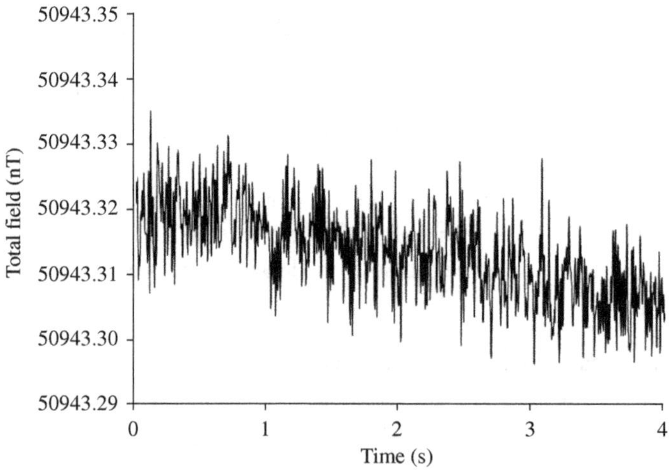

Figure 20.1 Four seconds of magnetic field as a function of time, measured inside high-quality shields. The magnetic field was updated every 4 ms, resulting in a Nyquist limit of 125 Hz.

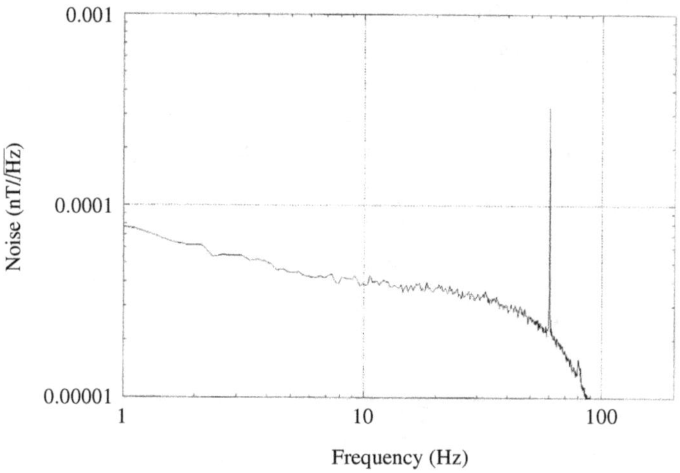

Figure 20.2 The power spectral distribution determined from 450 seconds of field data including the portion shown in Fig. 20.1. Note the 60 Hz noise spike. An additional spike at 90 Hz arises from the presence within the shields of a second magnetometer. Gradients within the shields result in a difference in Larmor frequencies of 90 Hz. (Data provided courtesy of Geometrics.)

studies, and space measurements, on the other hand, long-term stability can be very important. Long-term stability may differ substantially from the short-term sensitivity discussed above.

20.2.2 Resolution

The resolution of a magnetometer is the smallest change in magnetic field the magnetometer can resolve or report. The resolution may not be equal to the least-significant digit of the instrument's readout at all settings. The user should choose a magnetometer which has a counter resolution smaller than the smallest change in magnetic field being detected, or "quantization errors" will result, improperly defining the shape and character of the anomaly. Because the resolution of a magnetometer usually varies with sample rate, it is important to make sure that the resolution specification applies to the intended sample rate.

20.2.3 Sensitivity

The term "sensitivity" is a common specification offered for commercial magnetometers but because manufacturers define the term differently it makes for a difficult comparison. Sensitivity can be defined as the larger of either the resolution or the noise. With this definition, the sensitivity specification provides a single number that can be used for performance comparison between magnetometers that are limited by their resolution and those that are limited by their noise. The manufacturer should specify sensitivity for all expected speeds

Table 20.1. *Typical sensitivity at 1 Hz bandwidth for various commercial magnetometers.*

Atom	Sample size (mm)	Sensitivity (pT)
K	35	0.5
K	70	0.1
K	150	0.03
Cs	25	0.5
He-4	20	0.3
He-4	75	0.04
He-3	20	1.0

of reading and associated bandwidths over a range of magnetic fields characteristic of the Earth and perhaps a range of temperatures.

Instrumental factors that affect the sensitivity include the size of the atomic vapor sample and the atomic medium. Larger samples result in a measurement that makes use of a greater number of spins. Optical depth can approach unity without introducing spin-relaxing collisions. Together these two factors reduce shot noise and improve sensitivity. Large sample volumes also reduce wall relaxation processes, which can result in narrower resonances and thus improved sensitivity. Table 20.1 shows representative sensitivity in a 1 Hz bandwidth and the sensor size for a number of vendors.

20.2.4 Sample rate and cycle time

Sample rate is defined as the number of readings per second generated by the magnetometer. It is often specified as a frequency of reading in Hz. Cycle time, the number of seconds per reading, is the inverse of the sample rate. When constructing a magnetic-anomaly map, the field has to be sampled at a suitable spatial scale. This spatial scale should be small compared to the depth of the targets in order for the survey to locate the target accurately. The cycle time of the magnetometer and the speed with which it is moved over the ground determine the distance between measurements. In hand-carried devices, the user can walk slowly to increase the data density, at the expense of reduced production. In airborne and marine applications, this option is often impractical, as the vehicle may have some minimum operating speed. In any case, the desired sample rate will affect the choice of bandwidth.

20.2.5 Bandwidth

The bandwidth of a magnetometer is a measure of how well it can respond to rapid changes in the measured field. Higher bandwidth allows a magnetometer to faithfully reproduce a rapidly fluctuating actual field. All digital magnetometers report a value based on an average of the observed precession frequency during a "read" period. This value is normally the average value over a period less than or equal to the sample update time. In such cases, it is the sample rate and not any internal averaging that limits the frequency content of the data. In other cases where there are filters that average the frequency over longer times, these filters can become the limiting factor of frequency content.

The bandwidth of the magnetometer should be matched to the bandwidth of the magnetic field fluctuations of interest. When mapping out magnetic anomalies, the measurement bandwidth must be large enough to not distort the anomalies. If any significant processing of the data has been done inside the magnetometer, it is important to know the resulting bandwidth of the system because this determines the basic utility of the system. It is possible for example, to make smooth, low-variability magnetic data by applying a digital filter that removes all high frequencies from the output. Although the system appears to offer both low noise and rapid updates of the magnetic field, in fact such a filter removes both noise and any high-frequency content of the magnetic signal. In such cases it is the filter and not sample rate that determines the fastest speed at which a survey should be done. Over-filtering limits the performance of software that calculates the depth of targets and can mask the anomalies caused by small objects.

Commercial magnetometers are mixed analog-digital devices. The analog end can also include filtering. Narrow-band notch filters are not typically used, due to the wide range of frequencies over which the magnetometer must operate. A phase-locked loop can be used to reduce the noise bandwidth while preserving frequency agility. The time constant of the loop determines the noise bandwidth. The output can be digitized. Typically, the output is over-sampled to reduce aliasing. If the magnetic field has strong components near the Nyquist limit, then these may still be aliased down to the detection band. This problem can be detected as an anomalous increase of noise with the increase of the rate of readings.

20.2.6 Absolute error and drift

The absolute error is defined as the difference between the average of the readings of the magnetometer and the average of the field it measures. Drift is defined as the change in the absolute error with time. All magnetometers will have some absolute error and drift in their measurements. In most cases the drift will be much less than the absolute error.

For producing maps of magnetic anomalies, the drift specification is more important than the absolute error. For most surveys, a drift of less than 0.1 nT per day is acceptable. Greater drift than this causes noticeable artifacts such as survey block edge effects (i.e, discontinuities in the magnetic field map arising when adjacent regions are sampled at significantly different times). Because other survey errors (positional inaccuracies, heading errors due to internal instrument and mobile platform effects, diurnal Earth's field variations) usually dominate and are cumulative, absolute accuracies in the range of 2 to 4 nT are widely deemed acceptable. When a magnetometer is being used as a primary standard or in an observatory, then absolute accuracy and drift specifications are primary factors and sub-nT absolute errors are required. Criteria for achieving good absolute accuracy are numerous [2]. The gyromagnetic ratio is known precisely for protons in water at 25°C. The International Association of Geomagnetism and Aeronomy recommends using the latest CODATA values for all measurements after 1992. The 2010 CODATA value for the proton gyromagnetic ratio [3] is $\gamma_p = 2.675\,222\,005(63) \times 10^8\ \mathrm{s^{-1}\,T^{-1}}$ (or $0.267\,522\,2005\ \mathrm{rad\,s^{-1}\,nT}$). The gyromagnetic ratio of the proton in water is slightly different due to the shielding effects of the induced magnetic fields. At 25°C in a spherical sample surrounded by vacuum, the accepted value for water [3] is $2.675\,153\,268(66) \times 10^8\ \mathrm{s^{-1}\,T^{-1}}$. An accurate determination of the magnetic field from the precession of Cs depends on the hyperfine level and the nonlinear Zeeman effect [4]. Although the underlying spectroscopy is almost as well known (the fine-structure g-factor is known to about 1 part in 10^7), atomic magnetometers that measure a poorly defined blend of hyperfine lines will not achieve the highest absolute accuracy.

20.2.7 Gradient tolerance

Gradient tolerance is defined as the ability of the magnetometer to obtain a meaningful reading in a given gradient, not necessarily with the highest sensitivity. When the field has large gradients, errors in assessing the position of the sensor may dominate the error budget, so an increase in noise may be tolerable. Gradient tolerance depends on both the spectral line width and the size of sensor – smaller sensors tend to have high gradient tolerance. Wider spectral lines (Cs, Rb, ^{4}He) tend to show better gradient tolerance, although if a buffer gas is used then collisions with the buffer gas slow the movement of the atoms and limit the magnetometer's tolerance for gradients. The effect of gradients has been studied in the laboratory on paraffin-coated cells [5]. For a potassium-based sensor without buffer gas, a cell with a diameter of 35 mm has been shown to operate at gradients of over 50 000 nT/m. Gradient tolerance falls off sharply with increased sensor size. A similar potassium cell with a diameter of 70 mm can operate only in gradients of up to 1 000 nT/m, while 150 mm cells cannot withstand any movement of the cell and tolerate perhaps a 100 nT/m gradient at maximum.

Gradient tolerance also depends on the distance the atoms can diffuse over the time scale of the coherence. Proton-precession and Overhauser-type magnetometers generate the Larmor signal from a liquid sample such as water or hydrocarbons. Typical gradient tolerance for proton magnetometers is in the range of 1 000 nT/m but may be less depending on the absolute field strength (lower fields produce lower signals in proton and Overhauser magnetometers, lowering their gradient tolerance in low-field areas such as Brazil).

Magnetic gradients reduce the amplitude of the magnetometer signal. This can result in increased noise, lower sensitivity, and increased dead zone. In applications such as airborne surveys, the sensor is far from field sources, so gradient tolerance is not an important specification. In surveys of unexploded ordnance or landfills, gradients can be very large, requiring a magnetometer that tolerates large gradients.

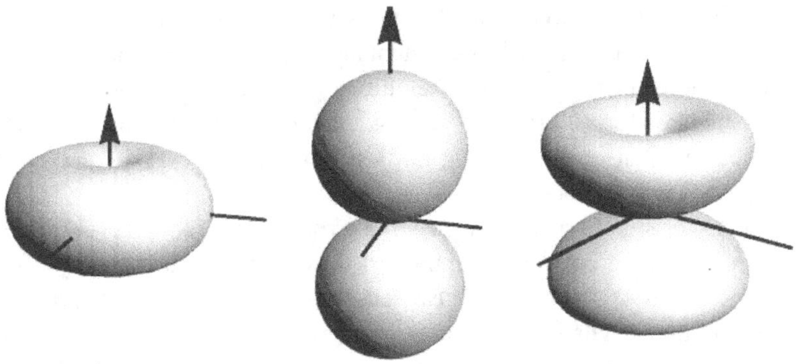

(a) (b) (c)

Figure 20.3 Polar plots showing signal strength for various sensor orientations with different types of dead zones: (a) a polar dead zone, (b) an equatorial dead zone, and (c) both (courtesy of Geometrics).

20.2.8 Dead zones

Atomic magnetometers have dead zones which occur when the orientation of the magnetometer is such that the signal is so weak that the magnetometer ceases to operate. The range of solid angles of the dead zone depends on the details of construction of the magnetometer, but the basic topology is a function of the underlying symmetry of the electromagnetic-field/atom system. Figure 20.3 shows dead zones associated with some commercial cesium magnetometer designs.

A number of ways exist to mitigate the problems with dead zones. In optically pumped systems an adjustable-orientation fixture can hold the sensor so that the sensor can be positioned such that it will not enter a dead zone during survey maneuvers. In proton-precession and Overhauser systems, dead zones can be eliminated with appropriate design. In some applications, the orientation of the optical sensor cannot be controlled. In such cases, it is necessary to have duplicate sensors oriented so that their dead zones do not overlap, or to choose a different measurement technology. ^{4}He magnetometers normally employ a design consisting of three integrated orthogonal sensor axes to achieve maximum sensitivity on all headings.

20.2.9 Heading error

Although scalar atomic magnetometers are designed to measure the total magnetic field, the field reported by a sensor depends slightly on its orientation in the field. This (unwanted) dependence of the measurement on orientation is termed heading error. Heading error is most important in airborne and marine systems and less so in hand-carried systems. This is because in airborne systems the aircraft must constantly change its orientation angle to maintain the correct path over the ground. In marine systems, the sensor undergoes some change in orientation as it is towed. In land systems, heading error is less apparent as a person tends to hold a sensor at a more or less constant angle. When generating magnetic contour maps, the magnetometer typically reverses direction on each pass across the measurement zone; heading error will produce "striping" in the final result. There are methods for post-acquisition removal of this artifact, but a better course is to minimize the error by careful design of the instrument.

Manufacturers commonly specify the heading error for a sensor as a peak-to-peak value. Heading errors are measured by placing the test sensor on a rotation stage surrounded by one or two fixed witness sensors. The test sensor is rotated through its operating range while monitoring the signals of the rotating and fixed sensors. Then the deviation of the test sensor from the witness sensors is plotted as a function of angle. Ideally, the witness sensors should be close enough to one another and the test sensor that gradient noise does not degrade the measurements, but far enough apart that the stray fields of the test sensor do not significantly change the measurements of the witness sensors.

The materials used to make the magnetometer must be carefully selected to minimize heading error. One technical source of heading error is the field generated by ferromagnetic particles within the sensor. As the sensor's orientation is rotated, the permanent magnetic field of the particles also rotates, resulting in a variation of the total scalar field in the sensed volume. Manufacturers select nonmagnetic parts and then test the components for residual magnetism prior to construction. Another, generally smaller heading error arises from the diamagnetic materials and possibly also paramagnetic materials used to build the sensor and housing. These are polarized by the local magnetic field, and the induced field contributes to the field in the sensed volume. If these materials are unevenly distributed around the measurement cell, they will give rise to a net field which changes with heading, hence a heading error. Likewise these materials displace air, which includes paramagnetic oxygen. Displacing the oxygen changes the field in the sensed volume. Magnetic fields can also be generated by the sensor electronics. For example, a current loop of 10 mA with an area of 1 cm^2 generates a magnetic field of 7 pT at a distance of 30 cm. Electronic components may also be ferromagnetic. Careful construction of detection and amplification electronics is essential. Usually, the sensor electronics are separated from the atomic sensor itself, but on some measurement platforms it is impossible to separate the measurement from the electronics by more than 1 m.

Changes in the phase of the magnetometer's response result in heading errors when the magnetometer is operated in self-oscillating configuration. One source of heading error is the geometric-phase shift of the self-oscillating signal. Larmor precession is the result of the torque on the magnetic moment associated with the atomic spins, $\boldsymbol{\tau} = \boldsymbol{\mu} \times \mathbf{B}$ (Chapter 1). From the magnetometer's point of view, a 180° rotation is equivalent to a reversal of **B**, hence a reversal of the precession. Such a phase reversal would change positive feedback needed to sustain self-oscillation into negative feedback. Commercial magnetometers automatically sense and correct electronically for the phase reversal to achieve continuous self-oscillation. However, at angles between 0° and 180°, the system can exhibit a smaller phase shift. This phase shift becomes a heading error as the feedback network dynamically changes the self-oscillation frequency to maintain a zero net phase change around the feedback loop.

The nonlinear Zeeman (NLZ) effect can be a source of heading error (see Chapter 1, Section 1.1.2). The linear Zeeman effect provides a good approximation at Earth's field (about 3.5 Hz/nT for Cs, or 175 kHz at 50 μT), so that a single Larmor precession frequency is a useful concept. The NLZ effect causes the precession frequency to depend slightly (~7 Hz for Cs at 50 μT) on the direction of the atom's spin relative to the magnetic field; the NLZ effect becomes more pronounced at higher fields. The amplitudes for pumping and probing the atoms also depend on the direction of the atom's spin as well as the angles between the optical field and the magnetic field, so that as the heading changes, the relative strengths of the different components making up the signal change, thereby shifting the frequency at which the system self-oscillates. The net result is a heading error.

Another source of heading error is the inherent asymmetry of the signal when it is a composite of several hyperfine transitions (see also Chapter 4), for instance when atoms are probed with atomic emission lamps that contain a mix of optical wavelengths. The relative intensities of the unresolved hyperfine peaks depend on the heading angle. The result is that the self-oscillation frequency changes by ~10 nT. Balancing two beams with opposite circular polarization suppresses this heading error due to the virtual light shift to a fraction of a nT. Helium and potassium magnetometers operate on a single spectral line; hence they do not suffer from hyperfine-structure heading error. The heading error of commercial potassium magnetometers is typically 50 pT. Commercial He magnetometers have an intrinsic heading error of 600 pT, which is compensated to 80 pT.

20.2.10 Range of measurement

Commercial magnetometers are designed to measure the Earth's field over the entire surface of the Earth. The Earth's field varies from approximately 67 μT in Antarctica to as little as 22 μT in Brazil. The magnetometer must measure the total field value over this entire range (plus anomalies, e.g., 20 to 80 μT) with no increased drift or heading errors due to changes in the absolute field values. Some naturally occurring anomalies can exceed this range. For instance, the field in some parts of Brazil approaches zero, while ferrous outcroppings can result in surface fields of 500 μT.

Factors that affect the operation over this range of fields include changes in the spectroscopy of the atoms and performance issues related to the electronics. For the electronics, the main issue is balancing the wide bandwidth needed to handle a wide range of oscillation frequencies against the need to reduce noise by limiting the bandwidth. The primary physics issue is the NLZ splitting. If NLZ splitting is less than the linewidth of the magnetic resonance, then the NLZ effect will broaden the line. The broadening can lead to reduced signal-to-noise and possibly to heading errors as discussed above. For instance, when operating potassium magnetometers at fields as low as 20 μT, operation on one line is only possible by reducing the bandwidth of the measurement to ~85 Hz, as a result of the decrease in the spread of the spectral lines. If the NLZ splitting is greater than the linewidth, then separate resonances can be observed, and it would be possible for the magnetometer to operate on any one of them. The range of Earth's field is sufficiently broad that some atomic systems can change from NLZ-unresolved to NLZ-resolved. Potassium and helium-4 magnetometers always operate with resolved NLZ (see Chapters 4 and 10, respectively). The magnetometer must operate properly over the full Earth-field range.

20.3 History of commercial magnetometry

20.3.1 Fluxgate magnetometers

Although neither a quantum nor an optical magnetometer, the fluxgate magnetometer merits discussion in this chapter because it has historically been used for similar applications as optical magnetometers. The fluxgate magnetometer was originally developed in the 1940s for use in anti-submarine warfare and after World War II was adopted for airborne geophysical surveys for natural-resource exploration.

A fluxgate magnetometer usually consists of a series of windings around a ferromagnetic core in which the permeability is a function of field strength. The windings are used to impress both an AC and a DC field on the core, in addition to the ambient field. The AC driving signal is used to drive the core through saturation around its hysteresis loop. The ambient field adds to the AC field in one half-cycle and subtracts in the other, creating a difference in the interval of saturation, detectable at twice the driving frequency using lock-in detection techniques for added sensitivity. This rectified asymmetry signal is proportional to the vector component of the ambient field along the AC field, although the constant of proportionality depends on the details of the nonlinear response of the core. The applied DC field is adjusted to null the ambient field. The current used to produce the DC field measures the ambient DC field and is independent of the core nonlinearity.

A number of fluxgate instruments have been developed. In its various configurations, the instrument is known as a second-harmonic, ring core, thin film, or oriented fluxgate magnetometer. The DC current used to null the ambient field is a source of some instability at long periods but careful design allows sensitivity of perhaps 0.05 nT in recently developed instruments, 0.1 nT in some of the airborne and monitoring versions, and 0.5 nT for hand-held portable instruments.

All fluxgate magnetometers measure a relative rather than absolute value of the magnetic field and as such must be calibrated in some way. A reference atomic magnetometer is commonly used to calibrate the fluxgate in airborne, satellite, and some ground survey applications. Since the fluxgate measures a vector component of the field, survey applications require that the instrument be referenced to some spatial coordinate system such as vertical, as in the case of ground portable surveys, or electromechanically oriented along the Earth field for airborne surveys. Archaeological survey in many parts of the world employ high-performance, low-drift multiple vertical fluxgate gradiometer arrays, providing excellent near-surface (short-range) anomaly detection.

Oriented fluxgate systems utilize three mutually orthogonal fluxgate sensors that are continuously oriented by a remote motor so that two of the sensors measure zero field. The third or primary sensor thus measures the maximum vector, otherwise referred to as total magnetic intensity. The precision of the sensors and the electromechanical orientation system are some of the principal limitations on sensitivity or resolution, particularly on a rapidly moving platform. Most satellites, a few airborne research survey aircraft, and some borehole probes utilize three mutually orthogonal fluxgate sensors referenced to an inertial or other spatial coordinate system in order to obtain vector information of the field. The periodic rotation of a satellite can be used to process fluxgate readings to reduce the effect of errors from drifting calibrations in the x, y, and z axes.

20.3.2 SQUID magnetometers

The superconducting quantum interference device (SQUID) magnetometer has the highest sensitivity of any commercial magnetometer. The SQUID magnetometer is based on Josephson's discovery in the early 1960s that small gaps in a circulating superconducting current loop were sensitive to magnetic flux perpendicular to the loop causing changes in the current flow. A resonant circuit is used to monitor the current changes and obtain vector-field information. SQUIDs are very sensitive ($\sim 3 \times 10^{-14}$ T), high-bandwidth (up to 50 kHz) devices [6]. The advent of high-temperature superconductors increased the usefulness of SQUIDs for geophysical surveys, although these devices have higher noise than low-temperature SQUID magnetometers. SQUID sensors have commercial applications in biomagnetics (heart and brain monitoring), magnetotellurics (measuring electrical currents flowing in the earth), and in paleomagnetics (where rock samples are measured for remanent field direction and density, demagnetized, then re-measured to determine the sample's chronological magnetic history). Recently SQUID sensors have also been used for stationary electromagnetic surveys as magnetic current sensors. However, SQUIDs are rarely used with mobile platforms due to a variety of practical limitations. The need for cryogenic cooling creates logistic difficulties for deploying SQUID sensors, because cryogenic fluids such as liquid nitrogen or liquid helium are not widely available. Motion of the sensor platform reduces the usefulness of vector sensors. Using three orthogonal SQUID sensors could in principle provide the total scalar field but, in practice, maintaining the relative orientation of the three sensors to the accuracy required to match the extraordinary sensitivity of the SQUID is a formidable challenge. Finally, SQUID sensors are several times more expensive than proton or alkali-vapor magnetometers. Thus, SQUID sensors tend to dominate only in markets where the highest sensitivity is critical for successful measurements.

20.3.3 Proton-precession and Overhauser magnetometers

The proton magnetometer was developed by Varian and Packard in 1954 as an outgrowth of research involving nuclear magnetic resonance [7]. The proton magnetometer, also called nuclear-precession, proton-precession, or proton free-precession magnetometer, consists of a coil of wire surrounding a fluid containing hydrogen atoms such as water, kerosene, etc. The proton is itself a magnetic dipole that can be polarized by applying a magnetic field at right angles to the field to be measured, which is produced by a DC current in the coil surrounding the sample. After the current in the coil is abruptly terminated, these spinning protons precess about the direction of the ambient magnetic field, at a frequency that is proportional to the field intensity. The precessing protons induce a signal, typically, 2 kHz, in the polarizing coil. The polarization usually decays in a matter of a few seconds, requiring a repolarization between measurements, and the measurements are thus discontinuous. Commercial magnetometers are operated at a rate of up to several readings per second.

Proton magnetometers, in common with atomic magnetometers, measure spin precession to determine the total scalar magnetic field. This characteristic facilitates magnetic survey applications on a moving platform because the sensor measures magnetic field independently of its orientation. As a result, precession-based magnetometers are widely used for ground, marine, and airborne surveys.

A variation of the proton magnetometer, the Overhauser or Abragam magnetometer, uses an organic hydrocarbon fluid containing free radicals that can be polarized by a (linearly polarized) radiofrequency exciting field whose frequency is appropriately adjusted. The free radical transfers its polarization to the proton sample. The Overhauser-effect magnetometer provides a continuous measurement. Because Overhauser magnetometers do not require a DC polarizing field, the exciting field (about 60 MHz) does not degrade the absolute accuracy of Overhauser magnetometers.

Commercial proton-precession magnetometers appeared in the early 1960s. Breakthroughs led to sensitivities of single-digit nT and then fractions of a nT; the improvements were immediately applied in mineral exploration. Geometrics [8] optimized the technology and used it in airborne surveys. A team in Grenoble, France, developed a continuous proton oscillator based on the Overhauser effect for military magnetic-anomaly detection [9]. Canadian development followed with similar continuous-precession magnetometers by Scintrex in 1970 [10]. GEM Systems [11] developed continuous and pulsed versions in the early 1980s. Sensors from these manufacturers achieved up to 5 readings/s and about 0.02 nT sensitivity.

The maximum sensitivity of commercial magnetometers based on proton precession of either type is generally 0.01 nT. The absolute accuracy of the free-precession magnetometer, however, is approximately 0.1 nT. This represents an absolute error of about 2 ppm in a field of 50 μT. The accuracy is limited primarily by the magnetic properties of the sensor, accuracy of the frequency-counting circuits, and known precision of an atomic constant. At present the proton-precession magnetometer provides the most accurate measurement of the absolute value of the Earth's magnetic field of any commercial magnetometers. The standard for magnetic observatories for scalar total-field measurement is an Overhauser proton-precession magnetometer with sensitivity of about 10–20 pT, absolute accuracies of ±0.1 nT, and a 1 Hz measurement rate. The frequency response of commercial proton magnetometers is limited by the highest cycle rate to a few Hz for free-precession or

Overhauser instruments. Proton magnetometers will not operate in high gradients, which limit their use in some ground surveys, such as over iron formations.

20.3.4 Alkali metal magnetometers: rubidium, cesium, and potassium

The optically pumped rubidium magnetometer was first used for geophysical measurements in 1957 [12]. Manufacturers included Varian, which later sold its magnetometer product line to CAE for military use and to Scintrex for commercial use. Early optically pumped magnetometers utilized a lamp containing the alkali metal. Light from the lamp passed through a glass cell containing the vapor of the same element and was focused on a photodetector on the other side of the vapor cell. In recent decades, diode lasers of the proper frequency have been developed to excite the atoms in the vapor cell, but many commercial instruments still use a lamp as the light source. The output of the photodetector is connected through an amplifier to a coil surrounding the vapor cell. This electro-optic system, through resonant absorption and re-radiation of energy, constitutes an oscillator whose frequency is directly proportional to the scalar magnetic field intensity in much the same way as for the proton-precession magnetometer. But where the proton-precession magnetometer operates in pulsed mode, optically pumped magnetometers typically operate with a continuous signal. The signal frequency is higher, roughly as the ratio of the electron mass to the proton mass: several hundred kHz for the alkali metals and 2 MHz for helium. The time response of the magnetometer is governed by the time interval over which the oscillation frequency is counted; practical devices can have kHz update rates.

The sensitivity of a cesium magnetometer has been shown to be 0.5 pT in a bandwidth of approximately 0.1 Hz through 10 Hz and possibly 1 pT between 10^{-3} Hz and 100 Hz. The absolute accuracy of a conventional optically pumped magnetometer, however, is generally 1 nT to several nT due to the complexity of various resonant lines on which it operates that in turn are a function of orientation in the Earth's magnetic field and several operating characteristics of the exciting lamp and amplifier.

Potassium narrow-spectral-line magnetometers appeared as a result of cooperation between the St. Petersburg research group led by Dr. E. B. Alexandrov and the Canadian manufacturer GEM Systems. The hyperfine structure of potassium (K) reduces certain heading-error effects and results in better absolute accuracy. The vapor pressure of potassium is lower than that of cesium at a given temperature, so typically potassium-based sensors must be heated, resulting in somewhat higher power consumption. Development has now resulted in a single-spectral-line operation and a sensitivity of some 30 fT once per second in stationary installations and about 0.1 pT in mobile applications. High absolute accuracy is obtained, similar to proton/Overhauser magnetometers.

The smaller intrinsic heading error of K means split-beam operation is not necessary to cancel it. This results in somewhat simpler electronics. The high sensitivity and small heading error make the K magnetometer a good fit for multiple-sensor applications. Potassium is now used in helicopter-borne and fixed-wing-craft-borne gradiometers and in ground surveys requiring superior sensitivity and gradient tolerance.

20.3.5 Helium-3 and helium-4 magnetometers

The innovation leading to helium magnetometers began with the first optical pumping of the metastable level of helium by Colegrove and Franken at the University of Michigan in 1958 [13]. The Franken patent [14] was purchased by Texas Instruments in 1959 leading to the first optically pumped scalar helium magnetometer at Texas Instruments (TI) in 1960 [15]. The TI team produced more than 1500 scalar He-4 magnetometers based on this technology for use by the US Navy for airborne magnetic-anomaly detection (MAD) of submarines. Lamp-pumped He-4 magnetometers were independently developed and manufactured beginning in the 1960s by Sinclair Research (later a part of ARCO) [16] and in the 1990s by China's Aero Geophysical Survey & Remote Sensing Center for Land and Resources (AGRS) [17].

Colegrove, Walters, and Schearer demonstrated a method for spin-exchange optical pumping of metastable helium atoms to create spin-polarized He-3 nuclei [18, 19], which led to the development of the He-3 nuclear free-precession magnetometer by Slocum and Marton in 1974 [20]. The first space vector-helium magnetometer was developed at TI in 1961 and is described in Chapter 10 [21]; a series of seven space magnetometers for Earth, planetary, and solar missions, built at the NASA Jet Propulsion Laboratory (JPL), followed. The ARCO–Sinclair magnetometer was not marketed but was used for in-house aeromagnetic surveys for oil and mineral exploration. AGRS also used helium magnetometers for aerial magnetic surveys.

All He-4 magnetometers manufactured in the twentieth century were lamp-pumped instruments from TI, JPL, and two geophysical-exploration organizations. The switch to laser pumping resulted in a significant advance in sensitivity of He magnetometers. In the USA, Polatomic [22] was founded by R. E. Slocum in 1982 and pursued practical application of 1083 nm semiconductor DBR laser pump sources for high-performance He-4 and He-3 magnetometers. One important application was the development of the next-generation laser-pumped AN/ASQ-233 He-4 system for US Navy anti-submarine aircraft, with research and development funding provided through the Small Business Innovation Research (SBIR) Program. The development of laser-pumped helium magnetometers over the last two decades resulted from a collaboration of four institutions: Polatomic, Inc., L. D. Schearer of the University of Missouri-Rolla in the USA, M. Leduc and F. Laloe of the Ecole Normale Superieure, Paris, and J. M. Hamel of the ISMRA at Caen, France. This effort included development of laser pump sources for He isotopes [23], applications of laser-pumped helium isotopes, and magnetometers using He-3 and He-4. In 1996, LETI-CEA Advanced Technologies, Grenoble, France, was selected to design and build a laser-pumped He-4 magnetometer for the SWARM satellites. At the time of writing this was scheduled for launch no earlier than November 2012. The SWARM instrument is discussed in detail in Chapter 15.

Helium-magnetometer technology has tended to emerge from government and corporate-sponsored R&D activities without a commercial outlet in North America and Europe. Alkali-vapor magnetometers have seen multiple commercial outlets as well as government-sponsored development. There has been surprisingly little interaction between the two groups over the past 50 years, with an intangible "repulsion factor" that seems to be based on research culture as much as technology. Following the invention of optical pumping of He-4 by Colegrove and Franken, the helium optical pumping patent was offered to Varian by Peter Franken. He reported that his friends at Varian would be happy to buy it but that they would only put it "on the shelf," because they were pleased with and committed to Rb and other alkali-vapor instruments. Franken immediately sold the patent in 1959 to Texas Instruments, who ten years later won the competition with Varian for the next-generation US Navy airborne AN/ASQ-81 MAD system, a contract which totaled approximately $1 billion in sales in today's dollars. In 1981 Varian made an offer to sell its alkali-vapor magnetometer business to TI, which would bring the manufacture of alkali and helium magnetometers under one roof, but the offer was declined. Meanwhile in the 1980s, EG&G owned Geometrics and considered purchasing a startup laser-helium-magnetometer company, Polatomic, but instead made an offer to sell Geometrics' alkali-vapor-magnetometer business to Polatomic, an offer that also was declined. Following a merger of its business unit with Raytheon in 1997, the Texas Instruments Defense Group exited the magnetometer-manufacturing business, whereupon a large part of the helium magnetometer technical team migrated from TI to Polatomic. Because the technology for making helium and alkali-vapor magnetometers has remained in separate companies, there is no optical magnetometer "superpower." A healthy magnetometer competition continues today.

20.4 Military applications

Important military applications that use atomic magnetometers include magnetic-anomaly detection (MAD) for both anti-submarine warfare (ASW) and for locating mines and unexploded ordnance (UXO). Other applications where atomic magnetometers might be useful include perimeter monitoring and detection of improvised explosive devices (IEDs).

The US Navy flies magnetic-anomaly-detection systems in the tail stinger of a P-3 Orion aircraft based on the AN/ASQ-81 and AN/ASQ-208 magnetometers. Combined sales amounted to about 1500 units at a unit cost of $350 000 (1991 dollars) [24]. Typically the MAD system will include one or more atomic magnetometers, a less-accurate vector magnetometer for correcting heading errors that arise in the atomic magnetometer and from the magnetic field of the aircraft, and motion sensors that can be used to compensate for the apparent signal due to spatial variations in Earth's field. The P-3 flies localization patterns at about a hundred meters above sea level. The Navy prefers scalar magnetometers for airborne magnetic-anomaly detection because the scalar measurement is less susceptible to orientation errors arising from the motion of the aircraft.

The P-3 platform was introduced in the late 1950s. The US Navy is presently investigating other options for airborne magnetometry, such as unmanned aerial vehicles (UAVs). Unmanned platforms can stay on patrol for longer periods of time and relieve flight crews of the drudgery of flying search patterns. Typically the craft are much smaller than the P-3, so instruments must be lighter and consume less electrical power. The smaller physical separation of the sensor from the body of the UAV may require improvements in algorithms for compensating for platform noise.

Underwater surveys of the seafloor are important in military and commercial applications. Although an unmanned underwater vehicle (UUV) requires an up-front capital cost, it provides significant savings in operating cost for near-seafloor surveys, especially in deep water. Taking into account the time it takes to turn around a deep-tow marine magnetometer system, a UUV can complete a deep-water survey in one quarter of the time it takes a surface ship towing an array to cover the same area. The UUV also can operate at greater depth, closer to targets buried on the seafloor [25].

20.5 Anticipated improvements

The future market for quantum magnetometers will be shaped by several forces. First, users will increasingly migrate to small, automated platforms such as UAVs for airborne sensing and UUVs for underwater exploration. These platforms in turn require smaller, lower-power sensors and improved rejection of platform noise. Second, new technologies may allow significant improvement in sensitivity or other performance parameters [26]. A new company, Twinleaf [27], was formed to commercialize some of these improvements. Another company, Southwest Sciences [28], has developed fiberized magnetometers for characterizing magnetic fields inside magnetic shields. Existing companies continue to upgrade their products by introducing promising new technologies.

New applications will demand improved performance. High-performance potassium gradiometers have recently been installed in Mexico, in Israel, and at a fixed magnetic observatory in Austria. Other high-sensitivity gradiometers are used for earthquake studies, magnetotellurics, and ionospheric measurements. In addition to high sensitivity, magnetometers for such gradiometers require high immunity from heading error, so that lower-order fields do not contaminate the measurement of gradients. Another new application uses an array of sensors to produce magnetic maps much more quickly. Commercial success hinges on reducing the cost of the sensor array significantly below the cost of a similar number of stand-alone instruments.

Surveys for unexploded ordnance (UXO) or improvised explosive devices (IEDs) are clear examples of dangerous endeavors. Robots carrying high-sensitivity magnetometers, perhaps configured in multiple-sensor arrays, can conduct these type of surveys efficiently with no danger to human life. As with all robotic magnetometer installations, careful and innovative compensation techniques will be required to minimize not only the induced, remanent, and eddy-current sources of noise but also noise generated by the control electronics, other sensors, or motive power and propulsion-system components. Currently work is underway to automate compensation for all these noise sources.

References

[1] J. M. Higbie, E. Corsini, and D. Budker, Rev. Sci. Instrum. **77**, 113106 (2006).

[2] I. Hrvoic, in *Proceedings of the VIth workshop on geomagnetic observatory instruments, data acquisition and processing*, edited by J. L. Rasson (Institut Meteorologique du Belgique, Brussels, 1994), pp. 70–72; available at gemsys.ca.

[3] P. J. Mohr, B. N. Taylor, and D. B. Newell, Rev. Mod. Phys. **80**, 633 (2008); values are tabulated in NIST Reference on Constants, Units and Uncertainty, http://physics.nist.gov/cgi-bin/cuu/Value?gammap.
[4] D. A. Steck, *Cesium D Line Data*, published online at http://steck.us/alkalidata/cesium-numbers.pdf.
[5] S. Pustelny, D. F. Jackson Kimball, S. M. Rochester, V. V. Yashchuk, and D. Budker, Phys. Rev. A **74**, 063406 (2006).
[6] M. N. Nabighian, V. J. S. Grauch, R. O. Hansen, T. R. LaFehr, Y. Li, J. W. Peirce, J. D. Phillips, and M. E. Ruder, Geophysics **70**, 33ND (2005).
[7] M. Packard and R. Varian, Phys. Rev. **93**, 941 (1954).
[8] Geometrics, San Jose, CA, USA; www.geometrics.com.
[9] G. Bonnet, J. L. Laffon, and P. Servez-Gavin, United States Patent 3,133,243 (May 12, 1964).
[10] Scintrex Limited, Concord, ON, Canada, www.scintrexltd.com.
[11] GEM Systems, Markham, ON, Canada, www.gemsys.ca.
[12] T. L. Skillman and P. L. Bender, J. Geophys. Res. **63**, 513 (1958).
[13] F. D. Colegrove and P. A. Franken, Phys. Rev. **119**, 680 (1960).
[14] P. A. Franken, United States Patent 3,122,702 (February 25, 1964).
[15] A. R. Keyser, J. A. Rice and L. D. Schearer, J. Geophys. Res. **66**, 4163 (1961).
[16] R. E. Hartline and F. H. Meister, US Patent 3,513,383 (May 19, 1970).
[17] Aero Geophysical Survey and Remote Sensing Center, China (AGRS), http://www.agrs.cn/cms/website/english/.
[18] G. K. Walters, F. D. Colegrove, and L. D. Schearer, Phys. Rev. Lett. **8**, 439 (1962).
[19] F. D. Colegrove, L. D. Schearer, and G. K. Walters, Phys. Rev. **132**, 2561 (1963).
[20] R. Slocum and B. Marton, IEEE Trans. Magn. **10**, 528 (1974).
[21] R. E. Slocum and F. N. Reilly, IEEE Trans. Nucl. Sci. **NS-10**, 165 (1963).
[22] Polatomic, Inc., Richardson, TX, USA, www.polatomic.com.
[23] L. Schearer, M. Leduc, F. Laloe, and J. M. Hamel, United States Patent 4,806,864 (February 21, 1989).
[24] Forecast International, *ASQ-81(V)/ASQ-208(V) ASW forecast* (February 2003).
[25] G. I. Allen, G. Sulzberger, J. T. Bono, J. S. Pray, and T. R. Clem, Proc. MTS/IEEE **3**, 1956 (2005).
[26] D. Budker and M. Romalis, Nature Physics **3**, 227 (2007).
[27] Twinleaf, Princeton, NJ, USA, www.twinleaf.com.
[28] Southwest Sciences, Inc., Santa Fe, NM, USA, www.swsciences.com.

Index

For EU product safety concerns, contact us at Calle de José Abascal, 56–1°,
28003 Madrid, Spain or eugpsr@cambridge.org.

www.ingramcontent.com/pod-product-compliance
Lightning Source LLC
Chambersburg PA
CBHW080205010726
47440CB00005B/282
* 9 7 8 1 1 0 7 0 1 0 3 5 2 *